Lectures on Dynamics of Stochastic Systems

Lectures on Dynamics of Stochastic Systems

Edited by

Valery I. Klyatskin

Translated from Russian by

A. Vinogradov

ELSEVIER

AMSTERDAM • BOSTON • HEIDELBERG • LONDON • NEW YORK OXFORD • PARIS
SAN DIEGO • SAN FRANCISCO • SINGAPORE • SYDNEY • TOKYO

Elsevier
32 Jamestown Road London NW1 7BY
30 Corporate Drive, Suite 400, Burlington, MA 01803, USA

First edition 2011

British Library Cataloguing in Publication Data
A catalogue record for this book is available from the British Library

Library of Congress Cataloging-in-Publication Data
A catalog record for this book is available from the Library of Congress

ISBN: 978-0-12-384966-3

For information on all Elsevier publications
visit our website at *www.elsevierdirect.com*

Typeset by: diacriTech, India

This book has been manufactured using Print On Demand technology. Each copy is produced to order and is limited to black ink. The online version of this book will show color figures where appropriate.

Contents

Preface

I keep six honest serving-men
(They taught me all I knew);
Their names are What and Why and When
And How and Where and Who.

R. Kipling

This book is a revised and more comprehensive re-edition of my book *Dynamics of Stochastic Systems*, Elsevier, Amsterdam, 2005.

Writing this book, I sourced from the series of lectures that I gave to scientific associates at the Institute of Calculus Mathematics, Russian Academy of Sciences. In the book, I use the functional approach to uniformly formulate general methods of statistical description and analysis of dynamic systems. These are described in terms of different types of equations with fluctuating parameters, such as ordinary differential equations, partial differential equations, boundary-value problems, and integral equations. Asymptotic methods of analyzing stochastic dynamic systems – the delta-correlated random process (field) approximation and the diffusion approximation – are also considered. General ideas are illustrated using examples of coherent phenomena in stochastic dynamic systems, such as clustering of particles and passive tracer in random velocity field, dynamic localization of plane waves in randomly layered media and caustic structure of wavefield in random media.

Working at this edition, I tried to take into account opinions and wishes of readers about both the style of the text and the choice of specific problems. Various mistakes and misprints have been corrected.

The book is aimed at scientists dealing with stochastic dynamic systems in different areas, such as acoustics, hydrodynamics, magnetohydrodynamics, radiophysics, theoretical and mathematical physics, and applied mathematics; it may also be useful for senior and postgraduate students.

The book consists of three parts.

The first part is, in essence, an introductory text. It raises a few typical physical problems to discuss their solutions obtained under random perturbations of parameters affecting system behavior. More detailed formulations of these problems and relevant statistical analysis can be found in other parts of the book.

The second part is devoted to the general theory of statistical analysis of dynamic systems with fluctuating parameters described by differential and integral equations. This theory is illustrated by analyzing specific dynamic systems. In addition, this part considers asymptotic methods of dynamic system statistical analysis, such as the delta-correlated random process (field) approximation and diffusion approximation.

The third part deals with the analysis of specific physical problems associated with coherent phenomena. These are clustering and diffusion of particles and passive tracer (density and magnetic fields) in a random velocity field, dynamic localization of plane waves propagating in layered random media and caustic structure of wavefield in random multidimensional media. These phenomena are described by ordinary differential equations and partial differential equations.

The material is supplemented with sections concerning dynamical and statistical descriptions of simplest hydrodynamic-type systems, the relationship between conventional methods (based on the Lyapunov stability analysis of stochastic dynamic systems), methods of statistical topography and statistical description of magnetic field generation in the random velocity field (stochastic [turbulent] dynamo).

Each lecture is appended with problems for the reader to solve on his or her own, which will be a good training for independent investigations.

V. I. Klyatskin
Moscow, Russia

Introduction

Different areas of physics pose statistical problems in ever greater numbers. Apart from issues traditionally obtained in statistical physics, many applications call for including fluctuation effects. While fluctuations may stem from different sources (such as thermal noise, instability, and turbulence), methods used to treat them are very similar. In many cases, the statistical nature of fluctuations may be deemed known (either from physical considerations or from problem formulation) and the physical processes may be modeled by differential, integro-differential or integral equations.

We will consider a statistical theory of dynamic and wave systems with fluctuating parameters. These systems can be described by ordinary differential equations, partial differential equations, integro-differential equations and integral equations. A popular way to solve such systems is by obtaining a closed system of equations for statistical characteristics, to study their solutions as comprehensively as possible.

We note that wave problems are often boundary-value problems. When this is the case, one may resort to the imbedding method to reformulate the equations at hand to initial-value problems, thus considerably simplifying the statistical analysis.

The purpose of this book is to demonstrate how different physical problems described by stochastic equations can be solved on the basis of a general approach.

In stochastic problems with fluctuating parameters, the variables are functions. It would be natural therefore to resort to functional methods for their analysis. We will use a *functional method* devised by Novikov [1] for Gaussian fluctuations of parameters in a turbulence theory and developed by the author of this book [2] for the general case of dynamic systems and fluctuating parameters of an arbitrary nature.

However, only a few dynamic systems lend themselves to analysis yielding solutions in a general form. It proved to be more efficient to use an asymptotic method where the statistical characteristics of dynamic problem solutions are expanded in powers of a small parameter. This is essentially a ratio of the random impact's correlation time to the time of observation or to another characteristic time scale of the problem (in some cases, these may be spatial rather than temporal scales). This method is essentially a generalization of the theory of Brownian motion. It is termed the *delta-correlated random process (field) approximation*.

For dynamic systems described by ordinary differential stochastic equations with Gaussian fluctuations of parameters, this method leads to a Markovian problem solving model, and the respective equation for transition probability density has the form of the *Fokker–Planck equation*. In this book, we will consider in depth the methods of analysis available for this equation and its boundary conditions. We will analyze solutions and validity conditions by way of integral transformations. In more complicated problems described by partial differential equations, this method leads to a generalized equation of the Fokker–Planck type in which variables are the derivatives of the solution's characteristic functional. For dynamic problems with non-Gaussian fluctuations of parameters, this method also yields Markovian type solutions. Under the circumstances, the probability density of respective dynamic stochastic equations satisfies a closed operator equation.

In physical investigations, Fokker–Planck and similar equations are usually set up from rule-of-thumb considerations, and dynamic equations are invoked only to calculate the coefficients of these equations. This approach is inconsistent, generally speaking. Indeed, the statistical problem is completely defined by dynamic equations and assumptions on the statistics of random impacts. For example, the Fokker–Planck equation must be a logical sequence of the dynamic equations and some assumptions on the character of random impacts. It is clear that not all problems lend themselves for reduction to a Fokker–Planck equation. The functional approach allows one to derive a Fokker–Planck equation from the problem's dynamic equation along with its applicability conditions. In terms of formal mathematics, our approach corresponds to that of R.L. Stratonovich (see, e.g., [3]).

For a certain class of Markovian random process (telegrapher's processes, Gaussian process and the like), the developed functional approach also yields closed equations for the solution probability density with allowance for a finite correlation time of random interactions.

For processes with Gaussian fluctuations of parameters, one may construct a better physical approximation than the delta-correlated random process (field) approximation – the *diffusion approximation* that allows for finiteness of correlation time radius. In this approximation, the solution is Markovian and its applicability condition has transparent physical meaning, namely, the statistical effects should be small within the correlation time of fluctuating parameters. This book treats these issues in depth from a general standpoint and for some specific physical applications.

Recently, the interest of both theoreticians and experimenters has been attracted to the relation of the behavior of average statistical characteristics of a problem solution with the behavior of the solution in certain happenings (realizations). This is especially important for geophysical problems related to the atmosphere and ocean where, generally speaking, a respective averaging ensemble is absent and experimenters, as a rule, deal with individual observations.

Seeking solutions to dynamic problems for these specific realizations of medium parameters is almost hopeless due to the extreme mathematical complexity of these problems. At the same time, researchers are interested in the main characteristics of these phenomena without much need to know specific details. Therefore, the idea of using a well-developed approach to random processes and fields based on ensemble averages rather than separate observations proved to be very fruitful. By way of example, almost all physical problems of the atmosphere and ocean to some extent are treated by statistical analysis.

Randomness in medium parameters gives rise to a stochastic behavior of physical fields. Individual samples of scalar two-dimensional fields $\rho(\boldsymbol{R}, t), \boldsymbol{R} = (x, y)$, say, recall a rough mountainous terrain with randomly scattered peaks, troughs, ridges and

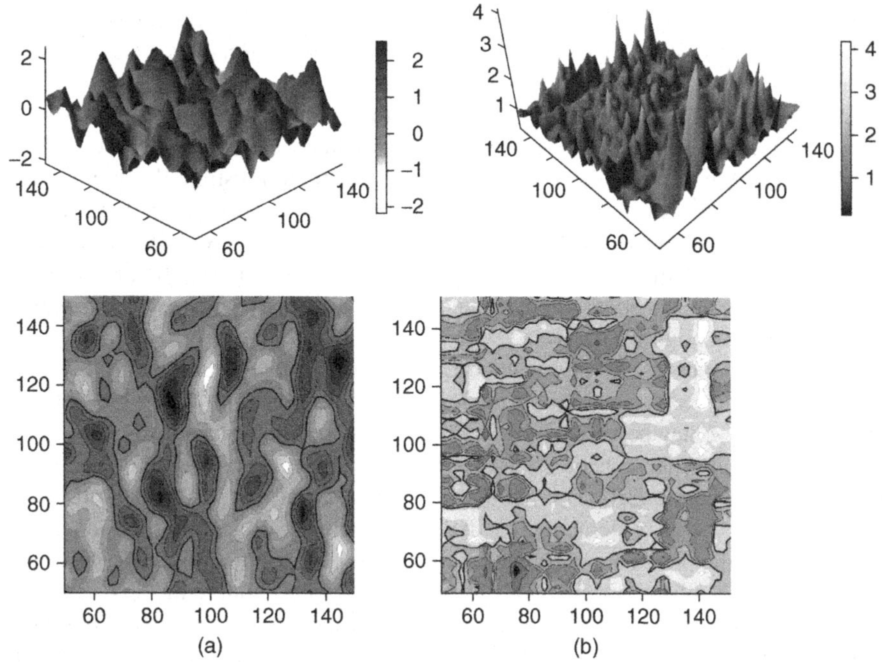

Figure 0.1 Realizations of the fields governed by (a) Gaussian and (b) lognormal distribution and the corresponding topographic level lines. The bold curves in the bottom patterns show level lines corresponding to levels 0 (a) and 1 (b).

saddles. Figure 0.1 shows examples of realizations of two random fields characterized by different statistical structures.

Common methods of statistical averaging (computing mean-type averages – $\langle \rho(\boldsymbol{R}, t) \rangle$, space-time correlation function – $\left\langle \rho(\boldsymbol{R}, t)\, \rho\left(\boldsymbol{R}', t'\right) \right\rangle$ etc., where $\langle \cdots \rangle$ implies averaging over an ensemble of random parameter samples) smooth the qualitative features of specific samples. Frequently, these statistical characteristics have nothing in common with the behavior of specific samples, and at first glance may even seem to be at variance with them. For example, the statistical averaging over all observations makes the field of average concentration of a passive tracer in a random velocity field ever more smooth, whereas each realization sample tends to be more irregular in space due to the mixture of areas with substantially different concentrations.

Thus, these types of statistical average usually characterize 'global' space–time dimensions of the area with stochastic processes, but give no details about the process behavior inside the area. For this case, details heavily depend on the velocity field pattern, specifically, on whether it is divergent or solenoidal. Thus, the first case will show with total probability that *clusters* will be formed, i.e. compact areas of enhanced concentration of tracer surrounded by vast areas of low-concentration tracer. In the circumstances, all statistical moments of the distance between the particles will grow with time exponentially; that is, on average, a statistical recession of particles will take place [4].

In a similar way, in the case of waves propagating in random media, an exponential spread of the rays will take place on average; but simultaneously, with total probability, *caustics* will form at finite distances. One more example to illustrate this point is the *dynamic localization* of plane waves in layered randomly inhomogeneous media. In this phenomenon, the wave field intensity exponentially decays inward to the medium with the probability equal to unity when the wave is incident on the half-space of such a medium, while all statistical moments increase exponentially with distance from the boundary of the medium [5].

These physical processes and phenomena occurring with the probability equal to unity will be referred to as *coherent* processes and phenomena [6]. This type of *statistical coherence* may be viewed as some organization of the complex dynamic system, and retrieval of its *statistically stable characteristics* is similar to the concept of *coherence* as *self-organization* of multicomponent systems that evolve from the random interactions of their elements [7]. In the general case, it is rather difficult to say whether or not the phenomenon occurs with the probability equal to unity. However, for a number of applications amenable to treatment with the simple models of fluctuating parameters, this may be handled by analytical means. In other cases, one may verify this by performing numerical modeling experiments or analyzing experimental findings.

The complete statistic (say, the whole body of all n-point space-time moment functions), would undoubtedly contain all the information about the investigated dynamic system. In practice, however, one may succeed only in studying the simplest statistical characteristics associated mainly with simultaneous and one-point probability distributions. It would be reasonable to ask how with these statistics on hand one would look into the quantitative and qualitative behavior of some system happenings?

This question is answered by *methods of statistical topography* [8]. These methods were highlighted by Ziman [9], who seems to have coined this term. Statistical topography yields a different philosophy of statistical analysis of dynamic stochastic systems, which may prove useful for experimenters planning a statistical processing of experimental data. These issues are treated in depth in this book.

More details about the material of this book and more exhaustive references can be found in the textbook quoted in reference [2] and recent reviews [6, 10–16].

Part I

Dynamical Description of Stochastic Systems

Lecture 1

Examples, Basic Problems, Peculiar Features of Solutions

In this chapter, we consider several dynamic systems described by differential equations of different types and discuss the features in the behaviors of solutions to these equations under random disturbances of parameters. Here, we content ourselves with the problems in the simplest formulation. More complete formulations will be discussed below, in the sections dealing with statistical analysis of the corresponding systems.

1.1 Ordinary Differential Equations: Initial-Value Problems

1.1.1 Particles Under the Random Velocity Field

In the simplest case, a particle under random velocity field is described by the system of ordinary differential equations of the first order

$$\frac{d}{dt}\boldsymbol{r}(t) = \boldsymbol{U}(\boldsymbol{r}, t), \quad \boldsymbol{r}(t_0) = \boldsymbol{r}_0, \tag{1.1}$$

where $\boldsymbol{U}(\boldsymbol{r}, t) = \boldsymbol{u}_0(\boldsymbol{r}, t) + \boldsymbol{u}(\boldsymbol{r}, t)$, $\boldsymbol{u}_0(\boldsymbol{r}, t)$ is the deterministic component of the velocity field (mean flow), and $\boldsymbol{u}(\boldsymbol{r}, t)$ is the random component. In the general case, field $\boldsymbol{u}(\boldsymbol{r}, t)$ can have both nondivergent (solenoidal, for which $\operatorname{div} \boldsymbol{u}(\boldsymbol{r}, t) = 0$) and divergent (for which $\operatorname{div} \boldsymbol{u}(\boldsymbol{r}, t) \neq 0$) components.

We dwell on stochastic features of the solution to problem (1.1) for a system of particles in the absence of mean flow ($\boldsymbol{u}_0(\boldsymbol{r}, t) = 0$). From Eq. (1.1) formally follows that every particle moves independently of other particles. However, if random field $\boldsymbol{u}(\boldsymbol{r}, t)$ has a finite spatial correlation radius l_{cor}, particles spaced by a distance shorter than l_{cor} appear in the common zone of infection of random field $\boldsymbol{u}(\boldsymbol{r}, t)$ and the behavior of such a system can show new collective features.

For steady velocity field $\boldsymbol{u}(\boldsymbol{r}, t) \equiv \boldsymbol{u}(\boldsymbol{r})$, Eq. (1.1) reduces to

$$\frac{d}{dt}\boldsymbol{r}(t) = \boldsymbol{u}(\boldsymbol{r}), \quad \boldsymbol{r}(0) = \boldsymbol{r}_0. \tag{1.2}$$

Lectures on Dynamics of Stochastic Systems. DOI: 10.1016/B978-0-12-384966-3.00001-5

This equation clearly shows that steady points $\tilde{\boldsymbol{r}}$ (at which $\boldsymbol{u}(\tilde{\boldsymbol{r}}) = 0$) remain the fixed points. Depending on whether these points are stable or unstable, they will attract or repel nearby particles. In view of randomness of function $\boldsymbol{u}(\boldsymbol{r})$, points $\tilde{\boldsymbol{r}}$ are random too.

It is expected that the similar behavior will be also characteristic of the general case of space-time random velocity field $\boldsymbol{u}(\boldsymbol{r},t)$.

If some points $\tilde{\boldsymbol{r}}$ remain stable during sufficiently long time, then clusters of particles (i.e., compact regions with elevated particle concentration, which occur merely in rarefied zones) must arise around these points in separate realizations of random field $\boldsymbol{u}(\boldsymbol{r}, t)$. On the contrary, if the stability of these points alternates with instability sufficiently rapidly and particles have no time for significant rearrangement, no clusters of particles will occur.

Simulations [17, 18] show that the behavior of a system of particles essentially depends on whether the random field of velocities is nondivergent or divergent. By way of example, Fig. 1.1*a* shows a schematic of the evolution of the two-dimensional system of particles uniformly distributed within the circle for a particular realization of the nondivergent steady field $\boldsymbol{u}(\boldsymbol{r})$.

Here, we use the dimensionless time related to statistical parameters of field $\boldsymbol{u}(\boldsymbol{r})$. In this case, the area of surface patch within the contour remains intact and particles relatively uniformly fill the region within the deformed contour. The only feature consists in the fractal-type irregularity of the deformed contour. This phenomenon – called chaotic advection – is under active study now (see, e.g., [19]).

On the contrary, in the case of the divergent velocity field $\boldsymbol{u}(\boldsymbol{r})$, particles uniformly distributed in the square at the initial instant will form clusters during the temporal evolution. Results simulated for this case are shown in Fig. 1.1*b*. We emphasize that the formation of clusters is purely a kinematic effect. This feature of particle dynamics disappears on averaging, over an ensemble of realizations of random velocity field.

To demonstrate the process of particle clustering, we consider the simplest problem [11], in which random velocity field $\boldsymbol{u}(\boldsymbol{r}, t)$ has the form

$$\boldsymbol{u}(\boldsymbol{r}, t) = \boldsymbol{v}(t) f(\boldsymbol{k}\boldsymbol{r}), \tag{1.3}$$

where $\boldsymbol{v}(t)$ is the random vector process and

$$f(\boldsymbol{k}\boldsymbol{r}) = \sin 2(\boldsymbol{k}\boldsymbol{r}) \tag{1.4}$$

is the deterministic function of one variable. This form of function $f(\boldsymbol{k}\boldsymbol{r})$ corresponds to the first term of the expansion in harmonic components and is commonly used in numerical simulations [17, 18].

In this case, Eq. (1.1) can be written in the form

$$\frac{d}{dt}\boldsymbol{r}(t) = \boldsymbol{v}(t) \sin 2(\boldsymbol{k}\boldsymbol{r}), \quad \boldsymbol{r}(0) = \boldsymbol{r}_0.$$

In the context of this model, motions of a particle along vector $\boldsymbol{k}$ and in the plane perpendicular to vector $\boldsymbol{k}$ are independent and can be separated. If we direct the x-axis

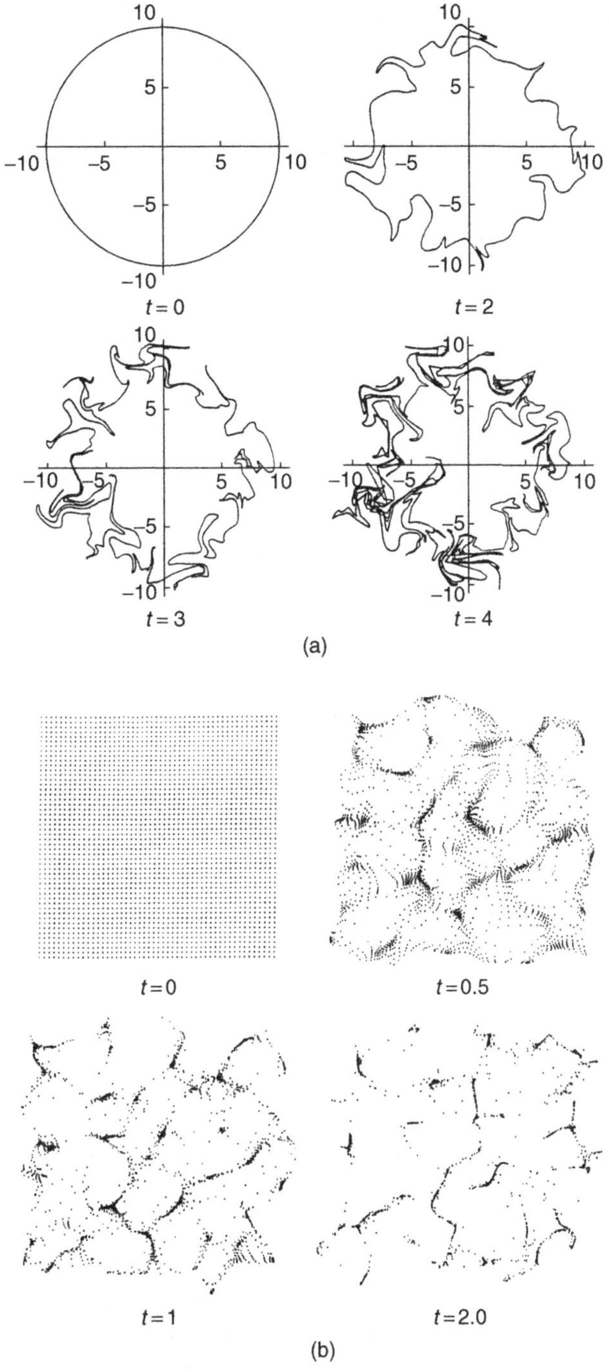

Figure 1.1 Diffusion of a system of particles described by Eqs. (1.2) numerically simulated for (*a*) solenoidal and (*b*) nondivergent random steady velocity field $\boldsymbol{u}(\boldsymbol{r})$.

along vector $\boldsymbol{k}$, then the equations assume the form

$$\begin{aligned}&\frac{d}{dt}x(t) = v_x(t)\sin(2kx), \qquad x(0) = x_0,\\&\frac{d}{dt}\boldsymbol{R}(t) = \boldsymbol{v}_{\boldsymbol{R}}(t)\sin(2kx), \quad \boldsymbol{R}(0) = \boldsymbol{R}_0.\end{aligned} \tag{1.5}$$

The solution of the first equation in system (1.5) is

$$x(t) = \frac{1}{k}\arctan\left[e^{T(t)}\tan(kx_0)\right], \tag{1.6}$$

where

$$T(t) = 2k\int_0^t d\tau v_x(\tau). \tag{1.7}$$

Taking into account the equalities following from Eq. (1.6)

$$\begin{aligned}\sin(2kx) &= \frac{\sin(2kx_0)}{e^{-T(t)}\cos^2(kx_0) + e^{T(t)}\sin^2(kx_0)},\\ \cos(2kx) &= \frac{1 - e^{2T(t)}\tan^2(kx_0)}{1 + e^{2T(t)}\tan^2(kx_0)},\end{aligned} \tag{1.8}$$

we can rewrite the second equation in Eqs. (1.5) in the form

$$\frac{d}{dt}\boldsymbol{R}(t|\boldsymbol{r}_0) = \frac{\sin(2kx_0)\boldsymbol{v}_{\boldsymbol{R}}(t)}{e^{-T(t)}\cos^2(kx_0) + e^{T(t)}\sin^2(kx_0)}.$$

As a result, we have

$$\boldsymbol{R}(t|\boldsymbol{r}_0) = \boldsymbol{R}_0 + \int_0^t d\tau \frac{\sin(2kx_0)\boldsymbol{v}_{\boldsymbol{R}}(\tau)}{e^{-T(\tau)}\cos^2(kx_0) + e^{T(\tau)}\sin^2(kx_0)}. \tag{1.9}$$

Consequently, if the initial particle position x_0 is such that

$$kx_0 = n\frac{\pi}{2}, \tag{1.10}$$

where $n = 0, \pm 1, \ldots$, then the particle will be the fixed particle and $\boldsymbol{r}(t) \equiv \boldsymbol{r}_0$.

Equalities (1.10) define planes in the general case and points in the one-dimensional case. They correspond to zeros of the field of velocities. Stability of these points depends on the sign of function $\boldsymbol{v}(t)$, and this sign changes during the evolution process. It can be expected that particles will gather around these points if $v_x(t) \neq 0$, which just corresponds to clustering of particles.

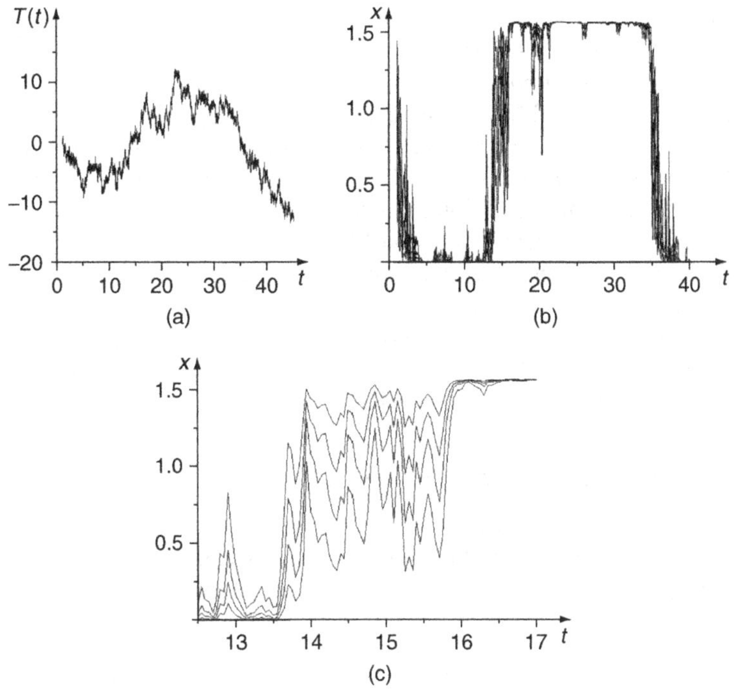

Figure 1.2 (*a*) Segment of a realization of random process $T(t)$ obtained by numerically integrating Eq. (1.7) for a realization of random process $v_x(t)$; (*b*), (*c*) x-coordinates simulated with this segment for four particles versus time.

In the case of a nondivergent velocity field, $v_x(t) = 0$ and, consequently, $T(t) \equiv 0$; as a result, we have

$$x(t|x_0) \equiv x_0, \quad \boldsymbol{R}(t|\boldsymbol{r}_0) = \boldsymbol{R}_0 + \sin 2(kx_0) \int_0^t d\tau \boldsymbol{v}_{\boldsymbol{R}}(\tau),$$

which means that no clustering occurs.

Figure 1.2*a* shows a fragment of the realization of random process $T(t)$ obtained by numerical integration of Eq. (1.7) for a realization of random process $v_x(t)$; we used this fragment for simulating the temporal evolution of coordinates of four particles $x(t)$, $x \in (0, \pi/2)$ initially located at coordinates $x_0(i) = \dfrac{\pi}{2}\dfrac{i}{5}$ $(i = 1, 2, 3, 4)$ (see Fig. 1.2*b*). Figure 1.2*b* shows that particles form a cluster in the vicinity of point $x = 0$ at the dimensionless time $t \approx 4$. Further, at time $t \approx 16$ the initial cluster disappears and new one appears in the vicinity of point $x = \pi/2$. At moment $t \approx 40$, the cluster appears again in the vicinity of point $x = 0$, and so on. In this process, particles in clusters remember their past history and significantly diverge during intermediate temporal segments (see Fig. 1.2*c*).

Thus, we see in this example that the cluster does not move from one region to another; instead, it first collapses and then a new cluster appears. Moreover, the lifetime of clusters significantly exceeds the duration of intermediate segments. It seems that this feature is characteristic of the specific model of velocity field and follows from steadiness of points (1.10).

As regards the particle diffusion along the y-direction, no cluster occurs there.

Note that such clustering in a system of particles was found, to all appearance for the first time, in papers [20, 21] as a result of simulating the so-called *Eole experiment* with the use of the simplest equations of atmospheric dynamics.

In this global experiment, 500 constant-density balloons were launched in Argentina in 1970–1971; these balloons traveled at a height of about 12 km and spread along the whole of the southern hemisphere.

Figure 1.3 shows the balloon distribution over the southern hemisphere for day 105 from the beginning of this process simulation [20]; this distribution clearly shows that balloons are concentrated in groups, which just corresponds to clustering.

1.1.2 Particles Under Random Forces

The system of equations (1.1) describes also the behavior of a particle under the field of random external forces $\boldsymbol{f}(\boldsymbol{r}, t)$. In the simplest case, the behavior of a particle in the presence of linear friction is described by the differential equation of the second order (Newton equation)

$$\frac{d^2}{dt^2}\boldsymbol{r}(t) = -\lambda\frac{d}{dt}\boldsymbol{r}(t) + \boldsymbol{f}(\boldsymbol{r}, t), \qquad \boldsymbol{r}(0) = \boldsymbol{r}_0, \quad \frac{d}{dt}\boldsymbol{r}(0) = \boldsymbol{v}_0, \tag{1.11}$$

or the systems of differential equation of the first order

$$\frac{d}{dt}\boldsymbol{r}(t) = \boldsymbol{v}(t), \quad \frac{d}{dt}\boldsymbol{v}(t) = -\lambda\boldsymbol{v}(t) + \boldsymbol{f}(\boldsymbol{r}, t), \qquad \boldsymbol{r}(0) = \boldsymbol{r}_0, \quad \boldsymbol{v}(0) = \boldsymbol{v}_0. \tag{1.12}$$

Results of numerical simulations of stochastic system (1.12) can be found in Refs [22, 23]. Stability of the system was studied in these papers by analyzing Lyapunov's characteristic parameters.

The behavior of a particle under the deterministic potential field in the presence of linear friction and random forces is described by the system of equations

$$\frac{d}{dt}\boldsymbol{r}(t) = \boldsymbol{v}(t), \quad \frac{d}{dt}\boldsymbol{v}(t) = -\lambda\boldsymbol{v}(t) - \frac{\partial U(\boldsymbol{r}, t)}{\partial \boldsymbol{r}} + \boldsymbol{f}(\boldsymbol{r}, t), \qquad \boldsymbol{r}(0) = \boldsymbol{r}_0, \quad \boldsymbol{v}(0) = \boldsymbol{v}_0. \tag{1.13}$$

which is the simplest example of *Hamiltonian systems* with linear friction. If friction and external forces are absent and function U is independent of time, $U(\boldsymbol{r}, t) = U(\boldsymbol{r})$,

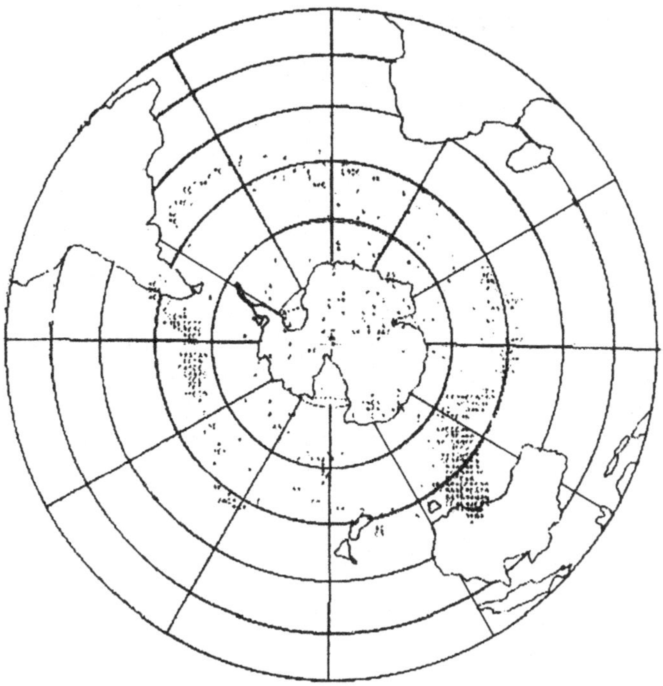

Figure 1.3 Balloon distribution in the atmosphere for day 105 from the beginning of process simulation.

the system has an integral of motion

$$\frac{d}{dt}E(t) = \text{const}, \quad E(t) = \frac{\boldsymbol{v}^2}{2} + U(\boldsymbol{r})$$

expressing energy conservation.

In statistical problems, equations of type (1.12), (1.13) are widely used to describe the *Brownian motion* of particles.

In the general case, Hamiltonian systems are described by the system of equations

$$\frac{d}{dt}\boldsymbol{r}(t) = \frac{\partial H(\boldsymbol{r}, \boldsymbol{v}, t)}{\partial \boldsymbol{v}}, \qquad \frac{d}{dt}\boldsymbol{v}(t) = -\frac{\partial H(\boldsymbol{r}, \boldsymbol{v}, t)}{\partial \boldsymbol{r}},$$
$$\boldsymbol{r}(0) = \boldsymbol{r}_0, \qquad \boldsymbol{v}(0) = \boldsymbol{v}_0, \tag{1.14}$$

where $H(\boldsymbol{r}, \boldsymbol{v}, t) = H(\boldsymbol{r}(t), \boldsymbol{v}(t), t)$ is the Hamiltonian function. In the case of conservative Hamiltonian systems, function $H(\boldsymbol{r}, \boldsymbol{v}, t)$ has no explicit dependence on time, $H(\boldsymbol{r}, \boldsymbol{v}, t) = H(\boldsymbol{r}, \boldsymbol{v})$, and the system has the integral of motion

$$H(\boldsymbol{r}, \boldsymbol{v}) = \text{const}.$$

1.1.3 The Hopping Phenomenon

Now, we dwell on another stochastic aspect related to dynamic equations of type (1.1); namely, we consider the *hopping phenomenon* caused by random fluctuations.

Consider the one-dimensional nonlinear equation

$$\frac{d}{dt}x(t) = x\left(1 - x^2\right) + f(t), \quad x(0) = x_0, \tag{1.15}$$

where $f(t)$ is the random function of time. In the absence of randomness ($f(t) \equiv 0$), the solution of Eq. (1.15) has two stable steady states $x = \pm 1$ and one instable state $x = 0$. Depending on the initial-value, solution of Eq. (1.15) arrives at one of the stable states. However, in the presence of small random disturbances $f(t)$, dynamic system (1.15) will first approach the vicinity of one of the stable states and then, after the lapse of certain time, it will be transferred into the vicinity of another stable state.

Note that Eq. (1.15) corresponds to limit process $\lambda \to \infty$ in the equation

$$\frac{d^2}{dt^2}x(t) + \lambda\frac{d}{dt}x(t) - \lambda\left\{\frac{dU(x)}{dx} + f(t)\right\} = 0,$$

that is known as the *Duffing equation* and is the special case of the one-dimensional Hamiltonian system (1.13)

$$\frac{d}{dt}x(t) = v(t), \quad \frac{d}{dt}v(t) = -\lambda\left\{v(t) - \frac{dU(x)}{dx} - f(t)\right\} \tag{1.16}$$

with the potential function

$$U(x) = \frac{x^2}{2} - \frac{x^4}{4}.$$

In other words, Eq. (1.15) corresponds to great friction coefficients λ.

Statistical description of this problem will be considered in Sect. 8.4.1, page 211. Additionally, we note that, in the context of statistical description, reduction of the Hamiltonian system (1.16) to the 'short-cut equation' is called the *Kramers problem*.

It is clear that the similar behavior can occur in more complicated situations.

Hydrodynamic-Type Nonlinear Systems

An important problem of studying large-scale processes in the atmosphere considered as a single physical system consists in revealing the mechanism of energy exchange between different 'degrees of freedom'. The analysis of such nonlinear processes on the base of simple models described by a small number of parameters (degrees of freedom) is recently of great attention. In this connection, A.M. Obukhov (see, for example [24]) introduced the concept of *hydrodynamic-type systems* (*HTS*). These systems have a finite number of parameters $v_1, \ldots, v_n$, but the general features of the dynamic equations governing system motions coincide with those characteristic of the hydrodynamic equations of perfect incompressible liquid, including quadratic nonlinearity, energy conservation, and regularity (phase volume invariance during system motions). The general description of HTS is given in Sect. 8.3.3, page 200. Here, we dwell only on the dynamic description of simplest systems.

The simplest system of this type (S_3) is equivalent to the *Euler equations* in the dynamics of solids; it describes the known problem on liquid motions in an ellipsoidal cavity [24]. Any finite-dimensional approximation of hydrodynamic equations also belongs to the class of HTS if it possesses the above features.

To model the cascade mechanism of energy transformation in a turbulent flow, Obukhov [25] suggested a multistage HTS. Each stage of this system consists of identical-scale triplets; at every next stage, the number of triplets is doubled and the scale is decreased in geometrical progression with ratio $Q \gtrsim 1$. As a result, this model describes interactions between the motions of different scales.

The first stage consists of a singe triplet whose unstable mode v_{01} is excited by an external force $f_0(t)$ applied to the system (Fig. 1.4*a*). The stable modes of this triplet $v_{1,1}$ and $v_{1,2}$ are the unstable modes of two triplets of the second stage; their stable modes $v_{2,1}$, $v_{2,2}$, $v_{2,3}$, and $v_{2,4}$ are, in turn, the unstable modes of four triplets of the third stage; and so on (Fig. 1.4*b*).

It should be noted however that physical processes described in terms of macroscopic equations occur in actuality against the background of processes characterized by shorter scales (noises). Among these processes are, for example, the molecular noises (in the context of macroscopic hydrodynamics), microturbulence (against the large-scale motions), and the effect of truncated (small-scale) terms in the finite-dimensional approximation of hydrodynamic equations. The effect of these small-scale noises should be analyzed in statistical terms. Such a consideration can be performed in terms of macroscopic variables. With this goal, one must include external random forces with certain statistical characteristics in the corresponding macroscopic equations. The models considered here require additionally the inclusion of dissipative terms in the equations of motion to ensure energy outflow to smaller-scale modes.

Accordingly, the simplest hydrodynamic models that allow simulating actual processes are the HTS with random forces and linear friction.

An important problem which appeared, for example, in the theory of climate consists in the determination of a possibility of significantly different circulation processes

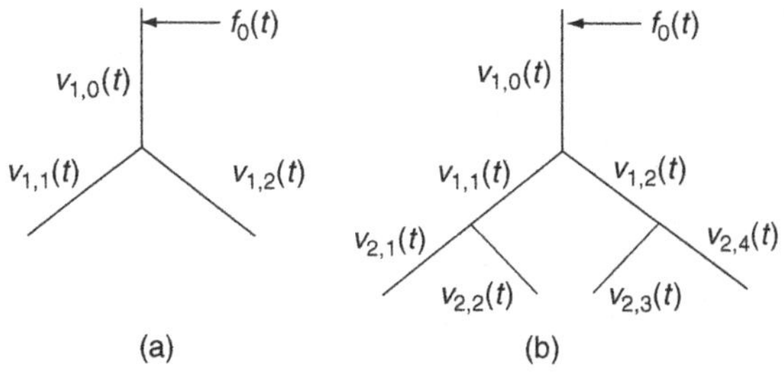

Figure 1.4 Diagrams of (*a*) three- and (*b*) seven-mode HTS.

which occurred under the same distribution of incoming heat, i.e., the problem of the existence of different motion regimes of a given hydrodynamic system under the same 'external conditions'. A quite natural phenomenon that should be considered in this context is the 'hopping' phenomenon that consists in switching between such regimes of motion. Characteristic of these regimes is the fact that the duration of switching is small in comparison with the lifetime of the corresponding regimes.

It is expedient to study these problems using the above simple models. The corresponding systems of quadratically nonlinear ordinary differential equations may generally have several stable regimes, certain parameters describing external conditions being identical. The hopping events are caused by variations of these conditions in time and by the effect of random noises. If the system has no stable regimes, its behavior can appear extremely difficult and require statistical description as it is the case in the *Lorentz model* [26]. In the case of availability of indifferent equilibrium states, the system may allow quasisteady-state regimes of motion. Switching between such regimes can be governed by the dynamic structure of the system, and system's behavior can appear 'stochastic' in this case, too.

In what follows, we study these hopping phenomena with the use of the simplest hydrodynamic models.

Dynamics of a Triplet (Gyroscope)

Consider first the case of a single stage, i.e., the triplet in the regime of forced motion (Fig. 1.4*a*). With this goal, we set force $f_0(t) = f_0 = \text{const}$ and assume the availability of dissipative forces acting on the stable modes of the triplet. The corresponding equations of motion have the form

$$\begin{aligned}
\frac{d}{dt}v_{1,0}(t) &= \mu\left(v_{1,1}^2(t) - v_{1,2}^2(t)\right) + f_0,\\
\frac{d}{dt}v_{1,1}(t) &= -\mu v_{1,0}(t)v_{1,1}(t) - \lambda v_{1,1}(t),\\
\frac{d}{dt}v_{1,2}(t) &= \mu v_{1,0}(t)v_{1,2}(t) - \lambda v_{1,2}(t).
\end{aligned} \tag{1.17}$$

If $f_0 > 0$, component $v_{1,1}(t)$ vanishes with time, so that the motion of the triplet (1.17) is described in the dimensionless variables

$$x = \sqrt{\frac{\mu}{f_0}}v_{1,0} - \frac{\lambda}{\sqrt{\mu f_0}}, \quad y = \sqrt{\frac{\mu}{f_0}}v_{1,2}, \quad \tau = \sqrt{\mu f_0}t, \tag{1.18}$$

by the two-mode system

$$\frac{d}{dt}x(t) = -y^2(t) + 1, \quad \frac{d}{dt}y(t) = x(t)y(t). \tag{1.19}$$

This system has the integral of motion $H_1 = x^2(t) + y^2(t) - 2\ln y(t)$. The change of variables $p(t) = x(t)$, $q(t) = \ln y(t)$ reduces the system to the Hamiltonian form

$$\frac{d}{dt}p(t) = -\frac{\partial \mathfrak{H}(p,q)}{\partial q}, \quad \frac{d}{dt}q(t) = \frac{\partial \mathfrak{H}(p,q)}{\partial p},$$

with the Hamiltonian $\mathfrak{H}(p,q) = \dfrac{p^2(t)}{2} + \dfrac{1}{2}e^{2q(t)} - q(t)$.

Thus, the behavior of a system with friction (1.17) under the action of a constant external force is described in terms of the Hamiltonian system.

Stationary points (0, 1) and (0, −1) of system (1.19) are the centers. If $H_1 - 1 \ll 1$, the period T_1 of motion along closed trajectories around each of these singular points is determined by the asymptotic formula (it is assumed that the sign of $y(t)$ remains intact during this motion)

$$T_1 \approx \sqrt{2}\pi\left[1 + \frac{H_1 - 1}{12}\right].$$

In the opposite limiting case $H_1 \gg 1$ (the trajectories are significantly distant from the mentioned centers), we obtain

$$T_1 \approx \frac{1}{\sqrt{H_1}}\left[2H_1 + \ln H_1\right].$$

Supplement now the dynamic system (1.17) with the linear friction acting on component $v_{1,0}(t)$:

$$\begin{aligned}
&\frac{d}{dt}v_{1,0}(t) = \mu\left(v_{1,1}^2(t) - v_{1,2}^2(t)\right) - \lambda v_{1,0}(t) + f_0,\\
&\frac{d}{dt}v_{1,1}(t) = -\mu v_{1,0}(t)v_{1,1}(t) - \lambda v_{1,1}(t),\\
&\frac{d}{dt}v_{1,2}(t) = \mu v_{1,0}(t)v_{1,2}(t) - \lambda v_{1,2}(t).
\end{aligned} \tag{1.20}$$

Introducing again the dimensionless variables

$$t \to t/\lambda, \quad v_{1,0}(t) \to \frac{\lambda}{\mu}v_0(t), \quad v_{1,2}(t) \to \frac{\lambda}{\mu}v_1(t), \quad v_{1,1}(t) \to \frac{\lambda}{\mu}v_2(t),$$

we arrive at the system of equations

$$\begin{aligned}
&\frac{d}{dt}v_0(t) = v_2^2(t) - v_1^2(t) - v_0(t) + R,\\
&\frac{d}{dt}v_1(t) = v_0(t)v_1(t) - v_1(t),\\
&\frac{d}{dt}v_2(t) = -v_0(t)v_2(t) - v_2(t),
\end{aligned} \tag{1.21}$$

where quantity $R = \dfrac{\mu f_0}{\lambda^2}$ is the analog of the Reynolds number.

Dynamic system (1.21) has steady-state solutions that depend now on parameter R, and $R = R_{\rm cr} = 1$ is the critical value.

For $R < 1$, the system has the stable steady-state solution

$$v_1 = v_2 = 0, \quad v_0 = R.$$

For $R > 1$, this solution becomes unstable with respect to small disturbances of parameters, and the steady-state regimes

$$v_0 = 1, \quad v_2 = 0, \quad v_1 = \pm\sqrt{R-1}, \tag{1.22}$$

become available. Here, we have an element of randomness because component v_1 can be either positive or negative, depending on the amplitude of small disturbance.

Assume now that all components of the triplet are acted on by random forces. This means that system (1.21) is replaced with the system of equations

$$\begin{aligned}
\frac{d}{dt}v_0(t) &= v_2^2(t) - v_1^2(t) - v_0(t) + R + f_0(t),\\
\frac{d}{dt}v_1(t) &= v_0(t)v_1(t) - v_1(t) + f_1(t),\\
\frac{d}{dt}v_2(t) &= -v_0(t)v_2(t) - v_2(t) + f_2(t).
\end{aligned} \tag{1.23}$$

This system describes the motion of a triplet (gyroscope) with the linear isotropic friction, which is driven by the force acting on the instable mode and having both regular (R) and random ($f(t)$) components. Such a situation occurs, for example, for a liquid moving in the ellipsoidal cavity.

For $R > 1$, dynamic system (1.23) under the action of random disturbances will first reach the vicinity of one of the stable states (1.22), and then, after the lapse of certain time, it will be hoppingly set into the vicinity of the other stable state. Figure 1.5

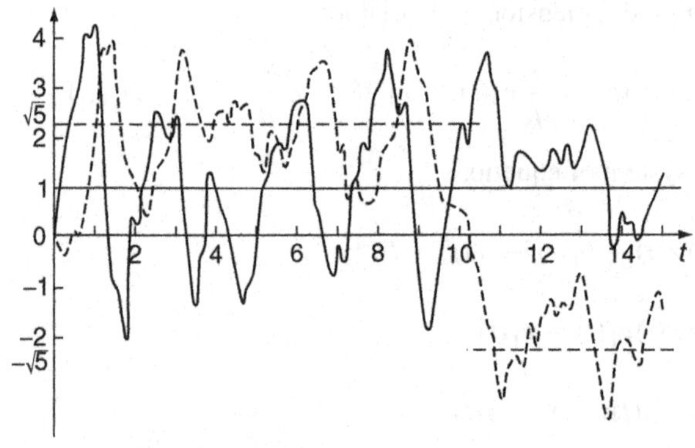

Figure 1.5 Hopping phenomenon simulated from system (1.23) for $R = 6$ and $\sigma = 0.1$ (the solid and dashed lines show components $v_0(t)$ and $v_1(t)$, respectively).

shows the results of simulations of this phenomenon for $R = 6$ and different realizations of random force $\boldsymbol{f}(t)$, whose components were simulated as the Gaussian random processes. (See Sect. 8.3.3, page 200 for the statistical description of this problem.)

Thus, within the framework of the dynamics of the first stage, hopping phenomena can occur only due to the effect of external random forces acting on all modes.

Hopping Between Quasisteady-State Regimes

The simplest two-stage system can be represented in the form

$$
\begin{aligned}
&\frac{d}{dt}v_{1,0}(t) = v_{1,1}^2(t) - v_{1,2}^2(t) + 1,\\
&\frac{d}{dt}v_{1,1}(t) = -v_{1,0}(t)v_{1,1}(t) + Q\left(v_{2,1}^2(t) - v_{2,2}^2(t)\right),\\
&\frac{d}{dt}v_{1,2}(t) = -v_{1,0}(t)v_{1,2}(t) + Q\left(v_{2,3}^2(t) - v_{2,4}^2(t)\right),\\
&\frac{d}{dt}v_{2,1}(t) = -Qv_{1,1}(t)v_{2,1}(t), \quad \frac{d}{dt}v_{2,2}(t) = Qv_{1,1}(t)v_{2,2}(t),\\
&\frac{d}{dt}v_{2,3}(t) = -Qv_{1,2}(t)v_{2,3}(t), \quad \frac{d}{dt}v_{2,4}(t) = Qv_{1,2}(t)v_{2,4}(t),
\end{aligned}
\tag{1.24}
$$

where we used the dimensionless variables similar to (1.18).

Only the components $v_{1,0}(t)$, $v_{1,2}(t)$, $v_{2,3}(t)$ and $v_{2,4}(t)$ survive for $f_0 > 0$ (see Fig. 1.6). These components satisfy the system of equations

$$
\begin{aligned}
&\frac{d}{dt}v_{1,0}(t) = -v_{1,2}^2(t) + 1,\\
&\frac{d}{dt}v_{1,2}(t) = -v_{1,0}(t)v_{1,2}(t) + Q\left(v_{2,3}^2(t) - v_{2,4}^2(t)\right),\\
&\frac{d}{dt}v_{2,3}(t) = -Qv_{1,2}(t)v_{2,3}(t), \quad \frac{d}{dt}v_{2,4}(t) = Qv_{1,2}(t)v_{2,4}(t),
\end{aligned}
$$

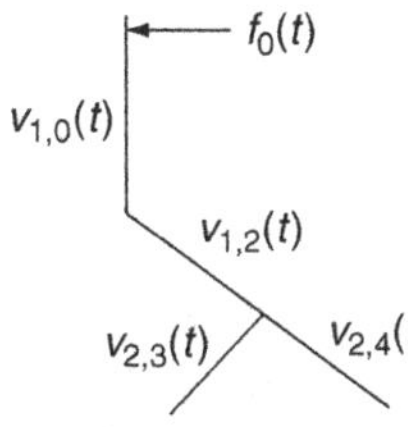

Figure 1.6 Diagram of the excited seven-mode HTS.

that have the integral of motion

$$\nu_{2,3}(t)\nu_{2,4}(t) = I = \text{const}.$$

Introducing notation

$$x(t) = \nu_{1,0}(t), \quad y(t) = \nu_{1,2}(t), \quad z(t) = \frac{\nu_{2,4}(t)}{\nu_{2,3}(t)},$$

we arrive at the system of three equations

$$\begin{aligned}
&\frac{d}{dt}x(t) = -y^2(t) + 1,\\
&\frac{d}{dt}y(t) = x(t)y(t) + QI\left(\frac{1}{z(t)} - z(t)\right),\\
&\frac{d}{dt}z(t) = y(t)z(t),
\end{aligned} \tag{1.25}$$

that describes the behavior of the seven-mode model (1.24).

Inclusion of the second stage significantly changes the dynamics of the first stage. Indeed, the initial-values of components $\nu_{2,3}$ and $\nu_{2,4}$ being arbitrarily small, they nevertheless determine certain value of the integral of motion I, and, by virtue of Eqs. (1.25), variable y will repeatedly change the sign.

Consider the case of small values of constant I in more detail. Figure 1.7 shows the numerical solution of Eqs. (1.25) with the initial conditions $x = 0.05$, $y = 1$, and $z = 1$ at $Q = \sqrt{8}$ and $I = 10^{-20}$. As may be seen, two types of motion are characteristic of system (1.25) for small values of constant I; namely, 'fast' motions occur in a small vicinity of either closed trajectory of the Hamiltonian system (1.19) near the plane $z = 1$, and relatively rare hopping events occur due to the changes of sign of variable y at $z \sim I$ or $z \sim -\frac{1}{I}$. Every such hopping event significantly changes the parameters of fast motion trajectories of system (1.25), so that the motion of the system acquires the features characteristic of the dynamic systems with strange attractors (see, e.g., [26]).

To describe 'slow' motions, we introduce variables X and Y by the formulas

$$x(t) = X(t)x_1(t), \quad y(t) = Y(t)y_1(t),$$

where $(x_1(t),\ y_1(t))$ is a solution to Eqs. (1.19). According to Eqs. (1.25), $X(t)$, $Y(t)$, and $z(t)$ satisfy the equations

$$\begin{aligned}
&\frac{d}{dt}X(t) = \frac{1}{x_1(t)}\left[\left(X^2(t) - Y^2(t)\right)y_1^2(t) + 1 - X(t)\right],\\
&\frac{d}{dt}Y(t) = (X(t) - 1)\,Y(t)x_1(t) + \frac{QI}{y_1(t)}\left(\frac{1}{z(t)} - z(t)\right),\\
&\frac{d}{dt}z(t) = 2QY(t)y_1(t)z(t).
\end{aligned}$$

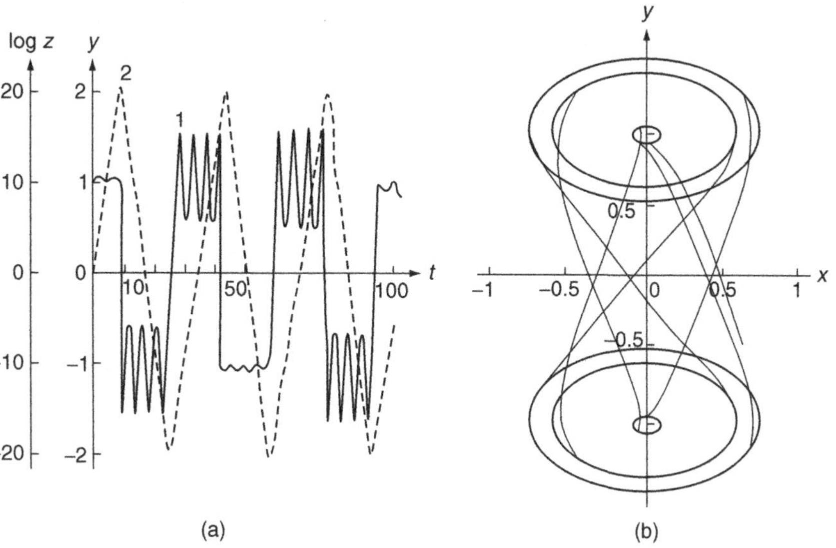

Figure 1.7 (*a*) Time-dependent behavior of components of system (1.25) (curve *1* for y and curve *2* for $\log z$) and (*b*) projection of the phase trajectory of system (1.25) on plane $(x,\ y)$.

Averaging these equations in slow varying quantities over the period[1] $T_1 = \sqrt{2}\pi$, we obtain

$$\frac{d}{dt}X(t) = 0, \quad \frac{d}{dt}Y(t) = QI\left(\frac{1}{Z(t)} - Z(t)\right), \quad \frac{d}{dt}Z(t) = 2QY(t)Z(t), \qquad (1.26)$$

where $Z = \widetilde{z}(t)$. The system (1.26) has the integral of motion

$$H_2 = Y^2(t) + I\left(\frac{1}{Z(t)} + Z(t)\right), \qquad (1.27)$$

so that we can use the corresponding variables to rewrite it in the Hamiltonian form, as it was done for the system (1.19). The motion of system (1.26) along closed trajectories around steady point (0, 1) is characterized by the half-period T_2 (the time between hopping events) given by the formula

$$T_2 = \frac{2}{Q}\int_{Z_1}^{Z_2} \frac{dZ}{Z\sqrt{H_2 - I\left(\frac{1}{Z} + Z\right)}} = \frac{1}{Q\sqrt{I\left(r + \sqrt{r^2-1}\right)}}K\left(\sqrt{\frac{2\sqrt{r^2-1}}{r+\sqrt{r^2-1}}}\right),$$

[1] This expression for T_1 appears to be quite adequate for the solution shown in Fig. 1.7. On closed trajectories of system (1.19) near which this solution passes, the values of Hamiltonian do not exceed 2. Because these trajectories are located in small vicinities of critical points (0, 1) and (0, −1) of this system, we performed averaging assuming that $\widetilde{y}_1(t) = 1$ and $\widetilde{1/y_1(t)} = 1$.

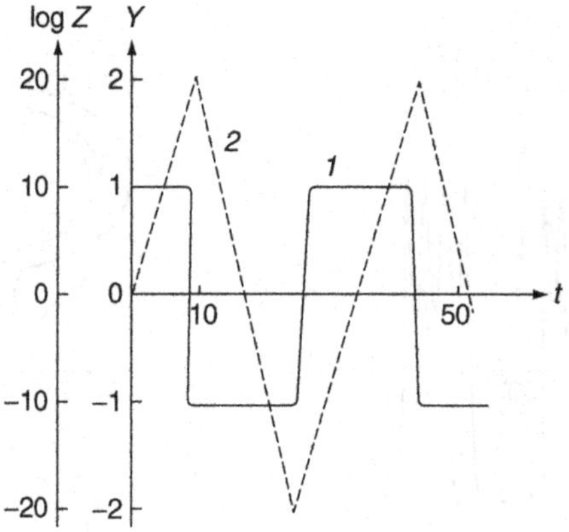

Figure 1.8 Time-dependent behavior of components of system (1.26) (curve *1* for Y and curve *2* for $\log Z$).

where Z_1 and Z_2 are the roots of Eq. (1.27) at $Y = 0$, $r = 2H_2/I$, and $K(z)$ is the complete elliptic integral of the first kind. For small I, we have

$$T_2 \approx \frac{1}{Q\sqrt{H_2}} \ln\left(\frac{4H_2}{I}\right). \tag{1.28}$$

Figure 1.8 shows the numerical solution of Eqs. (1.26) with the initial conditions $Y = 1$, $Z = 1$ (they correspond to the initial conditions used earlier for solving Eqs. (1.25)), constants Q and I also being coincident with those used for solving Eqs. (1.25). The comparison of curves in Figs. 1.7 and 1.8 shows that system (1.26) satisfactorily describes hopping events in system (1.25). The values of half-period T_2 determined by Eq. (1.28) and obtained from the numerical integration of Eqs. (1.25) are 33.54 and 33.51, respectively. We note that system (1.25) has an additional characteristic time $T_3 \sim 1/Q$, whose meaning is the duration of the hopping event.

1.1.4 Systems with Blow-Up Singularities

The simplest stochastic system showing singular behavior in time is described by the equation commonly used in the statistical theory of waves,

$$\frac{d}{dt}x(t) = -\lambda x^2(t) + f(t), \quad x(0) = x_0, \quad \lambda > 0, \tag{1.29}$$

where $f(t)$ is the random function of time.

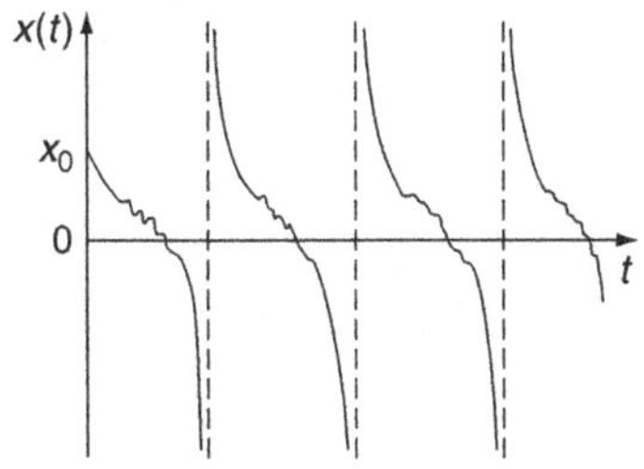

Figure 1.9 Typical realization of the solution to Eq. (1.29).

In the absence of randomness ($f(t) = 0$), the solution to Eq. (1.29) has the form

$$x(t) = \frac{1}{\lambda\,(t - t_0)}, \qquad t_0 = -\frac{1}{\lambda x_0}.$$

For $x_0 > 0$, we have $t_0 < 0$, and solution $x(t)$ monotonically tends to zero with increasing time. On the contrary, for $x_0 < 0$, solution $x(t)$ reaches $-\infty$ within a finite time $t_0 = -1/\lambda x_0$, which means that the solution becomes *singular* and shows the *blow-up behavior*. In this case, random force $f(t)$ has insignificant effect on the behavior of the system. The effect becomes significant only for positive parameter x_0.

Here, the solution, slightly fluctuating, decreases with time as long as it remains positive. On reaching sufficiently small value $x(t)$, the impact of force $f(t)$ can cause the solution to hop into the region of negative values of x, where it reaches the value of $-\infty$ within a certain finite time.

Thus, in the stochastic case, the solution to problem (1.29) shows the blow-up behavior for arbitrary values of parameter x_0 and always reaches $-\infty$ within a finite time t_0. Figure 1.9 schematically shows the temporal realization of the solution $x(t)$ to problem (1.29) for $t > t_0$; its behavior resembles a *quasi-periodic* structure.

1.1.5 Oscillator with Randomly Varying Frequency (Stochastic Parametric Resonance)

In the above stochastic examples, we considered the effect of additive random impacts (forces) on the behavior of systems. The simplest nontrivial system with multiplicative (parametric) impact can be illustrated by the example of *stochastic parametric resonance*. Such a system is described by the second-order equation

$$\frac{d^2}{dt^2}x(t) + \omega_0^2[1 + z(t)]x(t) = 0,$$
$$x(0) = x_0, \qquad \frac{d}{dt}x(0) = v_0, \qquad (1.30)$$

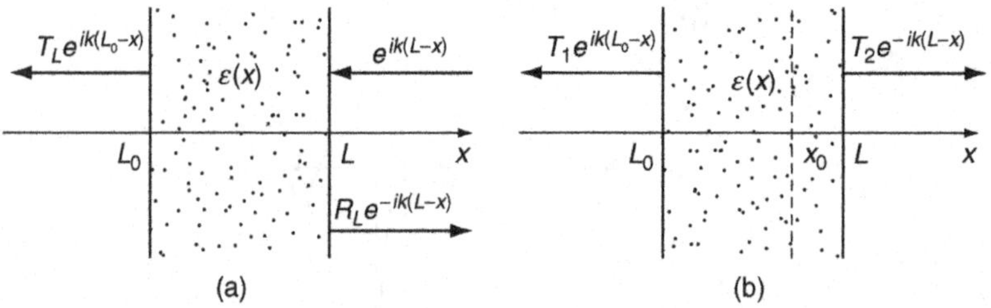

Figure 1.10 (*a*) Plane wave incident on the medium layer and (*b*) source inside the medium layer.

where $z(t)$ is the random function of time. This equation is characteristic of almost all fields of physics. It is physically obvious that dynamic system (1.30) is capable of parametric excitation, because random process $z(t)$ has harmonic components of all frequencies, including frequencies $2\omega_0/n\, (n = 1, 2, \ldots)$ that exactly correspond to the frequencies of parametric resonance in the system with periodic function $z(t)$, as, for example, in the case of the *Mathieu equation*.

1.2 Boundary-Value Problems for Linear Ordinary Differential Equations (Plane Waves in Layered Media)

In the previous section, we considered several dynamic systems described by a system of ordinary differential equations with given initial-values. Now, we consider the simplest linear boundary-value problem, namely, the steady one-dimensional wave problem.

Assume the layer of inhomogeneous medium occupies part of space $L_0 < x < L$, and let the unit-amplitude plane wave $u_0(x) = e^{-ik(x-L)}$ is incident on this layer from the region $x > L$ (Fig. 1.10*a*).

The wavefield satisfies the *Helmholtz equation*,

$$\frac{d^2}{dx^2}u(x) + k^2(x)u(x) = 0, \tag{1.31}$$

where

$$k^2(x) = k^2[1 + \varepsilon(x)],$$

and function $\varepsilon(x)$ describes medium inhomogeneities. We assume that $\varepsilon(x) = 0$, i.e., $k(x) = k$ outside the layer; inside the layer, we set $\varepsilon(x) = \varepsilon_1(x) + i\gamma$, where the real part $\varepsilon_1(x)$ is responsible for wave scattering in the medium and the imaginary part $\gamma \ll 1$ describes absorption of the wave in the medium.

In region $x > L$, the wavefield has the structure

$$u(x) = e^{-ik(x-L)} + R_L e^{ik(x-L)},$$

where R_L is the complex reflection coefficient. In region $x < L_0$, the structure of the wavefield is

$$u(x) = T_L e^{ik(L_0 - x)},$$

where T_L is the complex transmission coefficient. Boundary conditions for Eq. (1.31) are the continuity conditions for the field and the field derivative at layer boundaries; they can be written as follows

$$u(L) + \frac{i}{k}\frac{du(x)}{dx}\bigg|_{x=L} = 2, \qquad u(L_0) - \frac{i}{k}\frac{du(x)}{dx}\bigg|_{x=L_0} = 0. \tag{1.32}$$

Thus, the wavefield in the layer of an inhomogeneous medium is described by the boundary-value problems (1.31), (1.32). Dynamic equation (1.31) coincides in form with Eq. (1.30). Note that the problem under consideration assumes that function $\varepsilon(x)$ is discontinuous at layer boundaries. We will call the boundary-value problem (1.31), (1.32) the unmatched boundary-value problem. In such problems, wave scattering is caused not only by medium inhomogeneities, but also by discontinuities of function $\varepsilon(x)$ at layer boundaries.

If medium parameters (function $\varepsilon_1(x)$) are specified in the statistical form, then solving the stochastic problem (1.31), (1.32) consists in obtaining statistical characteristics of the reflection and transmission coefficients, which are related to the wavefield values at layer boundaries by the relationships

$$R_L = u(L) - 1, \quad T_L = u(L_0),$$

and the wavefield intensity

$$I(x) = |u(x)|^2$$

inside the inhomogeneous medium. Determination of these characteristics constitutes the subject of the *statistical theory of radiative transfer*.

Note that, for $x < L$, from (1.31) follows the equality

$$k\gamma I(x) = \frac{d}{dx}S(x),$$

where energy-flux density $S(x)$ is determined by the relationship

$$S(x) = \frac{i}{2k}\left[u(x)\frac{d}{dx}u^*(x) - u^*(x)\frac{d}{dx}u(x)\right].$$

By virtue of boundary conditions, we have $S(L) = 1 - |R_L|^2$ and $S(L_0) = |T_L|^2$.

For non-absorptive media ($\gamma = 0$), conservation of energy-flux density is expressed by the equality

$$|R_L|^2 + |T_L|^2 = 1. \tag{1.33}$$

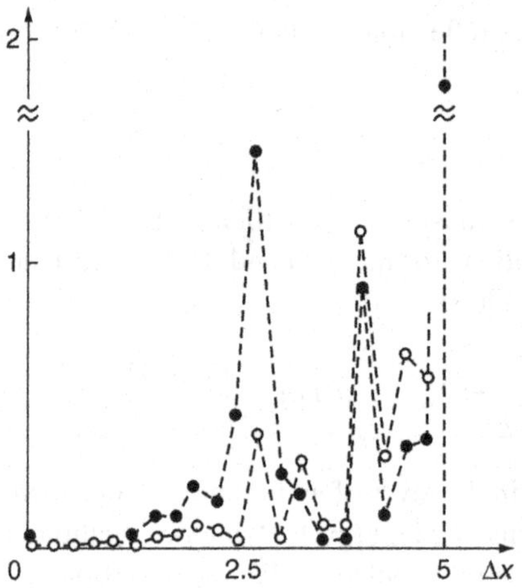

Figure 1.11 Dynamic localization phenomenon simulated for two realizations of medium inhomogeneities.

Consider some features characteristic of solutions to the stochastic boundary-value problem (1.31), (1.32). On the assumption that medium inhomogeneities are absent ($\varepsilon_1(x) = 0$) and absorption γ is sufficiently small, the intensity of the wavefield in the medium slowly decays with distance according to the exponential law

$$I(x) = |u(x)|^2 = e^{-k\gamma(L-x)}. \tag{1.34}$$

Figure 1.11 shows two realizations of the intensity of a wave in a sufficiently thick layer of medium. These realizations were simulated for two realizations of medium inhomogeneities. The difference between them consists in the fact that the corresponding functions $\varepsilon_1(x)$ have different signs in the middle of the layer at a distance of the wavelength. This offers a possibility of estimating the effect of a small medium mismatch on the solution of the boundary problem. Omitting the detailed description of problem parameters, we mention only that this figure clearly shows the prominent tendency of a sharp exponential decay (accompanied by significant spikes toward both higher and nearly zero-valued intensity values), which is caused by multiple reflections of the wave in the chaotically inhomogeneous random medium (the phenomenon of *dynamic localization*). Recall that absorption is small ($\gamma \ll 1$), so that it cannot significantly affect the dynamic localization.

The *imbedding method* offers a possibility of reformulating boundary-value problem (1.31), (1.32) to the dynamic initial-value problem with respect to parameter L (this parameter is the geometrical position of the layer right-hand boundary) by considering the solution to the boundary-value problem as a function of parameter L (see the next lecture). On such reformulation, the reflection coefficient R_L satisfies the Riccati equation

$$\frac{d}{dL}R_L = 2ikR_L + \frac{ik}{2}\varepsilon(L)\,(1 + R_L)^2, \quad R_{L_0} = 0, \tag{1.35}$$

and the wavefield in the medium layer $u(x) \equiv u(x; L)$ satisfies the linear equation

$$\begin{aligned} \frac{\partial}{\partial L} u(x; L) &= iku(x; L) + \frac{ik}{2}\varepsilon(L)\,(1 + R_L)\,u(x; L), \\ u(x; x) &= 1 + R_x, \end{aligned} \tag{1.36}$$

The equation for the reflection coefficient squared modulus $W_L = |R_L|^2$ for absent attenuation (i.e., at $\gamma = 0$) follows from Eq. (1.35)

$$\frac{d}{dL} W_L = -\frac{ik}{2}\varepsilon_1(L)\left(R_L - R_L^*\right)(1 - W_L), \quad W_{L_0} = 0. \tag{1.37}$$

Note that condition $W_{L_0} = 1$ will be the initial condition to Eq. (1.37) in the case of totally reflecting boundary at L_0. In this case, the wave incident on the layer of a non-absorptive medium ($\gamma = 0$) is totally reflected from the layer, i.e., $W_L = 1$.

In the general case of arbitrarily reflecting boundary L_0, the steady-state (independent of L) solution $W_L = 1$ corresponding to the total reflection of incident wave formally exists for a half-space ($L_0 \to -\infty$) filled with non-absorptive random medium, too. This solution, as it will be shown later, is actually realized in the statistical problem with a probability equal to unity.

If, in contrast to the above problem, we assume that function $k(x)$ is continuous at boundary $x = L$, i.e., if we assume that the wave number in the free half-space $x > L$ is equal to $k(L)$, then boundary conditions (1.32) of problem (1.31) will be replaced with the conditions

$$u(L) + \frac{i}{k(L)} \frac{du(x)}{dx}\bigg|_{x=L} = 2, \qquad u(L_0) - \frac{i}{k(L_0)} \frac{du(x)}{dx}\bigg|_{x=L_0} = 0. \tag{1.38}$$

We will call the boundary-value problem (1.31), (1.38) the *matched boundary-value problem.*

The field of a point source located in the layer of random medium is described by the similar boundary-value problem for Green's function of the Helmholtz equation:

$$\frac{d^2}{dx^2} G(x; x_0) + k^2[1 + \varepsilon(x)]G(x; x_0) = 2ik\delta(x - x_0),$$

$$G(L; x_0) + \frac{i}{k} \frac{dG(x; x_0)}{dx}\bigg|_{x=L} = 0, \quad G(L_0; x_0) - \frac{i}{k} \frac{dG(x; x_0)}{dx}\bigg|_{x=L_0} = 0.$$

Outside the layer, the solution has the form of outgoing waves (Fig. 1.10*b*)

$$G(x; x_0) = T_1 e^{ik(x-L)} \quad (x \geq L), \qquad G(x; x_0) = T_2 e^{-ik(x-L_0)} \quad (x \leq L_0).$$

Note that, for the source located at the layer boundary $x_0 = L$, this problem coincides with the boundary-value problem (1.31), (1.32) on the wave incident on the layer, which yields

$$G(x; L) = u(x; L).$$

1.3 Partial Differential Equations

Consider now several dynamic systems (dynamic fields) described by partial differential equations.

1.3.1 Linear First-Order Partial Differential Equations

Diffusion of Density Field Under Random Velocity Field

In the context of linear first-order partial differential equations, the simplest problems concern the equation of continuity for the concentration of a conservative tracer and the equation of transfer of a nonconservative passive tracer by random velocity field $\boldsymbol{U}(\boldsymbol{r}, t)$:

$$\left(\frac{\partial}{\partial t} + \frac{\partial}{\partial \boldsymbol{r}} \boldsymbol{U}(\boldsymbol{r}, t)\right) \rho(\boldsymbol{r}, t) = 0, \qquad \rho(\boldsymbol{r}, 0) = \rho_0(\boldsymbol{r}), \tag{1.39}$$

$$\left(\frac{\partial}{\partial t} + \boldsymbol{U}(\boldsymbol{r}, t) \frac{\partial}{\partial \boldsymbol{r}}\right) q(\boldsymbol{r}, t) = 0, \qquad q(\boldsymbol{r}, 0) = q_0(\boldsymbol{r}). \tag{1.40}$$

The conservative tracer is a tracer whose total mass remains intact

$$M_0 = \int d\boldsymbol{r} \rho(\boldsymbol{r}, t) = \int d\boldsymbol{r} \rho_0(\boldsymbol{r}) \tag{1.41}$$

We can use the method of characteristics to solve the linear first-order partial differential equations (1.39), (1.40). Introducing *characteristic curves* (particles)

$$\frac{d}{dt} \boldsymbol{r}(t) = \boldsymbol{U}(\boldsymbol{r}, t), \qquad \boldsymbol{r}(0) = \boldsymbol{r}_0, \tag{1.42}$$

we can write these equations in the form

$$\begin{aligned} &\frac{d}{dt} \rho(t) = -\frac{\partial \boldsymbol{U}(\boldsymbol{r}, t)}{\partial \boldsymbol{r}} \rho(t), \qquad \rho(0) = \rho_0(\boldsymbol{r}_0), \\ &\frac{d}{dt} q(t) = 0, \qquad q(0) = q_0(\boldsymbol{r}_0). \end{aligned} \tag{1.43}$$

This formulation of the problem corresponds to the *Lagrangian description*, while the initial dynamic equations (1.39), (1.40) correspond to the *Eulerian description*.

Here, we introduced the characteristic vector parameter $\boldsymbol{r}_0$ in the system of equations (1.42), (1.43). With this parameter, Eq. (1.42) coincides with Eq. (1.1) that describes particle dynamics under random velocity field.

The solution of system of equations (1.42), (1.43) depends on initial-value $\boldsymbol{r}_0$,

$$\boldsymbol{r}(t) = \boldsymbol{r}(t|\boldsymbol{r}_0), \qquad \rho(t) = \rho(t|\boldsymbol{r}_0), \tag{1.44}$$

which we will isolate in the argument list by the vertical bar.

The first equality in Eq. (1.44) can be considered as the algebraic equation in characteristic parameter; the solution of this equation

$$\boldsymbol{r}_0 = \boldsymbol{r}_0(\boldsymbol{r}, t)$$

exists because *divergence* $j(t|\boldsymbol{r}_0) = \det \|\partial r_i(t|\boldsymbol{r}_0)/\partial r_{0k}\|$ is different from zero. Consequently, we can write the solution of the initial equation (1.39) in the form

$$\rho(\boldsymbol{r}, t) = \rho(t|\boldsymbol{r}_0(\boldsymbol{r}, t)) = \int d\boldsymbol{r}_0 \rho(t|\boldsymbol{r}_0) j(t|\boldsymbol{r}_0) \delta\left(\boldsymbol{r}(t|\boldsymbol{r}_0) - \boldsymbol{r}\right).$$

Integrating this expression over $\boldsymbol{r}$, we obtain, in view of Eq. (1.41), the relationship between functions $\rho(t|\boldsymbol{r}_0)$ and $j(t|\boldsymbol{r}_0)$

$$\rho(t|\boldsymbol{r}_0) = \frac{\rho_0(\boldsymbol{r}_0)}{j(t|\boldsymbol{r}_0)}, \tag{1.45}$$

and, consequently, the density field can be rewritten in the form of equality

$$\rho(\boldsymbol{r}, t) = \int d\boldsymbol{r}_0 \rho(t|\boldsymbol{r}_0) j(t|\boldsymbol{r}_0) \delta\left(\boldsymbol{r}(t|\boldsymbol{r}_0) - \boldsymbol{r}\right) = \int d\boldsymbol{r}_0 \rho_0(\boldsymbol{r}_0) \delta\left(\boldsymbol{r}(t|\boldsymbol{r}_0) - \boldsymbol{r}\right) \tag{1.46}$$

that ascertains the relationship between the Lagrangian and Eulerian characteristics. For the position of the Lagrangian particle, the delta-function appeared in the right-hand side of this equality is the *indicator function* (see Lecture 3).

For a nondivergent velocity field ($\operatorname{div} \boldsymbol{U}(\boldsymbol{r}, t) = 0$), both particle divergence and particle density are conserved, i.e.,

$$j(t|\boldsymbol{r}_0) = 1, \qquad \rho(t|\boldsymbol{r}_0) = \rho_0(\boldsymbol{r}_0), \qquad q(t|\boldsymbol{r}_0) = q_0(\boldsymbol{r}_0).$$

Consider now the stochastic features of solutions to problem (1.39). A convenient way of analyzing dynamics of a random field consists in using *topographic* concepts. Indeed, in the case of the nondivergent velocity field, temporal evolution of the contour of constant concentration $\rho = \text{const}$ coincides with the dynamics of particles under this velocity field and, consequently, coincides with the dynamics shown in Fig. 1.1*a*. In this case, the area within the contour remains constant and, as it is seen from Fig. 1.1*a*, the pattern becomes highly indented, which is manifested in gradient sharpening and the appearance of contour dynamics for progressively shorter scales. In the other limiting case of a divergent velocity field, the area within the contour tends to zero, and the concentration field condenses in clusters. One can find examples simulated for this case in papers [17, 18]. These features of particle dynamics disappear on averaging over an ensemble of realizations.

Cluster formation in the Eulerian description can be traced using the random velocity field of form (1.3), (1.4), page 4. If $v_x(t) \neq 0$, then concentration field in

Lagrangian description $\rho(t|\boldsymbol{r}_0)$ in the particular case of uniform (independent of $\boldsymbol{r}$) initial distribution $\rho_0(\boldsymbol{r}) = \rho_0$ can be described by the following equation

$$\frac{d}{dt}\rho(t) = -2kv_x(t)\cos(2kx)\rho(t), \quad \rho(0) = \rho_0,$$

which can be rewritten, by virtue of Eq. (1.6), page 6, in the form

$$\frac{d}{dt}\rho(t|\boldsymbol{r}_0) = -2kv_x(t)\frac{1 - e^{2T(t)}\tan^2(kx_0)}{1 + e^{2T(t)}\tan^2(kx_0)}\rho(t|\boldsymbol{r}_0), \tag{1.47}$$

where function $T(t)$ is given by Eq. (1.7), page 6. Integrating Eq. (1.47), we obtain the Lagrangian representation of the velocity field in the framework of the model under consideration

$$\rho(t|x_0)/\rho_0 = \left[e^{-T(t)}\cos^2(kx_0) + e^{T(t)}\sin^2(kx_0)\right].$$

Eliminating characteristic parameter x_0 with the use of equalities

$$\begin{aligned}\sin^2(kx_0) &= \frac{e^{-T(t)}\sin^2(kx(t))}{e^{T(t)}\cos^2(kx(t)) + e^{-T(t)}\sin^2(kx(t))},\\ \cos^2(kx_0) &= \frac{e^{T(t)}\cos^2(kx(t))}{e^{T(t)}\cos^2(kx(t)) + e^{-T(t)}\sin^2(kx(t))},\end{aligned} \tag{1.48}$$

following from Eq. (1.6), page 6, we pass to the Eulerial description

$$\rho(\boldsymbol{r}, t)/\rho_0 = \frac{1}{e^{T(t)}\cos^2(kx) + e^{-T(t)}\sin^2(kx)}. \tag{1.49}$$

Expression (1.49) shows that the density field is low everywhere excluding the neighborhoods of points $kx = n\frac{\pi}{2}$, where $\rho(x,t)/\rho_0 = e^{\pm T(t)}$ and is sufficiently high if random factor $T(t)$ has appropriate sign.

Thus, in the problem under consideration, the cluster structure of the density field in the Eulerian description is formed in the neighborhoods of points

$$kx = n\frac{\pi}{2} \quad (n = 0, \pm 1, \pm 2, \ldots).$$

Note that Eulerian density field (1.49) averaged over spatial variables is independent of random factor $T(t)$,

$$\overline{\rho(x,t)/\rho_0} = 1,$$

and the average square of the density mainly grows with time

$$\overline{(\rho(x,t)/\rho_0)^2} = \frac{1}{2}\left(e^{T(t)} + e^{-T(t)}\right).$$

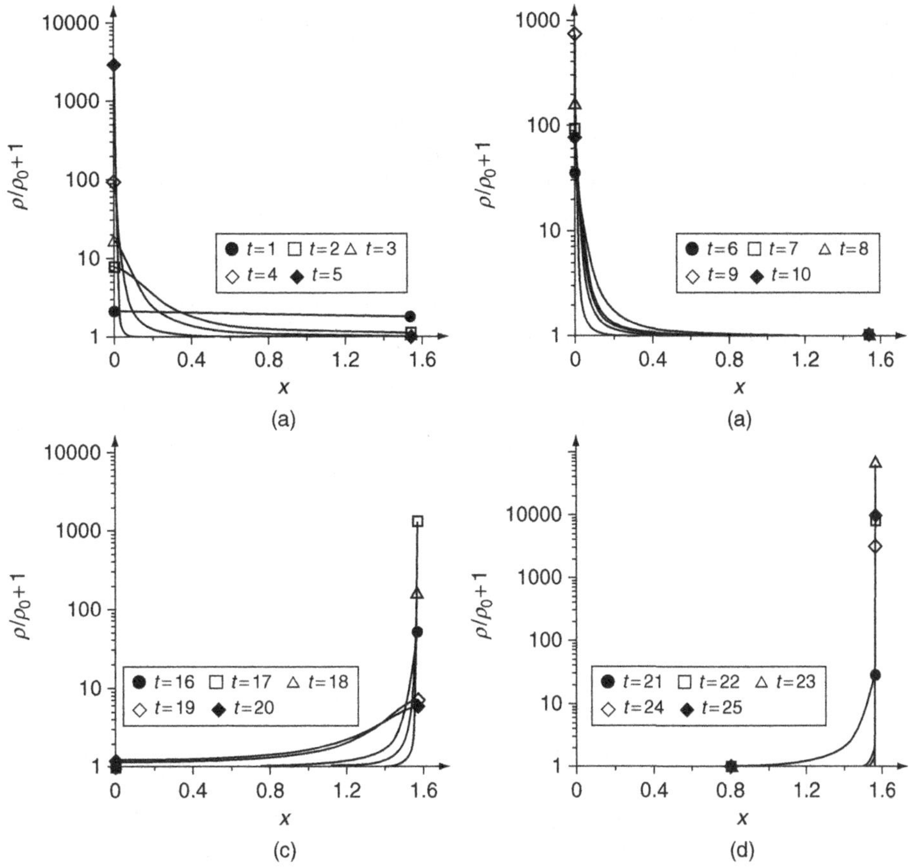

Figure 1.12 Space-time evolution of the Eulerian density field given by Eq. (1.49).

Figure 1.12 shows the Eulerian concentration field $1 + \rho(\boldsymbol{r}, t)/\rho_0$ and its space-time evolution calculated by Eq. (1.49) in the dimensionless space-time variables (the density field is added with a unity to avoid the difficulties of dealing with nearly zero-valued concentrations in the logarithmic scale). This figure shows successive patterns of concentration field rearrangement toward narrow neighborhoods of points $x \approx 0$ and $x \approx \pi/2$, i.e., the formation of clusters, in which relative density is as high as $10^3 - 10^4$, while relative density is practically zero in the whole other space. Note that the realization of the density field passes through the initial homogeneous state at the instants t such that $T(t) = 0$. As is seen from figures, the lifetimes of such clusters coincide on the order of magnitude with the time of cluster formation.

This model provides an insight into the difference between the diffusion processes in divergent and nondivergent velocity fields. In nondivergent (incompressible) velocity fields, particles (and, consequently, density field) have no time for attracting to stable centers of attraction during the lifetime of these centers, and particles slightly fluctuate relative to their initial location. On the contrary, in the divergent (compressible) velocity field, lifetime of stable centers of attraction is sufficient for particles to attract to them, because the speed of attraction increases exponentially, which is clearly seen from Eq. (1.49).

From the above description, it becomes obvious that dynamic equation (1.39) considered as a model equation describing actual physical phenomena can be used only on finite temporal intervals. A more complete analysis assumes the consideration of the field of tracer concentration gradient $\boldsymbol{p}(\boldsymbol{r}, t) = \nabla \rho(\boldsymbol{r}, t)$ that satisfies the equation (repeating indices assume summation)

$$\left(\frac{\partial}{\partial t} + \frac{\partial}{\partial \boldsymbol{r}} \boldsymbol{U}(\boldsymbol{r}, t)\right) p_i(\boldsymbol{r}, t) = -p_k(\boldsymbol{r}, t) \frac{\partial U_k(\boldsymbol{r}, t)}{\partial r_i} - \rho(\boldsymbol{r}, t) \frac{\partial^2 U_k(\boldsymbol{r}, t)}{\partial r_i \partial r_k}, \tag{1.50}$$
$$\boldsymbol{p}(\boldsymbol{r}, 0) = \boldsymbol{p}_0(\boldsymbol{r}) = \nabla \rho_0(\boldsymbol{r}).$$

In addition, one should also include the effect of the dynamic molecular diffusion (with the dynamic diffusion coefficient μ_ρ) that smooths the mentioned sharpen of the gradient; this effect is described by the linear second-order partial differential equation

$$\left(\frac{\partial}{\partial t} + \frac{\partial}{\partial \boldsymbol{r}} \boldsymbol{U}(\boldsymbol{r}, t)\right) \rho(\boldsymbol{r}, t) = \mu_\rho \Delta \rho(\boldsymbol{r}, t), \quad \rho(\boldsymbol{r}, 0) = \rho_0(\boldsymbol{r}). \tag{1.51}$$

Diffusion of Magnetic Field in a Random Velocity Field

The diffusion of such passive fields as the scalar density (particle concentration) field and the magnetic field is an important problem of the theory of turbulence in magnetohydrodynamics. Here, the basic stochastic equations are the continuity equation (1.39) for density field $\rho(\boldsymbol{r}, t)$ (see previous section) and the induction equation for nondivergent magnetic field $\boldsymbol{H}(\boldsymbol{r}, t)$ ($\operatorname{div} \boldsymbol{H}(\boldsymbol{r}, t) = 0$) [27]

$$\frac{\partial}{\partial t} \boldsymbol{H}(\boldsymbol{r}, t) = \operatorname{curl}\left[\boldsymbol{U}(\boldsymbol{r}, t) \times \boldsymbol{H}(\boldsymbol{r}, t)\right], \quad \boldsymbol{H}(\boldsymbol{r}, 0) = \boldsymbol{H}_0(\boldsymbol{r}). \tag{1.52}$$

In Eq. (1.52), $\boldsymbol{U}(\boldsymbol{r}, t)$ is the hydrodynamic velocity field and pseudovector $\boldsymbol{C} = \boldsymbol{A} \times \boldsymbol{B}$ is the vector product of vectors $\boldsymbol{A}$ and $\boldsymbol{B}$,

$$C_i = \varepsilon_{ijk} A_j B_k,$$

where ε_{ijk} is the pseudotensor such that $\varepsilon_{ijk} = 0$ if indices i, j, and k are not all different and $\varepsilon_{ijk} = 1$ or $\varepsilon_{ijk} = -1$, if indices i, j, and k are all different and form cyclic or anticyclic sequence (see, e.g., [28]). The operator

$$\operatorname{curl} \boldsymbol{C}(\boldsymbol{r}, t) = [\nabla \times \boldsymbol{C}(\boldsymbol{r}, t)], \quad \operatorname{curl} \boldsymbol{C}(\boldsymbol{r}, t)|_i = \varepsilon_{ijk} \frac{\partial}{\partial r_j} C_k(\boldsymbol{r}, t)$$

is called the *vortex* of field $\boldsymbol{C}(\boldsymbol{r}, t)$.

In magnetohydrodynamics, velocity field $\boldsymbol{U}(\boldsymbol{r}, t)$ is generally described by the Navier–Stokes equation complemented with the density of extraneous electromagnetic forces

$$f(\boldsymbol{r}, t) = \frac{1}{4\pi}\left[\operatorname{curl}\boldsymbol{H}(\boldsymbol{r}, t) \times \boldsymbol{H}(\boldsymbol{r}, t)\right].$$

Nevertheless we, as earlier, will consider velocity field $\boldsymbol{U}(\boldsymbol{r}, t)$ as random field whose statistical parameters are given.

The product of two pseudotensors is a tensor, and, in the case of pseudotensors ε, we have the following equality

$$\varepsilon_{ilm}\varepsilon_{jpq} = \delta_{ij}\delta_{lp}\delta_{mq} + \delta_{ip}\delta_{lq}\delta_{mj} + \delta_{iq}\delta_{lj}\delta_{mp} - \delta_{ij}\delta_{lq}\delta_{mp} - \delta_{ip}\delta_{lj}\delta_{mq} - \delta_{iq}\delta_{lp}\delta_{mj}. \tag{1.53}$$

Setting $j = m$ (repeated indices assume summation), we obtain

$$\varepsilon_{ilm}\varepsilon_{mpq} = (d-2)(\delta_{ip}\delta_{lq} - \delta_{iq}\delta_{lp}), \tag{1.54}$$

where d is the dimension of space, so that the above convolution becomes zero in the two-dimensional case.

Thus, the double vector product is given by the formula

$$[\boldsymbol{C} \times [\boldsymbol{A} \times \boldsymbol{B}]]_i = \varepsilon_{ilm}\varepsilon_{mpq}C_l A_p B_q = C_q A_i B_q - C_p A_p B_i. \tag{1.55}$$

If fields $\boldsymbol{C}$, $\boldsymbol{A}$, and $\boldsymbol{B}$ are the conventional vector fields, Eq. (1.55) assumes the form

$$[\boldsymbol{C} \times [\boldsymbol{A} \times \boldsymbol{B}]] = (\boldsymbol{C} \cdot \boldsymbol{B})\boldsymbol{A} - (\boldsymbol{C} \cdot \boldsymbol{A})\boldsymbol{B}. \tag{1.56}$$

In the case operator vector field $\boldsymbol{C} = \nabla = \frac{\partial}{\partial \boldsymbol{r}}$, Eq. (1.55) results in the expression

$$\operatorname{curl}\left[\boldsymbol{A}(\boldsymbol{r}) \times \boldsymbol{B}(\boldsymbol{r})\right] = \left(\frac{\partial}{\partial \boldsymbol{r}} \cdot \boldsymbol{B}(\boldsymbol{r})\right)\boldsymbol{A}(\boldsymbol{r}) - \left(\frac{\partial}{\partial \boldsymbol{r}} \cdot \boldsymbol{A}(\boldsymbol{r})\right)\boldsymbol{B}(\boldsymbol{r}). \tag{1.57}$$

Note that, if vector field $\boldsymbol{A}$ is an operator in Eq. (1.55), $\boldsymbol{A} = \nabla = \frac{\partial}{\partial \boldsymbol{r}}$, then we have

$$[\boldsymbol{C}(\boldsymbol{r}) \times \operatorname{curl}\boldsymbol{B}(\boldsymbol{r})] = C_q(\boldsymbol{r})\frac{\partial}{\partial \boldsymbol{r}}B_q(\boldsymbol{r}) - \left(\boldsymbol{C}(\boldsymbol{r}) \cdot \frac{\partial}{\partial \boldsymbol{r}}\right)\boldsymbol{B}(\boldsymbol{r})$$

and, in particular,

$$[\boldsymbol{B}(\boldsymbol{r}) \times \operatorname{curl}\boldsymbol{B}(\boldsymbol{r})] = \frac{1}{2}\frac{\partial}{\partial \boldsymbol{r}}\boldsymbol{B}^2(\boldsymbol{r}) - \left(\boldsymbol{B}(\boldsymbol{r}) \cdot \frac{\partial}{\partial \boldsymbol{r}}\right)\boldsymbol{B}(\boldsymbol{r}). \tag{1.58}$$

Using Eq. (1.57), we can rewrite Eq. (1.52) in the form

$$\left(\frac{\partial}{\partial t}+\frac{\partial}{\partial \boldsymbol{r}}\boldsymbol{u}(\boldsymbol{r},t)\right)\boldsymbol{H}(\boldsymbol{r},t)=\left(\boldsymbol{H}(\boldsymbol{r},t)\cdot\frac{\partial}{\partial \boldsymbol{r}}\right)\boldsymbol{u}(\boldsymbol{r},t),$$
$$\boldsymbol{H}(\boldsymbol{r},0)=\boldsymbol{H}_0(\boldsymbol{r}). \tag{1.59}$$

Dynamic system (1.59) is a conservative system, and magnetic field flux $\int d\boldsymbol{r}\,\boldsymbol{H}(\boldsymbol{r},t)$ remains constant during evolution.

We are interested in the evolution of magnetic field in space and time from given smooth initial distributions and, in particular, simply homogeneous ones, $\boldsymbol{H}_0(\boldsymbol{r}) = \boldsymbol{H}_0$. Clearly, at the initial evolutionary stages of the process, the effects of dynamic diffusion are insignificant, and Eq. (1.59) describes namely this case. Further stages of evolution require consideration of the effects of molecular diffusion; these effects are described by the equation

$$\left(\frac{\partial}{\partial t}+\frac{\partial}{\partial \boldsymbol{r}}\boldsymbol{u}(\boldsymbol{r},t)\right)\boldsymbol{H}(\boldsymbol{r},t)=\left(\boldsymbol{H}(\boldsymbol{r},t)\cdot\frac{\partial}{\partial \boldsymbol{r}}\right)\boldsymbol{u}(\boldsymbol{r},t)+\mu_{\boldsymbol{H}}\Delta\boldsymbol{H}(\boldsymbol{r},t), \tag{1.60}$$

where $\mu_{\boldsymbol{H}}$ is the dynamic diffusion coefficient for the magnetic field.

Note that, similarly to the case of tracer density [16], velocity filed model (1.3), (1.4), page 4, allows obtaining the magnetic field in an explicit form if molecular diffusion can be neglected. With the use of this model, the induction equation for homogeneous initial condition (1.59) assumes the form

$$\left(\frac{\partial}{\partial t}+v_x(t)\sin 2(kx)\frac{\partial}{\partial x}\right)\boldsymbol{H}(x,t)$$
$$=2k\cos 2(kx)\,[\boldsymbol{v}(t)H_x(x,t)-v_x(t)\boldsymbol{H}(x,t)],\quad \boldsymbol{H}(x,0)=\boldsymbol{H}_0,$$

from which follows that the x-component of the magnetic field remains constant ($H_x(x,t)=H_{x0}$), and the existence of this component H_{x0} causes the appearance of an additional source of magnetic field in the transverse (y,z)-plane

$$\left(\frac{\partial}{\partial t}+v_x(t)\sin 2(kx)\frac{\partial}{\partial x}\right)\boldsymbol{H}_\perp(\boldsymbol{r},t)$$
$$=2k\cos 2(kx)\,[\boldsymbol{v}_\perp(t)H_{x0}-v_x(t)\boldsymbol{H}_\perp(x,t)],\quad \boldsymbol{H}_\perp(x,0)=\boldsymbol{H}_{\perp 0}. \tag{1.61}$$

Equation (1.61) is a partial differential equation, and we can solve it using the method of characteristics (the Lagrangian description). The characteristics satisfy the equations

$$\frac{d}{dt}x(t|x_0)=v_x(t)\sin 2(kx(t|x_0)),$$
$$\frac{d}{dt}\boldsymbol{H}_\perp(t|x_0)=2k\cos 2(kx|x_0)\,[\boldsymbol{v}_\perp(t)H_{x0}-v_x(t)\boldsymbol{H}_\perp(t|x_0)],$$
$$x(0|x_0)=x_0,\quad \boldsymbol{H}_\perp(0|x_0)=\boldsymbol{H}_{\perp 0}, \tag{1.62}$$

where the vertical bar separates the dependence on characteristic parameter x_0.

The first equation in Eqs. (1.62) describes particle diffusion, and its solution has the form Eq. (1.6)

$$x(t|x_0) = \frac{1}{k} \arctan\left[e^{T(t)} \tan(kx_0)\right],$$

where function $T(t)$ is given by Eq. (1.7), page 6. The solution of the equation in the magnetic field has the form

$$\begin{aligned} \boldsymbol{H}_{\perp}(t|x_0) &= \left[e^{-T(t)}\cos^2(kx_0) + e^{T(t)}\sin^2(kx_0)\right]\boldsymbol{H}_{\perp 0} \\ &\quad + 2k\left[e^{-T(t)}\cos^2(kx_0) + e^{T(t)}\sin^2(kx_0)\right] \\ &\quad \times \int_0^t d\tau \frac{\left[e^{-T(\tau)}\cos^2(kx_0) - e^{T(\tau)}\sin^2(kx_0)\right]}{\left[e^{-T(\tau)}\cos^2(kx_0) + e^{T(\tau)}\sin^2(kx_0)\right]^2} v_x(\tau)\boldsymbol{v}_{\perp}(\tau)H_{x0}. \end{aligned}$$

Eliminating now characteristic parameter x_0 with the use of Eqs. (1.48), page 26, we pass to the Eulerian description

$$\begin{aligned} \boldsymbol{H}_{\perp}(x,t) &= \frac{\rho(x,t)}{\rho_0}\boldsymbol{H}_{\perp 0} \\ &\quad + 2kH_{x0}\int_0^t d\tau \frac{\left[e^{T(t)-T(\tau)}\cos^2(kx) - e^{-T(t)+T(\tau)}\sin^2(kx)\right]}{\left[e^{T(t)-T(\tau)}\cos^2(kx) + e^{-T(t)+T(\tau)}\sin^2(kx)\right]^2} v_x(\tau)\boldsymbol{v}_{\perp}(\tau), \end{aligned} \tag{1.63}$$

where the density of passive tracer $\rho(x,t)$ is described by Eq. (1.49). Making now the change of integration variables $t - \tau = \lambda$ in Eq. (1.63), we can rewrite it as

$$\begin{aligned} \boldsymbol{H}_{\perp}(x,t) &= \frac{\rho(x,t)}{\rho_0}\boldsymbol{H}_{\perp 0} + 2kH_{x0}\int_0^t d\lambda \frac{\left[e^{T(t)-T(\tau)}\cos^2(kx) - e^{-T(t)+T(\tau)}\sin^2(kx)\right]}{\left[e^{T(t)-T(\tau)}\cos^2(kx) + e^{-T(t)+T(\tau)}\sin^2(kx)\right]^2} \\ &\quad \times v_x(t-\lambda)\boldsymbol{v}_{\perp}(t-\lambda), \end{aligned}$$

where

$$T(t) - T(\tau) = \int_\tau^t d\xi v_x(\xi) = \int_0^{t-\tau} d\eta v_x(t-\eta) = \int_0^{\lambda} d\eta v_x(t-\eta).$$

Hence, dealing with the one-time statistical characteristics of magnetic field, we can replace $v_x(t-\lambda)$ with $v_x(\lambda)$ in view of stationarity of the velocity field (see Section 4.2.1, page 95) and rewrite Eq. (1.63) in a statistically equivalent form,

$$H_\perp(x,t) = \frac{\rho(x,t)}{\rho_0} H_{\perp 0} + 2kH_{x0}\int_0^t d\tau \frac{\left[e^{T(\tau)}\cos^2(kx) - e^{-T(\tau)}\sin^2(kx)\right]}{\left[e^{T(\tau)}\cos^2(kx) + e^{-T(\tau)}\sin^2(kx)\right]^2} v_x(\tau) v_\perp(\tau). \quad (1.64)$$

The first term describes magnetic field clustering like the density field clustering if $H_{\perp 0} \neq 0$. The second term describes the generation of a magnetic field in the transverse (y,z)-plane due to the presence of an initial field H_{x0}. At $H_{\perp 0} = 0$, this term, proportional to the square of the random velocity field, determines the situation. Like the density field, the structure of this field is also clustered.

Figures 1.13, 1.14 present the calculated space–time evolution of a realization of energy of the magnetic field generated in the transverse plane, $E(x,t) = H_\perp^2(x,t)$, in dimensionless variables (see Problem 8.8, page 226) at $H_{\perp 0} = 0$ for the same realization of the random process $T(t)$ as that presented previously in Fig. 1.2*a*.

First of all, we note that the total energy of the generated magnetic field concentrated in the segment $[0, \pi/2]$ increases rapidly with time (Fig. 1.13*a*). A general space–time structure of the magnetic energy clustering is shown in Fig. 1.13*b*. This structure was calculated in the following way. The coordinates of points x_i are plotted along the x-axis and the time is plotted along the t-axis (at 0.1 steps). The points are marked as white squares (unseen) if they contain less than 1% of the energy available in the entire layer at current t and as black squares if they contain more than 1% of the energy available in the entire layer at the time in question. There are a total of 40,000 points (100 steps in x and 400 steps in time).

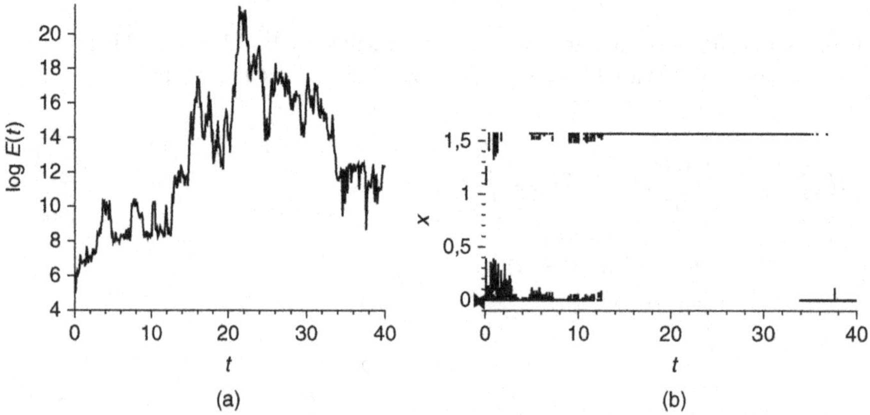

Figure 1.13 (*a*) Temporal evolution of the total magnetic energy in segment $[0, \pi/2]$ and (*b*) cluster structure in the (x,t)-plane.

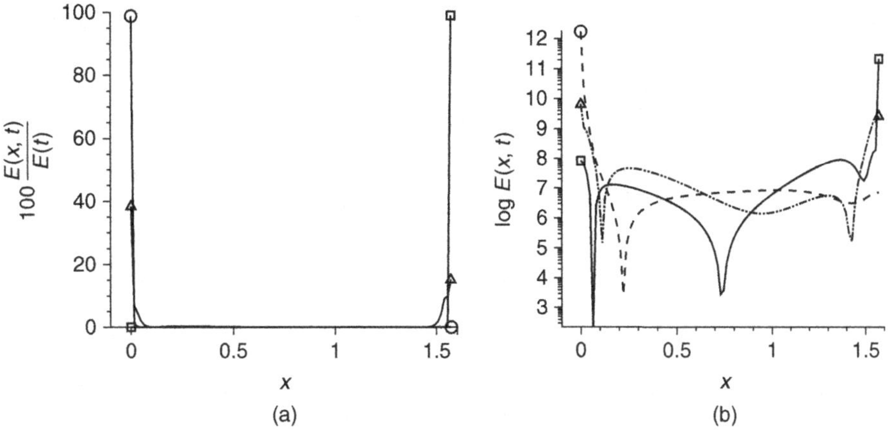

Figure 1.14 Dynamics of disappearance of a cluster at point 0 and appearance of a cluster at point $\pi/2$. The circles, triangles, and squares mark the curves corresponding to time instants $t = 10.4$, 10.8, and 11.8, respectively.

A more detailed pattern of the evolution of clusters with time is presented in Fig. 1.14. Figure 1.14*a* shows the percentage ratio of the generated magnetic energy field contained in a cluster to the total energy in the layer at the time under consideration; Fig. 1.14*b* shows the dynamics of the flow of magnetic energy perturbations from one boundary of the region to the other.

1.3.2 Quasilinear and Nonlinear First-Order Partial Differential Equations

Consider now the simplest quasilinear equation for scalar quantity $q(\boldsymbol{r}, t)$, which we write in the form

$$\left(\frac{\partial}{\partial t} + \boldsymbol{U}(t, q)\frac{\partial}{\partial \boldsymbol{r}}\right) q(\boldsymbol{r}, t) = Q(t, q), \qquad q(\boldsymbol{r}, 0) = q_0(\boldsymbol{r}), \tag{1.65}$$

where we assume for simplicity that functions $\boldsymbol{U}(t, q)$ and $Q(t, q)$ have no explicit dependence on spatial variable $\boldsymbol{r}$.

Supplement Eq. (1.65) with the equation for the gradient $\boldsymbol{p}(\boldsymbol{r}, t) = \nabla q(\boldsymbol{r}, t)$, which follows from Eq. (1.65), and the equation of continuity for conserved quantity $I(\boldsymbol{r}, t)$:

$$\begin{gathered}\left(\frac{\partial}{\partial t} + \boldsymbol{U}(t, q)\frac{\partial}{\partial \boldsymbol{r}}\right) \boldsymbol{p}(\boldsymbol{r}, t) + \frac{\partial \{\boldsymbol{U}(t, q)\boldsymbol{p}(\boldsymbol{r}, t)\}}{\partial q}\boldsymbol{p}(\boldsymbol{r}, t) = \frac{\partial Q(t, q)}{\partial q}\boldsymbol{p}(\boldsymbol{r}, t), \\ \frac{\partial}{\partial t}I(\boldsymbol{r}, t) + \frac{\partial}{\partial \boldsymbol{r}}\{\boldsymbol{U}(t, q)I(\boldsymbol{r}, t)\} = 0.\end{gathered} \tag{1.66}$$

From Eqs. (1.66) it follows that

$$\int d\boldsymbol{r} I(\boldsymbol{r}, t) = \int d\boldsymbol{r} I_0(\boldsymbol{r}). \tag{1.67}$$

In terms of characteristic curves determined by the system of ordinary differential equations, Eqs. (1.65) and (1.66) can be written in the form

$$\begin{aligned}
&\frac{d}{dt}\boldsymbol{r}(t) = \boldsymbol{U}(t, q), \quad \frac{d}{dt}q(t) = Q(t, q), \quad \boldsymbol{r}(0) = \boldsymbol{r}_0, \quad q(0) = q_0(\boldsymbol{r}_0), \\
&\frac{d}{dt}\boldsymbol{p}(t) = -\frac{\partial\{\boldsymbol{U}(t, q)\boldsymbol{p}(t)\}}{\partial q}\boldsymbol{p}(t) + \frac{\partial Q(t, q)}{\partial q}\boldsymbol{p}(t), \quad \boldsymbol{p}(0) = \frac{\partial q_0(\boldsymbol{r}_0)}{\partial \boldsymbol{r}_0}, \\
&\frac{d}{dt}I(t) = -\frac{\partial\{\boldsymbol{U}(t, q)\boldsymbol{p}(t)\}}{\partial q}I(t), \quad I(0) = I_0(\boldsymbol{r}_0).
\end{aligned} \tag{1.68}$$

Thus, the Lagrangian description considers the system (1.68) as the initial-value problem. In this description, the two first equations form a closed system that determines characteristic curves.

Expressing now characteristic parameter $\boldsymbol{r}_0$ in terms of t and $\boldsymbol{r}$, one can write the solution to Eqs. (1.65) and (1.66) in the Eulerian description as

$$\begin{aligned}
q(\boldsymbol{r}, t) &= \int d\boldsymbol{r}_0 q(t|\boldsymbol{r}_0) j(t|\boldsymbol{r}_0) \delta(\boldsymbol{r}(t|\boldsymbol{r}_0) - \boldsymbol{r}), \\
I(\boldsymbol{r}, t) &= \int d\boldsymbol{r}_0 I(t|\boldsymbol{r}_0) j(t|\boldsymbol{r}_0) \delta(\boldsymbol{r}(t|\boldsymbol{r}_0) - \boldsymbol{r}).
\end{aligned} \tag{1.69}$$

The feature of the transition from the Lagrangian description (1.68) to the Eulerian description (1.69) consists in the general appearance of ambiguities, which results in discontinuous solutions. These ambiguities are related to the fact that the divergence–Jacobian

$$j(t|\boldsymbol{r}_0) = \det\left\| \frac{\partial}{\partial r_{0k}} r_i(t|\boldsymbol{r}_0) \right\|$$

– can vanish at certain moments.

Quantities $I(t|\boldsymbol{r}_0)$ and $j(t|\boldsymbol{r}_0)$ are not independent. Indeed, integrating $I(\boldsymbol{r}, t)$ in Eq. (1.69) over $\boldsymbol{r}$ and taking into account Eq. (1.67), we see that there is the evolution integral

$$j(t|\boldsymbol{r}_0) = \frac{I_0(\boldsymbol{r}_0)}{I(t|\boldsymbol{r}_0)}, \tag{1.70}$$

from which follows that zero-valued divergence $j(t|\boldsymbol{r}_0)$ is accompanied by the infinite value of conservative quantity $I(t|\boldsymbol{r}_0)$.

It is obvious that all these results can be easily extended to the case in which functions $\boldsymbol{U}(\boldsymbol{r}, t, q)$ and $Q(\boldsymbol{r}, t, q)$ explicitly depend on spatial variable $\boldsymbol{r}$ and Eq. (1.65) itself is the vector equation.

As a particular physical example, we consider the equation for the velocity field $V(r, t)$ of low-inertia particles moving in the hydrodynamic flow whose velocity field is $u(r, t)$ (see, e.g., [29])

$$\left(\frac{\partial}{\partial t} + V(r, t)\frac{\partial}{\partial r}\right) V(r, t) = -\lambda\,[V(r, t) - u(r, t)]. \tag{1.71}$$

We will assume this equation the phenomenological equation.

In the general case, the solution to Eq. (1.71) can be non-unique, it can have discontinuities, etc. However, in the case of asymptotically small inertia property of particles (parameter $\lambda \to \infty$), which is of our concern here, the solution will be unique during reasonable temporal intervals. Note that, in the right-hand side of Eq. (1.71), term $F(r, t) = \lambda V(r, t)$ linear in the velocity field $V(r, t)$ is, according to the known *Stokes formula*, the resistance force acting on a slowly moving particle. If we approximate the particle by a sphere of radius a, parameter λ will be $\lambda = 6\pi a\eta/m_{\rm p}$, where η is the dynamic viscosity coefficient and $m_{\rm p}$ is the mass of the particle (see, e.g., [30]).

From Eq. (1.71) follows that velocity field $V(r, t)$ is the divergent field ($\operatorname{div} V(r, t) \neq 0$) even if hydrodynamic flow $u(r, t)$ is the nondivergent field ($\operatorname{div} u(r, t) = 0$). As a consequence, particle number density $n(r, t)$ in nondivergent hydrodynamic flows, which satisfies the linear equation of continuity

$$\left(\frac{\partial}{\partial t} + \frac{\partial}{\partial r}V(r, t)\right) n(r, t) = 0, \quad n(r, 0) = n_0(r), \tag{1.72}$$

similar to Eq. (1.39), shows the cluster behavior [31].

For large parameters $\lambda \to \infty$ (inertia-less particles), we have

$$V(r, t) \approx u(r, t), \tag{1.73}$$

and particle number density $n(r, t)$ in nondivergent hydrodynamic flows shows no cluster behavior.

The first-order partial differential equation (1.71) (the Eulerian description) is equivalent to the system of ordinary differential characteristic equations (the Lagrangian description)

$$\begin{aligned} &\frac{d}{dt}r(t) = V(r(t), t), \quad r(0) = r_0, \\ &\frac{d}{dt}V(t) = -\lambda\,[V(t) - u(r(t), t)], \quad V(0) = V_0(r_0), \end{aligned} \tag{1.74}$$

that describe diffusion of a particle under random external force and linear friction and coincide with Eq. (1.12).

Conventional statistical description usually assumes that fluctuations of the hydrodynamic velocity field are sufficiently small. For this reason, we can linearize the system of equations (1.74) and rewrite it in the form (for simplicity, we assume that the mean flow is absent and use the zero-valued initial conditions)

$$\frac{d}{dt}\boldsymbol{r}(t) = \boldsymbol{v}(t), \quad \frac{d}{dt}\boldsymbol{v}(t) = -\lambda\left[\boldsymbol{v}(t) - \boldsymbol{f}(t)\right],$$
$$\boldsymbol{r}(0) = 0, \quad \boldsymbol{v}(0) = 0, \tag{1.75}$$

the stochastic solution to which has the form

$$\boldsymbol{v}(t) = \lambda \int_0^t d\tau e^{-\lambda(t-\tau)} \boldsymbol{f}(\tau), \quad \boldsymbol{r}(t) = \int_0^t d\tau \left[1 - e^{-\lambda(t-\tau)}\right] \boldsymbol{f}(\tau).$$

Note that the closed linear equation of the first order in velocity $\boldsymbol{v}(t)$ is called the *Langevin equation.*

Now, we turn back to Eq. (1.71). Setting $\lambda = 0$, we arrive at the equation

$$\left(\frac{\partial}{\partial t} + \boldsymbol{V}(\boldsymbol{r}, t)\frac{\partial}{\partial \boldsymbol{r}}\right)\boldsymbol{V}(\boldsymbol{r}, t) = 0, \quad \boldsymbol{V}(\boldsymbol{r}, 0) = \boldsymbol{V}_0(\boldsymbol{r}), \tag{1.76}$$

called the *Riemann equation.* It describes free propagation of the *nonlinear Riemann wave.* The solution to this equation obviously satisfies the transcendental equation

$$\boldsymbol{V}(\boldsymbol{r}, t) = \boldsymbol{V}_0\left(\boldsymbol{r} - t\boldsymbol{V}(\boldsymbol{r}, t)\right).$$

In the special case of the one-dimensional equation

$$\left(\frac{\partial}{\partial t} + q(x, t)\frac{\partial}{\partial x}\right)q(x, t) = 0, \quad q(x, 0) = q_0(x), \tag{1.77}$$

the solution can be drawn in both implicit and explicit form. This equation coincides with Eq. (1.65) at $G(t, q) = 0$, $U(t, q) = q(x, t)$.

The method of characteristics applied to Eq. (1.77) gives

$$q(t|x_0) = q_0(x_0), \quad x(t|x_0) = x_0 + tq_0(x_0), \tag{1.78}$$

so that the solution of Eq. (1.77) can be written in the form of the transcendental equation

$$q(x, t) = q_0\left(x - tq(x, t)\right),$$

from which follows an expression for spatial derivative

$$p(x, t) = \frac{\partial}{\partial x}q(x, t) = \frac{q_0'(x_0)}{1 + tq_0'(x_0)}, \tag{1.79}$$

where

$$x_0 = x - tq(x,t) \quad \text{and} \quad q_0'(x_0) = \frac{d}{dx_0} q_0(x_0).$$

The function $p(x,t)$ by itself satisfies here the equation

$$\left(\frac{\partial}{\partial t} + q(x,t)\frac{\partial}{\partial x}\right) p(x,t) = -p^2(x,t), \quad p(x,0) = p_0(x) = q_0'(x). \tag{1.80}$$

For completeness, we give the equation of continuity for the density field $\rho(x,t)$

$$\left(\frac{\partial}{\partial t} + q(x,t)\frac{\partial}{\partial x}\right) \rho(x,t) = -p(x,t)\rho(x,t), \quad \rho(x,0) = \rho_0(x), \tag{1.81}$$

and its logarithm $\chi(x,t) = \ln \rho(x,t)$

$$\left(\frac{\partial}{\partial t} + q(x,t)\frac{\partial}{\partial x}\right) \chi(x,t) = -p(x,t), \quad \chi(x,0) = \chi_0(x), \tag{1.82}$$

which are related to the Riemann equation (1.77). The solution to Eq. (1.77) has the form

$$\rho(x,t) = \frac{\rho_0(x_0)}{1 + t p_0(x_0)} = \frac{\rho_0(x - tq(x,t))}{1 + t p_0(x - tq(x,t))}. \tag{1.83}$$

If $q_0'(x_0) < 0$, then derivative $\frac{\partial}{\partial x} q(x,t)$ and solution of Eq. (1.77) becomes discontinuous. For times prior to t_0, the solution is unique and representable in the form of a quadrature. To show this fact, we calculate the variational derivative (for variational derivative definitions and the corresponding operation rules, see Lecture 2)

$$\frac{\delta q(x,t)}{\delta q_0(x_0)} = \frac{1}{1 + t q_0'(x_0)} \delta(x - tq(x,t) - x_0).$$

Because $q(x,t) = q_0(x_0)$ and $x - tq_0(x_0) = x_0$, the argument of delta function vanishes at $x = F(x_0, t) = x_0 + tq_0(x_0)$. Consequently, we have

$$\frac{\delta q(x,t)}{\delta q_0(x_0)} = \delta(x - F(x_0,t)) = \frac{1}{2\pi} \int_{-\infty}^{\infty} dk\, e^{ik(x-x_0) - iktq_0(x_0)}.$$

We can consider this equality as the functional equation in variable $q_0(x_0)$. Then, integrating this equation with the initial value

$$q(x,t)\big|_{q_0(x_0)=0} = 0,$$

in the functional space, we obtain the solution of the Riemann equation in the form of the quadrature (see e.g., [32])

$$q(x,t) = \frac{i}{2\pi t}\int_{-\infty}^{\infty}\frac{dk}{k}\int_{-\infty}^{\infty} d\xi\, e^{ik(x-x_0)}\left[e^{-iktq_0(\xi)} - 1\right].$$

The mentioned ambiguity can be eliminated by considering the *Burgers equation*

$$\frac{\partial}{\partial t}q(x,t) + q(x,t)\frac{\partial}{\partial x}q(x,t) = \mu\frac{\partial^2}{\partial x^2}q(x,t), \quad q(x,0) = q_0(x),$$

(it includes the molecular viscosity and also can be solved in quadratures) followed by the limit process $\mu \to 0$.

In the general case, a nonlinear scalar first-order partial differential equation can be written in the form

$$\frac{\partial}{\partial t}q(\boldsymbol{r},t) + H(\boldsymbol{r},t,q,\boldsymbol{p}) = 0, \quad q(\boldsymbol{r},0) = q_0(\boldsymbol{r}), \tag{1.84}$$

where $\boldsymbol{p}(\boldsymbol{r},t) = \nabla q(\boldsymbol{r},t)$.

In terms of the Lagrangian description, this equation can be rewritten in the form of the system of characteristic equations:

$$\begin{aligned}
&\frac{d}{dt}\boldsymbol{r}(t|\boldsymbol{r}_0) = \frac{\partial}{\partial \boldsymbol{p}}H(\boldsymbol{r},t,q,\boldsymbol{p}), \quad \boldsymbol{r}(0|\boldsymbol{r}_0) = \boldsymbol{r}_0;\\
&\frac{d}{dt}\boldsymbol{p}(t|\boldsymbol{r}_0) = -\left(\frac{\partial}{\partial \boldsymbol{r}} + \boldsymbol{p}\frac{\partial}{\partial q}\right)H(\boldsymbol{r},t,q,\boldsymbol{p}), \quad \boldsymbol{p}(0|\boldsymbol{r}_0) = \boldsymbol{p}_0(\boldsymbol{r}_0);\\
&\frac{d}{dt}q(t|\boldsymbol{r}_0) = \left(\boldsymbol{p}\frac{\partial}{\partial \boldsymbol{p}} - 1\right)H(\boldsymbol{r},t,q,\boldsymbol{p}), \quad q(0|\boldsymbol{r}_0) = q_0(\boldsymbol{r}_0).
\end{aligned} \tag{1.85}$$

Now, we supplement Eq. (1.84) with the equation for the conservative quantity $I(\boldsymbol{r},t)$

$$\frac{\partial}{\partial t}I(\boldsymbol{r},t) + \frac{\partial}{\partial \boldsymbol{r}}\left\{\frac{\partial H(\boldsymbol{r},t,q,\boldsymbol{p})}{\partial \boldsymbol{p}}I(\boldsymbol{r},t)\right\} = 0, \quad I(\boldsymbol{r},0) = I_0(\boldsymbol{r}). \tag{1.86}$$

From Eq. (1.86) follows that

$$\int d\boldsymbol{r}I(\boldsymbol{r},t) = \int d\boldsymbol{r}I_0(\boldsymbol{r}). \tag{1.87}$$

Then, in the Lagrangian description, the corresponding quantity satisfies the equation

$$\frac{d}{dt}I(t|\boldsymbol{r}_0) = -\frac{\partial^2 H(\boldsymbol{r},t,q,\boldsymbol{p})}{\partial \boldsymbol{r}\partial \boldsymbol{p}}I(\boldsymbol{r},t), \quad I(0|\boldsymbol{r}_0) = I_0(\boldsymbol{r}_0),$$

so that the solution to Eq. (1.86) has the form

$$I(\boldsymbol{r},t) = I(t|\boldsymbol{r}_0(t,\boldsymbol{r})) = \int d\boldsymbol{r}_0 I(t|\boldsymbol{r}_0)\, j(t|\boldsymbol{r}_0)\, \delta(\boldsymbol{r}(t|\boldsymbol{r}_0) - \boldsymbol{r}), \tag{1.88}$$

where $j(t|\boldsymbol{r}_0) = \det \left\| \partial r_i(t|\boldsymbol{r}_0) / \partial r_{0j} \right\|$ is the divergence (Jacobian).

Quantities $I(t|\boldsymbol{r}_0)$ and $j(t|\boldsymbol{r}_0)$ are related to each other. Indeed, substituting Eq. (1.88) for $I(\boldsymbol{r},t)$ in Eq. (1.87), we see that there is the evolution integral

$$j(t|\boldsymbol{r}_0) = \frac{I_0(\boldsymbol{r}_0)}{I(t|\boldsymbol{r}_0)},$$

and Eq. (1.88) assumes the form

$$I(\boldsymbol{r},t) = \int d\boldsymbol{r}_0 I_0(\boldsymbol{r}_0)\, \delta(\boldsymbol{r}(t|\boldsymbol{r}_0) - \boldsymbol{r}).$$

1.3.3 Parabolic Equation of Quasioptics (Waves in Randomly Inhomogeneous Media)

We will describe propagation of a monochromatic wave $U(t,x,\boldsymbol{R}) = e^{-i\omega t}u(x,\boldsymbol{R})$ in the media with large-scale three-dimensional inhomogeneities responsible for small-angle scattering on the base of the complex *parabolic equation of quasioptics*,

$$\frac{\partial}{\partial x}u(x,\boldsymbol{R}) = \frac{i}{2k}\Delta_{\boldsymbol{R}}u(x,\boldsymbol{R}) + \frac{ik}{2}\varepsilon(x,\boldsymbol{R})u(x,\boldsymbol{R}), \qquad u(0,\boldsymbol{R}) = u_0(\boldsymbol{R}), \tag{1.89}$$

where $k = \omega/c$ is the wave number, c is the velocity of wave propagation, x-axis is directed along the initial direction of wave propagation, vecor $\boldsymbol{R}$ denotes the coordinates in the transverse plane, $\Delta_{\boldsymbol{R}} = \partial^2/\partial\boldsymbol{R}^2$, and function $\varepsilon(x,\boldsymbol{R})$ is the deviation of the refractive index (or dielectric permittivity) from unity. This equation was successfully used in many problems on wave propagation in Earth's atmosphere and oceans [33, 34].

Introducing the amplitude–phase representation of the wavefield in Eq. (1.89) by the formula

$$u(x,\boldsymbol{R}) = A(x,\boldsymbol{R})e^{iS(x,\boldsymbol{R})},$$

we can write the equation for the wavefield intensity

$$I(x,\boldsymbol{R}) = u(x,\boldsymbol{R})u^*(x,\boldsymbol{R}) = |u(x,\boldsymbol{R})|^2$$

in the form

$$\frac{\partial}{\partial x}I(x,\boldsymbol{R}) + \frac{1}{k}\nabla_{\boldsymbol{R}}\left\{\nabla_{\boldsymbol{R}}S(x,\boldsymbol{R})I(x,\boldsymbol{R})\right\} = 0, \qquad I(0,\boldsymbol{R}) = I_0(\boldsymbol{R}). \tag{1.90}$$

From this equation follows that the power of a wave in plane $x = \text{const}$ is conserved in the general case of arbitrary incident wave beam:

$$E_0 = \int I(x,\boldsymbol{R})d\boldsymbol{R} = \int I_0(\boldsymbol{R})d\boldsymbol{R}.$$

Equation (1.90) coincides in form with Eq. (1.39). Consequently, we can treat it as the transport equation for conservative tracer in the potential velocity field. However, this tracer can be considered the passive tracer only in the *geometrical optics approximation*, in which case the phase of the wave $S(x,\boldsymbol{R})$, the transverse gradient of the phase

$$\boldsymbol{p}(x,\boldsymbol{R}) = \frac{1}{k}\nabla_{\boldsymbol{R}}S(x,\boldsymbol{R}),$$

and the matrix of the phase second derivatives

$$u_{ij}(x,\boldsymbol{R}) = \frac{1}{k}\frac{\partial^2}{\partial R_i \partial R_j}S(x,\boldsymbol{R})$$

characterizing the *curvature of the phase front* $S(x,\boldsymbol{R}) = \text{const}$ satisfy the closed system of equations

$$\begin{aligned}
&\frac{\partial}{\partial x}S(x,\boldsymbol{R}) + \frac{k}{2}\boldsymbol{p}^2(x,\boldsymbol{R}) = \frac{k}{2}\varepsilon(x,\boldsymbol{R}),\\
&\left(\frac{\partial}{\partial x} + \boldsymbol{p}(x,\boldsymbol{R})\nabla_{\boldsymbol{R}}\right)\boldsymbol{p}(x,\boldsymbol{R}) = \frac{1}{2}\nabla_{\boldsymbol{R}}\varepsilon(x,\boldsymbol{R}),\\
&\left(\frac{\partial}{\partial x} + \boldsymbol{p}(x,\boldsymbol{R})\nabla_{\boldsymbol{R}}\right)u_{ij}(x,\boldsymbol{R}) + u_{ik}(x,\boldsymbol{R})u_{kj}(x,\boldsymbol{R}) = \frac{1}{2}\frac{\partial^2}{\partial R_i \partial R_j}\varepsilon(x,\boldsymbol{R}).
\end{aligned} \tag{1.91}$$

In the general case, i.e., with the inclusion of diffraction effects, this tracer is the active tracer.

According to the material of Sect. 1.3.1, realizations of intensity must show the cluster behavior, which manifests itself in the appearance of *caustic* structures.

An example demonstrating the appearance of the wavefield caustic structures is given in Fig. 1.15, which is a fragment of the photo on the back of the cover – the flyleaf – of book [33] that shows the transverse section of a laser beam in turbulent medium (laboratory measurements).

A photo of the pool in Fig. 1.16 also shows the prominent caustic structure of the wavefield on the pool bottom. Such structures appear due to light refraction and reflection by rough water surface, which corresponds to scattering by the so-called *phase screen*.

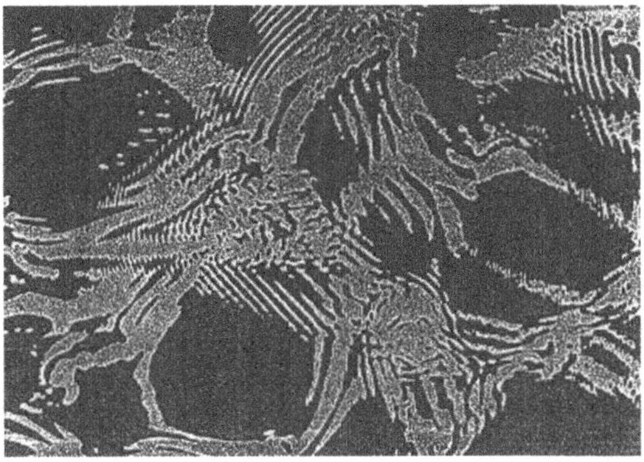

Figure 1.15 Transverse section of a laser beam in turbulent medium.

Figure 1.16 Caustics in a pool.

Note that Eq. (1.89) with parameter x concidered as time coincides in form with the *Schrödinger equation*. In a similar way, the *nonlinear parabolic equation* describing self-action of a harmonic wave field in multidimensional random media,

$$\frac{\partial}{\partial x}u(x,\boldsymbol{R}) = \frac{i}{2k}\Delta_{\boldsymbol{R}}u(x,\boldsymbol{R}) + \frac{ik}{2}\varepsilon(x,\boldsymbol{R};I(x,\boldsymbol{R}))u(x,\boldsymbol{R}), \qquad u(0,\boldsymbol{R}) = u_0(\boldsymbol{R}),$$

coincides in form with the *nonlinear Schrödinger equation*. Consequently, clustering of wave field energy must occur in this case too, because Eq. (1.90) is formally independent of the shape of function $\varepsilon(x,\boldsymbol{R};I(x,\boldsymbol{R}))$.

Consider now geometrical optics approximation (1.91) for parabolic equation (1.89). In this approximation, the equation for the wave phase is the Hamilton–Jacobi equation and the equation for the transverse gradient of the phase (1.91) is the closed quasilinear first-order partial differential equation, and we can solve it by the method of characteristics. Equations for the characteristic curves (*rays*) have the form

$$\frac{d}{dx}\boldsymbol{R}(x) = \boldsymbol{p}(x), \quad \frac{d}{dx}\boldsymbol{p}(x) = \frac{1}{2}\nabla_{\boldsymbol{R}}\varepsilon(x, \boldsymbol{R}), \tag{1.92}$$

and the wavefield intensity and matrix of the phase second derivatives along the characteristic curves will satisfy the equations

$$\begin{aligned} \frac{d}{dx}I(x) &= -I(x)u_{ii}(x), \\ \frac{d}{dx}u_{ij}(x) + u_{ik}(x)u_{kj}(x) &= \frac{1}{2}\frac{\partial^2}{\partial R_i \partial R_j}\varepsilon(x, \boldsymbol{R}). \end{aligned} \tag{1.93}$$

Equations (1.92) coincide in appearance with the equations for a particle under random external forces in the absence of friction (1.12) and form the system of the Hamilton equations.

In the two-dimensional case ($R = y$), Eqs. (1.92), (1.93) become significantly simpler and assume the form

$$\begin{aligned} &\frac{d}{dx}y(x) = p(x), \quad \frac{d}{dx}p(x) = \frac{1}{2}\frac{\partial}{\partial y}\varepsilon(x, y), \\ &\frac{d}{dx}I(x) = -I(x)u(x), \quad \frac{d}{dx}u(x) + u^2(x) = \frac{1}{2}\frac{\partial^2}{\partial y^2}\varepsilon(x, y). \end{aligned} \tag{1.94}$$

The last equation for $u(x)$ in (1.94) is similar to Eq. (1.29) whose solution shows the singular behavior. The only difference between these equations consists in the random term that has now a more complicated structure. Nevertheless, it is quite clear that solutions to stochastic problem (1.94) will show the blow-up behavior; namely, function $u(x)$ will reach minus infinity and intensity will reach plus infinity at a finite distance. Such a behavior of a wavefield in randomly inhomogeneous media corresponds to *random focusing*, i.e., to the formation of caustics, which means the appearance of points of multivaluedness (and discontinuity) in the solutions of quasilinear equation (1.91) for the transverse gradient of the wavefield phase.

1.3.4 Navier–Stokes Equation: Random Forces in Hydrodynamic Theory of Turbulence

Consider now the turbulent motion model that assumes the presence of external forces $\boldsymbol{f}(\boldsymbol{r}, t)$ acting on the liquid. Such a model is evidently only imaginary, because there are no actual analogues for these forces. However, assuming that forces $\boldsymbol{f}(\boldsymbol{r}, t)$ on average ensure an appreciable energy income only to large-scale velocity components, we can expect that, within the concepts of the theory of local isotropic turbulence, the imaginary nature of field $\boldsymbol{f}(\boldsymbol{r}, t)$ will only slightly affect statistical properties of small-scale turbulent components [35]. Consequently, this model is quite appropriate for describing small-scale properties of turbulence.

Motion of an incompressible liquid under external forces is governed by the *Navier–Stokes equation*

$$\left(\frac{\partial}{\partial t} + \boldsymbol{u}(\boldsymbol{r}, t)\frac{\partial}{\partial \boldsymbol{r}}\right)\boldsymbol{u}(\boldsymbol{r}, t) = -\frac{1}{\rho_0}\frac{\partial}{\partial \boldsymbol{r}}p(\boldsymbol{r}, t) + \nu\Delta\boldsymbol{u}(\boldsymbol{r}, t) + \boldsymbol{f}(\boldsymbol{r}, t),$$
$$\operatorname{div}\boldsymbol{u}(\boldsymbol{r}, t) = 0, \quad \operatorname{div}\boldsymbol{f}(\boldsymbol{r}, t) = 0. \tag{1.95}$$

Here, ρ_0 is the density of the liquid, ν is the kinematic viscosity, and pressure field $p(\boldsymbol{x}, t)$ is expressed in terms of the velocity field at the same instant by the relationship

$$p(\boldsymbol{r}, t) = -\rho_0 \int \Delta^{-1}\left(\boldsymbol{r}, \boldsymbol{r}'\right) \frac{\partial^2 \left(u_i(\boldsymbol{r}', t)u_j(\boldsymbol{r}', t)\right)}{\partial r_i' \partial r_j'} d\boldsymbol{r}', \tag{1.96}$$

where $\Delta^{-1}(\boldsymbol{r}, \boldsymbol{r}')$ is the integral operator inverse to the Laplace operator (repeated indices assume summation).

Note that the linearized equation (1.95)

$$\frac{\partial}{\partial t}\boldsymbol{u}(\boldsymbol{r}, t) = \nu\Delta\boldsymbol{u}(\boldsymbol{r}, t) + \boldsymbol{f}(\boldsymbol{r}, t), \quad \boldsymbol{u}(\boldsymbol{r}, 0) = 0, \tag{1.97}$$

describes the problem on turbulence degeneration for $t \to \infty$. The solution to this problem can be drawn in an explicit form:

$$\begin{aligned}
\boldsymbol{u}(\boldsymbol{r}, t) &= \int_0^t d\tau\, e^{\nu(t-\tau)\Delta} f(\boldsymbol{r}, \tau) = \int_0^t d\tau\, e^{\nu\tau\Delta} \int d\boldsymbol{r}'\, \delta(\boldsymbol{r} - \boldsymbol{r}') f(\boldsymbol{r}', t - \tau) \\
&= \frac{1}{(2\pi)^3} \int_0^t d\tau\, e^{\nu\tau\Delta} \int d\boldsymbol{r}' \int d\boldsymbol{q}\, e^{i\boldsymbol{q}(\boldsymbol{r}-\boldsymbol{r}')} f(\boldsymbol{r}', t - \tau) \\
&= \frac{1}{(2\pi)^3} \int_0^t d\tau \int d\boldsymbol{r}' \int d\boldsymbol{q}\, e^{-\nu\tau \boldsymbol{q}^2} e^{i\boldsymbol{q}\boldsymbol{r}'} f(\boldsymbol{r} - \boldsymbol{r}', t - \tau) \\
&= \int d\boldsymbol{r}' \int_0^t \frac{d\tau}{(4\pi\nu\tau)^{3/2}} \exp\left\{-\frac{\boldsymbol{r}'^2}{4\nu\tau}\right\} f(\boldsymbol{r} - \boldsymbol{r}', t - \tau).
\end{aligned} \tag{1.98}$$

The linear equation (1.97) is an extension of the Langevin equation in velocity (1.75), page 36 to random fields.

Neglecting the effect of viscosity and external random forces in Eq. (1.95), we arrive at the equation

$$\frac{\partial}{\partial t}\boldsymbol{u}(\boldsymbol{r},t)+(\boldsymbol{u}(\boldsymbol{r},t)\cdot\nabla)\,\boldsymbol{u}(\boldsymbol{r},t)=-\frac{1}{\rho}\nabla p(\boldsymbol{r},t) \tag{1.99}$$

that describes dynamics of *perfect liquid* and is called the *Euler equation*. Using Eq. (1.58), page 29, this equation can be rewritten in terms of the velocity field vortex $\boldsymbol{\omega}=\operatorname{curl}\boldsymbol{u}(\boldsymbol{r},t)$ as

$$\frac{\partial}{\partial t}\boldsymbol{\omega}(\boldsymbol{r},t)=\operatorname{curl}\left[\boldsymbol{u}(\boldsymbol{r},t)\times\boldsymbol{\omega}(\boldsymbol{r},t)\right],\quad \boldsymbol{\omega}(\boldsymbol{r},0)=\boldsymbol{\omega}_0(\boldsymbol{r}), \tag{1.100}$$

or

$$\left(\frac{\partial}{\partial t}+\frac{\partial}{\partial \boldsymbol{r}}\boldsymbol{u}(\boldsymbol{r},t)\right)\boldsymbol{\omega}(\boldsymbol{r},t)=\left(\boldsymbol{\omega}(\boldsymbol{r},t)\cdot\frac{\partial}{\partial \boldsymbol{r}}\right)\boldsymbol{\omega}(\boldsymbol{r},t),\quad \boldsymbol{\omega}(\boldsymbol{r},0)=\boldsymbol{\omega}_0(\boldsymbol{r}). \tag{1.101}$$

These equations coincide with Eqs. (1.52) and (1.59), pages 28 and 30, and appeared in the problem on the diffusion of magnetic field.

However, this coincidence is only formal, because different boundary conditions to these problems results in drastically different behaviors of fields $\boldsymbol{\omega}(\boldsymbol{r},t)$ and $\boldsymbol{H}(\boldsymbol{r},t)$ as functions of the velocity field $\boldsymbol{u}(\boldsymbol{r},t)$.

If we substitute Eq. (1.96) in Eq. (1.95) to exclude the pressure field, then we obtain in the three-dimensional case that the Fourier transform of the velocity field with respect to spatial coordinates

$$\widehat{u}_i(\boldsymbol{k},t)=\int d\boldsymbol{r}u_i(\boldsymbol{r},t)e^{-i\boldsymbol{kr}},\quad u_i(\boldsymbol{r},t)=\frac{1}{(2\pi)^3}\int d\boldsymbol{k}\widehat{u}_i(\boldsymbol{k},t)e^{i\boldsymbol{kr}},$$

($\widehat{u}_i^*(\boldsymbol{k},t)=\widehat{u}_i(-\boldsymbol{k},t)$) satisfies the nonlinear integro-differential equation

$$\begin{aligned}&\frac{\partial}{\partial t}\widehat{u}_i(\boldsymbol{k},t)+\frac{i}{2}\int d\boldsymbol{k}_1\int d\boldsymbol{k}_2\Lambda_i^{\alpha\beta}(\boldsymbol{k}_1,\boldsymbol{k}_2,\boldsymbol{k})\,\widehat{u}_\alpha(\boldsymbol{k}_1,t)\,\widehat{u}_\beta(\boldsymbol{k}_2,t)\\&\quad-\nu k^2\widehat{u}_i(\boldsymbol{k},t)=\widehat{f}_i(\boldsymbol{k},t),\end{aligned} \tag{1.102}$$

where

$$\Lambda_i^{\alpha\beta}(\boldsymbol{k}_1,\boldsymbol{k}_2,\boldsymbol{k})=\frac{1}{(2\pi)^3}\left\{k_\alpha\Delta_{i\beta}(\boldsymbol{k})+k_\beta\Delta_{i\alpha}(\boldsymbol{k})\right\}\delta(\boldsymbol{k}_1+\boldsymbol{k}_2-\boldsymbol{k}),$$

$$\Delta_{ij}(\boldsymbol{k})=\delta_{ij}-\frac{k_ik_j}{\boldsymbol{k}^2}\quad(i,\,\alpha,\,\beta\,=\,1,\,2,\,3),$$

and $\widehat{\boldsymbol{f}}(\boldsymbol{k},t)$ is the spatial Fourier harmonics of external forces,

$$\widehat{\boldsymbol{f}}(\boldsymbol{k},t)=\int d\boldsymbol{r}\boldsymbol{f}(\boldsymbol{r},t)e^{-i\boldsymbol{kr}},\quad \boldsymbol{f}(\boldsymbol{r},t)=\frac{1}{(2\pi)^3}\int d\boldsymbol{k}\widehat{\boldsymbol{f}}(\boldsymbol{k},t)e^{i\boldsymbol{kr}}.$$

A specific feature of the three-dimensional hydrodynamic motions consists in the fact that the absence of external forces and viscosity-driven effects is sufficient for energy conservation.

It appears convenient to describe the stationary turbulence in terms of the space-time Fourier harmonics of the velocity field

$$\widehat{u}_i(\boldsymbol{K}) = \int d\boldsymbol{x} \int_{-\infty}^{\infty} dt u_i(\boldsymbol{x}, t) e^{-i(\boldsymbol{kx}+\omega t)},$$

$$u_i(\boldsymbol{x}, t) = \frac{1}{(2\pi)^4} \int d\boldsymbol{k} \int_{-\infty}^{\infty} d\omega \widehat{u}_i(\boldsymbol{K}) e^{i(\boldsymbol{kx}+\omega t)},$$

where $\boldsymbol{K}$ is the four-dimensional wave vector $\{\boldsymbol{k}, \omega\}$ and field $\widehat{u}_i^*(\boldsymbol{K}) = \widehat{u}_i(-\boldsymbol{K})$ because field $u_i(\boldsymbol{r}, t)$ is real. In this case, we obtain the equation for component $\widehat{u}_i(\boldsymbol{K})$ by performing the Fourier transformation of Eq. (1.102) with respect to time:

$$(i\omega + \nu \boldsymbol{k}^2)\widehat{u}_i(\boldsymbol{K}) + \frac{i}{2} \int d^4\boldsymbol{K}_1 \int d^4\boldsymbol{K}_2 \Lambda_i^{\alpha\beta}(\boldsymbol{K}_1, \boldsymbol{K}_2, \boldsymbol{K}) \widehat{u}_\alpha(\boldsymbol{K}_1) \widehat{u}_\beta(\boldsymbol{K}) = \widehat{f}_i(\boldsymbol{K}), \tag{1.103}$$

where

$$\Lambda_i^{\alpha\beta}(\boldsymbol{K}_1, \boldsymbol{K}_2, \boldsymbol{K}) = \frac{1}{2\pi} \Lambda_i^{\alpha\beta}(\boldsymbol{k}_1, \boldsymbol{k}_2, \boldsymbol{k}) \, \delta(\omega_1 + \omega_2 - \omega),$$

and $\widehat{f}_i(\boldsymbol{K})$ are the space-time Fourier harmonics of external forces. The obtained Eq. (1.103) is now the integral (and not integro-differential) nonlinear equation.

Plane Motion Under the Action of a Periodic Force

Consider now the two-dimensional motion of an incompressible viscous liquid $\boldsymbol{U}(\boldsymbol{r}, t) = \{u(\boldsymbol{r}, t), v(\boldsymbol{r}, t)\}$ in plane $\boldsymbol{r} = \{x, y\}$ under the action of a spatially periodic force directed along the x-axis $f_x(\boldsymbol{r}, t) = \gamma \sin py$ $(\gamma > 0)$. Such a flow is usually called the *Kolmogorov flow* (*stream*). The corresponding motion is described by the system of equations

$$\begin{aligned}
\frac{\partial u}{\partial t} + \frac{\partial u^2}{\partial x} + \frac{\partial uv}{\partial y} &= -\frac{1}{\rho}\frac{\partial P}{\partial x} + \nu \Delta u + \gamma \sin py, \\
\frac{\partial v}{\partial t} + \frac{\partial uv}{\partial x} + \frac{\partial v^2}{\partial y} &= -\frac{1}{\rho}\frac{\partial P}{\partial y} + \nu \Delta v, \\
\frac{\partial u}{\partial x} + \frac{\partial v}{\partial y} &= 0,
\end{aligned} \tag{1.104}$$

where $P(\boldsymbol{r}, t)$ is the pressure, ρ is the density, and ν is the kinematic viscosity.

The system of the Navier–Stokes and continuity equations (1.104) has the steady-state solution that corresponds to the laminar flow at constant pressure along the x-axis and has the following form

$$u_{\text{s-s}}(\boldsymbol{r}) = \frac{\gamma}{\nu p^2} \sin py, \quad v_{\text{s-s}}(\boldsymbol{r}) = 0, \quad P_{\text{s-s}}(\boldsymbol{r}) = \text{const.} \tag{1.105}$$

Introducing scales of length p^{-1}, velocity $p^{-2}\upsilon^{-1}\gamma$, and time $p\upsilon\gamma^{-1}$ and using dimensionless variables, we reduce system (1.104) to the form

$$\begin{aligned}
\frac{\partial u}{\partial t} + \frac{\partial u^2}{\partial x} + \frac{\partial uv}{\partial y} &= -\frac{\partial P}{\partial x} + \frac{1}{R}\Delta u + \frac{1}{R}\sin y,\\
\frac{\partial v}{\partial t} + \frac{\partial uv}{\partial x} + \frac{\partial v^2}{\partial y} &= -\frac{\partial P}{\partial y} + \frac{1}{R}\Delta v,\\
\frac{\partial u}{\partial x} + \frac{\partial v}{\partial y} &= 0,
\end{aligned} \tag{1.106}$$

where $R = \dfrac{\gamma}{\nu^2 p^3}$ is the Reynolds number. In these variables, the steady-state solution has the form

$$u_{\text{s-s}}(\boldsymbol{r}) = \sin y, \quad v_{\text{s-s}}(\boldsymbol{r}) = 0, \quad P_{\text{s-s}}(\boldsymbol{r}) = \text{const}.$$

Introducing flow function $\psi(\boldsymbol{r}, t)$ by the relationship

$$u(\boldsymbol{r}, t) = \frac{\partial}{\partial y}\psi(\boldsymbol{r}, t), \quad v(\boldsymbol{r}, t) = -\frac{\partial}{\partial x}\psi(\boldsymbol{r}, t),$$

we obtain that it satisfies the equation

$$\left(\frac{\partial}{\partial t} - \frac{\Delta}{R}\right)\Delta\psi - \frac{\partial\psi}{\partial x}\frac{\partial\Delta\psi}{\partial y} + \frac{\partial\psi}{\partial y}\frac{\partial\Delta\psi}{\partial x} = \frac{1}{R}\cos y, \tag{1.107}$$

and

$$\psi_{\text{s-s}}(\boldsymbol{r}) = -\cos y.$$

It was shown [43, 44] that, in the linear problem formulation, the steady-state solution (1.105) corresponding to the laminar flow is unstable with respect to small disturbances for certain values of parameter R. These disturbances rapidly increase in time getting the energy from the flow (1.105); this causes the Reynolds stresses described by the nonlinear terms in Eq. (1.107) to increase, which results in decreasing the amplitude of laminar flow until certain new steady-state flow (called usually the *secondary flow*) is formed.

We represent the hydrodynamic fields in the form

$$\begin{aligned}
&u(\boldsymbol{r}, t) = U(y, t) + \widetilde{u}(\boldsymbol{r}, t), \quad v(\boldsymbol{r}, t) = \widetilde{v}(\boldsymbol{r}, t),\\
&P(\boldsymbol{r}, t) = P_0 + \widetilde{P}(\boldsymbol{r}, t), \quad \psi(\boldsymbol{r}, t) = \Psi(y, t) + \widetilde{\psi}(\boldsymbol{r}, t).
\end{aligned}$$

Here, $U(y, t)$ is the new profile of the steady-state flow to be determined together with the Reynolds stresses and the tilde denotes the corresponding ultimate disturbances. Abiding by the cited works, we will consider disturbances harmonic in variable x with wavelength $2\pi/\alpha$ ($\alpha > 0$). The new flow profile $U(y, t)$ is the result of averaging with respect to x over distances of about a wavelength.

One can easily see that, for $\alpha \geq 1$, the laminar flow (1.105) is unique and stable for all R [44] and instability can appear only for disturbances with $\alpha < 1$.

According to the linear stability theory, we will consider first the nonlinear interaction only between the first harmonic of disturbances and the mean flow and neglect the generation of higher harmonics and their mutual interactions and interaction with the mean flow.

We represent all disturbances in the form

$$\widetilde{\varphi}(\boldsymbol{r}, t) = \varphi^{(1)}(y, t)e^{i\alpha x} + \varphi^{(-1)}(y, t)e^{-i\alpha x},$$
$$\left(\widetilde{\varphi}(\boldsymbol{r}, t) = \widetilde{u}(\boldsymbol{r}, t),\ \widetilde{v}(\boldsymbol{r}, t),\ \widetilde{P}(\boldsymbol{r}, t),\ \widetilde{\psi}(\boldsymbol{r}, t)\right),$$

where quantity $\varphi^{(-1)}(y, t)$ is the complex conjugated of $\varphi^{(1)}(y, t)$. Then, using this representation in system (1.106) and eliminating quantities $\widetilde{P}(\boldsymbol{r}, t)$ and $\widetilde{u}(\boldsymbol{r}, t)$, we obtain the system of equations in mean flow $U(y, t)$ and disturbances $v^{(1)}(y, t)$ [24, 25]:

$$\frac{\partial}{\partial t}U + \frac{i}{\alpha}\left(v^{(-1)}\frac{\partial^2 v^{(1)}}{\partial y^2} - v^{(1)}\frac{\partial^2 v^{(-1)}}{\partial y^2}\right) = \frac{1}{R}\frac{\partial^2 U}{\partial y^2} + \frac{1}{R}\sin y,$$
$$\left(\frac{\partial}{\partial t} - \frac{\Delta}{R}\right)\Delta v^{(1)} + i\alpha\left[U\Delta v^{(1)} - v^{(1)}\frac{\partial^2 U}{\partial y^2}\right] = 0. \tag{1.108}$$

The second equation in system (1.108) is the known *Orr–Sommerfeld equation*. A similar system can be derived for the flow function.

To examine the stability of the laminar regime (1.105), we set

$$U(y) = \sin y$$

in the second equation of system (1.108) to obtain

$$\left(\frac{\partial}{\partial t} - \frac{\Delta}{R}\right)\Delta v^{(1)}(y, t) + i\alpha \sin y[1 + \Delta]v^{(1)}(y, t) = 0. \tag{1.109}$$

Representing disturbances $v^{(1)}(y, t)$ in the form

$$v^{(1)}(y, t) = \sum_{n=-\infty}^{\infty} v_n^{(1)} e^{\sigma t + iny}, \tag{1.110}$$

and substituting this representation in Eq. (1.109), we arrive at the recurrent system in quantities $v_n^{(1)}$

$$\frac{2}{\alpha}\left(\alpha^2 + n^2\right)\left[\sigma + \frac{\alpha^2 + n^2}{R}\right]v_n^{(1)} + v_{n-1}^{(1)}\left[\alpha^2 - 1 + (n-1)^2\right]$$
$$- v_{n+1}^{(1)}\left[\alpha^2 - 1 + (n+1)^2\right] = 0, \qquad n = -\infty, \ldots, +\infty. \tag{1.111}$$

The analysis of system (1.111) showed [43, 44] that, under certain restrictions on wave number α and Reynolds number R, positive values of parameter σ can exist, i.e., the solutions are unstable. The corresponding dispersion equation in σ has the form of an infinite continued fraction, and the critical Reynolds number is $R_{\text{cr}} = \sqrt{2}$ for $\alpha \to 0$. In other words, long-wave disturbances along the applied force appear to be most unstable. For this reason, we can consider parameter α as a small parameter of the problem at hand and integrate the Orr–Sommerfeld equation asymptotically. We will not dwell on details of this solution. Note only that the components of eigenvector $\left\{v_n^{(1)}\right\}$ of problem (1.111) have different orders of magnitude in parameter α. For example, all components of vector $\left\{v_n^{(1)}\right\}$ with $n = \pm 2, \pm 3, \ldots$ will have an order of α^4 at least. As a result, we can confine ourselves to the most significant harmonics with $n = 0, \pm 1$, which, in essence, is equivalent to the *Galerkin method* with the trigonometric coordinate functions. In this case,

$$U(y,t) = U(t)\sin y,$$

and the equation in $v^{(1)}$ assumes the form

$$\left(\frac{\partial}{\partial t} - \frac{\Delta}{R}\right)\Delta v^{(1)}(y,t) + i\alpha U(t)\sin y[1+\Delta]v^{(1)}(y,t) = 0.$$

Substituting the expansion

$$v^{(1)}(y,t) = \sum_{n=-1}^{1} v_n^{(1)}(t)e^{iny},$$

in Eqs. (1.108), we obtain that functions

$$U(t), \quad z_0(t) = v_0^{(1)}(t), \quad z_+(t) = v_1^{(1)}(t) + v_{-1}^{(1)}(t),$$

$$z_-(t) = \frac{v_1^{(1)}(t) - v_{-1}^{(1)}(t)}{2},$$

satisfy the system of equations [24, 45]

$$\begin{aligned}
&\left(\frac{d}{dt} + \frac{1}{R}\right)U(t) = \frac{1}{R} - \frac{4}{\alpha}z_0(t)z_-(t),\\
&\left(\frac{d}{dt} + \frac{\alpha^2}{R}\right)z_0(t) = \alpha U(t)z_-(t),\\
&\left(\frac{d}{dt} + \frac{1}{R}\right)z_-(t) = \frac{\alpha}{2}U(t)z_0(t), \quad \left(\frac{d}{dt} + \frac{1}{R}\right)z_+(t) = 0.
\end{aligned} \tag{1.112}$$

Equation in quantity $z_+(t)$ is independent of other equations; as a consequence, the corresponding disturbances can only decay with time. Three remaining equations form the simplest three-component hydrodynamic-type system (see Sect. 1.1.3, page 10). As was mentioned earlier, this system is equivalent to the dynamic system describing the motion of a gyroscope with anisotropic friction under the action of an external moment of force relative to the unstable axis. An analysis of system (1.112) shows that, for $R < R_{cr} = \sqrt{2}$, it yields the laminar regime with $U = 1, z_i = 0$. For $R > \sqrt{2}$, this regime becomes unstable, and new regime (the secondary flow) is formed that corresponds to the mean flow profile and steady-state Reynolds stresses

$$U = \frac{\sqrt{2}}{R}, \quad \frac{4}{\alpha} z_0 z_- = \frac{R-\sqrt{2}}{R^2}, \quad z_+ = 0,$$
$$\left[v_0^{(1)}\right]^2 = \frac{R-\sqrt{2}}{2\sqrt{2}R^2}, \quad v_1^{(1)} = \frac{\alpha}{\sqrt{2}} v_0^{(1)}, \quad \alpha \ll 1, \quad R \geq \sqrt{2}.$$

Turning back to the dimensional quantities, we obtain

$$U(y) = \sqrt{2}\upsilon p \sin py, \quad \langle \widetilde{u}\widetilde{v} \rangle = -\frac{\gamma}{p} \frac{R-\sqrt{2}}{R} \cos py. \tag{1.113}$$

Note that the amplitude of the steady-state mean flow is independent of the amplitude of exciting force. Moreover, quantity $v_0^{(1)}$ can be both positive and negative, depending on the signs of the amplitudes of small initial disturbances.

Flow function of the steady-state flow has the form

$$\psi_1(x, y) = -\frac{\sqrt{2}}{R} \cos y - \frac{2}{\alpha} v_0^{(1)} \left[\sqrt{2}\alpha \sin y \cos \alpha x + \sin \alpha x\right].$$

Figure 1.17 shows the current lines

$$\alpha \cos y + \sqrt{2}\alpha \sin y \cos \alpha x + \sin \alpha x = C$$

of flow (1.113) at $R = 2R_{cr} = 2\sqrt{2}$ ($v_0^{(1)} > 0$).

In addition, Fig. 1.17 shows schematically the profile of the mean flow. As distinct from the laminar solution, systems of spatially periodic vortices appear here, and the tilt of longer axes of these vortices is determined by the sign of the derivative of the mean flow profile with respect to y.

Flow (1.113) was derived under the assumption that the nonlinear interactions between different harmonics of the disturbance are insignificant in comparison with their interactions with the mean flow. This assumption will hold if flow (1.113) is, in turn, stable with respect to small disturbances. The corresponding stability analysis can be carried out by the standard procedure, i.e., by linearizing the equation for flow function (1.107) relative to flow (1.113) [45]. The analysis shows that flow (1.113) is stable

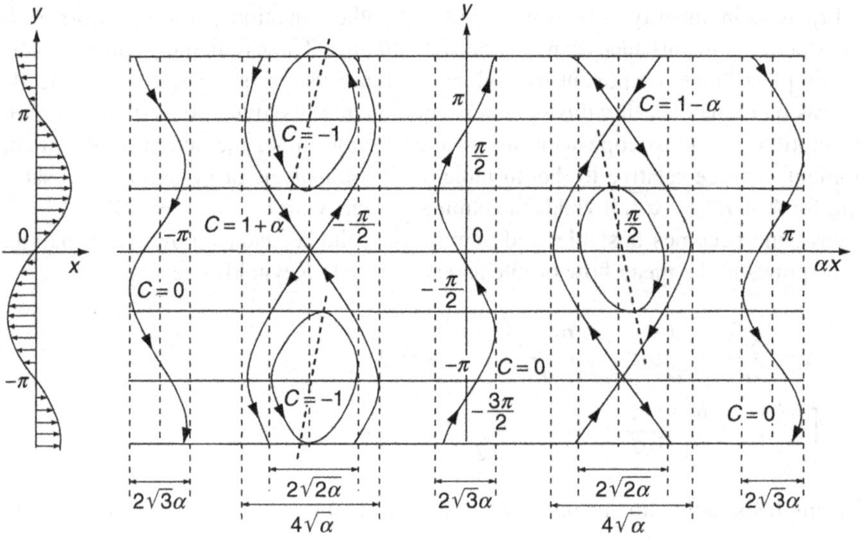

Figure 1.17 Mean velocity and current lines of the secondary flow at $R = 2R_{\text{cr}} = 2\sqrt{2}$ ($v_0^{(1)} > 0$).

if we restrict ourselves to the harmonics of the types same as in solution (1.113). However, the solution appears to be unstable with respect to small-scale disturbances. In this case, the nonlinear interaction of infinitely small disturbances governs the motion along with the interaction of the disturbances with the mean flow. Moreover, we cannot here content ourselves with a finite number of harmonics in x-coordinate and need to consider the infinite series. As regards the harmonics in y-coordinate, we, as earlier, can limit the consideration to the harmonics with $n = 0, \pm 1$.

Problem

Problem 1.1

Derive the system of equations which takes into account the whole infinite series of harmonics in x-coordinate and extends the system of gyroscopic-type equations (1.112) to the infinite-dimensional case.

Solution Represent the flow function in the form

$$\psi(x, y, t) = \psi_{-1}(x, t)e^{-iy} + \psi_0(x, t) + \psi_1(x, t)e^{iy} \\ (\psi_1^*(x, t) = \psi_{-1}(x, t)), \tag{1.114}$$

where $\psi_i(x, t)$ are the periodic functions in x-coordinate with a period $2\pi/\alpha$. Substituting Eq. (1.114) in Eq. (1.107), neglecting the terms of order α^3 in interactions of harmonics and the terms of order α^2 in the dissipative terms of harmonics $\psi_{\pm 1}$, and introducing new functions

$$\psi_+(x,t) = \frac{\psi_{-1} + \psi_1}{2}, \quad \psi_-(x,t) = \frac{\psi_{-1} - \psi_1}{2i},$$

we obtain the system of equations

$$\left(\frac{\partial}{\partial t} + \frac{1}{R}\right)\psi_+ - \psi_- \frac{\partial \psi_0}{\partial x} = -\frac{1}{2R},$$

$$\left(\frac{\partial}{\partial t} + \frac{1}{R}\right)\psi_- + \psi_+ \frac{\partial \psi_0}{\partial x} = 0,$$

$$\left(\frac{\partial}{\partial t} - \frac{1}{R}\frac{\partial^2}{\partial x^2}\right)\psi_0 + 2\left(\psi_- \frac{\partial \psi +}{\partial x} - \psi_+ \frac{\partial \psi_-}{\partial x}\right) = 0.$$

Its characteristic feature consists in the absence of steady-state solutions periodic in x-coordinate (except the solution corresponding to the laminar flow).

Lecture 2

Solution Dependence on Problem Type, Medium Parameters, and Initial Data

Below, we consider a number of dynamic systems described by both ordinary and partial differential equations. Many applications concerning research of statistical characteristics of the solutions to these equations require the knowledge of solution dependence (generally, in the functional form) on the medium parameters appearing in the equation as coefficients and initial values. Some properties appear common to all such dependencies, and two of them are of special interest in the context of statistical descriptions. We illustrate these dependencies by the example of the simplest problem, namely, the system of ordinary differential equations (1.1) that describes particle dynamics under random velocity field and which we reformulate in the form of the nonlinear integral equation

$$\boldsymbol{r}(t) = \boldsymbol{r}_0 + \int_{t_0}^{t} d\tau \boldsymbol{U}(\boldsymbol{r}(\tau), \tau). \tag{2.1}$$

The solution to Eq. (2.1) functionally depends on vector field $\boldsymbol{U}(\boldsymbol{r}', \tau)$ and initial values $\boldsymbol{r}_0$, t_0.

2.1 Functional Representation of Problem Solution

2.1.1 Variational (Functional) Derivatives

Recall first the general definition of a functional. One says that a functional is given if a rule is fixed that associates a number to every function from a certain function family. Below, we give some examples of functionals:

$$\textbf{(a)} \quad F[\varphi(\tau)] = \int_{t_1}^{t_2} d\tau a(\tau)\varphi(\tau),$$

Lectures on Dynamics of Stochastic Systems. DOI: 10.1016/B978-0-12-384966-3.00002-7

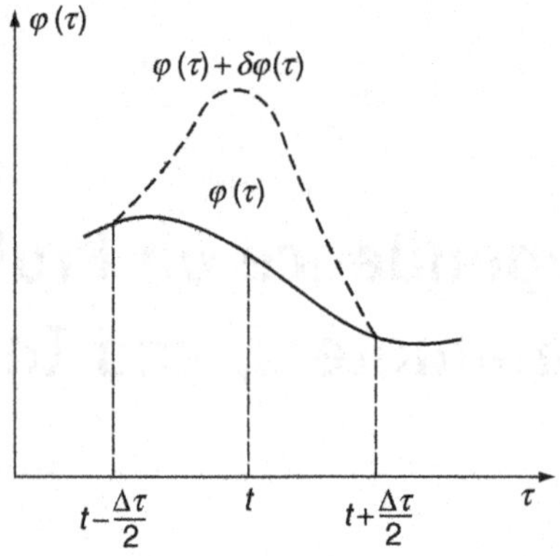

Figure 2.1 To definition of variational derivative.

where $a(t)$ is the given (fixed) function and limits t_1 and t_2 can be both finite and infinite. This is the linear functional.

$$\textbf{(b)}\quad F[\varphi(\tau)] = \int\limits_{t_1}^{t_2}\int\limits_{t_1}^{t_2} d\tau_1 d\tau_2 B(\tau_1, \tau_2)\varphi(\tau_1)\varphi(\tau_2),$$

where $B(t_1, t_2)$ is the given (fixed) function. This is the quadratic functional.

$$\textbf{(c)}\quad F[\varphi(\tau)] = f\left(\Phi[\varphi(\tau)]\right),$$

where $f(x)$ is the given function and quantity $\Phi[\varphi(\tau)]$ is the functional.

Estimate the difference between the values of a functional calculated for functions $\varphi(\tau)$ and $\varphi(\tau) + \delta\varphi(\tau)$ for $t - \dfrac{\Delta\tau}{2} < \tau < t + \dfrac{\Delta\tau}{2}$ (see Fig. 2.1).

The variation of a functional is defined as the linear (in $\delta\varphi(\tau)$) portion of the difference

$$\delta F[\varphi(\tau)] = \{F\left[\varphi(\tau) + \delta\varphi(\tau)\right] - F[\varphi(\tau)]\}.$$

The limit

$$\frac{\delta F[\varphi(\tau)]}{\delta\varphi(t)dt} = \lim_{\Delta\tau\to 0}\frac{\delta F[\varphi(\tau)]}{\int\limits_{\Delta\tau} d\tau\delta\varphi(\tau)}, \tag{2.2}$$

is called the *variational* (or *functional*) *derivative* (see, e.g., [35]).

For short, we will use notation $\dfrac{\delta F[\varphi(\tau)]}{\delta\varphi(t)}$ instead of $\dfrac{\delta F[\varphi(\tau)]}{\delta\varphi(t)dt}$.

Note that, if we use function $\delta\varphi(\tau) = \alpha\delta(\tau)$, where $\delta(\tau)$ is the Dirac delta function, then Eq. (2.2) can be represented in the form of the ordinary derivative

$$\frac{\delta F[\varphi(\tau)]}{\delta\varphi(t)} = \lim_{\alpha\to 0}\frac{d}{d\alpha}F[\varphi(\tau) + \alpha\delta(\tau - t)].$$

The variational derivative of functional $F[\varphi(\tau)]$ is again the functional of $\varphi(\tau)$, which depends additionally on point t as a parameter. As a result, this variational derivative will have two types of derivatives; one can differentiate it in the ordinary sense with respect to parameter t and in the functional sense with respect to $\varphi(\tau)$ at point $\tau = t'$, thus obtaining the second variational derivative of the initial functional

$$\frac{\delta^2 F[\varphi(\tau)]}{\delta\varphi(t')\delta\varphi(t)} = \frac{\delta}{\delta\varphi(t')}\left[\frac{\delta F[\varphi(\tau)]}{\delta\varphi(t)}\right].$$

The second variational derivative will now be the functional of $\varphi(\tau)$ dependent on two points t and t', and so forth.

Determine the variational derivatives of functionals (**a**), (**b**), and (**c**).

In case (**a**), we have

$$\delta F[\varphi(\tau)] = F[\varphi(\tau) + \delta\varphi(\tau)] - F[\varphi(\tau)] = \int\limits_{t-\frac{\Delta\tau}{2}}^{t+\frac{\Delta\tau}{2}} d\tau a(\tau)\delta\varphi(\tau).$$

If function $a(t)$ is continuous on segment $\Delta\tau$, then, by the average theorem,

$$\delta F[\varphi(\tau)] = a(t') \int\limits_{\Delta\tau} d\tau \delta\varphi(\tau),$$

where point t' belongs to segment $\left[t - \frac{\Delta\tau}{2}, t + \frac{\Delta\tau}{2}\right]$. Consequently,

$$\frac{\delta F[\varphi(\tau)]}{\delta\varphi(t)} = \lim_{\Delta\tau\to 0} a(t') = a(t). \tag{2.3}$$

In case (**b**), we obtain similarly

$$\frac{\delta F[\varphi(\tau)]}{\delta\varphi(t)} = \int\limits_{t_1}^{t_2} d\tau\, [B(\tau,t)+B(t,\tau)]\, \varphi(\tau) \quad (t_1 < t < t_2).$$

Note that function $B(\tau_1, \tau_2)$ can always be assumed a symmetric function of its arguments here.

In case (**c**), we have

$$\begin{aligned} F[\varphi(\tau) + \delta\varphi(\tau)] &= f\,(\Phi[\varphi(\tau)]) + \frac{\partial f(\Phi[\varphi(\tau)])}{\partial\Phi}\delta\Phi[\varphi(\tau)] + \cdots \\ &= F[\varphi(\tau)] + \frac{\partial f(\Phi[\varphi(\tau)])}{\partial\Phi}\delta\Phi[\varphi(\tau)] + \cdots \end{aligned}$$

and, consequently,

$$\frac{\delta}{\delta\varphi(t)} f\,(\Phi[\varphi(\tau)]) = \frac{\partial f(\Phi[\varphi(\tau)])}{\partial\Phi}\frac{\delta}{\delta\varphi(t)}\Phi[\varphi(\tau)]. \tag{2.4}$$

Consider now functional $\Phi[\varphi(\tau)] = F_1[\varphi(\tau)]F_2[\varphi(\tau)]$. We have

$$\delta\Phi[\varphi(\tau)] = F_1[\varphi(\tau) + \delta\varphi(\tau)]F_2[\varphi(\tau) + \delta\varphi(\tau)] - F_1[\varphi(\tau)]F_2[\varphi(\tau)]$$

$$= F_1[\varphi(\tau)]\delta F_2[\varphi(\tau)] + F_2[\varphi(\tau)]\delta F_1[\varphi(\tau)]$$

and, consequently,

$$\frac{\delta}{\delta\varphi(t)}F_1[\varphi(\tau)]F_2[\varphi(\tau)] = F_1[\varphi(\tau)]\frac{\delta}{\delta\varphi(t)}F_2[\varphi(\tau)] + F_2[\varphi(\tau)]\frac{\delta}{\delta\varphi(t)}F_1[\varphi(\tau)]. \tag{2.5}$$

We can define the expression for the variational derivative of functional $\varphi(\tau_0)$ with respect to function $\varphi(t)$ by the formal relationship

$$\frac{\delta\varphi(\tau_0)}{\delta\varphi(t)} = \delta(\tau_0 - t). \tag{2.6}$$

Formula (2.6) can be proved, for example, by considering the linear functional of the form

$$F[\varphi(\tau)] = \frac{1}{\sqrt{2\pi}\sigma}\int_{-\infty}^{\infty} d\tau\varphi(\tau)\exp\left\{-\frac{(\tau-\tau_0)^2}{2\sigma^2}\right\}. \tag{2.7}$$

According to Eq. (2.3), the variational derivative of this functional has the form

$$\frac{\delta}{\delta\varphi(t)}F[\varphi(\tau)] = \frac{1}{\sqrt{2\pi}\sigma}\exp\left\{-\frac{(t-\tau_0)^2}{2\sigma^2}\right\}. \tag{2.8}$$

Performing now formal limit process $\sigma \to 0$ in Eq. (2.7)) and (2.8), we obtain the desired formula (2.6). Moreover,

$$\frac{\delta F[\varphi(\tau)]}{\delta\varphi(t)} = \frac{\partial F[\varphi(\tau)]}{\partial\varphi(\tau)}\frac{\delta\varphi(\tau)}{\delta\varphi(t)} = \frac{\partial F[\varphi(\tau)]}{\partial\varphi(\tau)}\delta(\tau - t).$$

Formula (2.6) is very convenient for functional differentiation of functionals explicitly dependent on $\varphi(\tau)$. Indeed, for the quadratic functional (**b**), we have

$$\frac{\delta}{\delta\varphi(t)}\int_{t_1}^{t_2}\int_{t_1}^{t_2} d\tau_1 d\tau_2 B(\tau_1,\tau_2)\varphi(\tau_1)\varphi(\tau_2)$$

$$\overset{(2.5)}{=}\int_{t_1}^{t_2}\int_{t_1}^{t_2} d\tau_1 d\tau_2 B(\tau_1,\tau_2)\left[\frac{\delta\varphi(\tau_1)}{\delta\varphi(t)}\varphi(\tau_2)+\varphi(\tau_1)\frac{\delta\varphi(\tau_2)}{\delta\varphi(t)}\right]$$

$$\overset{(2.6)}{=}\int_{t_1}^{t_2} d\tau\,[B(t,\tau)+B(\tau,t)]\,\varphi(\tau)\quad (t_1<t<t_2).$$

Consider the functional

$$F[\varphi(\tau)]=\int_{t_1}^{t_2} d\tau L\left(\tau,\varphi(\tau),\frac{d\varphi(\tau)}{d\tau}\right)$$

as another example. In this case,

$$\frac{\delta}{\delta\varphi(t)}F[\varphi(\tau)]$$

$$\overset{(2.4)}{=}\int_{t_1}^{t_2} d\tau\left[\frac{\partial L\left(\tau,\varphi(\tau),\frac{d\varphi(\tau)}{d\tau}\right)}{\partial\varphi(\tau)}+\frac{\partial L\left(\tau,\varphi(\tau),\frac{d\varphi(\tau)}{d\tau}\right)}{\partial\dot{\varphi}(\tau)}\frac{d}{d\tau}\right]\frac{\delta\varphi(\tau)}{\delta\varphi(t)}$$

$$\overset{(2.6)}{=}\int_{t_1}^{t_2} d\tau\left[\frac{\partial L\left(\tau,\varphi(\tau),\frac{d\varphi(\tau)}{d\tau}\right)}{\partial\varphi(\tau)}+\frac{\partial L\left(\tau,\varphi(\tau),\frac{d\varphi(\tau)}{d\tau}\right)}{\partial\dot{\varphi}(\tau)}\frac{d}{d\tau}\right]\delta(\tau-t)$$

$$=\left(-\frac{d}{dt}\frac{\partial}{\partial\dot{\varphi}(t)}+\frac{\partial}{\partial\varphi(t)}\right)L\left(t,\varphi(t),\frac{d\varphi(t)}{dt}\right),$$

where $\dot{\varphi}(t)=\frac{d}{dt}\varphi(t)$ if point t belongs to interval (t_1,t_2).

Just as a function can be expanded in the Taylor series, a functional $F[\varphi(\tau)+\eta(\tau)]$ can be expanded in the functional Taylor series in function $\eta(\tau)$ for $\eta(\tau)\sim 0$

$$F[\varphi(\tau)+\eta(\tau)] = F[\varphi(\tau)] + \int_{-\infty}^{\infty} dt \frac{\delta F[\varphi(\tau)]}{\delta\varphi(t)}\eta(t) + \frac{1}{2!}\int_{-\infty}^{\infty}\int_{-\infty}^{\infty} dt_1 dt_2 \frac{\delta^2 F[\varphi(\tau)]}{\delta\varphi(t_1)\delta\varphi(t_2)}\eta(t_1)\eta(t_2) + \cdots . \tag{2.9}$$

Note that the operator expression

$$1 + \int_{-\infty}^{\infty} dt\eta(t)\frac{\delta}{\delta\varphi(t)} + \frac{1}{2!}\int_{-\infty}^{\infty}\int_{-\infty}^{\infty} dt_1 dt_2 \eta(t_1)\eta(t_2)\frac{\delta^2}{\delta\varphi(t_1)\delta\varphi(t_2)} + \cdots$$
$$= 1 + \int_{-\infty}^{\infty} dt\eta(t)\frac{\delta}{\delta\varphi(t)} + \frac{1}{2!}\left[\int_{-\infty}^{\infty} dt\eta(t)\frac{\delta}{\delta\varphi(t)}\right]^2 + \cdots \tag{2.10}$$

can be written shortly as the operator

$$\exp\left\{\int_{-\infty}^{\infty} dt\eta(t)\frac{\delta}{\delta\varphi(t)}\right\}, \tag{2.11}$$

whose action should be treated precisely in the sense of expansion (2.10). Using this operator, we can rewrite Eq. (2.9) in the form

$$F[\varphi(\tau)+\eta(\tau)] = e^{\int_{-\infty}^{\infty} dt\eta(t)\frac{\delta}{\delta\varphi(t)}} F[\varphi(\tau)], \tag{2.12}$$

which enables us to interpret operator (2.11) as the functional shift operator.

Consider now functional $F[t;\varphi(\tau)]$ dependent on parameter t. We can differentiate this functional with respect to t and determine its variational derivative with respect to $\varphi(t')$, as well. One can easily see that these operations commute, i.e., the equality

$$\frac{\partial}{\partial t}\frac{\delta F[t;\varphi(\tau)]}{\delta\varphi(t')} = \frac{\delta}{\delta\varphi(t')}\frac{\partial F[t;\varphi(\tau)]}{\partial t} \tag{2.13}$$

holds. If the domain of τ is independent of t, the validity of Eq. (2.13) is obvious. Otherwise, for example, for functionals $F[t;\varphi(\tau)]$ with $0 \le \tau \le t$, the validity of Eq. (2.13) can be checked on by expanding functional $F[t;\varphi(\tau)]$ in the functional Taylor series.

2.1.2 Principle of Dynamic Causality

Vary Eq. (2.1) with respect to field $\boldsymbol{U}(\boldsymbol{r}, t)$. Assuming that the initial position $\boldsymbol{r}_0$ is independent of field $\boldsymbol{U}$, we obtain the equation linear in variational derivative (the linear variational differential equation)

$$\frac{\delta r_i(t)}{\delta U_j(\boldsymbol{r}, t')} = \delta_{ij}\delta\left(\boldsymbol{r} - \boldsymbol{r}(t')\right)\theta\left(t' - t_0\right)\theta\left(t - t'\right) + \int_{t_0}^{t} d\tau \frac{\partial U_i(\boldsymbol{r}(\tau), \tau)}{\partial r_k} \frac{\delta r_k(\tau)}{\delta U_j(\boldsymbol{r}, t')}, \tag{2.14}$$

where $\delta\left(\boldsymbol{r} - \boldsymbol{r}'\right)$ is the Dirac delta function, and $\theta(z)$ is the Heaviside step function. From Eq. (2.14) follows that

$$\frac{\delta r_i(t)}{\delta U_j(\boldsymbol{r}, t')} = 0 \quad \text{for } t' > t \text{ or } t' < t_0, \tag{2.15}$$

which means that solution $\boldsymbol{r}(t)$ to the dynamic problem (2.1) as a functional of field $\boldsymbol{U}\left(\boldsymbol{r}, t'\right)$ depends on $\boldsymbol{U}\left(\boldsymbol{r}, t'\right)$ only for $t_0 < t' < t$. Consequently, function $\boldsymbol{r}(t)$ will remain unchanged if field $\boldsymbol{U}\left(\boldsymbol{r}, t'\right)$ varies outside the interval (t_0, t), i.e., for $t' < t_0$ or $t' > t$. We will call condition (2.15) the *dynamic causality condition*.

Taking this condition into account, we can rewrite Eq. (2.14) in the form

$$\begin{aligned}\frac{\delta r_i(t)}{\delta U_j(\boldsymbol{r}, t')} &= \delta_{ij}\delta\left(\boldsymbol{r} - \boldsymbol{r}(t')\right)\theta\left(t' - t_0\right)\theta\left(t - t'\right) \\ &\quad + \int_{t'}^{t} d\tau \frac{\partial U_i(\boldsymbol{r}(\tau), \tau)}{\partial r_k} \frac{\delta r_k(\tau)}{\delta U_j(\boldsymbol{r}, t')}.\end{aligned} \tag{2.16}$$

As a consequence, limit $t \to t' + 0$ yields the equality

$$\left.\frac{\delta r_i(t)}{\delta U_j(\boldsymbol{r}, t')}\right|_{t=t'+0} = \delta_{ij}\delta\left(\boldsymbol{r} - \boldsymbol{r}(t')\right). \tag{2.17}$$

Integral equation (2.16) in variational derivative is obviously equivalent to the linear differential equation with the initial value

$$\begin{aligned}&\frac{\partial}{\partial t}\left(\frac{\delta r_i(t)}{\delta U_j(\boldsymbol{r}, t')}\right) = \frac{\partial U_i(\boldsymbol{r}(t), t)}{\partial r_k}\left(\frac{\delta r_k(t)}{\delta U_j(\boldsymbol{r}, t')}\right), \\ &\left.\frac{\delta r_i(t)}{\delta U_j(\boldsymbol{r}, t')}\right|_{t=t'} = \delta_{ij}\delta\left(\boldsymbol{r} - \boldsymbol{r}(t')\right).\end{aligned} \tag{2.18}$$

The dynamic causality condition is the general property of problems described by differential equations with initial values. The boundary-value problems possess no such property. Indeed, in the case of problem (1.31), (1.32), page 20–21 that describes propagation of a plane wave in a layer of inhomogeneous medium, wavefield $u(x)$ at point x and reflection and transmission coefficients depend functionally on function $\varepsilon(x)$ for all x of layer (L_0, L). However, using the imbedding method, we can convert this problem into the initial-value problem with respect to an auxiliary parameter L and make use the causality property in terms of the equations of the imbedding method.

2.2 Solution Dependence on Problem's Parameters

2.2.1 Solution Dependence on Initial Data

Here, we will use the vertical bar symbol to isolate the dependence of solution $\boldsymbol{r}(t)$ to Eq. (2.1) on the initial parameters $\boldsymbol{r}_0$ and t_0:

$$\boldsymbol{r}(t) = \boldsymbol{r}(t|\boldsymbol{r}_0, t_0), \quad \boldsymbol{r}_0 = \boldsymbol{r}(t_0|\boldsymbol{r}_0, t_0).$$

Let us differentiate Eq. (2.1) with respect to parameters r_{0k} and t_0. As a result, we obtain linear equations for Jacobi's matrix $\dfrac{\partial}{\partial r_{0k}} r_i(t|\boldsymbol{r}_0, t_0)$ and quantity $\dfrac{\partial}{\partial t_0} r_i(t|\boldsymbol{r}_0, t_0)$

$$\begin{aligned}
\frac{\partial r_i(t|\boldsymbol{r}_0, t_0)}{\partial r_{0k}} &= \delta_{ik} + \int_{t_0}^{t} d\tau \frac{\partial U_i(\boldsymbol{r}(\tau), \tau)}{\partial r_j} \frac{\partial r_j(\tau|\boldsymbol{r}_0, t_0)}{\partial r_{0k}}, \\
\frac{\partial r_i(t|\boldsymbol{r}_0, t_0)}{\partial t_0} &= -U_i(\boldsymbol{r}_0(t_0), t_0) + \int_{t_0}^{t} d\tau \frac{\partial U_i(\boldsymbol{r}(\tau), \tau)}{\partial r_j} \frac{\partial r_j(\tau|\boldsymbol{r}_0, t_0)}{\partial t_0}.
\end{aligned} \tag{2.19}$$

Multiplying now the first of these equations by $U_k(\boldsymbol{r}_0(t), t)$, summing over index k, adding the result to the second equation, and introducing the vector function

$$F_i(t|\boldsymbol{r}_0, t_0) = \left(\frac{\partial}{\partial t_0} + \boldsymbol{U}(\boldsymbol{r}_0, t_0)\frac{\partial}{\partial \boldsymbol{r}_0}\right) r_i(t|\boldsymbol{r}_0, t_0),$$

we obtain that this function satisfies the linear homogeneous equation

$$F_i(t|\boldsymbol{r}_0, t_0) = \int_{t_0}^{t} d\tau \frac{\partial U_i(\boldsymbol{r}(\tau), \tau)}{\partial r_k} F_k(\tau|\boldsymbol{r}_0, t_0). \tag{2.20}$$

Differentiating this equation with respect to time, we arrive at the ordinary differential equation

$$\frac{\partial}{\partial t} F_i(t|\boldsymbol{r}_0, t_0) = \frac{\partial U_i(\boldsymbol{r}(t), t)}{\partial r_k} F_k(t|\boldsymbol{r}_0, t_0)$$

with the initial condition $F_i(t_0|\boldsymbol{r}_0, t_0) = 0$ at $t = t_0$, which follows from Eq. (2.20); as a consequence, we have $F_i(t|\boldsymbol{r}_0, t_0) \equiv 0$. Therefore, we obtain the equality

$$\left(\frac{\partial}{\partial t_0} + \boldsymbol{U}(\boldsymbol{r}_0, t_0)\frac{\partial}{\partial \boldsymbol{r}_0}\right) r_i(t|\boldsymbol{r}_0, t_0) = 0, \tag{2.21}$$

which can be considered as the linear partial differential equation with the derivatives with respect to variables $\boldsymbol{r}_0$, t_0 and the initial value at $t_0 = t$

$$\boldsymbol{r}(t|\boldsymbol{r}_0, t) = \boldsymbol{r}_0. \tag{2.22}$$

The variable t appears now in problem (2.21), (2.22) as a parameter.

Equation (2.21) is solved using the time direction inverse to that used in solving problem (1.1), pages 50–51; for this reason, we will call it the *backward equation*.

Equation (2.21) with the initial condition (2.22) can be rewritten as the integral equation

$$\boldsymbol{r}(t|\boldsymbol{r}_0, t_0) = \boldsymbol{r}_0 + \int\limits_{t_0}^{t} d\tau \left(\boldsymbol{U}(\boldsymbol{r}_0, \tau)\frac{\partial}{\partial \boldsymbol{r}_0}\right)\boldsymbol{r}(t|\boldsymbol{r}_0, \tau). \tag{2.23}$$

Varying now Eq. (2.23) with respect to function $U_j(\boldsymbol{r}', t')$, we obtain the integral equation

$$\begin{aligned}\frac{\delta r_i(t|\boldsymbol{r}_0, t_0)}{\delta U_j(\boldsymbol{r}', t')} &= \delta(\boldsymbol{r}_0 - \boldsymbol{r}')\theta(t' - t_0)\theta(t - t')\frac{\partial r_i(t|\boldsymbol{r}_0, t')}{\partial r_{j0}} \\ &+ \int\limits_{t_0}^{t} d\tau \left(\boldsymbol{U}(\boldsymbol{r}_0, \tau)\frac{\partial}{\partial \boldsymbol{r}_0}\right)\frac{\delta r_i(t|\boldsymbol{r}_0, \tau)}{\delta U_j(\boldsymbol{r}', t')},\end{aligned} \tag{2.24}$$

from which follows that

$$\frac{\delta r_i(t|\boldsymbol{r}_0, t_0)}{\delta U_j(\boldsymbol{r}', t')} = 0, \quad \text{if } t' > t \quad \text{or} \quad t' < t_0,$$

which means that function $\boldsymbol{r}(t|\boldsymbol{r}_0, t_0)$ also possesses the property of dynamic causality with respect to parameter t_0 (2.15) – which is quite natural – and Eq. (2.24) can be rewritten in the form (for $t_0 < t' < t$)

$$\frac{\delta r_i(t|\boldsymbol{r}_0, t_0)}{\delta U_j(\boldsymbol{r}', t')} = \delta(\boldsymbol{r}_0 - \boldsymbol{r}')\frac{\partial r_i(t|\boldsymbol{r}_0, t')}{\partial r_{j0}} + \int\limits_{t_0}^{t'} d\tau \left(\boldsymbol{U}(\boldsymbol{r}_0, \tau)\frac{\partial}{\partial \boldsymbol{r}_0}\right)\frac{\delta r_i(t|\boldsymbol{r}_0, \tau)}{\delta U_j(\boldsymbol{r}', t')}. \tag{2.25}$$

Setting now $t' \to t_0 + 0$, we obtain the equality

$$\left.\frac{\delta r_i(t|\boldsymbol{r}_0, t_0)}{\delta U_j(\boldsymbol{r}, t')}\right|_{t'=t_0+0} = \delta(\boldsymbol{r}_0 - \boldsymbol{r})\frac{\partial r_i(t|\boldsymbol{r}_0, t_0)}{\partial r_{0j}}. \tag{2.26}$$

2.2.2 Imbedding Method for Boundary-Value Problems

Consider first boundary-value problems formulated in terms of ordinary differential equations. The *imbedding method* (or *invariant imbedding method*, as it is usually called in mathematical literature) offers a possibility of reducing boundary-value problems at hand to the evolution-type initial-value problems possessing the property of dynamic causality with respect to an auxiliary parameter.

The idea of this method was first suggested by V.A. Ambartsumyan (the so-called *Ambartsumyan invariance principle*) [36–38] for solving the equations of linear theory of radiative transfer. Further, mathematicians grasped this idea and used it to convert boundary-value (nonlinear, in the general case) problems into evolution-type initial-value problems that are more convenient for simulations. Several monographs (see, e.g., [39–41]) deal with this method and consider both physical and computational aspects.

Consider the dynamic system described in terms of the system of ordinary differential equations

$$\frac{d}{dt}\boldsymbol{x}(t) = \boldsymbol{F}(t, \boldsymbol{x}(t)), \tag{2.27}$$

defined on segment $t \in [0, T]$ with the boundary conditions

$$g\boldsymbol{x}(0) + h\boldsymbol{x}(T) = \boldsymbol{v}, \tag{2.28}$$

where g and h are the constant matrixes.

Dynamic problem (2.27), (2.28) possesses no dynamic causality property, which means that the solution $\boldsymbol{x}(t)$ to this problem at instant t functionally depends on external forces $\boldsymbol{F}(\tau, \boldsymbol{x}(\tau))$ for all $0 \le \tau \le T$. Moreover, even boundary values $\boldsymbol{x}(0)$ and $\boldsymbol{x}(T)$ are functionals of field $\boldsymbol{F}(\tau, \boldsymbol{x}(\tau))$. The absence of dynamic causality in problem (2.27), (2.28) prevents us from using the known statistical methods of analyzing statistical characteristics of the solution to Eq. (2.27) if external force functional $\boldsymbol{F}(t, \boldsymbol{x})$ is the random space- and time-domain field. Introducing the one-time probability density $P(t; \boldsymbol{x})$ of the solution to Eq. (2.27), we can easily see that condition (2.28) is insufficient for determining the value of this probability at any point. The boundary condition imposes only certain functional restriction.

Note that the solution to problem (2.27), (2.28) parametrically depends on T and $\boldsymbol{v}$, i.e., $\boldsymbol{x}(t) = \boldsymbol{x}(t; T, \boldsymbol{v})$. Abiding by paper [42], we introduce functions

$$\boldsymbol{R}(T, \boldsymbol{v}) = \boldsymbol{x}(T; T, \boldsymbol{v}), \quad \boldsymbol{S}(T, \boldsymbol{v}) = \boldsymbol{x}(0; T, \boldsymbol{v})$$

that describe the boundary values of the solution to Eq. (2.27).

Differentiate Eq. (2.27) with respect to T and $\boldsymbol{v}$. We obtain two linear equations in the corresponding derivatives

$$\begin{aligned} \frac{d}{dt}\frac{\partial x_i(t; T, \boldsymbol{v})}{\partial T} &= \frac{\partial F_i(t, \boldsymbol{x})}{\partial x_l}\frac{\partial x_l(t; T, \boldsymbol{v})}{\partial T}, \\ \frac{d}{dt}\frac{\partial x_i(t; T, \boldsymbol{v})}{\partial v_k} &= \frac{\partial F_i(t, \boldsymbol{x})}{\partial x_l}\frac{\partial x_l(t; T, \boldsymbol{v})}{\partial v_k}. \end{aligned} \tag{2.29}$$

These equations are identical in form; consequently, we can expect that their solutions are related by the linear expression

$$\frac{\partial x_i(t; T, \boldsymbol{v})}{\partial T} = \lambda_k(T, \boldsymbol{v}) \frac{\partial x_i(t; T, \boldsymbol{v})}{\partial v_k}, \tag{2.30}$$

if vector quantity $\boldsymbol{\lambda}(T, \boldsymbol{v})$ is such that boundary conditions (2.28) are satisfied and the solution is unique. To determine vector quantity $\boldsymbol{\lambda}(T, \boldsymbol{v})$, we first set $t = 0$ in Eq. (2.30) and multiply the result by matrix g; then, we set $t = T$ and multiply the result by matrix h; and, finally, we combine the obtained expressions. Taking into account Eq. (2.28), we obtain

$$g \frac{\partial \boldsymbol{x}(0; T, \boldsymbol{v})}{\partial T} + h \left. \frac{\partial \boldsymbol{x}(t; T, \boldsymbol{v})}{\partial T} \right|_{t=T} = \boldsymbol{\lambda}(T, \boldsymbol{v}).$$

In view of the fact that

$$\left. \frac{\partial \boldsymbol{x}(t; T, \boldsymbol{v})}{\partial T} \right|_{t=T} = \frac{\partial \boldsymbol{x}(T; T, \boldsymbol{v})}{\partial T} - \left. \frac{\partial \boldsymbol{x}(t; T, \boldsymbol{v})}{\partial t} \right|_{t=T} = \frac{\partial \boldsymbol{R}(T, \boldsymbol{v})}{\partial T} - \boldsymbol{F}(T, \boldsymbol{R}(T, \boldsymbol{v}))$$

(with allowance for Eq. (2.27)), we obtain the desired expression for quantity $\boldsymbol{\lambda}(T, \boldsymbol{v})$,

$$\boldsymbol{\lambda}(T, \boldsymbol{v}) = -h\boldsymbol{F}(T, \boldsymbol{R}(T, \boldsymbol{v})). \tag{2.31}$$

Expression (2.30) with parameter $\boldsymbol{\lambda}(T, \boldsymbol{v})$ defined by Eq. (2.31), i.e., the expression

$$\frac{\partial x_i(t; T, \boldsymbol{v})}{\partial T} = -h_{kl} F_l(T, \boldsymbol{R}(T, \boldsymbol{v})) \frac{\partial x_i(t; T, \boldsymbol{v})}{\partial v_k}, \tag{2.32}$$

can be considered as the linear differential equation; one needs only to supplement it with the corresponding initial condition

$$\boldsymbol{x}(t; T, \boldsymbol{v})|_{T=t} = \boldsymbol{R}(t, \boldsymbol{v})$$

assuming that function $\boldsymbol{R}(T, \boldsymbol{v})$ is known.

The equation for this function can be obtained from the equality

$$\frac{\partial \boldsymbol{R}(T, \boldsymbol{v})}{\partial T} = \left. \frac{\partial \boldsymbol{x}(t; T, \boldsymbol{v})}{\partial t} \right|_{t=T} + \left. \frac{\partial \boldsymbol{x}(t; T, \boldsymbol{v})}{\partial T} \right|_{t=T}. \tag{2.33}$$

The right-hand side of Eq. (2.33) is the sum of the right-hand sides of Eq. (2.27) and (2.30) at $t = T$. As a result, we obtain the closed nonlinear (quasilinear) equation

$$\frac{\partial \boldsymbol{R}(T, \boldsymbol{v})}{\partial T} = -h_{kl} F_l(T, \boldsymbol{R}(T, \boldsymbol{v})) \frac{\partial \boldsymbol{R}(T, \boldsymbol{v})}{\partial v_k} + \boldsymbol{F}(T, \boldsymbol{R}(T, \boldsymbol{v})). \tag{2.34}$$

The initial condition for Eq. (2.34) follows from Eq. (2.28) for $T \to 0$

$$\boldsymbol{R}(T,\boldsymbol{v})|_{T=0} = (g+h)^{-1}\boldsymbol{v}. \tag{2.35}$$

Setting now $t = 0$ in Eq. (2.29), we obtain for the secondary boundary quantity $\boldsymbol{S}(T,\boldsymbol{v}) = \boldsymbol{x}(0;T,\boldsymbol{v})$ the equation

$$\frac{\partial \boldsymbol{S}(T,\boldsymbol{v})}{\partial T} = -h_{kl}F_l\left(T,\boldsymbol{R}(T,\boldsymbol{v})\right)\frac{\partial \boldsymbol{S}(T,\boldsymbol{v})}{\partial v_k} \tag{2.36}$$

with the initial condition

$$\boldsymbol{S}(T,\boldsymbol{v})|_{T=0} = (g+h)^{-1}\boldsymbol{v}$$

following from Eq. (2.35).

Thus, the problem reduces to the closed quasilinear equation (2.34) with initial value (2.35) and linear equation (2.30) whose coefficients and initial value are determined by the solution of Eq. (2.34).

In the problem under consideration, input 0 and output T are symmetric. For this reason, one can solve it not only from T to 0, but also from 0 to T. In the latter case, functions $\boldsymbol{R}(T,\boldsymbol{v})$ and $\boldsymbol{S}(T,\boldsymbol{v})$ switch the places.

An important point consists in the fact that, despite the initial problem (2.27) is nonlinear, Eq. (2.30) is the linear equation, because it is essentially the equation in variations. It is Eq. (2.34) that is responsible for nonlinearity.

Note that the above technique of deriving imbedding equations for Eq. (2.27) can be easily extended to the boundary condition of the form [42]

$$\boldsymbol{g}\left(\boldsymbol{x}(0)\right) + \boldsymbol{h}\left(\boldsymbol{x}(T)\right) + \int_0^T d\tau \boldsymbol{K}\left(\tau,\boldsymbol{x}(\tau)\right) = \boldsymbol{v},$$

where $\boldsymbol{g}(\boldsymbol{x})$, $\boldsymbol{h}(\boldsymbol{x})$ and $\boldsymbol{K}(T,\boldsymbol{x})$ are arbitrary given vector functions.

If function $\boldsymbol{F}(t,\boldsymbol{x})$ is linear in $\boldsymbol{x}$, $F_i(t,\boldsymbol{x}) = A_{ij}(t)x_j(t)$, then boundary-value problem (2.27), (2.28) assumes the simpler form

$$\frac{d}{dt}\boldsymbol{x}(t) = A(t)\boldsymbol{x}(t), \quad g\boldsymbol{x}(0) + h\boldsymbol{x}(T) = \boldsymbol{v},$$

and the solution of Eq. (2.30), (2.34) and (2.36) will be the function linear in $\boldsymbol{v}$

$$\boldsymbol{x}(t;T,\boldsymbol{v}) = X(t;T)\boldsymbol{v}. \tag{2.37}$$

As a result, we arrive at the closed matrix Riccati equation for matrix $R(T) = X(T;T)$

$$\frac{d}{dT}R(T) = A(T)R(T) - R(T)hA(T)R(T), \quad R(0) = (g+h)^{-1}. \tag{2.38}$$

As regards matrix $X(t,T)$, it satisfies the linear matrix equation with the initial condition

$$\frac{\partial}{\partial T}X(t;T) = -X(t;T)hA(T)R(T), \quad X(t;T)_{T=t} = R(t). \tag{2.39}$$

Problems

Problem 2.1

Helmholtz equation with unmatched boundary. Derive the imbedding equations for the stationary wave boundary-value problem

$$\left(\frac{d^2}{dx^2} + k^2(x)\right) u(x) = 0,$$
$$\left(\frac{d}{dx} + ik_1\right) u(x)\Big|_{x=L_0} = 0, \quad \left(\frac{d}{dx} - ik_0\right) u(x)\Big|_{x=L} = -2ik_0.$$

Instruction *Reformulate this boundary-value problem as the initial-value in terms of functions* $u(x) = u(x; L)$ *and* $v(x; L) = \frac{\partial}{\partial x} u(x; L)$

$$\frac{d}{dx} u(x; L) = v(x; L), \quad \frac{d}{dx} v(x; L) = -k^2(x) u(x; L),$$
$$v(L_0; L) + ik_1 u(L_0; L) = 0, \quad v(L; L) - ik_0 u(L; L) = -2ik_0.$$

Solution

$$\frac{\partial}{\partial L} u(x; L) = \left\{ik_0 + \frac{i}{2k_0}\left[k^2(L) - k_0^2\right] u(L; L)\right\} u(x; L),$$
$$u(x; L)|_{L=x} = u(x; x),$$
$$\frac{d}{dL} u(L; L) = 2ik_0\left[u(L; L) - 1\right] + \frac{i}{2k_0}\left[k^2(L) - k_0^2\right] u(L; L)^2,$$
$$u(L_0; L_0) = \frac{2k_0}{k_0 + k_1}.$$

Problem 2.2

Helmholtz equation with matched boundary. Derive the imbedding equations for the stationary wave boundary-value problem

$$\left(\frac{d^2}{dx^2} + k^2(x)\right) u(x; L) = 0,$$
$$\left(\frac{d}{dx} + ik_1\right) u(x; L)\Big|_{x=L_0} = 0, \quad \left(\frac{d}{dx} - ik(L)\right) u(x; L)\Big|_{x=L} = -2ik(L).$$

Solution

$$\frac{\partial}{\partial L}u(x;L) = \left\{ik(L) + \frac{1}{2}\frac{k'(L)}{k(L)}\left[2 - u(L;L)\right]\right\} u(x;L),$$

$$u(x;L)|_{L=x} = u(x;x),$$

$$\frac{d}{dL}u(L;L) = ik(L)\left[u(L;L) - 1\right] + \frac{1}{2}\frac{k'(L)}{k(L)}\left[2 - u(L;L)\right]u(L;L),$$

$$u(L_0;L_0) = \frac{2k(L_0)}{k(L_0) + k_1}.$$

Problem 2.3

Derive the imbedding equations for the boundary-value problem

$$\left(\frac{d^2}{dx^2} - \frac{\rho'(x)}{\rho(x)}\frac{d}{dx} + p^2\left[1 + \varepsilon(x)\right]\right)u(x) = 0,$$

$$\left(\frac{1}{\rho(x)}\frac{d}{dx} + ip\right)u(x)\bigg|_{x=L_0} = 0, \quad \left(\frac{1}{\rho(x)}\frac{d}{dx} - ip\right)u(x)\bigg|_{x=L} = -2ip,$$

where $\rho'(x) = \dfrac{d\rho(x)}{dx}$

Solution The imbedding equations with respect to parameter L have the form

$$\frac{\partial}{\partial L}u(x;L) = \left\{ip\rho(L) + \varphi(L)u(L;L)\right\} u(x;L),$$

$$u(x;L)|_{L=x} = u(x;x);$$

$$\frac{d}{dL}u(L;L) = 2ip\rho(L)\left[u(L;L) - 1\right] + \varphi(L)u^2(L;L),$$

$$u(L;L)|_{L=L_0} = 1,$$

where

$$\varphi(x) = \frac{ip}{2\rho(x)}\left[1 + \varepsilon(x) - \rho^2(x)\right].$$

Remark The above boundary-value problem describes many physical processes, such as acoustic waves in a medium with variable density and some types of electromagnetic waves. In these cases, function $\varepsilon(x)$ describes inhomogeneities of the velocity of wave propagation (refractive index or dielectric permittivity) and function $\rho(x)$ appears only in the case of acoustic waves and characterize medium density.

Problem 2.4

Derive the imbedding equations for the matrix Helmholtz equation

$$\left(\frac{d^2}{dx^2}+\gamma(x)\frac{d}{dx}+K(x)\right)U(x)=0,$$

$$\left(\frac{d}{dx}+B\right)U(x)\bigg|_{x=L}=D,\qquad \left(\frac{d}{dx}+C\right)U(x)\bigg|_{x=0}=0,$$

where $\gamma(x)$, $K(x)$, and $U(x)$ are the variable matrixes, while B, C, and D are the constant matrixes.

Solution The imbedding equations with respect to parameter L have the form

$$\frac{\partial}{\partial L}U(x;L)=U(x;L)\Lambda(L),\qquad U(x;L)|_{L=x}=U(x;x),$$

$$\frac{d}{dL}U(L;L)=D-\left\{BU(L;L)+U(L;L)D^{-1}BD\right\}+U(L;L)D^{-1}\gamma(L)D$$
$$+U(L;L)D^{-1}\left\{K(L)-\gamma(L)B+B^2\right\}U(L;L),$$
$$U(0;0)=(B-C)^{-1}D,$$

where matrix $\Lambda(L)$ is given by the formula

$$\Lambda(L)=D^{-1}\left[\gamma(L)-B\right]D+D^{-1}\left[K(L)+B^2-\gamma(L)B\right]U(L;L).$$

Lecture 3

Indicator Function and Liouville Equation

Modern apparatus of the theory of random processes is able of constructing closed descriptions of dynamic systems if these systems meet the condition of dynamic causality and are formulated in terms of linear partial differential equations or certain types of integral equations. One can use indicator functions to perform the transition from the initial, generally nonlinear system to the equivalent description in terms of the linear partial differential equations. However, this approach results in increasing the dimension of the space of variables. Consider such a transition in the dynamic systems described in the Lecture 1 as examples.

3.1 Ordinary Differential Equations

Let a stochastic problem is described by the system of equations (1.1), which we'll rewrite in the form

$$\frac{d}{dt}\boldsymbol{r}(t) = \boldsymbol{U}\left(\boldsymbol{r}(t), t\right), \quad \boldsymbol{r}(t_0) = \boldsymbol{r}_0. \tag{3.1}$$

We introduce the scalar function

$$\varphi(\boldsymbol{r}, t) = \delta(\boldsymbol{r}(t) - \boldsymbol{r}), \tag{3.2}$$

which is concentrated on the section of the random process $\boldsymbol{r}(t)$ by a given plane $\boldsymbol{r}(t) = \boldsymbol{r}$ and is usually called the *indicator function*.

Differentiating Eq. (3.2) with respect to time t, we obtain, using Eq. (3.1), the equality

$$\frac{\partial}{\partial t}\varphi(\boldsymbol{r}, t) = -\frac{d\boldsymbol{r}(t)}{dt}\frac{\partial}{\partial \boldsymbol{r}}\delta(\boldsymbol{r}(t) - \boldsymbol{r}) = -\boldsymbol{U}\left(\boldsymbol{r}(t), t\right)\frac{\partial}{\partial \boldsymbol{r}}\delta(\boldsymbol{r}(t) - \boldsymbol{r})$$
$$= -\frac{\partial}{\partial \boldsymbol{r}}\left[\boldsymbol{U}\left(\boldsymbol{r}(t), t\right)\delta(\boldsymbol{r}(t) - \boldsymbol{r})\right].$$

Using then the *probing property* of the delta-function

$$\boldsymbol{U}\left(\boldsymbol{r}(t), t\right)\delta(\boldsymbol{r}(t) - \boldsymbol{r}) = \boldsymbol{U}\left(\boldsymbol{r}, t\right)\delta(\boldsymbol{r}(t) - \boldsymbol{r}),$$

Lectures on Dynamics of Stochastic Systems. DOI: 10.1016/B978-0-12-384966-3.00003-9

we obtain the linear partial differential equation

$$\left(\frac{\partial}{\partial t}+\frac{\partial}{\partial \boldsymbol{r}}\boldsymbol{U}(\boldsymbol{r},t)\right)\varphi(\boldsymbol{r},t)=0, \qquad \varphi(\boldsymbol{r},t_0)=\delta(\boldsymbol{r}_0-\boldsymbol{r}), \tag{3.3}$$

equivalent to the initial system. This equation is called the *Liouville equation*. In terms of phase points moving in phase space $\{\boldsymbol{r}\}$, this equation corresponds to the equation of continuity.

The above derivation procedure of the Liouville equation clearly shows the necessity of distinguishing the function $\boldsymbol{r}(t)$ and the parameter $\boldsymbol{r}$. In this connection, correct working with the delta-function assumes that the argument of function $\boldsymbol{r}(t)$ was explicitly written in all intermediate calculations (cf. the forms of Eqs. (3.1) and (1.1), page 3). Neglecting this rule usually results in errors.

The transition from system (3.1) to Liouville equation (3.3) entails enlargement of the phase space $(\boldsymbol{r}, t)$, which, however, has a finite dimension. Note that Eq. (3.3) coincides in form with the equation of tracer transfer by the velocity field $\boldsymbol{U}(\boldsymbol{r},t)$ (1.39); the only difference consists in the initial values.

Equation (3.3) can be rewritten in the form of the linear integral equation

$$\varphi(\boldsymbol{r},t)=\delta(\boldsymbol{r}-\boldsymbol{r}_0)-\int\limits_{t_0}^{t}d\tau\frac{\partial}{\partial \boldsymbol{r}}\{\boldsymbol{U}(\boldsymbol{r},\tau)\varphi(\boldsymbol{r},\tau)\}.$$

Varying this equation with respect to function $U_i(\boldsymbol{r}',t')$ with the use of Eq. (2.6), page 56, we obtain the integral equation

$$\begin{aligned}\frac{\delta\varphi(\boldsymbol{r},t)}{\delta U_j(\boldsymbol{r}',t')}&=-\int\limits_{t_0}^{t}d\tau\frac{\partial}{\partial r_i}\left\{\frac{\delta U_i(\boldsymbol{r},\tau)}{\delta U_j(\boldsymbol{r}',t')}\varphi(\boldsymbol{r},\tau)\right\}-\int\limits_{t_0}^{t}d\tau\frac{\partial}{\partial \boldsymbol{r}}\left\{\boldsymbol{U}(\boldsymbol{r},\tau)\frac{\delta\varphi(\boldsymbol{r},\tau)}{\delta U_j(\boldsymbol{r}',t')}\right\}\\&=-\theta(t-t')\theta(t'-t_0)\frac{\partial}{\partial r_j}\left\{\delta\left(\boldsymbol{r}-\boldsymbol{r}'\right)\varphi\left(\boldsymbol{r},t'\right)\right\}\\&\quad-\int\limits_{t_0}^{t}d\tau\frac{\partial}{\partial \boldsymbol{r}}\left\{\boldsymbol{U}(\boldsymbol{r},\tau)\frac{\delta\varphi(\boldsymbol{r},\tau)}{\delta U_j(\boldsymbol{r}',t')}\right\},\end{aligned} \tag{3.4}$$

where $\theta(t)$ is, as earlier, the Heaviside step function. As a result, we have

$$\frac{\delta\varphi(\boldsymbol{r},t)}{\delta U_j(\boldsymbol{r}',t')}\sim\theta(t-t')\theta(t'-t_0),$$

and the variational derivative satisfies the condition

$$\frac{\delta\varphi(\boldsymbol{r},t)}{\delta U_j(\boldsymbol{r}',t')}=0, \quad \text{if} \quad t'<t_0 \quad \text{or} \quad t'>t,$$

which expresses the condition of dynamic causality for the Liouville equation (3.3).

For $t_0 < t' < t$, Eq. (3.4) can be written in the form

$$\frac{\delta \varphi(\boldsymbol{r},t)}{\delta U_j(\boldsymbol{r}',t')} = -\frac{\partial}{\partial r_j}\left\{\delta\left(\boldsymbol{r}-\boldsymbol{r}'\right)\varphi\left(\boldsymbol{r},t'\right)\right\} - \int\limits_{t'}^{t} d\tau \frac{\partial}{\partial \boldsymbol{r}}\left\{\boldsymbol{U}(\boldsymbol{r},\tau)\frac{\delta \varphi(\boldsymbol{r},\tau)}{\delta U_j(\boldsymbol{r}',t')}\right\}, \tag{3.5}$$

from which follows that

$$\left.\frac{\delta \varphi(\boldsymbol{r},t)}{\delta U_i(\boldsymbol{r}',t')}\right|_{t=t'+0} = -\frac{\partial}{\partial r_i}\left\{\delta\left(\boldsymbol{r}-\boldsymbol{r}'\right)\varphi(\boldsymbol{r},t')\right\}. \tag{3.6}$$

Note that equality (3.6) can be derived immediately from Eq. (2.17), page 59. Indeed, by definition of function $\varphi(\boldsymbol{r},t)$, variational derivative $\delta\varphi(\boldsymbol{r},t)/\delta U_i(\boldsymbol{r}',t')$ has the form

$$\frac{\delta \varphi(\boldsymbol{r},t)}{\delta U_j(\boldsymbol{r}',t')} = \frac{\delta}{\delta U_j(\boldsymbol{r}',t')}\delta(\boldsymbol{r}(t)-\boldsymbol{r}) = -\frac{\partial}{\partial r_i}\delta(\boldsymbol{r}(t)-\boldsymbol{r})\frac{\delta r_i(t)}{\delta U_j(\boldsymbol{r}',t')}. \tag{3.7}$$

As a consequence, for $t = t' + 0$, we have the equality

$$\begin{aligned}\left.\frac{\delta \varphi(\boldsymbol{r},t)}{\delta U_j(\boldsymbol{r}',t')}\right|_{t=t'+0} &= -\frac{\partial}{\partial r_i}\delta(\boldsymbol{r}(t')-\boldsymbol{r})\left.\frac{\delta r_i(t)}{\delta U_j(\boldsymbol{r}',t')}\right|_{t=t'+0}\\ &= -\frac{\partial}{\partial r_j}\left\{\delta(\boldsymbol{r}(t')-\boldsymbol{r})\delta\left(\boldsymbol{r}'-\boldsymbol{r}\right)\right\} = -\frac{\partial}{\partial r_j}\left\{\varphi(\boldsymbol{r},t')\delta\left(\boldsymbol{r}'-\boldsymbol{r}\right)\right\}.\end{aligned}$$

The solution to Eq. (3.1) and, consequently, function (3.2) depends on the initial values $t_0, \boldsymbol{r}_0$. Indeed, function $\boldsymbol{r}(t) = \boldsymbol{r}(t|\boldsymbol{r}_0, t_0)$ as a function of variables $\boldsymbol{r}_0$ and t_0 satisfies the linear first-order partial differential equation (1.105). The equations of such type also allow the transition to the equations in the indicator function $\varphi(\boldsymbol{r},t|\boldsymbol{r}_0,t_0)$ (see the next section); in the case under consideration, this equation is again the linear first-order partial differential equation including the derivatives with respect to variables $\boldsymbol{r}_0$ and t_0

$$\left(\frac{\partial}{\partial t_0} + \boldsymbol{U}(\boldsymbol{r}_0,t_0)\frac{\partial}{\partial \boldsymbol{r}_0}\right)\varphi(\boldsymbol{r},t|\boldsymbol{r}_0,t_0) = 0, \quad \varphi(\boldsymbol{r},t|\boldsymbol{r}_0,t) = \delta(\boldsymbol{r}_0-\boldsymbol{r}). \tag{3.8}$$

Equation (3.8) can be called the *backward Liouville equation.* At the same time, as a consequence of the dynamic causality, the variational derivative of indicator function $\varphi(\boldsymbol{r},t|\boldsymbol{r}_0,t_0)$,

$$\frac{\delta \varphi(\boldsymbol{r},t|\boldsymbol{r}_0,t_0)}{\delta U_j(\boldsymbol{r}',t')} = \frac{\delta}{\delta U_j(\boldsymbol{r}',t')}\delta(\boldsymbol{r}(t|\boldsymbol{r}_0,t_0)-\boldsymbol{r}) = -\frac{\partial}{\partial r_i}\delta(\boldsymbol{r}(t|\boldsymbol{r}_0,t_0)-\boldsymbol{r})\frac{\delta r_i(t|\boldsymbol{r}_0,t_0)}{\delta U_j(\boldsymbol{r}',t')},$$

can be expressed in view of Eq. (2.26), page 61, as follows

$$\left.\frac{\delta\varphi(\boldsymbol{r},t|\boldsymbol{r}_0,t_0)}{\delta U_j(\boldsymbol{r}',t')}\right|_{t'=t_0+0} = -\frac{\partial}{\partial r_i}\delta(\boldsymbol{r}(t|\boldsymbol{r}_0,t_0)-\boldsymbol{r})\left.\frac{\delta r_i(t|\boldsymbol{r}_0,t_0)}{\delta U_j(\boldsymbol{r}',t')}\right|_{t'=t_0+0}$$
$$= -\delta(\boldsymbol{r}_0-\boldsymbol{r}')\frac{\partial r_i(t|\boldsymbol{r}_0,t_0)}{\partial r_{0j}}\frac{\partial\varphi(\boldsymbol{r},t|\boldsymbol{r}_0,t_0)}{\partial r_i}. \tag{3.9}$$

On the other hand,

$$\frac{\partial\varphi(\boldsymbol{r},t|\boldsymbol{r}_0,t_0)}{\partial r_{0j}} = -\frac{\partial r_i(t|\boldsymbol{r}_0,t_0)}{\partial r_{0j}}\frac{\partial\varphi(\boldsymbol{r},t|\boldsymbol{r}_0,t_0)}{\partial r_i},$$

and Eq. (3.9) assumes the final closed form

$$\left.\frac{\delta\varphi(\boldsymbol{r},t|\boldsymbol{r}_0,t_0)}{\delta U_j(\boldsymbol{r}',t')}\right|_{t'=t_0+0} = \delta(\boldsymbol{r}_0-\boldsymbol{r}')\frac{\partial\varphi(\boldsymbol{r},t|\boldsymbol{r}_0,t_0)}{\partial r_{0j}}. \tag{3.10}$$

3.2 First-Order Partial Differential Equations

If the initial-value problem is formulated in terms of partial differential equations, we always can convert it to the equivalent formulation in terms of the linear variational differential equation in the infinite-dimensional space (the *Hopf equation*). For some particular types of problems, this approach is simplified. Indeed, if the initial dynamic system satisfies the first-order partial differential equation (either *linear* as Eq. (1.39), page 24, or *quasilinear* as Eq. (1.65), page 33, or in the general case *nonlinear* as Eq. (1.84), page 38), then the phase space of the corresponding indicator function will be the finite-dimension space, which follows from the fact that first-order partial differential equations are equivalent to systems of ordinary (characteristic) differential equations. Consider these cases in more detail.

3.2.1 Linear Equations

Backward Liouville Equation

First of all, we consider the derivation of the backward Liouville equation in indicator function

$$\varphi(\boldsymbol{r},t|\boldsymbol{r}_0,t_0) = \delta\left(\boldsymbol{r}\left(t|\boldsymbol{r}_0,t_0\right)-\boldsymbol{r}\right)$$

considered as a function of parameters $(\boldsymbol{r}_0, t_0)$. Differentiating function $\varphi(\boldsymbol{r},t|\boldsymbol{r}_0,t_0)$ with respect to t_0 and r_{0j} and taking into account Eq. (2.21), page 61, we obtain the expressions

$$\begin{aligned}\frac{\partial\varphi(\boldsymbol{r},t|\boldsymbol{r}_0,t_0)}{\partial t_0} &= \frac{\partial\varphi(\boldsymbol{r},t|\boldsymbol{r}_0,t_0)}{\partial r_i}\boldsymbol{U}\left(\boldsymbol{r}_0,t_0\right)\frac{\partial r_i\left(t|\boldsymbol{r}_0,t_0\right)}{\partial\boldsymbol{r}_0},\\ \frac{\partial\varphi(\boldsymbol{r},t|\boldsymbol{r}_0,t_0)}{\partial\boldsymbol{r}_0} &= -\frac{\partial\varphi(\boldsymbol{r},t|\boldsymbol{r}_0,t_0)}{\partial r_i}\frac{\partial r_i(t|\boldsymbol{r}_0,t_0)}{\partial\boldsymbol{r}_0}.\end{aligned} \tag{3.11}$$

Taking scalar product of the second equation in Eq. (3.11) with $\boldsymbol{U}(\boldsymbol{r}_0, t_0)$ and summing the result with the first one, we arrive at the closed equation

$$\left(\frac{\partial}{\partial t_0} + \boldsymbol{U}(\boldsymbol{r}_0, t_0)\frac{\partial}{\partial \boldsymbol{r}_0}\right)\varphi(\boldsymbol{r}, t|\boldsymbol{r}_0, t_0) = 0, \quad \varphi(\boldsymbol{r}, t|\boldsymbol{r}_0, t) = \delta(\boldsymbol{r} - \boldsymbol{r}_0), \tag{3.12}$$

which is just the backward Liouville equation given earlier without derivation.

Rewrite Eq. (3.12) as the linear integral equation

$$\varphi(\boldsymbol{r}, t|\boldsymbol{r}_0, t_0) = \delta(\boldsymbol{r} - \boldsymbol{r}_0) + \int_{t_0}^{t} d\tau \boldsymbol{U}(\boldsymbol{r}_0, \tau)\frac{\partial}{\partial \boldsymbol{r}_0}\varphi(\boldsymbol{r}, t|\boldsymbol{r}_0, \tau). \tag{3.13}$$

According to this equation, variational derivative $\dfrac{\delta\varphi(\boldsymbol{r}, t|\boldsymbol{r}_0, t_0)}{\delta U_j(\boldsymbol{r}', t')}$ satisfies the linear integral equation

$$\frac{\delta\varphi(\boldsymbol{r}, t|\boldsymbol{r}_0, t_0)}{\delta U_j(\boldsymbol{r}', t')} = \delta(\boldsymbol{r}_0 - \boldsymbol{r}')\theta(t' - t_0)\theta(t - t')\frac{\partial\varphi(\boldsymbol{r}, t|\boldsymbol{r}_0, t')}{\partial r_{0j}} + \int_{t_0}^{t} d\tau \boldsymbol{U}(\boldsymbol{r}_0, \tau)\frac{\partial}{\partial \boldsymbol{r}_0}\frac{\delta\varphi(\boldsymbol{r}, t|\boldsymbol{r}_0, \tau)}{\delta U_j(\boldsymbol{r}', t')},$$

from which follows that

$$\frac{\delta\varphi(\boldsymbol{r}, t|\boldsymbol{r}_0, t_0)}{\delta U_j(\boldsymbol{r}', t')} \sim \theta(t' - t_0)\theta(t - t'),$$

i.e., the condition of dynamic causality holds. For $t_0 < t' < t$, this equation is simplified to the form

$$\frac{\delta\varphi(\boldsymbol{r}, t|\boldsymbol{r}_0, t_0)}{\delta U_j(\boldsymbol{r}', t')} = \delta(\boldsymbol{r}_0 - \boldsymbol{r}')\frac{\partial\varphi(\boldsymbol{r}, t|\boldsymbol{r}_0, t')}{\partial r_{0j}} + \int_{t_0}^{t'} d\tau \boldsymbol{U}(\boldsymbol{r}_0, \tau)\frac{\partial}{\partial \boldsymbol{r}_0}\frac{\delta\varphi(\boldsymbol{r}, t|\boldsymbol{r}_0, \tau)}{\delta U_j(\boldsymbol{r}', t')},$$

and we arrive, in the limit $t' = t_0 + 0$, at Eq. (3.10) derived earlier in a different way.

Density Field of Passive Tracer Under Random Velocity Field

Consider the problem on tracer transfer by random velocity field in more details. The problem is formulated in terms of Eq. (1.39), page 24, that we rewrite in the form

$$\left(\frac{\partial}{\partial t} + \boldsymbol{U}(\boldsymbol{r}, t)\frac{\partial}{\partial \boldsymbol{r}}\right)\rho(\boldsymbol{r}, t) + \frac{\partial \boldsymbol{U}(\boldsymbol{r}, t)}{\partial \boldsymbol{r}}\rho(\boldsymbol{r}, t) = 0, \quad \rho(\boldsymbol{r}, 0) = \rho_0(\boldsymbol{r}). \tag{3.14}$$

To describe the density field in the Eulerian description, we introduce the indicator function

$$\varphi(\boldsymbol{r}, t; \rho) = \delta(\rho(t, \boldsymbol{r}) - \rho), \tag{3.15}$$

which is similar to function (3.2) and is localized on surface

$$\rho(\boldsymbol{r}, t) = \rho = \text{const},$$

in the three-dimensional case or on a contour in the two-dimensional case. An equation for this function can be easily obtained either immediately from Eq. (3.14), or from the Liouville equation in the Lagrangian description. Indeed, differentiating Eq. (3.15) with respect to time t and using dynamic equation (3.14) and probing property of the delta-function, we obtain the equation

$$\frac{\partial}{\partial t}\varphi(\boldsymbol{r}, t; \rho) = \frac{\partial \boldsymbol{U}(\boldsymbol{r}, t)}{\partial \boldsymbol{r}} \frac{\partial}{\partial \rho}\rho\varphi(\boldsymbol{r}, t; \rho) + \boldsymbol{U}(\boldsymbol{r}, t)\frac{\partial \rho(\boldsymbol{r}, t)}{\partial \boldsymbol{r}} \frac{\partial}{\partial \rho}\varphi(\boldsymbol{r}, t; \rho). \tag{3.16}$$

However, this equation is unclosed because the right-hand side includes term $\partial\rho(\boldsymbol{r}, t)/\partial\boldsymbol{r}$ that cannot be explicitly expressed through $\rho(\boldsymbol{r}, t)$.

On the other hand, differentiating function (3.15) with respect to $\boldsymbol{r}$, we obtain the equality

$$\frac{\partial}{\partial \boldsymbol{r}}\varphi(\boldsymbol{r}, t; \rho) = -\frac{\partial \rho(\boldsymbol{r}, t)}{\partial \boldsymbol{r}} \frac{\partial}{\partial \rho}\varphi(\boldsymbol{r}, t; \rho). \tag{3.17}$$

Eliminating now the last term in Eq. (3.16) with the use of (3.17), we obtain the closed Liouville equation in the Eulerian description

$$\begin{gathered}\left(\frac{\partial}{\partial t} + \boldsymbol{U}(\boldsymbol{r}, t)\frac{\partial}{\partial \boldsymbol{r}}\right)\varphi(\boldsymbol{r}, t; \rho) = \frac{\partial \boldsymbol{U}(\boldsymbol{r}, t)}{\partial \boldsymbol{r}} \frac{\partial}{\partial \rho}\left[\rho\varphi(\boldsymbol{r}, t; \rho)\right], \\ \varphi(\boldsymbol{r}, 0; \rho) = \delta(\rho_0(\boldsymbol{r}) - \rho).\end{gathered} \tag{3.18}$$

Note that the Liouville equation (3.18) becomes the one-dimensional equation in the case of velocity field described by the simplest model Eq. (1.15), page 10, $\boldsymbol{U}(x, t) = \boldsymbol{v}(t)f(x)$ (x is the dimensionless variable),

$$\begin{gathered}\left(\frac{\partial}{\partial t} + v_x(t)f(x)\frac{\partial}{\partial x}\right)\varphi(x, t; \rho) = v_x(t)\frac{\partial f(x)}{\partial x} \frac{\partial}{\partial \rho}[\rho\varphi(x, t; \rho)], \\ \varphi(x, 0; \rho) = \delta(\rho_0(x) - \rho).\end{gathered} \tag{3.19}$$

To obtain more complete description, we consider the extended indicator function including both density field $\rho(\boldsymbol{r}, t)$ and its spatial gradient $\boldsymbol{p}(\boldsymbol{r}, t) = \nabla\rho(\boldsymbol{r}, t)$

$$\varphi(\boldsymbol{r}, t; \rho, \boldsymbol{p}) = \delta\left(\rho(\boldsymbol{r}, t) - \rho\right)\delta\left(\boldsymbol{p}(\boldsymbol{r}, t) - \boldsymbol{p}\right). \tag{3.20}$$

Differentiating Eq. (3.20) with respect to time, we obtain the equality

$$\frac{\partial}{\partial t}\varphi(\boldsymbol{r},t;\rho,\boldsymbol{p}) = -\left[\frac{\partial}{\partial \rho}\frac{\partial \rho(\boldsymbol{r},t)}{\partial t} + \frac{\partial}{\partial p_i}\frac{\partial p_i(\boldsymbol{r},t)}{\partial t}\right]\varphi(\boldsymbol{r},t;\rho,\boldsymbol{p}). \tag{3.21}$$

Using now dynamic equation (3.14) for density and Eq. (1.50), page 28 for the density spatial gradient, we can rewrite Eq. (3.21) as the equation

$$\begin{aligned}\frac{\partial}{\partial t}\varphi(\boldsymbol{r},t;\rho,\boldsymbol{p}) &= \frac{\partial}{\partial \rho}\left[\frac{\partial \boldsymbol{U}(\boldsymbol{r},t)}{\partial \boldsymbol{r}}\rho + \boldsymbol{U}(\boldsymbol{r},t)\boldsymbol{p}\right]\varphi(\boldsymbol{r},t;\rho,\boldsymbol{p})\\ &+ \frac{\partial}{\partial p_i}\left[\boldsymbol{U}(\boldsymbol{r},t)\frac{\partial p_i(\boldsymbol{r},t)}{\partial \boldsymbol{r}} + \frac{\partial \boldsymbol{U}(\boldsymbol{r},t)}{\partial \boldsymbol{r}}p_i\right.\\ &\left. + p_k\frac{\partial U_k(\boldsymbol{r},t)}{\partial r_i} + \rho\frac{\partial^2 \boldsymbol{U}(\boldsymbol{r},t)}{\partial r_i \partial \boldsymbol{r}}\right]\varphi(\boldsymbol{r},t;\rho,\boldsymbol{p}),\end{aligned} \tag{3.22}$$

which is unclosed because of the term $\partial p_i(\boldsymbol{r},t)/\partial \boldsymbol{r}$ in the right-hand side.

Differentiating function (3.20) with respect to $\boldsymbol{r}$, we obtain the equality

$$\frac{\partial}{\partial \boldsymbol{r}}\varphi(\boldsymbol{r},t;\rho,\boldsymbol{p}) = -\left[\boldsymbol{p}\frac{\partial}{\partial \rho} + \frac{\partial p_i(\boldsymbol{r},t)}{\partial \boldsymbol{r}}\frac{\partial}{\partial p_i}\right]\varphi(\boldsymbol{r},t;\rho,\boldsymbol{p}). \tag{3.23}$$

Now, multiplying Eq. (3.23) by $\boldsymbol{U}(\boldsymbol{r},t)$ and adding the result to Eq. (3.22), we obtain the closed Liouville equation for the extended indicator function

$$\begin{aligned}&\left(\frac{\partial}{\partial t} + \boldsymbol{U}(\boldsymbol{r},t)\frac{\partial}{\partial \boldsymbol{r}}\right)\varphi(\boldsymbol{r},t;\rho,\boldsymbol{p})\\ &= \left[\frac{\partial U_k(\boldsymbol{r},t)}{\partial r_i}\frac{\partial}{\partial p_i}p_k + \frac{\partial \boldsymbol{U}(\boldsymbol{r},t)}{\partial \boldsymbol{r}}\left(\frac{\partial}{\partial \rho}\rho + \frac{\partial}{\partial \boldsymbol{p}}\boldsymbol{p}\right) + \frac{\partial^2 \boldsymbol{U}(\boldsymbol{r},t)}{\partial r_i \partial \boldsymbol{r}}\frac{\partial}{\partial p_i}\rho\right] \times \varphi(\boldsymbol{r},t;\rho,\boldsymbol{p})\end{aligned} \tag{3.24}$$

with the initial value

$$\varphi(\boldsymbol{r},0;\rho,\boldsymbol{p}) = \delta\left(\rho_0(\boldsymbol{r}) - \rho\right)\delta\left(\boldsymbol{p}_0(\boldsymbol{r}) - \boldsymbol{p}\right).$$

In the case of nondivergent velocity field ($\operatorname{div}\boldsymbol{U}(\boldsymbol{r},t) = 0$), the Liouville equation (3.24) is simplified to the form

$$\begin{aligned}\left(\frac{\partial}{\partial t} + \boldsymbol{U}(\boldsymbol{r},t)\frac{\partial}{\partial \boldsymbol{r}}\right)\varphi(\boldsymbol{r},t;\rho,\boldsymbol{p}) &= \frac{\partial U_k(\boldsymbol{r},t)}{\partial r_i}\frac{\partial}{\partial p_i}p_k\varphi(\boldsymbol{r},t;\rho,\boldsymbol{p}),\\ \varphi(\boldsymbol{r},0;\rho,\boldsymbol{p}) &= \delta\left(\rho_0(\boldsymbol{r}) - \rho\right)\delta\left(\boldsymbol{p}_0(\boldsymbol{r}) - \boldsymbol{p}\right).\end{aligned} \tag{3.25}$$

Magnetic Field Diffusion Under Random Velocity Field

Let us introduce the indicator function of magnetic field $\boldsymbol{H}(\boldsymbol{r}, t)$,

$$\varphi(\boldsymbol{r}, t; \boldsymbol{H}) = \delta(\boldsymbol{H}(\boldsymbol{r}, t) - \boldsymbol{H}).$$

Differentiating this function with respect to time with the use of dynamic equation (1.59), page 30, and probing property of the delta function, we arrive at the equation

$$\begin{aligned}
\frac{\partial}{\partial t}\varphi(\boldsymbol{r}, t; \boldsymbol{H}) &= -\frac{\partial H_i(\boldsymbol{r}, t)}{\partial t}\frac{\partial}{\partial H_i}\varphi(\boldsymbol{r}, t; \boldsymbol{H}) \\
&= -\left[\boldsymbol{H}(\boldsymbol{r}, t)\frac{\partial u_i(\boldsymbol{r}, t)}{\partial \boldsymbol{r}} - H_i(\boldsymbol{r}, t)\frac{\partial \boldsymbol{u}(\boldsymbol{r}, t)}{\partial \boldsymbol{r}}\right]\frac{\partial}{\partial H_i}\varphi(\boldsymbol{r}, t; \boldsymbol{H}) \\
&\quad + \boldsymbol{u}(\boldsymbol{r}, t)\frac{\partial H_i(\boldsymbol{r}, t)}{\partial \boldsymbol{r}}\frac{\partial}{\partial H_i}\varphi(\boldsymbol{r}, t; \boldsymbol{H}) \\
&= -\frac{\partial}{\partial H_i}\left[\boldsymbol{H}\frac{\partial u_i(\boldsymbol{r}, t)}{\partial \boldsymbol{r}} - H_i\frac{\partial \boldsymbol{u}(\boldsymbol{r}, t)}{\partial \boldsymbol{r}}\right]\varphi(\boldsymbol{r}, t; \boldsymbol{H}) \\
&\quad + \boldsymbol{u}(\boldsymbol{r}, t)\frac{\partial H_i(\boldsymbol{r}, t)}{\partial \boldsymbol{r}}\frac{\partial}{\partial H_i}\varphi(\boldsymbol{r}, t; \boldsymbol{H}),
\end{aligned}$$

which is unclosed because of the last term. However, since

$$\frac{\partial}{\partial \boldsymbol{r}}\varphi(\boldsymbol{r}, t; \boldsymbol{H}) = -\frac{\partial H_i(\boldsymbol{r}, t)}{\partial \boldsymbol{r}}\frac{\partial}{\partial H_i}\delta(\boldsymbol{H}(\boldsymbol{r}, t) - \boldsymbol{H})$$

and, hence,

$$-\boldsymbol{u}(\boldsymbol{r}, t)\frac{\partial}{\partial \boldsymbol{r}}\varphi(\boldsymbol{r}, t; \boldsymbol{H}) = \boldsymbol{u}(\boldsymbol{r}, t)\frac{\partial H_i(\boldsymbol{r}, t)}{\partial \boldsymbol{r}}\frac{\partial}{\partial H_i}\delta(\boldsymbol{H}(\boldsymbol{r}, t) - \boldsymbol{H}),$$

we obtain the desired closed Liouville equation [16]

$$\left(\frac{\partial}{\partial t} + \boldsymbol{u}(\boldsymbol{r}, t)\frac{\partial}{\partial \boldsymbol{r}}\right)\varphi(\boldsymbol{r}, t; \boldsymbol{H}) = -\frac{\partial}{\partial H_i}\left[\boldsymbol{H}\frac{\partial u_i(\boldsymbol{r}, t)}{\partial \boldsymbol{r}} - H_i\frac{\partial \boldsymbol{u}(\boldsymbol{r}, t)}{\partial \boldsymbol{r}}\right]\varphi(\boldsymbol{r}, t; \boldsymbol{H}) \tag{3.26}$$

with the initial condition

$$\varphi(\boldsymbol{r}, 0; \boldsymbol{H}) = \delta(\boldsymbol{H}_0(\boldsymbol{r}) - \boldsymbol{H}).$$

Note that, in the case of velocity field described by the simplest model (1.15), page 10, $\boldsymbol{u}(x, t) = \boldsymbol{v}(t)f(x)$ (x is the dimensionless variable), the Liouville equation (3.26) with homogeneous initial conditions becomes the equation

$$\begin{aligned}
\frac{\partial}{\partial t}\varphi(x, t; \boldsymbol{H}) &= \left[-\frac{\partial}{\partial x}f(x) + f'(x)\left(1 + \frac{\partial}{\partial \boldsymbol{H}}\boldsymbol{H}\right)\right]v_x(t)\varphi(x, t; \boldsymbol{H}) \\
&\quad - f'(x)H_{x0}v_i(t)\frac{\partial}{\partial H_i}\varphi(x, t; \boldsymbol{H})
\end{aligned} \tag{3.27}$$

with the initial condition

$$\varphi(x, 0; \boldsymbol{H}) = \delta(\boldsymbol{H}_0 - \boldsymbol{H}).$$

3.2.2 Quasilinear Equations

Consider now the simplest quasilinear equation for scalar quantity $q(\boldsymbol{r}, t)$ (1.65), page 33,

$$\left(\frac{\partial}{\partial t} + \boldsymbol{U}(t, q)\frac{\partial}{\partial \boldsymbol{r}}\right) q(\boldsymbol{r}, t) = Q(t, q), \qquad q(\boldsymbol{r}, 0) = q_0(\boldsymbol{r}). \tag{3.28}$$

In this case, an attempt of deriving a closed equation for the indicator function $\varphi(\boldsymbol{r}, t; q) = \delta(q(\boldsymbol{r}, t) - q)$ on the analogy of the linear problem will fail. Here, we must supplement Eq. (3.28) with (1.66), page 33, for the gradient field $\boldsymbol{p}(\boldsymbol{r}, t) = \nabla q(\boldsymbol{r}, t)$

$$\left(\frac{\partial}{\partial t} + \boldsymbol{U}(t, q)\frac{\partial}{\partial \boldsymbol{r}}\right) \boldsymbol{p}(\boldsymbol{r}, t) + \frac{\partial \left\{\boldsymbol{U}(t, q)\boldsymbol{p}(\boldsymbol{r}, t)\right\}}{\partial q}\boldsymbol{p}(\boldsymbol{r}, t) = \frac{\partial Q(t, q)}{\partial q}\boldsymbol{p}(\boldsymbol{r}, t) \tag{3.29}$$

and consider the extended indicator function

$$\varphi(\boldsymbol{r}, t; q, \boldsymbol{p}) = \delta(q(\boldsymbol{r}, t) - q)\delta(\boldsymbol{p}(\boldsymbol{r}, t) - \boldsymbol{p}). \tag{3.30}$$

Differentiating Eq. (3.30) with respect to time and using Eqs. (3.28) and (3.29), we obtain the equation

$$\begin{aligned}
\frac{\partial}{\partial t}\varphi(\boldsymbol{r}, t; q, \boldsymbol{p}) &= -\left[\frac{\partial}{\partial q}\frac{\partial q(\boldsymbol{r}, t)}{\partial t} + \frac{\partial}{\partial p_k}\frac{\partial p_k(\boldsymbol{r}, t)}{\partial t}\right]\varphi(\boldsymbol{r}, t; q, \boldsymbol{p}) \\
&= \left\{\frac{\partial}{\partial q}\left[\boldsymbol{p}\boldsymbol{U}(t, q) - Q(t, q)\right] + \frac{\partial}{\partial p_k}\boldsymbol{U}(t, q)\frac{\partial p_k(\boldsymbol{r}, t)}{\partial \boldsymbol{r}}\right\}\varphi(\boldsymbol{r}, t; q, \boldsymbol{p}) \\
&+ \frac{\partial}{\partial p_k}\left[p_k\left(\boldsymbol{p}\frac{\partial \boldsymbol{U}(t, q)}{\partial q} - \frac{\partial Q(t, q)}{\partial q}\right)\right]\varphi(\boldsymbol{r}, t; q, \boldsymbol{p}),
\end{aligned} \tag{3.31}$$

which is unclosed, however, because of the term $\partial p_k(\boldsymbol{r}, t)/\partial \boldsymbol{r}$ in the right-hand side.

Differentiating function (3.30) with respect to $\boldsymbol{r}$, we obtain the equality

$$\frac{\partial}{\partial \boldsymbol{r}}\varphi(\boldsymbol{r}, t; q, \boldsymbol{p}) = -\left[\boldsymbol{p}\frac{\partial}{\partial q} + \frac{\partial p_k(\boldsymbol{r}, t)}{\partial \boldsymbol{r}}\frac{\partial}{\partial p_k}\right]\varphi(\boldsymbol{r}, t; q, \boldsymbol{p}), \tag{3.32}$$

from which follows that

$$\frac{\partial}{\partial p_k}\frac{\partial p_k(\boldsymbol{r}, t)}{\partial \boldsymbol{r}}\varphi(\boldsymbol{r}, t; q, \boldsymbol{p}) = -\left[\frac{\partial}{\partial \boldsymbol{r}} + \boldsymbol{p}\frac{\partial}{\partial q}\right]\varphi(\boldsymbol{r}, t; q, \boldsymbol{p}).$$

Consequently, Eq. (3.31) can be rewritten in the closed form

$$\left(\frac{\partial}{\partial t}+\boldsymbol{U}(t,q)\frac{\partial}{\partial \boldsymbol{r}}\right)\varphi(\boldsymbol{r},t;q,\boldsymbol{p})=\left[\boldsymbol{p}\frac{\partial \boldsymbol{U}(t,q)}{\partial q}-\frac{\partial}{\partial q}Q(t,q)\right]\varphi(\boldsymbol{r},t;q,\boldsymbol{p})$$
$$+\frac{\partial}{\partial p_k}\left[p_k\left(\boldsymbol{p}\frac{\partial \boldsymbol{U}(t,q)}{\partial q}-\frac{\partial Q(t,q)}{\partial q}\right)\right]\varphi(\boldsymbol{r},t;q,\boldsymbol{p}), \tag{3.33}$$

which is just the desired Liouville equation for the quasilinear equation (3.28) in the extended phase space $\{q,\boldsymbol{p}\}$ with the initial value

$$\varphi(\boldsymbol{r},0;q,\boldsymbol{p})=\delta(q_0(\boldsymbol{r})-q)\delta(\boldsymbol{p}_0(\boldsymbol{r})-\boldsymbol{p}).$$

Note that the equation of continuity for conserved quantity $I(\boldsymbol{r},t)$

$$\frac{\partial}{\partial t}I(\boldsymbol{r},t)+\frac{\partial}{\partial \boldsymbol{r}}\left\{\boldsymbol{U}(t,q)I(\boldsymbol{r},t)\right\}=0,\qquad I(\boldsymbol{r},0)=I_0(\boldsymbol{r})$$

can be combined with Eqs. (3.28), (3.29). In this case, the indicator function has the form

$$\varphi(\boldsymbol{r},t;q,\boldsymbol{p},I)=\delta(q(\boldsymbol{r},t)-q)\delta(\boldsymbol{p}(\boldsymbol{r},t)-\boldsymbol{p})\delta(I(\boldsymbol{r},t)-I)$$

and derivation of the closed equation for this function appears possible in space $\{q,\ \boldsymbol{p},\ I\}$, which follows from the fact that, in the Lagrangian description, quantity inverse to $I(\boldsymbol{r},t)$ coincides with the divergence.

Note that the Liouville equation in indicator function

$$\varphi(x,t;q,p,\chi)=\delta\left(q(x,t)-u\right)\delta\left(p(x,t)-p\right)\delta\left(\chi(x,t)-\chi\right)$$

in the case of the one-dimensional Riemann equation (1.77), page 36, in function $q(x,t)$ complemented with the equations for the gradient field $p(x,t)=\dfrac{\partial q(x,t)}{\partial x}$– (1.80) and the logarithm of the density field $\chi(x,t)=\ln\rho(x,t)$– (1.82) assumes the form of the equation

$$\left(\frac{\partial}{\partial t}+q\frac{\partial}{\partial x}\right)\varphi(x,t;q,p,\chi)=\left(p+\frac{\partial}{\partial p}p^2+p\frac{\partial}{\partial \chi}\right)\varphi(x,t;q,p,\chi) \tag{3.34}$$

with the initial condition

$$\varphi(x,0;q,p,\chi)=\delta(q_0(x)-q)\delta(p_0(x)-p)\delta\left(\chi_0(x)-\chi\right).$$

The solution to Eq. (3.34) can be represented in the form

$$\varphi(x,t;q,p,\chi)=\exp\left\{-t\left(q\frac{\partial}{\partial x}-p\frac{\partial}{\partial \chi}\right)\right\}\widetilde{\varphi}(x,t;q,p,\chi),$$

where function $\widetilde{\varphi}(x, t; q, p)$ satisfies the equation

$$\left(\frac{\partial}{\partial t} - p^2 \frac{\partial}{\partial p}\right) \widetilde{\varphi}(x, t; q, p, \chi) = 3p\widetilde{\varphi}(x, t; q, p, \chi) \tag{3.35}$$

with the initial condition

$$\varphi(x, 0; q, p, \chi) = \delta(q_0(x) - q)\delta(p_0(x) - p)\delta\left(\chi_0(x) - \chi\right).$$

Solving now Eq. (3.35) by the method of characteristics, we obtain the expression

$$\begin{aligned}\widetilde{\varphi}(x, t; q, p, \chi) = {} & \frac{1}{(1 - pt)^3}\delta(q_0(x) - q)\delta\left(p_0(x) - \frac{p}{1 - pt}\right) \\ & \times \delta\left(\chi_0(x) - \ln\left(1 + tp_0(x)\right) - \chi\right),\end{aligned}$$

that can be rewritten in the form

$$\begin{aligned}\widetilde{\varphi}(x, t; q, p, \chi) = {} & [1 + tp_0(x)]\,\delta(q_0(x) - q)\delta\left(\frac{p_0(x)}{1 + tp_0(x)} - p\right) \\ & \times \delta\left(\chi_0(x) - \ln\left(1 + tp_0(x)\right) - \chi\right).\end{aligned} \tag{3.36}$$

Thus, we obtain the solution to Eq. (3.34) in the final form

$$\begin{aligned}\varphi(x, t; q, p, \chi) = {} & \exp\left\{-qt\frac{\partial}{\partial x}\right\} [1 + tp_0(x)]\,\delta(q_0(x) - q)\delta\left(\frac{p_0(x)}{1 + tp_0(x)} - p\right) \\ & \times \delta\left(\chi_0(x) - \ln\left(1 + tp_0(x)\right) - \chi\right),\end{aligned} \tag{3.37}$$

in agreement with solutions (1.78), (1.79) and (1.83), pages 36 and 37.

3.2.3 General-Form Nonlinear Equations

Consider now the scalar nonlinear first-order partial differential equation in the general form (1.84), page 38,

$$\frac{\partial}{\partial t}q(\boldsymbol{r}, t) + H\left(\boldsymbol{r}, t, q, \boldsymbol{p}\right) = 0, \qquad q\left(\boldsymbol{r}, 0\right) = q_0\left(\boldsymbol{r}\right), \tag{3.38}$$

where $\boldsymbol{p}(\boldsymbol{r}, t) = \partial q(\boldsymbol{r}, t)/\partial \boldsymbol{r}$. In order to derive the closed Liouville equation in this case, we must supplement Eq. (3.38) with equations for vector $\boldsymbol{p}(\boldsymbol{r}, t)$ and second derivative matrix $U_{ik}(\boldsymbol{r}, t) = \partial^2 q(\boldsymbol{r}, t)/\partial r_i \partial r_k$.

Introduce now the extended indicator function

$$\varphi(\boldsymbol{r}, t; q, \boldsymbol{p}, U, I) = \delta(q(\boldsymbol{r}, t) - q)\delta(\boldsymbol{p}(\boldsymbol{r}, t) - \boldsymbol{p})\delta(U(\boldsymbol{r}, t) - U)\delta(I(\boldsymbol{r}, t) - I), \tag{3.39}$$

where we included for generality an additional conserved variable $I(\boldsymbol{r}, t)$ satisfying the equation of continuity (1.86)

$$\frac{\partial}{\partial t}I(\boldsymbol{r}, t) + \frac{\partial}{\partial \boldsymbol{r}}\left\{\frac{\partial H(\boldsymbol{r}, t, q, \boldsymbol{p})}{\partial \boldsymbol{p}}I(\boldsymbol{r}, t)\right\} = 0, \quad I(\boldsymbol{r}, 0) = I_0(\boldsymbol{r}). \tag{3.40}$$

Equations (3.38), (3.40) describe, for example, wave propagation in inhomogeneous media within the frames of the geometrical optics approximation of the parabolic equation of quasi-optics. Differentiating function (3.39) with respect to time and using dynamic equations for functions $q(\boldsymbol{r}, t)$, $\boldsymbol{p}(\boldsymbol{r}, t)$, $U(\boldsymbol{r}, t)$ and $I(\boldsymbol{r}, t)$, we generally obtain an unclosed equation containing third-order derivatives of function $q(\boldsymbol{r}, t)$ with respect to spatial variable $\boldsymbol{r}$. However, the combination

$$\left(\frac{\partial}{\partial t} + \frac{\partial H(\boldsymbol{r}, t, q, \boldsymbol{p})}{\partial \boldsymbol{p}}\frac{\partial}{\partial \boldsymbol{r}}\right)\varphi(\boldsymbol{r}, t; q, \boldsymbol{p}, U, I)$$

will not include the third-order derivatives; as a result, we obtain the closed Liouville equation in space $\{q, \boldsymbol{r}, U, I\}$.

3.3 Higher-Order Partial Differential Equations

If the initial dynamic system includes higher-order derivatives (e.g., Laplace operator), derivation of a closed equation for the corresponding indicator function becomes impossible. In this case, only the variational differential equation (the *Hopf equation*) can be derived in the closed form for the functional whose average over an ensemble of realizations coincides with the *characteristic functional* of the solution to the corresponding dynamic equation. Consider such a transition using the partial differential equations considered in Lecture 1 as examples.

3.3.1 Parabolic Equation of Quasi-Optics

The first example concerns wave propagation in random medium within the frames of linear parabolic equation (1.89), page 39,

$$\frac{\partial}{\partial x}u(x, \boldsymbol{R}) = \frac{i}{2k}\Delta_{\boldsymbol{R}}u(x, \boldsymbol{R}) + \frac{ik}{2}\varepsilon(x, \boldsymbol{R})u(x, \boldsymbol{R}), \quad u(0, \boldsymbol{R}) = u_0(\boldsymbol{R}). \tag{3.41}$$

Equation (3.41) as the initial-value problem possesses the property of dynamic causality with respect to parameter x. Indeed, rewrite this equation in the integral form

$$u(x, \boldsymbol{R}) = u_0(\boldsymbol{R}) + \frac{i}{2k}\int_0^x d\xi\, \Delta_{\boldsymbol{R}}u(\xi, \boldsymbol{R}) + \frac{ik}{2}\int_0^x d\xi\, \varepsilon(\xi, \boldsymbol{R})u(\xi, \boldsymbol{R}). \tag{3.42}$$

Function $u(x, \boldsymbol{R})$ is a functional of field $\varepsilon(x, \boldsymbol{R})$, i.e.

$$u(x, \boldsymbol{R}) = u\left[x, \boldsymbol{R};\ \varepsilon\left(\widetilde{x}, \widetilde{\boldsymbol{R}}\right)\right].$$

Varying Eq. (3.42) with respect to function $\varepsilon(x', \boldsymbol{R}')$, we obtain the integral equation in variational derivative $\dfrac{\delta u(x, \boldsymbol{R})}{\delta\varepsilon(x', \boldsymbol{R}')}$

$$\frac{\delta u(x, \boldsymbol{R})}{\delta\varepsilon(x', \boldsymbol{R}')} = \frac{ik}{2}\theta(x - x')\theta(x')\delta(\boldsymbol{R} - \boldsymbol{R}')u(x', \boldsymbol{R}) + \frac{ik}{2}\int\limits_0^x d\xi\,\varepsilon(\xi, \boldsymbol{R})\frac{\delta u(\xi, \boldsymbol{R})}{\delta\varepsilon(x', \boldsymbol{R}')} + \frac{i}{2k}\int\limits_0^x d\xi\,\Delta_{\boldsymbol{R}}\frac{\delta u(\xi, \boldsymbol{R})}{\delta\varepsilon(x', \boldsymbol{R}')}. \tag{3.43}$$

From this equation follows that

$$\frac{\delta u(x, \boldsymbol{R})}{\delta\varepsilon(x', \boldsymbol{R}')} = 0 \quad \text{if} \quad x < x' \quad \text{or} \quad x' < 0, \tag{3.44}$$

which just expresses the dynamic causality property. As a consequence of this fact, we can rewrite Eq. (3.43) in the form (for $0 < x' < x$)

$$\frac{\delta u(x, \boldsymbol{R})}{\delta\varepsilon(x', \boldsymbol{R}')} = \frac{ik}{2}\delta(\boldsymbol{R} - \boldsymbol{R}')u(x', \boldsymbol{R}) + \frac{ik}{2}\int\limits_{x'}^x d\xi\,\varepsilon(\xi, \boldsymbol{R})\frac{\delta u(\xi, \boldsymbol{R})}{\delta\varepsilon(x', \boldsymbol{R}')} + \frac{i}{2k}\int\limits_{x'}^x d\xi\,\Delta_{\boldsymbol{R}}\frac{\delta u(\xi, \boldsymbol{R})}{\delta\varepsilon(x', \boldsymbol{R}')}, \tag{3.45}$$

from which follows the equality

$$\frac{\delta u(x, \boldsymbol{R})}{\delta\varepsilon(x - 0, \boldsymbol{R}')} = \frac{ik}{2}\delta(\boldsymbol{R} - \boldsymbol{R}')u(x, \boldsymbol{R}). \tag{3.46}$$

Consider now the functional

$$\varphi[x; v(\boldsymbol{R}'), v^*(\boldsymbol{R}')] = \exp\left\{i\int d\boldsymbol{R}'\left[u(x, \boldsymbol{R}')v(\boldsymbol{R}') + u^*(x, \boldsymbol{R}')v^*(\boldsymbol{R}')\right]\right\}, \tag{3.47}$$

where wavefield $u(x, \boldsymbol{R})$ satisfies Eq. (3.41) and $u^*(x, \boldsymbol{R})$ is the complex conjugated function. Differentiating (3.47) with respect to x and using dynamic equation (3.41) and its complex conjugate, we obtain the equality

$$\frac{\partial}{\partial x}\varphi[x; v(\boldsymbol{R}'), v^*(\boldsymbol{R}')] = -\frac{1}{2k}\int d\boldsymbol{R}\,[v(\boldsymbol{R})\Delta_{\boldsymbol{R}}u(x, \boldsymbol{R}) - v^*(\boldsymbol{R})\Delta_{\boldsymbol{R}}u^*(x, \boldsymbol{R})]\,\varphi[x; v(\boldsymbol{R}'), v^*(\boldsymbol{R}')] - \frac{k}{2}\int d\boldsymbol{R}\varepsilon(x, \boldsymbol{R})\left[v(\boldsymbol{R})u(x, \boldsymbol{R}) - v^*(\boldsymbol{R})u^*(x, \boldsymbol{R})\right]\varphi[x; v(\boldsymbol{R}'), v^*(\boldsymbol{R}')],$$

which can be written as the variational differential equation

$$\frac{\partial}{\partial x}\varphi[x; v(\boldsymbol{R}'), v^*(\boldsymbol{R}')] = \frac{ik}{2}\int d\boldsymbol{R}\varepsilon(x,\boldsymbol{R})\widehat{M}(\boldsymbol{R})\varphi[x; v(\boldsymbol{R}'), v^*(\boldsymbol{R}')]$$
$$+\frac{i}{2k}\int d\boldsymbol{R}\left[v(\boldsymbol{R})\Delta_{\boldsymbol{R}}\frac{\delta}{\delta v(\boldsymbol{R})} - v^*(\boldsymbol{R})\Delta_{\boldsymbol{R}}\frac{\delta}{\delta v^*(\boldsymbol{R})}\right]\times\varphi[x; v(\boldsymbol{R}'), v^*(\boldsymbol{R}')], \quad (3.48)$$

with the initial value

$$\varphi[0; v(\boldsymbol{R}'), v^*(\boldsymbol{R}')] = \exp\left\{i\int d\boldsymbol{R}'\left[u_0(\boldsymbol{R}')v(\boldsymbol{R}') + u_0^*(\boldsymbol{R}')v^*(\boldsymbol{R}')\right]\right\},$$

and with the Hermitian operator

$$\widehat{M}(\boldsymbol{R}) = v(\boldsymbol{R})\frac{\delta}{\delta v(\boldsymbol{R})} - v^*(\boldsymbol{R})\frac{\delta}{\delta v^*(\boldsymbol{R})}.$$

Equation (3.48) is equivalent to the input equation (3.41).

Rewrite problem (3.48) with the initial condition in the integral form

$$\varphi[x; v(\widetilde{\boldsymbol{R}}), v^*(\widetilde{\boldsymbol{R}})] = \varphi[0; v(\widetilde{\boldsymbol{R}}), v^*(\widetilde{\boldsymbol{R}})]$$
$$+\frac{ik}{2}\int_0^x d\xi\int d\boldsymbol{R}\varepsilon(\xi,\boldsymbol{R})\widehat{M}(\boldsymbol{R})\varphi[\xi; v(\widetilde{\boldsymbol{R}}), v^*(\widetilde{\boldsymbol{R}})]$$
$$+\frac{i}{2k}\int_0^x d\xi\int d\boldsymbol{R}\left[v(\boldsymbol{R})\Delta_{\boldsymbol{R}}\frac{\delta}{\delta v(\boldsymbol{R})} - v^*(\boldsymbol{R})\Delta_{\boldsymbol{R}}\frac{\delta}{\delta v^*(\boldsymbol{R})}\right]\times\varphi[\xi; v(\widetilde{\boldsymbol{R}}), v^*(\widetilde{\boldsymbol{R}})],$$

and vary the resulting equation with respect to function $\varepsilon(x',\boldsymbol{R}')$. Then we obtain that variational derivative $\dfrac{\delta\varphi[x; v(\widetilde{\boldsymbol{R}}), v^*(\widetilde{\boldsymbol{R}})]}{\delta\varepsilon(x',\boldsymbol{R}')}$ satisfies the integral equation

$$\frac{\delta\varphi[x; v(\widetilde{\boldsymbol{R}}), v^*(\widetilde{\boldsymbol{R}})]}{\delta\varepsilon(x',\boldsymbol{R}')} = \theta(x-x')\theta(x')\frac{ik}{2}\widehat{M}(\boldsymbol{R}')\varphi[x'; v(\widetilde{\boldsymbol{R}}), v^*(\widetilde{\boldsymbol{R}})]$$
$$+\frac{ik}{2}\int_0^x d\xi\int d\boldsymbol{R}\varepsilon(\xi,\boldsymbol{R})\widehat{M}(\boldsymbol{R})\frac{\delta\varphi[\xi; v(\widetilde{\boldsymbol{R}}), v^*(\widetilde{\boldsymbol{R}})]}{\delta\varepsilon(x',\boldsymbol{R}')}$$
$$+\frac{i}{2k}\int_0^x d\xi\int d\boldsymbol{R}\left[v(\boldsymbol{R})\Delta_{\boldsymbol{R}}\frac{\delta}{\delta v(\boldsymbol{R})} - v^*(\boldsymbol{R})\Delta_{\boldsymbol{R}}\frac{\delta}{\delta v^*(\boldsymbol{R})}\right]\times\frac{\delta\varphi[\xi; v(\widetilde{\boldsymbol{R}}), v^*(\widetilde{\boldsymbol{R}})]}{\delta\varepsilon(x',\boldsymbol{R}')}.$$

In view of the dynamic causality property, we can rewrite it in the form of the equation (for $0 < x' < x$)

$$\frac{\delta\varphi[x; v(\widetilde{\boldsymbol{R}}), v^*(\widetilde{\boldsymbol{R}})]}{\delta\varepsilon(x', \boldsymbol{R}')} = \frac{ik}{2}\widehat{M}(\boldsymbol{R}')\varphi[x'; v(\widetilde{\boldsymbol{R}}), v^*(\widetilde{\boldsymbol{R}})]$$
$$+ \frac{ik}{2}\int\limits_{x'}^{x} d\xi \int d\boldsymbol{R}\varepsilon(\xi, \boldsymbol{R})\widehat{M}(\boldsymbol{R})\frac{\delta\varphi[\xi; v(\widetilde{\boldsymbol{R}}), v^*(\widetilde{\boldsymbol{R}})]}{\delta\varepsilon(x', \boldsymbol{R}')}$$
$$+ \frac{i}{2k}\int\limits_{x'}^{x} d\xi \int d\boldsymbol{R}\left[v(\boldsymbol{R})\Delta_{\boldsymbol{R}}\frac{\delta}{\delta v(\boldsymbol{R})} - v^*(\boldsymbol{R})\Delta_{\boldsymbol{R}}\frac{\delta}{\delta v^*(\boldsymbol{R})}\right] \times \frac{\delta\varphi[\xi; v(\widetilde{\boldsymbol{R}}), v^*(\widetilde{\boldsymbol{R}})]}{\delta\varepsilon(x', \boldsymbol{R}')},$$

from which follows the equality

$$\frac{\delta}{\delta\varepsilon(x-0, \boldsymbol{R})}\varphi[x; v(\widetilde{\boldsymbol{R}}), v^*(\widetilde{\boldsymbol{R}})] = \frac{ik}{2}\widehat{M}(\boldsymbol{R})\varphi[x; v(\widetilde{\boldsymbol{R}}), v^*(\widetilde{\boldsymbol{R}})]. \tag{3.49}$$

3.3.2 Random Forces in Hydrodynamic Theory of Turbulence

Consider now integro-differential equation (1.102), page 44, for the Fourier harmonics $\widehat{\boldsymbol{u}}(\boldsymbol{k}, t)$ of the solution to the Navier–Stokes equation (1.95)

$$\frac{\partial}{\partial t}\widehat{u}_i(\boldsymbol{k}, t) + \frac{i}{2}\int d\boldsymbol{k}_1 \int d\boldsymbol{k}_2 \Lambda_i^{\alpha\beta}(\boldsymbol{k}_1, \boldsymbol{k}_2, \boldsymbol{k})\widehat{u}_\alpha(\boldsymbol{k}_1, t)\widehat{u}_\beta(\boldsymbol{k}_2, t)$$
$$- \nu k^2\widehat{u}_i(\boldsymbol{k}, t) = \widehat{f}_i(\boldsymbol{k}, t), \tag{3.50}$$

where

$$\Lambda_i^{\alpha\beta}(\boldsymbol{k}_1, \boldsymbol{k}_2, \boldsymbol{k}) = \frac{1}{(2\pi)^3}\left\{k_\alpha\Delta_{i\beta}(\boldsymbol{k}) + k_\beta\Delta_{i\alpha}(\boldsymbol{k})\right\}\delta(\boldsymbol{k}_1 + \boldsymbol{k}_2 - \boldsymbol{k}),$$
$$\Delta_{ij}(\boldsymbol{k}) = \delta_{ij} - \frac{k_i k_j}{\boldsymbol{k}^2} \quad (i, \alpha, \beta = 1, 2, 3),$$

and $\widehat{\boldsymbol{f}}(\boldsymbol{k}, t)$ is the spatial Fourier harmonics of external forces.

Function $\widehat{u}_i(\boldsymbol{k}, t)$ satisfies the principle of dynamic causality, which means that the variational derivative of function

$$\widehat{u}_i(\boldsymbol{k}, t) = \widehat{u}_i\left[\boldsymbol{k}, t; \widehat{f}_\alpha\left(\widetilde{\boldsymbol{k}}, \widetilde{t}\right)\right]$$

(considered as a functional of field $\widehat{\boldsymbol{f}}(\boldsymbol{k}, t)$) with respect to field $\widehat{\boldsymbol{f}}\left(\boldsymbol{k}', t'\right)$ vanishes for $t' > t$ and $t' < 0$,

$$\frac{\delta\widehat{u}_i(\boldsymbol{k}, t)}{\delta\widehat{f}_j\left(\boldsymbol{k}', t'\right)} = 0, \quad \text{if} \quad t' > t \quad \text{or} \quad t' < 0,$$

and, moreover,

$$\frac{\delta \widehat{u}_i(\boldsymbol{k},t)}{\delta \widehat{f}_j(\boldsymbol{k}',t-0)} = \delta_{ij}\delta(\boldsymbol{k}-\boldsymbol{k}'). \tag{3.51}$$

Consider the functional

$$\varphi[t;z] = \varphi[t;z(\boldsymbol{k}')] = \exp\left\{i\int d\boldsymbol{k}'\widehat{\boldsymbol{u}}(\boldsymbol{k}',t)\boldsymbol{z}(\boldsymbol{k}')\right\}. \tag{3.52}$$

Differentiating this functional with respect to time t and using dynamic equation (3.50), we obtain the equality

$$\begin{aligned}\frac{\partial}{\partial t}\varphi[t;z] = &-i\int d\boldsymbol{k}z_i(\boldsymbol{k})\left\{\nu k^2\widehat{u}_i(\boldsymbol{k},t) - \widehat{f}_i(\boldsymbol{k},t)\right\}\varphi[t;z]\\ &+\frac{1}{2}\int d\boldsymbol{k}z_i(\boldsymbol{k})\int d\boldsymbol{k}_1\int d\boldsymbol{k}_2\Lambda_i^{\alpha\beta}(\boldsymbol{k}_1,\boldsymbol{k}_2,\boldsymbol{k})\widehat{u}_\alpha(\boldsymbol{k}_1,t)\widehat{u}_\beta(\boldsymbol{k}_2,t)\varphi[t;z],\end{aligned}$$

which can be rewritten in the functional space as the linear *Hopf equation* containing variational derivatives

$$\begin{aligned}\frac{\partial}{\partial t}\varphi[t;z] = &-\frac{1}{2}\int d\boldsymbol{k}z_i(\boldsymbol{k})\int d\boldsymbol{k}_1\int d\boldsymbol{k}_2\Lambda_i^{\alpha\beta}(\boldsymbol{k}_1,\boldsymbol{k}_2,\boldsymbol{k})\frac{\delta^2\varphi[t;z]}{\delta z_\alpha(\boldsymbol{k}_1)\delta z_\beta(\boldsymbol{k}_2)}\\ &-\int d\boldsymbol{k}z_i(\boldsymbol{k})\left\{\nu k^2\frac{\delta}{\delta z_i(\boldsymbol{k})} - i\widehat{f}_i(\boldsymbol{k},t)\right\}\varphi[t;z],\end{aligned} \tag{3.53}$$

with the given initial condition at $t = 0$.

This means that Eq. (3.53) satisfies the condition of dynamic causality; a consequence of this condition is the equality

$$\frac{\delta}{\delta \widehat{f}(\boldsymbol{k},t-0)}\varphi[t;z] = iz(\boldsymbol{k})\varphi[t;z]. \tag{3.54}$$

In a similar way, considering the space-time harmonics of the velocity field $\widehat{u}_i(\boldsymbol{K})$, where $\boldsymbol{K}$ is the four-dimensional wave vector $\{\boldsymbol{k},\omega\}$ and $\widehat{u}_i^*(\boldsymbol{K}) = \widehat{u}_i(-\boldsymbol{K})$ because field $u_i(\boldsymbol{r},t)$ is real, we obtain the nonlinear integral equation (1.103), page 45,

$$(i\omega+\nu k^2)\widehat{u}_i(\boldsymbol{K}) + \frac{i}{2}\int d^4\boldsymbol{K}_1\int d^4\boldsymbol{K}_2\Lambda_i^{\alpha\beta}(\boldsymbol{K}_1,\boldsymbol{K}_2,\boldsymbol{K})\widehat{u}_\alpha(\boldsymbol{K}_1)\widehat{u}_\beta(\boldsymbol{K}) = \widehat{f}_i(\boldsymbol{K}), \tag{3.55}$$

where

$$\Lambda_i^{\alpha\beta}(\boldsymbol{K}_1,\boldsymbol{K}_2,\boldsymbol{K}) = \frac{1}{2\pi}\Lambda_i^{\alpha\beta}(\boldsymbol{k}_1,\boldsymbol{k}_2,\boldsymbol{k})\delta(\omega_1+\omega_2-\omega),$$

and $\widehat{f}_i(\boldsymbol{K})$ are the space-time Fourier harmonics of external forces. In this case, dealing with the functional

$$\varphi[z] = \varphi[z(\boldsymbol{K}')] = \exp\left\{i\int d^4\boldsymbol{K}'\widehat{\boldsymbol{u}}(\boldsymbol{K}')\boldsymbol{z}(\boldsymbol{K}')\right\}, \tag{3.56}$$

we derive the linear variational integro-differential equation of the form

$$(i\omega + \nu \boldsymbol{k}^2)\frac{\delta\varphi[z]}{\delta z_i(\boldsymbol{K})} = \widehat{if_i}(\boldsymbol{K})\,\varphi[z] - \frac{1}{2}\int d^4\boldsymbol{K}_1 \int d^4\boldsymbol{K}_2 \Lambda_i^{\alpha\beta}(\boldsymbol{K}_1, \boldsymbol{K}_2, \boldsymbol{K}) \frac{\delta\varphi[z]}{\delta z_\alpha(\boldsymbol{K}_1)\,\delta z_\beta(\boldsymbol{K}_2)}. \tag{3.57}$$

Problems

Problem 3.1

Derive the Liouville equation for the Hamiltonian system of equations (1.14), page 9.

Solution Indicator function $\varphi(\boldsymbol{r}, \boldsymbol{v}, t) = \delta(\boldsymbol{r}(t) - \boldsymbol{r})\delta(\boldsymbol{v}(t) - \boldsymbol{v})$ satisfies the Liouville equation (equation of continuity in the $\{\boldsymbol{r}, \boldsymbol{v}\}$ space)

$$\frac{\partial\varphi(\boldsymbol{r}, \boldsymbol{v}, t)}{\partial t} = \left\{-\frac{\partial}{\partial r_k}\frac{\partial H(\boldsymbol{r}, \boldsymbol{v}, t)}{\partial v_k} + \frac{\partial}{\partial v_k}\frac{\partial H(\boldsymbol{r}, \boldsymbol{v}, t)}{\partial r_k}\right\}\varphi(\boldsymbol{r}, \boldsymbol{v}, t)$$

that can be rewritten in the form

$$\frac{\partial\varphi(\boldsymbol{r}, \boldsymbol{v}, t)}{\partial t} = \{H(\boldsymbol{r}, \boldsymbol{v}, t), \varphi(\boldsymbol{r}, \boldsymbol{v}, t)\}, \tag{3.58}$$

where

$$\{H, \varphi\} = \left(\frac{\partial H}{\partial r_k}\frac{\partial\varphi}{\partial v_k} - \frac{\partial H}{\partial v_k}\frac{\partial\varphi}{\partial r_k}\right)$$

is the Poisson bracket of functions $H(\boldsymbol{r}, \boldsymbol{v}, t)$ and $\varphi(\boldsymbol{r}, \boldsymbol{v}, t)$.

Problem 3.2

Determine the relationship between the Eulerian and Lagrangian indicator functions for linear problem (3.14), page 73.

Solution

$$\varphi(\boldsymbol{r}, t; \rho) = \int d\boldsymbol{r}_0 \int_0^\infty dj\, j\varphi_{\mathrm{Lag}}(\boldsymbol{r}, \rho, j, t|\boldsymbol{r}_0).$$

Problem 3.3

Derive Eq. (3.18), page 74, from the dynamic system (1.42), (1.43), page 24, in the Lagrangian description.

Part II

Statistical Description of Stochastic Systems

Lecture 4

Random Quantities, Processes, and Fields

Prior to consider statistical descriptions of the problems mentioned in Lecture 1, we discuss basic concepts of the theory of random quantities, processes, and fields.

4.1 Random Quantities and their Characteristics

The probability for a random quantity ξ to fall in interval $-\infty < \xi < z$ is the monotonic function

$$F(z) = P(-\infty < \xi < z) = \langle \theta(z - \xi) \rangle_{\xi}, \quad F(\infty) = 1, \tag{4.1}$$

where

$$\theta(z) = \begin{cases} 1, & \text{if } z > 0, \\ 0, & \text{if } z < 0, \end{cases}$$

is the Heaviside step function and $\langle \cdots \rangle_{\xi}$ denotes averaging over an ensemble of realizations of random quantity ξ. This function is called the *probability distribution function* or the *integral distribution function*. Definition (4.1) reflects the real-world procedure of finding the probability according to the rule

$$P(-\infty < \xi < z) = \lim_{N \to \infty} \frac{n}{N},$$

where n is the integer equal to the number of realizations of event $\xi < z$ in N independent trials. Consequently, the probability for a random quantity ξ to fall into interval $z < \xi < z + dz$, where dz is the infinitesimal increment, can be written in the form

$$P(z < \xi < z + dz) = P(z)dz,$$

where function $P(z)$ called the *probability density* is represented by the formula

$$P(z) = \frac{d}{dz} P(-\infty < \xi < z) = \langle \delta(z - \xi) \rangle_{\xi}, \tag{4.2}$$

Lectures on Dynamics of Stochastic Systems. DOI: 10.1016/B978-0-12-384966-3.00004-0

where $\delta(z)$ is the Dirac delta function. In terms of probability density $p(z)$, the integral distribution function is expressed by the formula

$$F(z) = P(-\infty < \xi < z) = \int_{-\infty}^{z} d\xi P(\xi), \tag{4.3}$$

so that

$$P(z) > 0, \qquad \int_{-\infty}^{\infty} dz P(z) = 1.$$

Multiplying Eq. (4.2) by arbitrary function $f(z)$ and integrating over the whole domain of variable z, we express the mean value of the arbitrary function of random quantity in the following form

$$\langle f(\xi) \rangle_{\xi} = \int_{-\infty}^{\infty} dz P(z) f(z). \tag{4.4}$$

The *characteristic function* defined by the equality

$$\Phi(v) = \left\langle e^{iv\xi} \right\rangle_{\xi} = \int_{-\infty}^{\infty} dz e^{ivz} P(z)$$

is a very important quantity that exhaustively describes all characteristics of random quantity ξ. The characteristic function being known, we can obtain the probability density (via the Fourier transform)

$$P(x) = \frac{1}{2\pi} \int_{-\infty}^{\infty} dv \Phi(v) e^{-ivx},$$

moments

$$M_n = \left\langle \xi^n \right\rangle = \int_{-\infty}^{\infty} dz p(z) z^n = \left(\frac{d}{idv} \right)^n \Phi(v) \bigg|_{v=0},$$

cumulants (or semi-invariants)

$$K_n = \left(\frac{d}{idv} \right)^n \Theta(v) \bigg|_{v=0},$$

where $\Theta(v) = \ln \Phi(v)$, and other statistical characteristics. In terms of moments and cumulants of random quantity ξ, functions $\Theta(v)$ and $\Phi(v)$ are the Taylor series

$$\Phi(v) = \sum_{n=0}^{\infty} \frac{i^n}{n!} M_n v^n, \quad \Theta(v) = \sum_{n=1}^{\infty} \frac{i^n}{n!} K_n v^n. \tag{4.5}$$

In the case of multidimensional random quantity $\boldsymbol{\xi} = \{z_1, \ldots, z_n\}$, the exhaustive statistical description assumes the multidimensional characteristic function

$$\Phi(\boldsymbol{v}) = \left\langle e^{i\boldsymbol{v}\boldsymbol{\xi}} \right\rangle_{\boldsymbol{\xi}}, \quad \boldsymbol{v} = \{v_1, \ldots, v_n\}. \tag{4.6}$$

The corresponding joined probability density for quantities $\xi_1, \ldots, \xi_n$ is the Fourier transform of characteristic function $\Phi(\boldsymbol{v})$, i.e.,

$$P(\boldsymbol{x}) = \frac{1}{(2\pi)^n} \int d\boldsymbol{v} \Phi(\boldsymbol{v}) e^{-i\boldsymbol{v}\boldsymbol{x}}, \quad \boldsymbol{x} = \{x_1, \ldots, x_n\}. \tag{4.7}$$

Substituting function $\Phi(\boldsymbol{v})$ defined by Eq. (4.6) in Eq. (4.7) and integrating the result over $\boldsymbol{v}$, we obtain the obvious equality

$$P(\boldsymbol{x}) = \langle \delta(\boldsymbol{\xi} - \boldsymbol{x}) \rangle_{\boldsymbol{\xi}} = \langle \delta(\xi_1 - x_1) \ldots \delta(\xi_n - x_n) \rangle \tag{4.8}$$

that can serve the definition of the probability density of random vector quantity $\boldsymbol{\xi}$.

In this case, the moments and cumulants of random quantity $\boldsymbol{\xi}$ are defined by the expressions

$$M_{i_1,\ldots,i_n} = \left. \frac{\partial^n}{i^n \partial v_{i_1} \ldots \partial v_{i_n}} \Phi(\boldsymbol{v}) \right|_{\boldsymbol{v}=0}, \quad K_{i_1,\ldots,i_n} = \left. \frac{\partial^n}{i^n \partial v_{i_1} \ldots \partial v_{i_n}} \Theta(\boldsymbol{v}) \right|_{\boldsymbol{v}=0},$$

where $\Theta(\boldsymbol{v}) = \ln \Phi(\boldsymbol{v})$, and functions $\Theta(\boldsymbol{v})$ and $\Phi(\boldsymbol{v})$ are expressed in terms of moments $M_{i_1,\ldots,i_n}$ and cumulants $K_{i_1,\ldots,i_n}$ via the Taylor series

$$\Phi(\boldsymbol{v}) = \sum_{n=0}^{\infty} \frac{i^n}{n!} M_{i_1,\ldots,i_n} v_{i_1} \ldots v_{i_n}, \quad \Theta(\boldsymbol{v}) = \sum_{n=1}^{\infty} \frac{i^n}{n!} K_{i_1,\ldots,i_n} v_{i_1} \ldots v_{i_n}. \tag{4.9}$$

Note that, for quantities ξ assuming only discrete values $\xi_i (i = 1, 2, \ldots)$ with probabilities p_i, formula (4.8) is replaced with its discrete analog

$$P_k = \left\langle \delta_{z,\xi_k} \right\rangle,$$

where $\delta_{i,k}$ is the Kronecker delta ($\delta_{i,k} = 1$ for $i = k$ and 0 otherwise).

Consider now statistical average $\langle \xi f(\xi) \rangle_{\xi}$, where $f(z)$ is arbitrary deterministic function such that the above average exists. We calculate this average using the procedure that will be widely used in what follows. Instead of $f(\xi)$, we consider function

$f(\xi+\eta)$, where η is arbitrary deterministic quantity. Expand function $f(\xi+\eta)$ in the Taylor series in powers of ξ, i.e., represent it in the form

$$f(\xi+\eta)=\sum_{n=0}^{\infty}\frac{1}{n!}f^{(n)}(\eta)\xi^n=\exp\left\{\xi\frac{d}{d\eta}\right\}f(\eta),$$

where we introduced the shift operator with respect to η. Then we can write the equality

$$\begin{aligned}\langle\xi f(\xi+\eta)\rangle_\xi&=\left\langle\xi\exp\left\{\xi\frac{d}{d\eta}\right\}\right\rangle_\xi f(\eta)\\&=\frac{\left\langle\xi\exp\left\{\xi\frac{d}{d\eta}\right\}\right\rangle_\xi}{\left\langle\exp\left\{\xi\frac{d}{d\eta}\right\}\right\rangle_\xi}\left\langle\exp\left\{\xi\frac{d}{d\eta}\right\}\right\rangle_\xi f(\eta)=\Omega\left(\frac{d}{id\eta}\right)\langle f(\xi+\eta)\rangle_\xi,\end{aligned} \tag{4.10}$$

where function

$$\Omega(v)=\frac{\left\langle\xi e^{i\xi v}\right\rangle_\xi}{\left\langle e^{i\xi v}\right\rangle_\xi}=\frac{d}{idv}\ln\Phi(v)=\frac{d}{idv}\Theta(v),$$

and $\Phi(v)$ is the characteristic function of random quantity ξ. Using now the Taylor series (4.5) for function $\Theta(v)$, we obtain function $\Omega(v)$ in the form of the series

$$\Omega(v)=\sum_{n=0}^{\infty}\frac{i^n}{n!}K_{n+1}v^n. \tag{4.11}$$

Because variable η appears in the right-hand side of Eq. (4.10) only as the term of sum $\xi+\eta$, we can replace differentiation with respect to η with differentiation with respect to ξ (in so doing, operator $\Omega(d/id\xi)$ should be introduced into averaging brackets) and set $\eta=0$. As a result, we obtain the equality

$$\langle\xi f(\xi)\rangle_\xi=\left\langle\Omega\left(\frac{d}{id\xi}\right)f(\xi)\right\rangle_\xi,$$

which can be rewritten, using expansion (4.11) for $\Omega(v)$, as the series in cumulants K_n

$$\langle\xi f(\xi)\rangle_\xi=\sum_{n=0}^{\infty}\frac{1}{n!}K_{n+1}\left\langle\frac{d^nf(\xi)}{d\xi^n}\right\rangle_\xi. \tag{4.12}$$

Note that, setting $f(\xi) = \xi^{n-1}$ in Eq. (4.12), we obtain the recurrence formula that relates moments and cumulants of random quantity ξ in the form

$$M_n = \sum_{k=1}^{n} \frac{(n-1)!}{(k-1)!(n-k)!} K_k M_{n-k} \quad (M_0 = 1, \quad n = 1, 2, \ldots). \tag{4.13}$$

In a similar way, we can obtain the following operator expression for statistical average $\langle g(\xi) f(\xi) \rangle_\xi$

$$\langle g(\xi + \eta_1) f(\xi + \eta_2) \rangle_\xi = \exp\left\{ \Theta\left(\frac{d}{id\eta_1} + \frac{d}{id\eta_2} \right) - \Theta\left(\frac{d}{id\eta_1} \right) - \Theta\left(\frac{d}{id\eta_2} \right) \right\} \times \langle g(\xi + \eta_1) \rangle_\xi \, \langle f(\xi + \eta_2) \rangle_\xi \, . \tag{4.14}$$

In the particular case $g(z) = e^{\omega z}$, where parameter ω assumes complex values too, we obtain the expression

$$\left\langle e^{\omega\xi} f(\xi + \eta) \right\rangle_\xi = \exp\left\{ \Theta\left(\frac{1}{i}\left(\omega + \frac{d}{d\eta} \right) \right) - \Theta\left(\frac{d}{id\eta} \right) \right\} \langle f(\xi + \eta) \rangle_\xi . \tag{4.15}$$

To illustrate practicability of the above formulas, we consider two types of random quantities ξ as examples.

1. Let ξ be the Gaussian random quantity with the probability density

$$P(z) = \frac{1}{\sqrt{2\pi}\sigma} \exp\left\{ -\frac{z^2}{2\sigma^2} \right\}.$$

Then, we have

$$\Phi(v) = \exp\left\{ -\frac{v^2\sigma^2}{2} \right\}, \quad \Theta(v) = -\frac{v^2\sigma^2}{2},$$

so that

$$M_1 = K_1 = \langle \xi \rangle = 0, \quad M_2 = K_2 = \sigma^2 = \left\langle \xi^2 \right\rangle, \quad K_{n>2} = 0.$$

In this case, the recurrence formula (4.13) assumes the form

$$M_n = (n-1)\sigma^2 M_{n-2}, \quad n = 2, \ldots, \tag{4.16}$$

from which follows that

$$M_{2n+1} = 0, \quad M_{2n} = (2n-1)!!\sigma^{2n}.$$

For averages (4.12) and (4.15), we obtain the expressions

$$\langle \xi f(\xi) \rangle_\xi = \sigma^2 \left\langle \frac{df(\xi)}{d\xi} \right\rangle_\xi, \quad \left\langle e^{\omega\xi} f(\xi) \right\rangle_\xi = \exp\left\{ \frac{\omega^2\sigma^2}{2} \right\} \left\langle f(\xi + \omega\sigma^2) \right\rangle_\xi . \tag{4.17}$$

Additional useful formulas can be derived from Eqs. (4.17); for example,

$$\left\langle e^{\omega\xi}\right\rangle_{\xi} = \exp\left\{\frac{\omega^2\sigma^2}{2}\right\}, \quad \left\langle \xi e^{\omega\xi}\right\rangle_{\xi} = \omega\sigma^2 \exp\left\{\frac{\omega^2\sigma^2}{2}\right\},$$

and so on.

If random quantity ξ has a nonzero mean value, $\langle\xi\rangle \neq 0$, the probability density is given by the expression

$$P(z) = \frac{1}{\sqrt{2\pi\sigma^2}} \exp\left\{-\frac{(\xi - \langle\xi\rangle)^2}{2\sigma^2}\right\}. \tag{4.18}$$

In this case, the characteristic function assumes the form

$$\Phi(v) = \exp\left\{iv\langle\xi\rangle - \frac{v^2\sigma^2}{2}\right\}, \qquad \Theta(v) = iv\langle\xi\rangle - \frac{v^2\sigma^2}{2},$$

and, consequently, moments and cumulants are given by the expressions

$$M_1 = K_1 = \langle\xi\rangle, \;\; M_2 = \langle\xi\rangle^2 + \sigma^2, \;\; K_2 = \sigma^2, \;\; K_{n>2} = 0.$$

The corresponding integral distribution function (4.3) assumes the form

$$\begin{aligned} F(z) &= \frac{1}{\sqrt{2\pi\sigma^2}} \int\limits_{-\infty}^{z} d\xi \, \exp\left\{-\frac{(\xi - \langle\xi\rangle)^2}{2\sigma^2}\right\} \\ &= \frac{1}{\sqrt{2\pi}} \int\limits_{-\infty}^{\frac{z-\langle\xi\rangle}{\sigma}} dy \, \exp\left\{-\frac{y^2}{2}\right\} = \Pr\left(\frac{z - \langle\xi\rangle}{\sigma}\right), \end{aligned} \tag{4.19}$$

where function

$$\Pr(z) = \frac{1}{\sqrt{2\pi}} \int\limits_{-\infty}^{z} dx \, \exp\left\{-\frac{x^2}{2}\right\} \tag{4.20}$$

is known as the *probability integral*. It is clear that

$$\Pr(\infty) = 1 \quad \text{and} \quad \Pr(0) = \frac{1}{2}. \tag{4.21}$$

For negative z ($z < 0$), this function satisfies the relationship

$$\Pr(-|z|) = \frac{1}{\sqrt{2\pi}} \int\limits_{|z|}^{\infty} dx \, \exp\left\{-\frac{x^2}{2}\right\} = 1 - \Pr(|z|). \tag{4.22}$$

Asymptotic behavior of probability integral for $z \to \pm\infty$ can be easily derived from Eqs. (4.20) and (4.22); namely,

$$\Pr(z) \approx \begin{cases} 1 - \dfrac{1}{z\sqrt{2\pi}} \exp\left\{-\dfrac{z^2}{2}\right\}, & z \to \infty, \\ \dfrac{1}{|z|\sqrt{2\pi}} \exp\left\{-\dfrac{z^2}{2}\right\} & z \to -\infty. \end{cases} \tag{4.23}$$

If we deal with the random Gaussian vector whose components are $\xi_i(\langle\xi_i\rangle_\xi = 0, i = 1, \ldots, n)$, then the characteristic function is given by the equality

$$\Phi(\mathbf{v}) = \exp\left\{-\frac{1}{2}B_{ij}v_iv_j\right\}, \quad \Theta(\mathbf{v}) = -\frac{1}{2}B_{ij}v_iv_j,$$

where matrix $B_{ij} = \langle\xi_i\xi_j\rangle$ and repeated indices assume summation. In this case, Eq. (4.17) is replaced with the equality

$$\langle \boldsymbol{\xi} f(\boldsymbol{\xi})\rangle_\xi = B\left\langle \frac{df(\boldsymbol{\xi})}{d\boldsymbol{\xi}} \right\rangle_\xi. \tag{4.24}$$

2. Let $\xi \equiv n$ be the integer random quantity governed by Poisson distribution

$$P_n = \frac{\bar{n}^n}{n!}e^{-\bar{n}},$$

where $\bar{n}$ is the average value of quantity n. In this case, we have

$$\Phi(v) = \exp\left\{\bar{n}\left(e^{iv} - 1\right)\right\}, \quad \Theta(v) = \bar{n}\left(e^{iv}-1\right).$$

The recurrence formula (4.13) and Eq. (4.12) assume for this random quantity the forms

$$M_l = \bar{n}\sum_{k=0}^{l-1} \frac{(l-1)!}{k!(l-1-k)!}M_k \equiv \bar{n}\left\langle(n+1)^{l-1}\right\rangle, \ \langle nf(n)\rangle = \bar{n}\,\langle f(n+1)\rangle. \tag{4.25}$$

4.2 Random Processes, Fields, and their Characteristics

4.2.1 General Remarks

If we deal with random function (random process) $z(t)$, then all statistical characteristics of this function at any fixed instant t are exhaustively described in terms of the one-time probability density

$$P(z,t) = \langle\delta\left(z(t) - z\right)\rangle, \tag{4.26}$$

dependent parametrically on time t by the following relationship

$$\langle f(z(t))\rangle = \int_{-\infty}^{\infty} dz f(z) P(z,t).$$

The integral distribution function for this process, i.e., the probability of the event that process $z(t) < Z$ at instant t, is calculated by the formula

$$F(t, Z) = P(z(t) < Z) = \int_{-\infty}^{Z} dz P(z, t),$$

from which follows that

$$F(t, Z) = \langle \theta(Z - z(t)) \rangle, \quad F(t, \infty) = 1, \tag{4.27}$$

where $\theta(z)$ is the Heaviside step function equal to zero for $z < 0$ and unity for $z > 0$.

Note that the singular Dirac delta function

$$\varphi(z, t) = \delta\left(z(t) - z\right)$$

appeared in Eq. (4.26) in angle brackets of averaging is called the *indicator function.*

Similar definitions hold for the two-time probability density

$$P(z_1, t_1; z_2, t_2) = \langle \varphi(z_1, t_1; z_2, t_2) \rangle,$$

and for the general case of the n-time probability density

$$P(z_1, t_1; \ldots; z_n, t_n) = \langle \varphi(z_1, t_1; \ldots; z_n, t_n) \rangle,$$

where

$$\varphi(z_1, t_1; \ldots; z_n, t_n) = \delta(z(t_1) - z_1) \ldots \delta(z(t_n) - z_n).$$

is the n-time indicator function.

Process $z(t)$ is called *stationary* if all its statistical characteristics are invariant with respect to arbitrary temporal shift, i.e., if

$$P(z_1, t_1 + \tau; \ldots; z_n, t_n + \tau) = P(z_1, t_1; \ldots; z_n, t_n).$$

In particular, the one-time probability density of stationary process is at all independent of time, and the correlation function depends only on the absolute value of difference of times,

$$B_z(t_1, t_2) = \langle z(t_1) z(t_2) \rangle = B_z(|t_1 - t_2|).$$

Temporal scale τ_0 characteristic of correlation function $B_z(t)$ is called the temporal correlation radius of process $z(t)$. We can determine this scale, say, by the equality

$$\int_{0}^{\infty} \langle z(t + \tau) z(t) \rangle \, d\tau = \tau_0 \left\langle z^2(t) \right\rangle. \tag{4.28}$$

Note that the Fourier transform of stationary process correlation function

$$\Phi_z(\omega) = \int_{-\infty}^{\infty} dt B_z(t) e^{i\omega t},$$

is called the *temporal spectral function* (or simply *temporal spectrum*).

For random field $f(\boldsymbol{x}, t)$, the one- and n-time probability densities are defined similarly

$$P(\boldsymbol{x}, t; f) = \langle \varphi(\boldsymbol{x}, t; f) \rangle, \tag{4.29}$$

$$P(\boldsymbol{x}_1, t_1, f_1; \ldots; \boldsymbol{x}_n, t_n, f_n) = \langle \varphi(\boldsymbol{x}_1, t_1, f_1; \ldots; \boldsymbol{x}_n, t_n, f_n) \rangle, \tag{4.30}$$

where the indicator functions are defined as follows:

$$\begin{aligned} &\varphi(\boldsymbol{x}, t; f) = \delta(f(\boldsymbol{x}, t) - f), \\ &\varphi(\boldsymbol{x}_1, t_1, f_1; \ldots; \boldsymbol{x}_n, t_n, f_n) = \delta\left(f(\boldsymbol{x}_1, t_1) - f_1\right) \ldots \delta\left(f(\boldsymbol{x}_n, t_n) - f_n\right). \end{aligned} \tag{4.31}$$

For clarity, we use here variables $\boldsymbol{x}$ and t as spatial and temporal coordinates; however, in many physical problems, some preferred spatial coordinate can play the role of the temporal coordinate.

Random field $f(\boldsymbol{x}, t)$ is called the spatially homogeneous field if all its statistical characteristics are invariant relative to spatial translations by arbitrary vector $\boldsymbol{a}$, i.e., if

$$P(\boldsymbol{x}_1 + \boldsymbol{a}, t_1, f_1; \ldots; \boldsymbol{x}_n + \boldsymbol{a}, t_n, f_n) = P(\boldsymbol{x}_1, t_1, f_1; \ldots; \boldsymbol{x}_n, t_n, f_n).$$

In this case, the one-point probability density $P(\boldsymbol{x}, t; f) = P(t; f)$ is independent of $\boldsymbol{x}$, and the spatial correlation function $B_f(\boldsymbol{x}_1, t_1; \boldsymbol{x}_2, t_2)$ depends on the difference $\boldsymbol{x}_1 - \boldsymbol{x}_2$:

$$B_f(\boldsymbol{x}_1, t_1; \boldsymbol{x}_2, t_2) = \langle f(\boldsymbol{x}_1, t_1) f(\boldsymbol{x}_2, t_2) \rangle = B_f(\boldsymbol{x}_1 - \boldsymbol{x}_2; t_1, t_2).$$

If random field $f(\boldsymbol{x}, t)$ is additionally invariant with respect to rotation of all vectors $\boldsymbol{x}_i$ by arbitrary angle, i.e., with respect to rotations of the reference system, then field $f(\boldsymbol{x}, t)$ is called the homogeneous isotropic random field. In this case, the correlation function depends on the length $|\boldsymbol{x}_1 - \boldsymbol{x}_2|$:

$$B_f(\boldsymbol{x}_1, t_1; \boldsymbol{x}_2, t_2) = \langle f(\boldsymbol{x}_1, t_1) f(\boldsymbol{x}_2, t_2) \rangle = B_f(|\boldsymbol{x}_1 - \boldsymbol{x}_2|; t_1, t_2).$$

The corresponding Fourier transform of the correlation function with respect to the spatial variable defines the *spatial spectral function* (called also the *angular spectrum*)

$$\Phi_f(\boldsymbol{k}, t) = \int d\boldsymbol{x} B_f(\boldsymbol{x}, t) e^{i\boldsymbol{k}\boldsymbol{x}},$$

and the Fourier transform of the correlation function of random field $f(\boldsymbol{x},t)$ stationary in time and homogeneous in space defines the space–time spectrum

$$\Phi_f(\boldsymbol{k},\omega) = \int d\boldsymbol{x} \int\limits_{-\infty}^{\infty} dt B_f(\boldsymbol{x},t) e^{i(\boldsymbol{kx}+\omega t)}.$$

In the case of isotropic random field $f(\boldsymbol{x},t)$, the space–time spectrum appears to be isotropic in the $\boldsymbol{k}$-space:

$$\Phi_f(\boldsymbol{k},\omega) = \Phi_f(k,\omega).$$

An exhaustive description of random function $z(t)$ can be given in terms of the characteristic functional

$$\Phi[v(\tau)] = \left\langle \exp\left\{ i \int\limits_{-\infty}^{\infty} d\tau v(\tau) z(\tau) \right\} \right\rangle,$$

where $v(t)$ is arbitrary (but sufficiently smooth) function. Functional $\Phi[v(\tau)]$ being known, one can determine such characteristics of random function $z(t)$ as mean value $\langle z(t) \rangle$, correlation function $\langle z(t_1) z(t_2) \rangle$, n-time moment function $\langle z(t_1) \ldots z(t_n) \rangle$, etc.

Indeed, expanding functional $\Phi[v(\tau)]$ in the functional Taylor series, we obtain the representation of characteristic functional in terms of the moment functions of process $z(t)$:

$$\Phi[v(\tau)] = \sum_{n=0}^{\infty} \frac{i^n}{n!} \int\limits_{-\infty}^{\infty} dt_1 \ldots \int\limits_{-\infty}^{\infty} dt_n M_n(t_1,\ldots,t_n) v(t_1) \ldots v(t_n),$$

$$M_n(t_1,\ldots,t_n) = \langle z(t_1) \ldots z(t_n) \rangle = \frac{1}{i^n} \frac{\delta^n}{\delta v(t_1) \ldots \delta v(t_n)} \Phi[v(\tau)] \bigg|_{v=0}.$$

Consequently, the moment functions of random process $z(t)$ are expressed in terms of variational derivatives of the characteristic functional.

Represent functional $\Phi[v(\tau)]$ in the form $\Phi[v(\tau)] = \exp\{\Theta[v(\tau)]\}$. Functional $\Theta[v(\tau)]$ also can be expanded in the functional Taylor series

$$\Theta[v(\tau)] = \sum_{n=1}^{\infty} \frac{i^n}{n!} \int\limits_{-\infty}^{\infty} dt_1 \ldots \int\limits_{-\infty}^{\infty} dt_n K_n(t_1,\ldots,t_n) v(t_1) \ldots v(t_n), \tag{4.32}$$

where function

$$K_n(t_1,\ldots,t_n) = \frac{1}{i^n} \frac{\delta^n}{\delta v(t_1) \ldots \delta v(t_n)} \Theta[v(\tau)] \bigg|_{v=0}$$

is called the n-th order *cumulant function* of random process $z(t)$.

The characteristic functional and the n-th order cumulant functions of scalar random field $f(\mathbf{x}, t)$ are defined similarly

$$\Phi[v(\mathbf{x}', \tau)] = \left\langle \exp\left\{ i \int d\mathbf{x} \int_{-\infty}^{\infty} dt v(\mathbf{x},t) f(\mathbf{x}, t) \right\} \right\rangle = \exp\left\{ \Theta[v(\mathbf{x}', \tau)] \right\},$$

$$M_n(\mathbf{x}_1, t_1, \ldots, \mathbf{x}_n, t_n) = \frac{1}{i^n} \frac{\delta^n}{\delta v(\mathbf{x}_1, t_1) \ldots \delta v(\mathbf{x}_n, t_n)} \Phi[v(\mathbf{x}', \tau)] \Bigg|_{v=0},$$

$$K_n(\mathbf{x}_1, t_1, \ldots, \mathbf{x}_n, t_n) = \frac{1}{i^n} \frac{\delta^n}{\delta v(\mathbf{x}_1, t_1) \ldots \delta v(\mathbf{x}_n, t_n)} \Theta[v(\mathbf{x}', \tau)] \Bigg|_{v=0}.$$

In the case of vector random field $\mathbf{f}(\mathbf{x}, t)$, we must assume that $\mathbf{v}(\mathbf{x},t)$ is the vector function.

As we noted earlier, characteristic functionals ensure the exhaustive description of random processes and fields. However, even one-time probability densities provide certain information about temporal behavior and spatial structure of random processes for arbitrary long temporal intervals. The ideas of statistical topography of random processes and fields can assist in obtaining this information.

4.2.2 Statistical Topography of Random Processes and Fields

Random Processes

We discuss first the concept of *typical realization curve* of random process $z(t)$. This concept concerns the fundamental features of the behavior of a separate process realization as a whole for temporal intervals of arbitrary duration.

Random Process Typical Realization Curve We will call the typical realization curve of random process $z(t)$ the deterministic curve $z^*(t)$, which is the *median of the integral distribution function* (4.27) and is determined as the solution to the algebraic equation

$$F\left(t, z^*(t)\right) = \int_{-\infty}^{z^*(t)} dz \, P(z, t) = \frac{1}{2}. \tag{4.33}$$

The reason for this definition rests on the median property consisting in the fact that, for any temporal interval (t_1, t_2), random process $z(t)$ entwines about curve $z^*(t)$ in a way to force the identity of average times during which the inequalities $z(t) > z^*(t)$ and $z(t) < z^*(t)$ hold (Fig. 4.1):

$$\left\langle T_{z(t)>z^*(t)} \right\rangle = \left\langle T_{z(t)<z^*(t)} \right\rangle = \frac{1}{2} (t_2 - t_1). \tag{4.34}$$

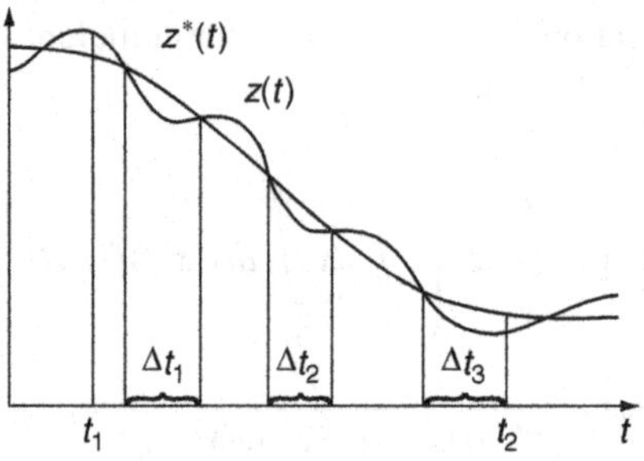

Figure 4.1 To the definition of the typical realization curve of a random process.

Indeed, integrating Eq. (4.33) over temporal interval (t_1, t_2), we obtain

$$\int_{t_1}^{t_2} dtF\left(t, z^*(t)\right) = \frac{1}{2}(t_2 - t_1). \tag{4.35}$$

On the other hand, in view of the definition of integral distribution function (4.27), the integral in the right-hand side of Eq. (4.35) can be represented as

$$\int_{t_1}^{t_2} dtF\left(t, z^*(t)\right) = \langle T(t_1, t_2)\rangle, \tag{4.36}$$

where $T(t_1, t_2) = \sum_{1}^{N} \Delta t_k$ is the combined time during which the realization of process $z(t)$ is above curve $z^*(t)$ in interval (t_1, t_2). Combining Eqs. (4.35) and (4.36), we obtain Eq. (4.34).

Curve $z^*(t)$ can significantly differ from any particular realization of process $z(t)$ and cannot describe possible magnitudes of spikes. Nevertheless, the definitional domain of typical realization curve $z^*(t)$ derived from the one-point probability density of random process $z(t)$ is the whole temporal axis $t \in (0, \infty)$.

Consideration of specific random processes allows additional information to be obtained concerning the realization's spikes relative to the typical realization curve.

Statistics of Random Process Cross Points with a Line The one-time probability density (4.26) of random process $z(t)$ is a result of averaging the singular indicator function over an ensemble of realizations of this process. This function is concentrated at the points at which process $z(t)$ crosses line $z = \text{const}$. Because the cross points are determined as roots of the algebraic equation

$$z(t_n) = z \quad (n = 0, 1, \ldots, \infty),$$

we can rewrite the indicator function in the following form

$$\varphi(z,t)=\sum_{k=1}^{n}\frac{1}{|p(t_k)|}\delta(t-t_k),$$

where $p(t)=\frac{d}{dt}z(t)$.

The number of cross points by itself is obviously a random quantity described by the formula

$$n(t,z)=\int_{-\infty}^{t}d\tau|p(\tau)|\varphi(\tau;z). \tag{4.37}$$

As a consequence, the average number of points where process $z(t)$ crosses line $z=$ const can be described in terms of the correlation between the process derivative with respect to time and the process indicator function, or in terms of joint one-time probability density of process $z(t)$ and its derivative with respect to time $\frac{d}{dt}z(t)$.

In a similar way, we can determine certain elements of statistics related to some other special points (such as points of maxima or minima) of random process $z(t)$.

Random Fields

Similar to topography of mountain ranges, statistical topography studies the systems of contours (level lines in the two-dimensional case and surfaces of constant values in the three-dimensional case) specified by the equality $f(\boldsymbol{r},t)=f=$ const.

For analyzing a system of contours (in this section, we will deal for simplicity with the two-dimensional case and assume $\boldsymbol{r}=\boldsymbol{R}$), we introduce the singular indicator function (4.31) concentrated on these contours.

The convenience of function (4.31) consists, in particular, in the fact that it allows simple expressions for quantities such as the total area of regions where $f(\boldsymbol{R},t)>f$ (i.e., within level lines $f(\boldsymbol{R},t)=f$) and the total mass of the field within these regions [4]

$$S(t;f)=\int\theta(f(\boldsymbol{R},t)-f)d\boldsymbol{R}=\int_{f}^{\infty}df'\int d\boldsymbol{R}\varphi(t,\boldsymbol{R};f'),$$

$$M(t;f)=\int f(\boldsymbol{R},t)\theta(f(\boldsymbol{R},t)-f)d\boldsymbol{R}=\int_{f}^{\infty}f'df'\int d\boldsymbol{R}\varphi(t,\boldsymbol{R};f').$$

As we mentioned earlier, the mean value of indicator function (4.31) over an ensemble of realizations determines the one-time (in time) and one-point (in space) probability density

$$P(\boldsymbol{R},t;f)=\langle\varphi(\boldsymbol{R},t;f)\rangle=\langle\delta\left(f(\boldsymbol{R},t)-f\right)\rangle.$$

Consequently, this probability density immediately determines ensemble-averaged values of the above expressions.

If we include into consideration the spatial gradient

$$p(\boldsymbol{R},t) = \nabla f(\boldsymbol{R},t),$$

we can obtain additional information on details of the structure of field $f(\boldsymbol{R},t)$. For example, quantity

$$l(t;f) = \int d\boldsymbol{R}\, |\boldsymbol{p}(\boldsymbol{R},t)|\, \delta(f(\boldsymbol{R},t) - f) = \oint dl, \tag{4.38}$$

is the total length of contours and extends formula (4.37) to random fields.

The integrand in Eq. (4.38) is described in terms of the extended indicator function

$$\varphi(\boldsymbol{R},t;f,\boldsymbol{p}) = \delta\left(f(\boldsymbol{R},t) - f\right)\delta\left(\boldsymbol{p}(\boldsymbol{R},t) - \boldsymbol{p}\right), \tag{4.39}$$

so that the average value of total length (4.38) is related to the joint one-time probability density of field $f(\boldsymbol{R},t)$ and its gradient $\boldsymbol{p}(\boldsymbol{R},t)$, which is defined as the ensemble average of indicator function (4.39), i.e., as the function

$$P(\boldsymbol{R},t;f,\boldsymbol{p}) = \langle \delta\left(f(\boldsymbol{R},t) - f\right)\delta\left(\boldsymbol{p}(\boldsymbol{R},t) - \boldsymbol{p}\right)\rangle.$$

Inclusion of second-order spatial derivatives into consideration allows us to estimate the total number of contours $f(\boldsymbol{R},t) = f = \text{const}$ by the approximate formula (neglecting unclosed lines)

$$N(t;f) = N_{\text{in}}(t;f) - N_{\text{out}}(t;f) = \frac{1}{2\pi}\int d\boldsymbol{R}\kappa(t,\boldsymbol{R};f)\, |\boldsymbol{p}(\boldsymbol{R},t)|\, \delta\left(f(\boldsymbol{R},t) - f\right),$$

where $N_{\text{in}}(t;f)$ and $N_{\text{out}}(t;f)$ are the numbers of contours for which vector $\boldsymbol{p}$ is directed along internal and external normals, respectively, and $\kappa(\boldsymbol{R},t;f)$ is the curvature of the level line.

Recall that, in the case of the spatially homogeneous field $f(\boldsymbol{R},t)$, the corresponding probability densities $P(\boldsymbol{R},t;f)$ and $P(\boldsymbol{R},t;f,\boldsymbol{p})$ are independent of $\boldsymbol{R}$. In this case, if statistical averages of the above expressions (without integration over $\boldsymbol{R}$) exist, they will characterize the corresponding specific (per unit area) values of these quantities. In this case, random field $f(\boldsymbol{R},t)$ is statistically equivalent to the random process whose statistical characteristics coincide with the spatial one-point characteristics of field $f(\boldsymbol{R},t)$.

Consider now several examples of random processes.

4.2.3 Gaussian Random Process

We start the discussion with the continuous processes; namely, we consider the Gaussian random process $z(t)$ with zero-valued mean ($\langle z(t)\rangle = 0$) and correlation function $B(t_1,t_2) = \langle z(t_1)z(t_2)\rangle$. The corresponding characteristic functional assumes the form

$$\Phi[v(\tau)] = \exp\left\{-\frac{1}{2}\int_{-\infty}^{\infty}\int_{-\infty}^{\infty} dt_1 dt_2 B(t_1, t_2) v(t_1) v(t_2)\right\}. \tag{4.40}$$

Only one cumulant function (the correlation function)

$$K_2(t_1, t_2) = B(t_1, t_2),$$

is different from zero for this process, so that

$$\Theta[v(\tau)] = -\frac{1}{2}\int_{-\infty}^{\infty}\int_{-\infty}^{\infty} dt_1 dt_2 B(t_1, t_2) v(t_1) v(t_2). \tag{4.41}$$

Consider the nth-order variational derivative of functional $\Phi[v(\tau)]$. It satisfies the following line of equalities:

$$\begin{aligned}\frac{\delta^n}{\delta v(t_1)\ldots\delta v(t_n)}\Phi[v(\tau)] &= \frac{\delta^{n-1}}{\delta v(t_2)\ldots\delta v(t_n)}\frac{\delta\Theta[v(\tau)]}{\delta v(t_1)}\Phi[v(\tau)] \\ &= \frac{\delta^2\Theta[v(\tau)]}{\delta v(t_1)\delta v(t_2)}\frac{\delta^{n-2}}{\delta v(t_3)\ldots\delta v(t_n)}\Phi[v(\tau)] \\ &\quad + \frac{\delta^{n-2}}{\delta v(t_3)\ldots\delta v(t_n)}\frac{\delta\Theta[v(\tau)]}{\delta v(t_1)}\frac{\delta\Phi[v(\tau)]}{\delta v(t_2)}.\end{aligned}$$

Setting now $v = 0$, we obtain that moment functions of the Gaussian process $z(t)$ satisfy the recurrence formula

$$M_n(t_1, \ldots, t_n) = \sum_{k=2}^{n} B(t_1, t_2) M_{n-2}(t_2, \ldots, t_{k-1}, t_{k+1}, \ldots, t_n). \tag{4.42}$$

From this formula it follows that, for the Gaussian process with zero-valued mean, all moment functions of odd orders are identically equal to zero and the moment functions of even orders are represented as sums of terms which are the products of averages of all possible pairs $z(t_i)z(t_k)$.

If we assume that function $v(\tau)$ in Eq. (4.41) is different from zero only in interval $0 < \tau < t$, the characteristic functional

$$\begin{aligned}\Phi[t; v(\tau)] &= \left\langle \exp\left(i\int_0^t d\tau z(\tau) v(\tau)\right)\right\rangle \\ &= \exp\left\{-\int_0^t d\tau_1 \int_0^{\tau_1} d\tau_2 B(\tau_1, \tau_2) v(\tau_1) v(\tau_2)\right\},\end{aligned} \tag{4.43}$$

becomes a function of time t and satisfies the ordinary differential equation

$$\frac{d}{dt}\Phi[t; v(\tau)] = -v(t)\int_0^t dt_1 B(t, t_1)v(t_1)\Phi[t; v(\tau)], \quad \Phi[0; v(\tau)] = 1. \tag{4.44}$$

To obtain the one-time characteristic function of the Gaussian random process at instant t, we specify function $v(\tau)$ in Eq. (4.40) in the form

$$v(\tau) = v\delta(\tau - t).$$

Then, we obtain

$$\Phi(v, t) = \left\langle e^{ivz(t)} \right\rangle = \int_{-\infty}^{\infty} dz\, P(z, t)e^{ivz} = \exp\left\{-\frac{1}{2}\sigma^2(t)v^2\right\}, \tag{4.45}$$

where $\sigma^2(t) = B(t, t)$. Using the inverse Fourier transform of (4.45), we obtain the one-time probability density of the Gaussian random process

$$P(z, t) = \frac{1}{2\pi}\int_{-\infty}^{\infty} dv\, \Phi(v, t)e^{-ivz} = \frac{1}{\sqrt{2\pi\sigma^2(t)}}\exp\left\{-\frac{z^2}{2\sigma^2(t)}\right\}. \tag{4.46}$$

Note that, in the case of stationary random process $z(t)$, variance $\sigma^2(t)$ is independent of time t, i.e., $\sigma^2(t) = \sigma^2 = \text{const}$.

Density $P(z, t)$ as a function of z is symmetric relative to point $z = 0$,

$$P(z, t) = P(-z, t).$$

If mean value of the Gaussian random process is different from zero, then we can consider process $z(t) - \langle z(t)\rangle$ to obtain instead of Eq. (4.46) the expression

$$P(z, t) = \frac{1}{\sqrt{2\pi\sigma^2(t)}}\exp\left\{-\frac{(z - \langle z(t)\rangle)^2}{2\sigma^2(t)}\right\}, \tag{4.47}$$

and the corresponding integral distribution function assumes the form

$$F(z, t) = \frac{1}{\sqrt{2\pi\sigma^2(t)}}\int_{-\infty}^{z} dz\, \exp\left\{-\frac{(z - \langle z(t)\rangle)^2}{2\sigma^2(t)}\right\} = \Pr\left(\frac{z - \langle z(t)\rangle}{\sigma(t)}\right),$$

where probability integral $\Pr(z)$ is defined by Eq. (4.20), page 94.

In view of Eq. (4.21), page 94, the typical realization curve (4.33) of the Gaussian random process $z(t)$ in this case coincides with the mean value of process $z(t)$

$$z^*(t) = \langle z(t)\rangle. \tag{4.48}$$

4.2.4 Gaussian Vector Random Field

Consider now the basic space–time statistical characteristics of the vector Gaussian random field $\boldsymbol{u}(\boldsymbol{r},t)$ with zero-valued mean and correlation tensor

$$B_{ij}(\boldsymbol{r},t;\boldsymbol{r}',t')=\left\langle u_i(\boldsymbol{r},t)u_j(\boldsymbol{r}',t')\right\rangle.$$

For definiteness, we will identify this vector field with the velocity field.

In the general case, random field $\boldsymbol{u}(\boldsymbol{r},t)$ is assumed to be the divergent ($\operatorname{div}\boldsymbol{u}(\boldsymbol{r},t)\neq 0$) Gaussian field statistically homogeneous and possessing spherical symmetry (but not possessing reflection symmetry) in space and stationary in time with correlation and spectral tensors ($\tau=t-t_1$)

$$B_{ij}(\boldsymbol{r}-\boldsymbol{r}_1,\tau)=\left\langle u_i(\boldsymbol{r},t)u_j(\boldsymbol{r}_1,t_1)\right\rangle=\int d\boldsymbol{k}\,E_{ij}(\boldsymbol{k},\tau)e^{i\boldsymbol{k}(\boldsymbol{r}-\boldsymbol{r}_1)},$$

$$E_{ij}(\boldsymbol{k},\tau)=\frac{1}{(2\pi)^d}\int d\boldsymbol{r}\,B_{ij}(\boldsymbol{r},\tau)e^{-i\boldsymbol{k}\boldsymbol{r}},$$

where d is the dimension of space. In view of the assumed symmetry conditions, correlation tensor $B_{ij}(\boldsymbol{r}-\boldsymbol{r}_1,\tau)$ has vector structure [28] ($\boldsymbol{r}-\boldsymbol{r}_1\to\boldsymbol{r}$)

$$B_{ij}(\boldsymbol{r},\tau)=B_{ij}^{\mathrm{iso}}(\boldsymbol{r},\tau)+C(r,\tau)\varepsilon_{ijk}r_k, \tag{4.49}$$

where the isotropic portion of the correlation tensor is expressed as follows

$$B_{ij}^{\mathrm{iso}}(\boldsymbol{r},\tau)=A(r,\tau)r_ir_j+B(r,\tau)\delta_{ij}.$$

Here, ε_{ijk} is the pseudotensor described in Sect. 1.3.1, page 24. This pseudotensor is used, for example, to determine the *velocity vortex field*

$$\boldsymbol{\omega}(\boldsymbol{r},t)=\operatorname{curl}\boldsymbol{u}(\boldsymbol{r},t)=\left[\nabla\times\boldsymbol{u}(\boldsymbol{r},t)\right], \tag{4.50}$$

whose components have the form

$$\omega_i(\boldsymbol{r},t)=\varepsilon_{ijk}\frac{\partial u_k(\boldsymbol{r},t)}{\partial r_j}.$$

Note that, in the two-dimensional case, vector $\boldsymbol{\omega}(\boldsymbol{r},t)$ has a single component orthogonal to velocity field $\boldsymbol{u}(\boldsymbol{r},t)$; as a result, quantity $\boldsymbol{u}(\boldsymbol{r},t)\cdot\boldsymbol{\omega}(\boldsymbol{r},t)$ called *velocity field helicity* vanishes,

$$\boldsymbol{u}(\boldsymbol{r},t)\cdot\boldsymbol{\omega}(\boldsymbol{r},t)=0.$$

Expression (1.53), page 29, allows obtaining the following representations for the vortex field correlation and spectral tensors [28]:

$$\left\langle \omega_i(\boldsymbol{r},t)\omega_j(\boldsymbol{r}',t')\right\rangle = -\delta_{ij}\Delta_{\boldsymbol{r}}B_{ll}(\boldsymbol{r}-\boldsymbol{r}',t-t') + \frac{\partial^2 B_{ll}(\boldsymbol{r}-\boldsymbol{r}',t-t')}{\partial r_i \partial r_j}$$
$$+ \Delta_{\boldsymbol{r}}B_{ij}(\boldsymbol{r}-\boldsymbol{r}',t-t'),$$
$$\Omega_{ij}(\boldsymbol{k},t-t') = (\delta_{ij}\boldsymbol{k}^2 - k_ik_j)E_{ll}(\boldsymbol{k},t-t') - \boldsymbol{k}^2E_{ij}(\boldsymbol{k},t-t').$$

Convolving them with respect to indices i and j and setting $\boldsymbol{r} = \boldsymbol{r}'$, we obtain the expressions for the correlation and spectral functions of the vortex field

$$\left\langle \omega_i(\boldsymbol{r},t)\omega_i(\boldsymbol{r},t')\right\rangle = -\Delta_{\boldsymbol{r}}B_{ll}(\boldsymbol{0},t-t') = -\left\langle \boldsymbol{u}(\boldsymbol{r},t)\Delta_{\boldsymbol{r}}\boldsymbol{u}(\boldsymbol{r},t')\right\rangle,$$
$$\Omega_{ii}(\boldsymbol{k},t-t') = \boldsymbol{k}^2E_{ii}(\boldsymbol{k},t-t'). \tag{4.51}$$

Note that the procedure similar to that used to calculate vortex field variance and spectrum allows additionally calculating average helicity of the velocity field

$$\chi(\boldsymbol{r},t) = \boldsymbol{u}(\boldsymbol{r},t)\boldsymbol{\omega}(\boldsymbol{r},t) = \boldsymbol{u}(\boldsymbol{r},t)\cdot \operatorname{curl}\boldsymbol{u}(\boldsymbol{r},t),$$

in the three-dimensional case in the absence of reflection symmetry. Indeed, the use of Eq. (1.54), page 29, results in the following expression for statistically averaged quantity $\left\langle \omega_i(\boldsymbol{r},t)u_j(\boldsymbol{r},t)\right\rangle$

$$\left\langle \omega_i(\boldsymbol{r},t)u_p(\boldsymbol{r}',t)\right\rangle_{\boldsymbol{r}=\boldsymbol{r}'} = \varepsilon_{ijk}\left\langle \frac{\partial u_k(\boldsymbol{r},t)}{\partial r_j}u_p(\boldsymbol{r}',t)\right\rangle_{\boldsymbol{r}=\boldsymbol{r}'} = \varepsilon_{ijk}\frac{\partial}{\partial r_j}B_{kp}(\boldsymbol{r},0)_{\boldsymbol{r}=\boldsymbol{0}}$$
$$= \lim_{\boldsymbol{r}\to\boldsymbol{0}} \varepsilon_{ijk}\frac{\partial}{\partial r_j}C(r,0)\varepsilon_{kpm}r_m = C(0,0)\varepsilon_{ijm}\varepsilon_{mpj} = 2C(0,0)\delta_{ip},$$

and, hence, the average helicity of the Gaussian random field in the three-dimensional case is

$$\langle \chi(\boldsymbol{r},t)\rangle = 6C(0,0). \tag{4.52}$$

The isotropic portion of correlation tensor corresponds to the spatial spectral tensor of the form

$$E_{ij}(\boldsymbol{k},\tau) = E^{\mathrm{s}}_{ij}(\boldsymbol{k},\tau) + E^{\mathrm{p}}_{ij}(\boldsymbol{k},\tau),$$

where the spectral components of the tensor of velocity field have the following structure

$$E^{\mathrm{s}}_{ij}(\boldsymbol{k},\tau) = E^{\mathrm{s}}(k,\tau)\left(\delta_{ij} - \frac{k_ik_j}{k^2}\right), \quad E^{\mathrm{p}}_{ij}(\boldsymbol{k},\tau) = E^{\mathrm{p}}(k,\tau)\frac{k_ik_j}{k^2}.$$

Here, $E^{\mathrm{s}}(k,\tau)$ and $E^{\mathrm{p}}(k,\tau)$ are the solenoidal and potential components of the spectral density of velocity field, respectively.

Define now function $B_{ij}(\boldsymbol{r})$ as the integral of correlation function (4.49) over time, i.e.,

$$B_{ij}(\boldsymbol{r}) = \int_0^\infty d\tau B_{ij}(\boldsymbol{r},\tau) = B_{ij}^{\mathrm{iso}}(\boldsymbol{r}) + C(r)\varepsilon_{ijk}r_k. \tag{4.53}$$

Then, $B_{ij}(\mathbf{0}) = D_0\delta_{ij}$ and, hence, quantity

$$B_{ii}(\mathbf{0}) = D_0 d = \tau_0\sigma_{\boldsymbol{u}}^2 = \int d\boldsymbol{k}\, \left[(d-1)\,E^{\mathrm{s}}(k) + E^{\mathrm{p}}(k)\right], \tag{4.54}$$

defines temporal correlation radius of the velocity field τ_0. Here, $\sigma_{\boldsymbol{u}}^2 = B_{ii}(\mathbf{0},0) = \langle \boldsymbol{u}^2(\boldsymbol{r},t)\rangle$ is the variance of the velocity field, and functions $E^{\mathrm{s}}(k)$ and $E^{\mathrm{p}}(k)$ are defined as follows

$$E^{\mathrm{s}}(k) = \int_0^\infty d\tau\; E^{\mathrm{s}}(k,\tau), \quad E^{\mathrm{p}}(k) = \int_0^\infty d\tau\; E^{\mathrm{p}}(k,\tau). \tag{4.55}$$

The use of correlation function (4.53) appreciably simplifies calculations dealing with spatial derivatives of the velocity field. Indeed, we have

$$\frac{\partial B_{ij}(\mathbf{0})}{\partial r_k} = C(0)\varepsilon_{ijk}, \tag{4.56}$$

$$\begin{aligned} -\frac{\partial^2 B_{ij}(0)}{\partial r_k \partial r_l} &= \frac{D^{\mathrm{s}}}{d(d+2)}\left[(d+1)\delta_{kl}\delta_{ij} - \delta_{ki}\delta_{lj} - \delta_{kj}\delta_{li}\right] \\ &\quad + \frac{D^{\mathrm{p}}}{d(d+2)}\left[\delta_{kl}\delta_{ij} + \delta_{ki}\delta_{lj} + \delta_{kj}\delta_{li}\right], \end{aligned} \tag{4.57}$$

$$\frac{\partial^3 B_{kp}(\mathbf{0})}{\partial r_n \partial r_m \partial r_j} = -2\alpha\left(\varepsilon_{kpj}\delta_{nm} + \varepsilon_{kpm}\delta_{nj} + \varepsilon_{kpn}\delta_{mj}\right), \tag{4.58}$$

and, consequently,

$$\frac{\partial^3 B_{kp}(\mathbf{0})}{\partial \boldsymbol{r}^2 \partial r_j} = -2\alpha(d+2)\varepsilon_{kpj},$$

where

$$
\begin{aligned}
D^{s} &= \int dk\, k^2 E^{s}(k) = \frac{1}{d-1}\int_0^\infty d\tau\, \langle \boldsymbol{\omega}(\boldsymbol{r}, t+\tau)\boldsymbol{\omega}(\boldsymbol{r}, t)\rangle, \\
D^{p} &= \int dk\, k^2 E^{p}(k) = \int_0^\infty d\tau \left\langle \frac{\partial \boldsymbol{u}(\boldsymbol{r}, t+\tau)}{\partial \boldsymbol{r}} \frac{\partial \boldsymbol{u}(\boldsymbol{r}, t)}{\partial \boldsymbol{r}} \right\rangle, \\
C(r) &= C(0) - \alpha r^2,
\end{aligned}
\tag{4.59}
$$

and $\boldsymbol{\omega}(\boldsymbol{r}, t)$ is the velocity field vortex.

4.2.5 Logarithmically Normal Random Process

Note that the so-called *logarithmically normal* (*lognormal*) random process $y(t)$

$$y(t) = e^{z(t)},$$

whose logarithm is the Gaussian random process, is described by the one-time probability density $P(y, t)$ of the form

$$P(y, t) = \frac{1}{y} P(z = \ln y, t) = \frac{1}{y\sqrt{2\pi\sigma^2(t)}} \exp\left\{-\frac{\ln^2\left[e^{-\langle z(t)\rangle} y\right]}{2\sigma^2(t)}\right\},$$

and, consequently, the corresponding integral distribution function assumes the form

$$
\begin{aligned}
F(y, t) &= \int_{-\infty}^{y} dy'\, P(y', t) = \frac{1}{\sqrt{2\pi\sigma^2(t)}} \int_{-\infty}^{y} \frac{dy'}{y'} \exp\left\{-\frac{\ln^2\left[e^{-\langle z(t)\rangle} y'\right]}{2\sigma^2(t)}\right\} \\
&= \frac{1}{\sqrt{2\pi\sigma^2(t)}} \int_{-\infty}^{\ln\left[e^{-\langle z(t)\rangle} y\right]} dz \exp\left\{-\frac{z^2}{2\sigma^2(t)}\right\} = \Pr\left(\frac{\ln\left[e^{-\langle z(t)\rangle} y\right]}{\sigma(t)}\right),
\end{aligned}
$$

where probability integral $\Pr(z)$ is defined by Eq. (4.20), page 94.

In view of Eq. (4.21), page 94, the right-hand side of this equality turns into $1/2$ if

$$e^{-\langle z(t)\rangle} y(t) = 1,$$

which means that the typical realization curve $y^*(t)$ of the lognormal random process $y(t)$ is given by the equality

$$y^*(t) = e^{\langle z(t)\rangle} = e^{\langle \ln y(t)\rangle}. \tag{4.60}$$

Note that, if we know the moment functions of random process $y(t)$ as functions of time, i.e., if we know functions $\langle y^n(t)\rangle$ $(n = 1, 2, \ldots)$, we know also the statistical characteristics of random process $z(t) = \ln y(t)$. Indeed,

$$\left\langle y^n(t)\right\rangle = \left\langle e^{n\ln y(t)}\right\rangle = \exp\left\{n\left\langle \ln y(t)\right\rangle + \frac{n^2}{2}\sigma^2_{\ln y}(t)\right\},$$

and, consequently,

$$\langle \ln y(t)\rangle = \lim_{n\to 0}\frac{1}{n}\ln\left\langle y^n(t)\right\rangle, \quad \sigma^2_{\ln y}(t) = \lim_{n\to\infty}\frac{2}{n^2}\ln\left\langle y^n(t)\right\rangle. \tag{4.61}$$

Further, analyzing statistical topography of random processes and fields, we will use, in line with the integral distribution function, functions

$$V(y,t) = \int_y^\infty dy' P(y',t) \quad \text{and} \quad Y(y,t) = \int_y^\infty y' dy' P(y',t),$$

closely related to the probability integral $\Pr(z)$.

For function $V(y,t)$, we have

$$\begin{aligned} V(y,t) &= \frac{1}{\sqrt{2\pi\sigma^2(t)}}\int_y^\infty \frac{dy'}{y'}\exp\left\{-\frac{\left(\ln y' - \langle z(t)\rangle\right)^2}{2\sigma^2(t)}\right\} \\ &= \frac{1}{\sqrt{2\pi}}\int_{\frac{\ln y - \langle z(t)\rangle}{\sigma(t)}}^\infty dz\exp\left\{-\frac{z^2}{2}\right\} = \Pr\left(-\frac{\ln y - \langle z(t)\rangle}{\sigma(t)}\right) \\ &= \Pr\left\{\frac{1}{\sigma(t)}\ln\left(\frac{1}{y}e^{\langle z(t)\rangle}\right)\right\}. \end{aligned} \tag{4.62}$$

For function $Y(y,t)$, we obtain the expression

$$\begin{aligned} Y(y,t) &= \frac{1}{\sqrt{2\pi\sigma^2(t)}}\int_y^\infty dy'\exp\left\{-\frac{\left(\ln y' - \langle z(t)\rangle\right)^2}{2\sigma^2(t)}\right\} \\ &= \frac{1}{\sqrt{2\pi}}e^{\langle z(t)\rangle + \sigma^2(t)/2}\int_{\frac{\ln y - \langle z(t)\rangle - \sigma^2(t)}{\sigma(t)}}^\infty dz\exp\left\{-\frac{z^2}{2}\right\} \\ &= e^{\langle z(t)\rangle + \sigma^2(t)/2}\Pr\left(-\frac{\ln y - \langle z(t)\rangle - \sigma^2(t)}{\sigma(t)}\right) \\ &= e^{\langle z(t)\rangle + \sigma^2(t)/2}\Pr\left\{\frac{1}{\sigma(t)}\ln\left(\frac{1}{y}e^{\langle z(t)\rangle + \sigma^2(t)}\right)\right\}. \end{aligned} \tag{4.63}$$

Note that for random process $\widetilde{y}(t) = y_0 y(t)$, i.e.,

$$\widetilde{y}(t) = y_0 e^{z(t)},$$

instead of Eqs. (4.62), (4.63) we have the expressions

$$\begin{aligned}\widetilde{V}(y,t) &= \Pr\left\{\frac{1}{\sigma(t)}\ln\left(\frac{y_0}{y}e^{\langle z(t)\rangle}\right)\right\},\\ \widetilde{Y}(y,t) &= y_0 e^{\langle z(t)\rangle+\sigma^2(t)/2}\Pr\left\{\frac{1}{\sigma(t)}\ln\left(\frac{y_0}{y}e^{\langle z(t)\rangle+\sigma^2(t)}\right)\right\}.\end{aligned} \tag{4.64}$$

4.2.6 Discontinuous Random Processes

Consider now some examples of discontinuous processes. The discontinuous processes are the random functions that change their time-dependent behavior at discrete instants $t_1, t_2, \ldots$ given statistically. The description of discontinuous processes requires first of all either the knowledge of the statistics of these instants, or the knowledge of the statistics of number $n(0,t)$ of instants t_i falling in time interval $(0,t)$. In the latter case, we have the equality

$$n(0,t) = n(0,t') + n(t',t), \quad 0 \le t' \le t.$$

The quantity $n(0,t)$ by itself is a random process, and Fig. 4.2 shows its possible realization.

The set of points of discontinuity $t_1, t_2, \ldots$ of process $z(t)$ is called the *stream of points*. In what follows, we will consider *Poisson's stationary stream of points* in which the probability of falling n points in interval (t_1, t_2) is specified by Poisson's formula

$$P_{n(t_1,t_2)=n} = \frac{\left[\overline{n(t_1,t_2)}\right]^n}{n!} e^{-\overline{n(t_1,t_2)}}, \tag{4.65}$$

with the mean number of points in interval (t_1, t_2) given by the formula

$$\overline{n(t_1,t_2)} = \nu |t_1 - t_2|,$$

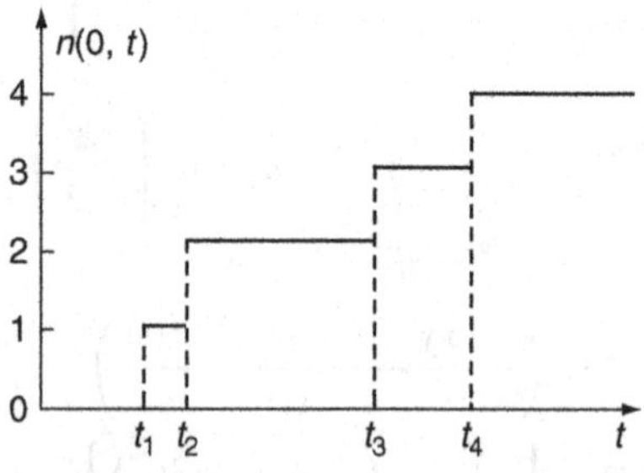

Figure 4.2 A possible realization of process $n(0,t)$.

where ν is the mean number of points per unit time. It is assumed here that the numbers of points falling in nonoverlapping intervals are statistically independent and the instants at which points were fallen in interval (t_1, t_2) under the condition that their total number was n are also statistically independent and uniformly distributed over the interval (t_1, t_2). The length of the interval between adjacent points of discontinuity satisfies the exponential distribution.

Poisson stream of points is an example of the Markovian processes.

Consider now random processes whose points of discontinuity form Poisson's streams of points. Currently, three types of such process – *Poisson's process*, *telegrapher's process*, and *generalized telegrapher's process* – are mainly used to model problems in physics. Below, we focus our attention on these processes.

Poisson's (Impulse) Random Process

Poisson's (impulse) random process $z(t)$ is the process described by the formula

$$z(t) = \sum_{i=1}^{n} \xi_i g(t - t_i), \tag{4.66}$$

where random quantities ξ_i are statistically independent and distributed with probability density $p(\xi)$; random points t_k are uniformly distributed on interval $(0, T)$, so that their number n obeys Poisson's law with parameter $\bar{n} = \nu T$; and function $g(t)$ is the deterministic function that describes the pulse envelope ($g(t) = 0$ for $t < 0$).

The characteristic functional of Poisson's random process $z(t)$ assumes the form

$$\Phi[t; v(\tau)] = \exp\left\{\nu \int_0^t dt' \left[W\left(\int_{t'}^{t} d\tau v(\tau) g(t - t')\right) - 1\right]\right\}, \tag{4.67}$$

where

$$W(v) = \int_{-\infty}^{\infty} d\xi p(\xi) e^{i\xi v},$$

is the characteristic function of random quantity ξ. Consequently, functional $\Theta[t; v(\tau)]$ is given by the formula

$$\begin{aligned}\Theta[t; v(\tau)] &= \ln \Phi[t; v(\tau)] \\ &= \nu \int_0^t dt' \int_{-\infty}^{\infty} d\xi p(\xi) \left\{\exp\left[i\xi \int_{t'}^{t} d\tau v(\tau) g(t - t')\right] - 1\right\},\end{aligned} \tag{4.68}$$

and cumulant functions assume the form

$$K_n(t_1,\ldots,t_n) = \nu\left\langle\xi^n\right\rangle \int\limits_0^{\min\{t_1,\ldots,t_n\}} dt' g(t-t')\ldots g(t_n-t'). \tag{4.69}$$

We consider two types of Poisson's processes important for applications.

1. Let $g(t) = \theta(t) = \begin{cases} 1, & t > 0, \\ 0, & t < 0, \end{cases}$, i.e., $z(t) = \sum\limits_{i=1}^{n} \xi_i \theta(t - t_i)$. In this case,

$$K_n(t_1,\ldots,t_n) = \nu\left\langle\xi^n\right\rangle \min\{t_1,\ldots,t_n\}.$$

If additionally $\xi = 1$, then process $z(t) \equiv n(0,t)$, and we have

$$K_n(t_1,\ldots,t_n) = \nu \min\{t_1,\ldots,t_n\},$$

$$\Theta[t;v(\tau)] = \nu \int\limits_0^t dt' \left\{\exp\left[i\int\limits_{t'}^t d\tau v(\tau)\right] - 1\right\}. \tag{4.70}$$

2. Let now $g(t) = \delta(t)$. In this case, process

$$z(t) = \sum_{i=1}^{n} \xi_i \delta(t - t_i)$$

is usually called the *shot noise process*. This process is a special case of the *delta-correlated processes* (see Sect. 5.5, page 130). For such a process, functional $\Theta[t;v(\tau)]$ is

$$\Theta[t;v(\tau)] = \nu \int\limits_0^t d\tau \int\limits_{-\infty}^{\infty} d\xi p(\xi) \left\{e^{i\xi v(\tau)} - 1\right\} \tag{4.71}$$

and cumulant functions assume the form

$$K_n(t_1,\ldots,t_n) = \nu\left\langle\xi^n\right\rangle \delta(t_1 - t_2)\delta(t_2 - t_3)\ldots\delta(t_{n-1} - t_n).$$

Note that Poisson's process $z(t)$ with arbitrary given impulse function $g(t)$ can be always expressed in terms of Poisson's delta-correlated random process $z_\delta(t)$ by the formula

$$z(t) = \int\limits_0^t d\tau g(t-\tau) z_\delta(\tau). \tag{4.72}$$

Telegrapher's Random Process

Consider now statistical characteristics of telegrapher's random process (Fig. 4.3) defined by the formula

$$z(t) = a(-1)^{n(0,t)} \quad (z(0) = a,\ z^2(t) \equiv a^2), \tag{4.73}$$

where $n(t_1, t_2)$ is the random sequence of integers equal to the number of points of discontinuity in interval (t_1, t_2).

We consider two cases.

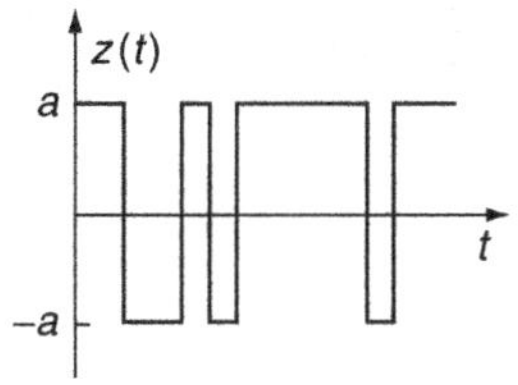

Figure 4.3 A possible realization of telegrapher's random process.

1. We will assume first that amplitude a is the deterministic quantity.

For the two first moment functions of process $z(t)$, we have the expressions

$$\begin{aligned}\langle z(t)\rangle &= a\sum_{n(0,t)=0}^{\infty}(-1)^{n(0,t)}P_{n(0,t)} = ae^{-2\bar{n}(0,t)} = ae^{-2\nu t},\\ \langle z(t_1)z(t_2)\rangle &= a^2\left\langle(-1)^{n(0,t_1)+n(0,t_2)}\right\rangle = a^2\left\langle(-1)^{n(t_2,t_1)}\right\rangle\\ &= a^2e^{-2\bar{n}(t_2,t_1)} = a^2e^{-2\nu(t_1-t_2)}\quad(t_1\geq t_2).\end{aligned}\tag{4.74}$$

The higher moment functions for $t_1 \geqslant t_2 \geqslant \ldots \geqslant t_n$ satisfy the recurrence relationship

$$\begin{aligned}M_n(t_1,\ldots,t_n) &= \langle z(t_1)\ldots z(t_n)\rangle\\ &= a^2\left\langle(-1)^{n(0,t_1)+n(0,t_2)+n(0,t_3)+\cdots+n(0,t_n)}\right\rangle\\ &= a^2\left\langle(-1)^{n(t_2,t_1)}\right\rangle\left\langle(-1)^{n(0,t_3)+\cdots+n(0,t_n)}\right\rangle\\ &= \langle z(t_1)z(t_2)\rangle\, M_{n-2}(t_3,\ldots,t_n).\end{aligned}\tag{4.75}$$

This relationship is very similar to Eq. (4.42) for the Gaussian process with correlation function $B(t_1, t_2)$. The only difference is that the right-hand side of Eq. (4.75) coincides with only one term of the sum in Eq. (4.42), namely, with the term that corresponds to the above order of times.

Consider now the characteristic functional of this process

$$\Phi_a[t;\nu(\tau)] = \left\langle\exp\left\{i\int_0^t d\tau z(\tau)\nu(\tau)\right\}\right\rangle,$$

where index a means that amplitude a is the deterministic quantity. Expanding the characteristic functional in the functional Taylor series and using recurrence formula (4.75), we obtain the integral equation

$$\begin{aligned}\Phi_a[t;\nu(\tau)] &= 1 + ia\int_0^t dt_1 e^{-2\nu t_1}\nu(t_1)\\ &\quad - a^2\int_0^t dt_1\int_0^{t_1} dt_2 e^{-2\nu(t_1-t_2)}\nu(t_1)\nu(t_2)\Phi_a[t_2;\nu(\tau)].\end{aligned}\tag{4.76}$$

Differentiating Eq. (4.76) with respect to t, we obtain the integro-differential equation

$$\frac{d}{dt}\Phi_a[t; \nu(\tau)] = iae^{-2\nu t}\nu(t) - a^2\nu(t)\int_0^t dt_1 e^{-2\nu(t-t_1)}\nu(t_1)\Phi_a[t_1; \nu(\tau)]. \tag{4.77}$$

2. Assume now amplitude a is the random quantity with probability density $p(a)$. To obtain the characteristic functional of process $z(t)$ in this case, we should average Eq. (4.77) with respect to random amplitude a. In the general case, such averaging cannot be performed analytically. Analytical averaging of Eq. (4.77) appears to be possible only if probability density of random amplitude a has the form

$$p(a) = \frac{1}{2}\left[\delta\left(a - a_0\right) + \delta\left(a + a_0\right)\right], \tag{4.78}$$

with $\langle a\rangle = 0$ and $\left\langle a^2\right\rangle = a_0^2$ (in fact, this very case is what is called usually telegrapher's process). As a result, we obtain the integro-differential equation

$$\frac{d}{dt}\Phi[t; \nu(\tau)] = -a_0^2\nu(t)\int_0^t dt_1 e^{-2\nu(t-t_1)}\nu(t_1)\Phi[t_1; \nu(\tau)]. \tag{4.79}$$

Now, we dwell on an important limiting theorem concerning telegrapher's random processes.

Consider the random process

$$\xi_N(t) = z_1(t) + \cdots + z_N(t),$$

where all $z_k(t)$ are statistically independent telegrapher's processes with zero-valued means and correlation functions

$$\langle z(t)z(t+\tau)\rangle = \frac{\sigma^2}{N}e^{-\alpha|\tau|}.$$

In the limit $N \to \infty$, we obtain that process $\xi(t) = \lim_{N\to\infty}\xi_N(t)$ is the Gaussian random process with the exponential correlation function

$$\langle \xi(t)\xi(t+\tau)\rangle = \sigma^2 e^{-\alpha|\tau|},$$

i.e., the Gaussian Markovian process. Thus, process $\xi_N(t)$ for finite N is the finite-number-of-states process approximating the Gaussian Markovian process.

Generalized Telegrapher's Random Process

Consider now generalized telegrapher's process defined by the formula

$$z(t) = a_{n(0,t)}. \tag{4.80}$$

Here, $n(0, t)$ is the sequence of integers described above and quantities a_k are assumed statistically independent with distribution function $p(a)$. Figure 4.4 shows a possible realization of such a process.

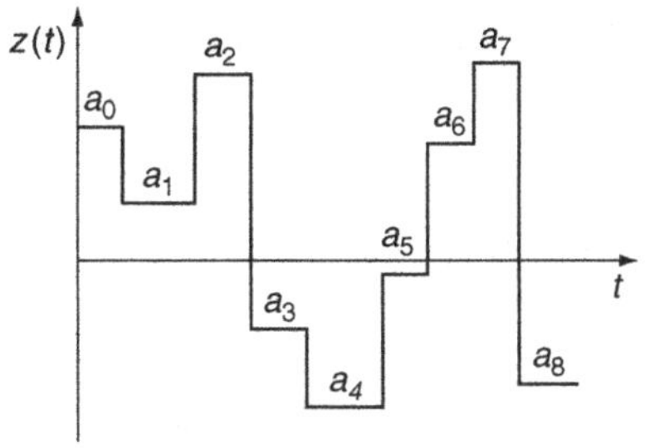

Figure 4.4 A possible realization of generalized telegrapher's random process.

For process $z(t)$, we have

$$\langle z(t)\rangle = \sum_{k=0}^{\infty}\left\langle a_k \delta_{k,n(0,t)}\right\rangle = \langle a\rangle,$$

$$\langle z(t_1)z(t_2)\rangle = \sum_{k=0}^{\infty}\sum_{k=0}^{\infty}\langle a_k a_l\rangle \left\langle \delta_{k,n(0,t_1)}\delta_{l,n(0,t_2)}\right\rangle$$

$$= \left\langle a^2\right\rangle \left\{\sum_{k=0}^{\infty}\left\langle \delta_{k,n(0,t_1)}\right\rangle\left\langle \delta_{0,n(t_2,t_1)}\right\rangle + 1 - \sum_{k=0}^{\infty}\left\langle \delta_{k,n(0,t_2)}\right\rangle\left\langle \delta_{0,n(t_2,t_1)}\right\rangle\right\}$$

$$= \left\langle a^2\right\rangle e^{-\nu(t_1-t_2)} + \left\langle a^2\right\rangle\left(1 - e^{-\nu(t_1-t_2)}\right) \quad (t_1 \geqslant t_2),$$

and so on. In addition, the probability of absence of points of discontinuity in interval (t_2, t_1) is given by the formula

$$P_{n(t_2,t_1)=0} = \left\langle \delta_{0,n(t_2,t_1)}\right\rangle = e^{-\nu|t_1-t_2|}.$$

Earlier, we mentioned that Poisson's stream of points and processes based on these streams are the Markovian processes. Below, we consider this important class of random processes in detail.

4.3 Markovian Processes

4.3.1 General Properties

In the foregoing section, we considered the characteristic functional that describes all statistical characteristics of random process $z(t)$. Specification of the argument of the functional in the form

$$v(t) = \sum_{k=1}^{n} v_k \delta(t - t_k)$$

transforms the characteristic functional into the joint characteristic function of random quantities $z_k = z(t_k)$

$$\Phi_n(v_1, \ldots, v_n) = \left\langle \exp\left\{ i \sum_{k=1}^{n} v_k z(t_k) \right\} \right\rangle,$$

whose Fourier transform is the joint probability density of process $z(t)$ at discrete instants

$$P_n(z_1, t_1; \ldots; z_n, t_n) = \langle \delta(z(t_1) - z_1) \ldots \delta(z(t_n) - z_n) \rangle . \tag{4.81}$$

Assume that the above instants are ordered according to the line of inequalities

$$t_1 \geq t_2 \geq \cdots \geq t_n.$$

Then, by definition of the conditional probability, we have

$$P_n(z_1, t_1; \ldots; z_n, t_n) = p_n(z_1, t_1 | z_2, t_2; \ldots; z_n, t_n) P_{n-1}(z_2, t_2; \ldots; z_n, t_n), \tag{4.82}$$

where p_n is the conditional probability density of the value of process $z(t)$ at instant t_1 under the condition that function $z(t)$ was equal to z_k at instants t_k for $k = 2, \ldots, n$ ($z(t_k) = z_k$, $k = 2, \ldots, n$). If process $z(t)$ is such that the conditional probability density for all $t_1 > t_2$ is unambiguously determined by the value z_2 of the process at instant t_2 and is independent of the previous history, i.e., if

$$p_n(z_1, t_1 | z_2, t_2; \ldots; z_n, t_n) = p(z_1, t_1 | z_2, t_2), \tag{4.83}$$

then this process is called the Markovian process, or the memoryless process. In this case, function

$$p(z, t | z_0, t_0) = \langle \delta(z(t) - z) | z(t_0) = z_0 \rangle \quad (t > t_0), \tag{4.84}$$

is called the *transition probability density*. Setting $t = t_0$ in Eq. (4.84), we obtain the equality

$$p(z, t_0 | z_0, t_0) = \delta(z - z_0).$$

Substituting expression (4.82) in Eq. (4.81), we obtain the recurrence formula for the n-time probability density of process $z(t)$. Iterating this formula, we find the expression of probability density P_n in terms of the one-time probability density $(t_1 \geq t_2 \geq \cdots \geq t_n)$

$$P_n(z_1, t_1; \ldots; z_n, t_n) = p(z_1, t_1|z_2, t_2) \ldots p(z_{n-1}, t_{n-1}|z_n, t_n)P(t_n, z_n). \tag{4.85}$$

Substituting expression (4.82) in Eq. (4.81), we obtain the recurrence formula for the n-time probability density of process $z(t)$. Iterating this formula, we find the relationship of probability density P_n with the one-time probability density $(t_1 \geq t_2 \geq \cdots \geq t_n)$

$$P_n(z_1, t_1; \ldots; z_n, t_n) = p(z_1, t_1|z_2, t_2) \ldots p(z_{n-1}, t_{n-1}|z_n, t_n)P(t_n, z_n). \tag{4.86}$$

Nevertheless, statistical analysis of stochastic equations often requires the knowledge of the characteristic functional of random process $z(t)$.

4.3.2 Characteristic Functional of the Markovian Process

For the Markovian process $z(t)$, no closed equation can be derived in the general case for the characteristic functional $\Phi[t; v(\tau)] = \langle\varphi[t; v(\tau)]\rangle$, where

$$\varphi[t; v(\tau)] = \exp\left\{i \int_0^t d\tau z(\tau)v(\tau)\right\}.$$

Instead, we can derive the closed equation for the functional

$$\Psi[z, t; v(\tau)] = \langle\delta(z(t) - z)\varphi[t; v(\tau)]\rangle, \tag{4.87}$$

describing correlations of process $z(t)$ with its prehistory. The characteristic functional $\Phi[t; v(\tau)]$ can be obtained from functional $\Psi[z, t; v(\tau)]$ by the formula

$$\Phi[t; v(\tau)] = \int_{-\infty}^{\infty} dz \Psi[z, t; v(\tau)]. \tag{4.88}$$

To derive the equation in functional $\Psi[z, t; v(\tau)]$, we note that the following equality

$$\varphi[t; v(\tau)] = 1 + i \int_0^t dt_1 z(t_1)v(t_1)\varphi[t_1; v(\tau)] \tag{4.89}$$

holds. Substituting Eq. (4.89) in Eq. (4.87), we obtain the expression

$$\Psi[z, t; v(\tau)] = P(z, t) + i \int_0^t dt_1 v(t_1) \langle\delta(z(t) - z)z(t_1)\varphi[t_1; v(\tau)]\rangle, \tag{4.90}$$

where $P(z, t) = \langle\delta(z(t) - z)\rangle$ is the one-time probability density of random quantity $z(t)$.

We rewrite Eq. (4.90) in the form

$$\Psi[z, t; v(\tau)] = P(z, t) + i\int_0^t dt_1 v(t_1) \int_{-\infty}^{\infty} dz_1 z_1 \langle\delta(z(t) - z)\delta(z(t_1) - z_1)\varphi[t_1; v(\tau)]\rangle. \tag{4.91}$$

Taking into account the fact that process $z(t)$ is the Markovian process, we can perform averaging in Eq. (4.91) to obtain the closed integral equation

$$\Psi[z, t; v(\tau)] = P(z, t) + i\int_0^t dt_1 v(t_1) \int_{-\infty}^{\infty} dz_1 z_1 p(z, t|z_1, t_1)\Psi[z_1, t_1; v(\tau)], \tag{4.92}$$

where $p(z, t; z_0, t_0)$ is the transition probability density.

Integrating Eq. (4.92) with respect to z, we obtain an additional relationship between the characteristic functional $\Phi[t; v(\tau)]$ and functional $\Psi[z, t; v(\tau)]$. This relationship has the form

$$\frac{1}{iv(t)}\frac{d}{dt}\Phi[t; v(\tau)] = \int_{-\infty}^{\infty} dz_1 z_1 \Psi[z_1, t; v(\tau)] = \Psi[t; v(\tau)]. \tag{4.93}$$

Multiplying Eq. (4.92) by z and integrating the result over z, we obtain the relationship between functionals $\Psi[t; v(\tau)]$ and $\Psi[z, t; v(\tau)]$

$$\Psi[t; v(\tau)] = \langle z(t)\rangle + i\int_0^t dt_1 v(t_1) \int_{-\infty}^{\infty} dz_1 \langle z(t)|z_1, t_1\rangle \, \Psi[z_1, t_1; v(\tau)]. \tag{4.94}$$

Equation (4.92) is generally a complicated integral equation whose explicit form depends on functions $P(z, t)$ and $p(z, t; z_0, t_0)$, i.e., on parameters of the Markovian process.

In some specific cases, the situation is simplified. For example, for telegrapher's process, we have from Eq. (4.74)

$$\langle z(t)|z_1, t_1\rangle = z_1 e^{-2\nu(t-t_1)}, \quad \langle z(t)\rangle = 0,$$

and we obtain Eq. (4.24), page 95.

Problems

Problem 4.1

Derive expression for the characteristic functional of Poisson's random process (4.67).

Solution The characteristic functional of Poisson's random process has the form

$$\Phi[t; v(\tau)] = \left\langle \exp\left\{ i \int_0^t d\tau\, v(\tau) z(\tau) \right\} \right\rangle_z. \tag{4.95}$$

We will average Eq. (4.95) in two stages. At first, we perform averaging over random quantity ξ and positions of random points t_k:

$$\begin{aligned} \Phi[t; v(\tau)] &= \left\langle \left\langle \exp\left\{ i \int_0^t d\tau\, v(\tau) \sum_{k=1}^{n} \xi_k g(\tau - t_k) \right\} \right\rangle_{\xi_i, t_k} \right\rangle_n \\ &= \left\langle \left[\frac{1}{T} \int_0^T dt' \int_{-\infty}^{\infty} d\xi \exp\left\{ i\xi \int_0^t d\tau\, v(\tau) g(\tau - t') \right\} \right]^n \right\rangle_n \\ &= \left\langle \left[\frac{1}{T} \int_0^T dt'\, W\left(\int_{t'}^{t} d\tau\, v(\tau) g(\tau - t') \right) \right]^n \right\rangle_n, \end{aligned} \tag{4.96}$$

where $W(v) = \int_{-\infty}^{\infty} d\xi\, p(\xi) e^{i\xi v}$ is the characteristic function of random quantity ξ. Let $t < T$. Then Eq. (4.96) can be rewritten in the form

$$\Phi[t; v(\tau)] = \left\langle \left[\frac{1}{T} \int_0^t dt'\, W\left(\int_{t'}^{t} d\tau\, v(\tau) g(\tau - t') \right) + \frac{T - t}{T} \right]^n \right\rangle_n. \tag{4.97}$$

Average now Eq. (4.97) over Poisson's distribution $p_n = \dfrac{e^{-\bar{n}}}{n!} (\bar{n})^n$ of random quantity n:

$$\begin{aligned} \Phi[t; v(\tau)] &= \sum_0^{\infty} \frac{e^{-\bar{n}}}{n!} (\bar{n})^n \left\langle \left[\frac{1}{T} \int_0^t dt'\, W\left(\int_{t'}^{t} d\tau\, v(\tau) g(\tau - t') \right) + \frac{T - t}{T} \right]^n \right\rangle_n \\ &= \exp\left\{ -\bar{n} + \frac{\bar{n}}{T} \left[\int_0^t dt'\, W\left(\int_{t'}^{t} d\tau\, v(\tau) g(\tau - t') \right) + T - t \right] \right\}. \end{aligned}$$

As a result, we obtain the expression (4.67) for the characteristic functional of Poisson's random process.

Problem 4.2

Show that random process

$$\xi_N(t) = z_1(t) + \cdots + z_N(t),$$

where all $z_k(t)$ are statistically independent telegrapher's processes with zero-valued means and correlation functions

$$\langle z(t)z(t+\tau)\rangle = \frac{\sigma^2}{N} e^{-\alpha|\tau|}$$

in the limit $N \to \infty$ is the Gaussian Markovian random process

$$\xi(t) = \lim_{N\to\infty} \xi_N(t)$$

with the exponential correlation function

$$\langle \xi(t)\xi(t+\tau)\rangle = \sigma^2 e^{-\alpha|\tau|}.$$

Remark *Consider the differential equation for the characteristic functional of random process $\xi_N(t)$*

$$\Phi_N[t; v(\tau)] = \left\langle \exp\left\{ i \int_0^t d\tau \xi_N(\tau) v(\tau) \right\} \right\rangle.$$

Solution The characteristic functional of process $z_k(t)$ satisfies Eq. (4.79)

$$\frac{d}{dt}\Phi[t; v(\tau)] = -\frac{\sigma^2}{N} v(t) \int_0^t dt_1 e^{-\alpha(t-t_1)} v(t_1)\Phi[t_1; v(\tau)],$$

from which follows that $\Phi[t; v(\tau)] \to 1$ for $N \to \infty$.

For the characteristic functional of random process $\xi_N(t)$, we have the expression

$$\Phi_N[t; v(\tau)] = \left\langle \exp\left\{ i \int_0^t d\tau \xi_N(\tau) v(\tau) \right\} \right\rangle = \{\Phi[t; v(\tau)]\}^N.$$

Consequently, it satisfies the equation

$$\frac{d}{dt}\ln\Phi_N[t;v(\tau)] = -\sigma^2 v(t)\int_0^t dt_1 e^{-\alpha(t-t_1)}v(t_1)\frac{\Phi[t_1;v(\tau)]}{\Phi[t;v(\tau)]}.$$

In the limit $N \to \infty$, we obtain the equation

$$\frac{d}{dt}\ln\Phi_\infty[t;v(\tau)] = -\sigma^2 v(t)\int_0^t dt_1 e^{-\alpha(t-t_1)}v(t_1),$$

which means that process $\xi(t)$ is the Gaussian random process.

Lecture 5

Correlation Splitting

5.1 General Remarks

For simplicity, we content ourselves here with the one-dimensional random processes (extensions to multidimensional cases are obvious). We need the ability of calculating correlation $\langle z(t)]R[z(\tau)]\rangle$, where $R[z(\tau)]$ is the functional that can depend on process $z(t)$ both explicitly and implicitly.

To calculate this average, we consider auxiliary functional $R[z(\tau)+\eta(\tau)]$, where $\eta(t)$ is arbitrary deterministic function, and calculate the correlation

$$\langle z(t)R[z(\tau)+\eta(\tau)]\rangle. \tag{5.1}$$

The correlation of interest will be obtained by setting $\eta(\tau)=0$ in the final result.

We can expand the above auxiliary functional $R[z(\tau)+\eta(\tau)]$ in the functional Taylor series with respect to $z(\tau)$. The result can be represented in the form

$$R\left[z(\tau)+\eta(\tau)\right]=\exp\left\{\int\limits_{-\infty}^{\infty} d\tau z(\tau)\frac{\delta}{\delta\eta(\tau)}\right\}R\left[\eta(\tau)\right],$$

where we introduced the functional shift operator. With this representation, we can obtain the following expression for correlation (5.1)

$$\langle z(t)R[z(\tau)+\eta(\tau)]\rangle=\Omega\left[t;\frac{\delta}{i\delta\eta(\tau)}\right]\langle R[z(\tau)+\eta(\tau)]\rangle, \tag{5.2}$$

where functional

$$\Omega[t;v(\tau)]=\frac{\left\langle z(t)\exp\left\{i\int\limits_{-\infty}^{\infty}d\tau z(\tau)v(\tau)\right\}\right\rangle}{\left\langle \exp\left\{i\int\limits_{-\infty}^{\infty}d\tau z(\tau)v(\tau)\right\}\right\rangle}=\frac{\delta}{i\delta v(t)}\Theta[v(\tau)]. \tag{5.3}$$

Lectures on Dynamics of Stochastic Systems. DOI: 10.1016/B978-0-12-384966-3.00005-2

Here $\Theta[v(\tau)] = \ln \Phi[v(\tau)]$ and $\Phi[v(\tau)]$ is the characteristic functional of random process $z(t)$.

Replacing differentiation with respect to $\eta(\tau)$ by differentiation with respect to $z(\tau)$ and setting $\eta(\tau) = 0$, we obtain the expression

$$\langle z(t)R[z(\tau)\rangle = \left\langle \Omega\left[t; \frac{\delta}{i\delta z(\tau)}\right] R[z(\tau)]\right\rangle. \tag{5.4}$$

If we expand functional $\Theta[v(\tau)]$ in the functional Taylor series (4.32), page 98, and differentiate the result with respect to $v(t)$, we obtain

$$\Omega[t; v(\tau)] = \sum_{n=0}^{\infty} \frac{i^n}{n!} \int_{-\infty}^{\infty} dt_1 \ldots \int_{-\infty}^{\infty} dt_n K_{n+1}(t, t_1, \ldots, t_n) v(t_1) \ldots v(t_n)$$

and expression (5.4) will assume the form

$$\begin{aligned} &\left\langle z(t')R[z(\tau)]\right\rangle \\ &= \sum_{n=0}^{\infty} \frac{1}{n!} \int_{-\infty}^{\infty} dt_1 \ldots \int_{-\infty}^{\infty} dt_n K_{n+1}(t', t_1, \ldots, t_n) \left\langle \frac{\delta^n R[z(\tau)]}{\delta z(t_1) \ldots \delta z(t_n)} \right\rangle. \end{aligned} \tag{5.5}$$

In physical problems satisfying the condition of dynamical causality in time, statistical characteristics of the solution at instant t depend on the statistical characteristics of process $z(\tau)$ for $0 \leq \tau \leq t$, which are completely described by the characteristic functional

$$\Phi[t; v(\tau)] = \exp\{\Theta[t; v(\tau)]\} = \left\langle \exp\left\{ i \int_0^t d\tau z(\tau) v(\tau) \right\} \right\rangle.$$

In this case, the obtained formulas hold also for calculating statistical averages $\left\langle z(t')R[t; z(\tau)]\right\rangle$ for $t' < t, \quad \tau \leq t$, i.e., we have the equality

$$\left\langle z(t')R[t; z(\tau)]\right\rangle = \left\langle \Omega\left[t', t; \frac{\delta}{i\delta z(\tau)}\right] R[t; z(\tau)]\right\rangle \quad (0 < t' < t), \tag{5.6}$$

where

$$\begin{aligned} \Omega[t', t; v(\tau)] &= \frac{\delta}{i\delta v(t')]} \Theta[t; v(\tau)] \\ &= \sum_{n=0}^{\infty} \frac{i^n}{n!} \int_0^t dt_1 \ldots \int_0^t dt_n K_{n+1}(t', t_1, \ldots, t_n) v(t_1) \ldots v(t_n). \end{aligned} \tag{5.7}$$

For $t' = t - 0$, formula (5.6) holds as before, i.e.,

$$\langle z(t)R[t; z(\tau)]\rangle = \left\langle \Omega\left[t, t; \frac{\delta}{i\delta z(\tau)}\right] R[t; z(\tau)]\right\rangle. \tag{5.8}$$

However, expansion (5.7) not always gives the correct result in the limit $t' \to t - 0$ (which means that the limiting process and the procedure of expansion in the functional Taylor series can be non-commutable). In this case,

$$\Omega[t, t; v(\tau)] = \frac{\left\langle z(t) \exp\left\{ i \int\limits_0^t d\tau z(\tau) v(\tau) \right\}\right\rangle}{\left\langle \exp\left\{ i \int\limits_0^t d\tau z(\tau) v(\tau) \right\}\right\rangle} = \frac{d}{iv(t)dt}\Theta[t; v(\tau)], \tag{5.9}$$

and statistical averages in Eqs. (5.6) and (5.8) can be discontinuous at $t' = t - 0$.

Consider several examples of random processes.

5.2 Gaussian Process

In the case of the Gaussian random process $z(t)$, all formulas obtained in the previous section become significantly simpler. In this case, the logarithm of characteristic functional $\Phi[v(\tau)]$ is given by Eq. (4.41), page 103, (we assume that the mean value of process $z(t)$ is zero), and functional $\Theta[t, v(\tau)]$ assumes the form

$$\Theta[t, v(\tau)] = -\frac{1}{2} \int\limits_{-\infty}^{\infty} \int\limits_{-\infty}^{\infty} d\tau_1 d\tau_2 B(\tau_1, \tau_2) v(\tau_1) v(\tau_2).$$

As a consequence, functional $\Omega[t; v(\tau)]$ (5.3) is the linear functional

$$\Omega[t; v(\tau)] = i \int\limits_{-\infty}^{\infty} d\tau_1 B(t, \tau_1) v(\tau_1), \tag{5.10}$$

and Eq. (5.2) assumes the form

$$\langle z(t)R[z(\tau) + \eta(\tau)]\rangle = \int\limits_{-\infty}^{\infty} d\tau_1 B(t, \tau_1) \frac{\delta}{\delta\eta(\tau_1)} \langle R[z(\tau) + \eta(\tau)]\rangle. \tag{5.11}$$

Replacing differentiation with respect to $\eta(\tau)$ by differentiation with respect to $z(\tau)$ and setting $\eta(\tau) = 0$, we obtain the equality

$$\langle z(t)R[z(\tau)]\rangle = \int\limits_{-\infty}^{\infty} d\tau_1 B(t, \tau_1)\left\langle \frac{\delta}{\delta z(\tau_1)} R[z(\tau)]\right\rangle \tag{5.12}$$

commonly known in physics as the *Furutsu–Novikov formula* [1, 46].

One can easily obtain the multi-dimensional extension of Eq. (5.12); it can be written in the form

$$\left\langle z_{i_1,\dots,i_n}(\boldsymbol{r})R[z]\right\rangle = \int d\boldsymbol{r}' \left\langle z_{i_1,\dots,i_n}(\boldsymbol{r}) z_{j_1,\dots,j_n}(\boldsymbol{r}')\right\rangle \left\langle \frac{\delta R[z]}{\delta z_{j_1,\dots,j_n}(\boldsymbol{r}')}\right\rangle, \tag{5.13}$$

where $\boldsymbol{r}$ stands for all continuous arguments of random vector field $z(\boldsymbol{r})$ and $i_1, \dots, i_n$ are the discrete (index) arguments. Repeated index arguments in the right-hand side of Eq. (5.13) assume summation.

Formulas (5.12) and (5.13) extend formulas (4.17), (4.24) to the Gaussian random processes.

If random process $z(\tau)$ is defined only on time interval $[0, t]$, then functional $\Theta[t, v(\tau)]$ assumes the form

$$\Theta[t, v(\tau)] = -\frac{1}{2}\int\limits_0^t \int\limits_0^t d\tau_1 d\tau_2 B(\tau_1, \tau_2) v(\tau_1) v(\tau_2), \tag{5.14}$$

and functionals $\Omega[t', t; v(\tau)]$ and $\Omega[t, t; v(\tau)]$ are the linear functionals

$$\begin{aligned}\Omega[t', t; v(\tau)] &= \frac{\delta}{i\delta v(t')}\Theta[t, v(\tau)] = i\int\limits_0^t d\tau B(t', \tau) v(\tau),\\ \Omega[t, t; v(\tau)] &= \frac{d}{iv(t)dt}\Theta[t, v(\tau)] = i\int\limits_0^t d\tau B(t, \tau) v(\tau).\end{aligned} \tag{5.15}$$

As a consequence, Eqs. (5.6), (5.8) assume the form

$$\left\langle z(t')R[t, z(\tau)]\right\rangle = \int\limits_0^t d\tau B(t', \tau)\left\langle \frac{\delta R[z(\tau)]}{\delta z(\tau)}\right\rangle \quad (t' \leqslant t) \tag{5.16}$$

that coincides with Eq. (5.12) if the condition

$$\frac{\delta R[t; z(\tau)]}{\delta z(\tau)} = 0 \quad \text{for} \quad \tau < 0, \quad \tau > t \tag{5.17}$$

holds.

5.3 Poisson's Process

The Poisson process is defined by Eq. (4.66), page 111, and its characteristic functional is given by Eq. (4.67), page 111. In this case, formulas (5.7) and (5.9) for functionals $\Omega[t', t; v(\tau)]$ and $\Omega[t, t; v(\tau)]$ assume the forms

$$\Omega[t', t; v(\tau)] = \frac{\delta}{i\delta v(t')}\Theta[t, v(\tau)]$$
$$= -i\int_0^{t'} d\tau g(t'-\tau)\dot{W}\left(\int_\tau^t d\tau_1 v(\tau_1)g(\tau_1-\tau)\right),$$
$$\Omega[t, t; v(\tau)] = \frac{d}{iv(t)dt}\Theta[t, v(\tau)]$$
$$= -i\int_0^{t} d\tau g(t-\tau)\dot{W}\left(\int_\tau^t d\tau_1 v(\tau_1)g(\tau_1-\tau)\right), \qquad (5.18)$$

where $\dot{W}(v) = \dfrac{dW(v)}{dv} = i\int_{-\infty}^{\infty} d\xi \xi p(\xi)e^{i\xi v}$.

Changing the integration order, we can rewrite equalities (5.18) in the form

$$\Omega[t', t; v(\tau)]$$
$$= \int_{-\infty}^{\infty} d\xi \xi p(\xi)\int_0^{t'} d\tau g(t'-\tau)\exp\left\{i\xi\int_\tau^t d\tau_1 v(\tau_1)g(\tau_1-\tau)\right\} \quad (t' \leqslant t). \quad (5.19)$$

As a result, we obtain that correlations of Poisson's random process $z(t)$ with functionals of this process are described by the expression

$$\langle z(t')R[t; z(\tau)]\rangle$$
$$= \nu\int_{-\infty}^{\infty} d\xi \xi p(\xi)\int_0^{t'} d\tau' g(t'-\tau')\langle R[t; z(\tau)+\xi g(\tau-\tau')]\rangle \quad (t' \leqslant t). \quad (5.20)$$

As we mentioned earlier, random process $n(0, t)$ describing the number of jumps during temporal interval $(0, t)$ is the special case of Poisson's process. In this case, $p(\xi) = \delta(\xi - 1)$ and $g(t) = \theta(t)$, so that Eq. (5.20) assumes the extra-simple form

$$\langle n(0, t)R[t; n(0, \tau)]\rangle = \nu\int_0^{t'} d\tau \langle R[t; n(0, \tau)+\theta(t'-\tau)]\rangle \quad (t' \leqslant t). \qquad (5.21)$$

Equality (5.21) extends formula (4.25) for Poisson's random quantities to Poisson's random processes.

5.4 Telegrapher's Random Process

Now, we dwell on telegrapher's random process defined by formula (4.73), page 112, $-z(t) = a(-1)^{n(0,t)}$, where a is the deterministic quantity.

As we mentioned earlier, calculation of correlator $\langle z(t)R[t; z(\tau)]\rangle$ for $\tau \le t$ assumes the knowledge of the characteristic functional of process $z(t)$; unfortunately, the characteristic functional is unavailable in this case. We know only Eqs. (4.77) and (4.79), page 114, that describe the relationship between the functional

$$\Psi[t; v(\tau)] = \frac{1}{iv(t)}\frac{d}{dt}\Phi[t; v(\tau)] = \left\langle z(t)\exp\left\{i\int\limits_0^t d\tau z(\tau)v(\tau)\right\}\right\rangle$$

and the characteristic functional $\Phi[t; v(\tau)]$ in the form

$$\left\langle z(t)\exp\left\{i\int\limits_0^t d\tau z(\tau)v(\tau)\right\}\right\rangle$$

$$= ae^{-2\nu t} + ia^2\int\limits_0^t dt_1 e^{-2\nu(t-t_1)}v(t_1)\left\langle\exp\left\{i\int\limits_0^{t_1} d\tau z(\tau)v(\tau)\right\}\right\rangle.$$

This relationship is sufficient to split correlator $\langle z(t)R[t; z(\tau)]\rangle$, i.e., to express it immediately in terms of the average functional $\langle R[t; z(\tau)]\rangle$.

Indeed, the equality

$$\begin{aligned}\langle z(t)R[t; z(\tau)+\eta(\tau)]\rangle &= \left\langle z(t)\exp\left\{\int\limits_0^t d\tau z(\tau)\frac{\delta}{\delta\eta(\tau)}\right\}\right\rangle R[t; \eta(\tau)] \\ &= aR[t; \eta(\tau)]e^{-2\nu t} \\ &\quad + a^2\int\limits_0^t dt_1 e^{-2\nu(t-t_1)}\frac{\delta}{\delta\eta(t_1)}\langle R[t; z(\tau)\theta(t_1-\tau)+\eta(\tau)]\rangle,\end{aligned} \tag{5.22}$$

where $\eta(t)$ is arbitrary deterministic function, holds for any functional $R[t; z(\tau)]$ for $\tau \le t$. The final result is obtained by the limit process $\eta \to 0$:

$$\begin{aligned}\langle z(t)R[t; z(\tau)]\rangle &= aR[t; 0]e^{-2\nu t} \\ &\quad + a^2\int\limits_0^t dt_1 e^{-2\nu(t-t_1)}\left\langle\frac{\delta}{\delta z(t_1)}\widetilde{R}[t, t_1; z(\tau)]\right\rangle,\end{aligned} \tag{5.23}$$

where functional $\widetilde{R}[t, t_1; z(\tau)]$ is defined by the formula

$$\widetilde{R}[t, t_1; z(\tau)] = R[t; z(\tau)\theta(t_1 - \tau + 0)]. \tag{5.24}$$

In the case of random quantity a distributed according to probability density distribution (4.78), page 114, additional averaging (5.23) over a results in the equality

$$\langle z(t)R[t; z(\tau)]\rangle = a_0^2 \int_0^t dt_1 e^{-2\nu(t-t_1)} \left\langle \frac{\delta}{\delta z(t_1)} \widetilde{R}[t, t_1; z(\tau)] \right\rangle. \tag{5.25}$$

Formula (5.25) is very similar to the formula for splitting the correlator of the Gaussian process $z(t)$ characterized by the exponential correlation function (i.e., the Gaussian Markovian process) with functional $R[t; z(\tau)]$. The difference consists in the fact that the right-hand side of Eq. (5.25) depends on the functional that is cut by the process $z(\tau)$ rather than on functional $R[t; z(\tau)]$ itself.

Let us differentiate Eq. (5.23) with respect to time t. Taking into account the fact that there is no need in differentiating with respect to the upper limit of the integral in the right-hand side of Eq. (5.23), we obtain the expression [47]

$$\left(\frac{d}{dt} + 2\nu\right) \langle z(t)R[t; z(\tau)]\rangle = \left\langle z(t)\frac{d}{dt}R[t; z(\tau)]\right\rangle \tag{5.26}$$

usually called the *differentiation formula*.

One essential point should be noticed. Functional $R[t; z(\tau)]$ in differentiation formula (5.26) is an arbitrary functional and can simply coincide with process $z(t-0)$. In the general case, the realization of telegrapher's process is the generalized function. The derivative of this process is also the generalized function (the sequence of delta-functions), so that

$$z(t)\frac{d}{dt}z(t) \neq \frac{1}{2}\frac{d}{dt}z^2(t) \equiv 0$$

in the general case. These generalized functions, as any generalized functions, are defined only in terms of functionals constructed on them. In the case of our interest, such functionals are the average quantities denoted by angle brackets $\langle \cdots \rangle$, and the above differentiation formula describes the differential constraint between different functionals related to random process $z(t)$ and its one-sided derivatives for $t \to t-0$, such as dz/dt, d^2z/dt^2. For example, formula (5.26) allows derivation of equalities, such as

$$\left.\left\langle z(t)\frac{d}{d\tau}z(\tau)\right\rangle\right|_{\tau=t-0} = 2\nu\left\langle z^2\right\rangle, \quad \left.\left\langle z(t)\frac{d^2}{d\tau^2}z(\tau)\right\rangle\right|_{\tau=t-0} = 4\nu^2\left\langle z^2\right\rangle.$$

It is clear that these formulas can be obtained immediately by differentiating the correlation function $\langle z(t)z(t')\rangle$ with respect to t' $(t' < t)$ followed by limit process $t' \to t-0$.

Now, we dwell on some extensions of the above formulas for the vector Markovian process $z(t) = \{z_1(t), \ldots, z_N(t)\}$. For example, all above Markovian processes having the exponential correlation function

$$\langle z(t)z(t+\tau)\rangle = \left\langle z^2\right\rangle e^{-\alpha|\tau|} \tag{5.27}$$

satisfy the equality

$$\left(\frac{\partial}{\partial t} + \alpha k\right)\langle z_1(t)\ldots z_k(t)R[t; z(\tau)]\rangle = \left\langle z_1(t)\ldots z_k(t)\frac{\partial}{\partial t}R[t; z(\tau)]\right\rangle. \tag{5.28}$$

Formula (5.28) defines the rule of factoring the operation of differentiating with respect to time out of angle brackets of averaging; in particular, we have

$$\left\langle z_1(t)\ldots z_k(t)\frac{\partial^n}{\partial t^n}R[t; z(\tau)]\right\rangle = \left(\frac{\partial}{\partial t} + \alpha k\right)^n \langle z_1(t)\ldots z_k(t)R[t; z(\tau)]\rangle, \tag{5.29}$$

where $k = 1, \ldots, N$.

5.5 Delta-Correlated Random Processes

Random processes $z(t)$ that can be treated as delta-correlated processes are of special interest in physics. The importance of this approximation follows first of all from the fact that it is physically obvious in the context of many physical problems and the corresponding dynamic systems allow obtaining closed equations for probability densities of the solutions to these systems.

For the Gaussian delta-correlated (in time) process, correlation function has the form

$$B(t_1, t_2) = \langle z(t_1)z(t_2)\rangle = B(t_1)\delta(t_1 - t_2), \quad (\langle z(t)\rangle = 0).$$

In this case, functionals $\Theta[t; v(\tau)]$, $\Omega[t', t; v(\tau)]$ and $\Omega[t, t; v(\tau)]$ introduced by Eqs. (5.14), (5.15) are

$$\Theta[t; v(\tau)] = -\frac{1}{2}\int_0^t d\tau B(\tau)v^2(\tau),$$

$$\Omega[t', t; v(\tau)] = iB(t')v(t'), \quad \Omega[t, t; v(\tau)] = \frac{i}{2}B(t)v(t),$$

and Eqs. (5.16) assume significantly simpler forms

$$\begin{aligned} \langle z(t')R[t; v(\tau)]\rangle &= B(t')\left\langle \frac{\delta}{\delta z(t')}R[t; v(\tau)]\right\rangle \quad (0 < t' < t), \\ \langle z(t)R[t; v(\tau)]\rangle &= \frac{1}{2}B(t)\left\langle \frac{\delta}{\delta z(t)}R[t; v(\tau)]\right\rangle. \end{aligned} \tag{5.30}$$

Formulas (5.30) show that statistical averages of the Gaussian delta-correlated process considered here are discontinuous at $t' = t$. This discontinuity is dictated entirely by the fact that this process is delta-correlated; if the process at hand is not delta-correlated, no discontinuity occurs (see Eq. (5.16)).

Poisson's delta-correlated random process corresponds to the limit process

$$g(t) \to \delta(t).$$

In this case, the logarithm of the characteristic functional has the simple form (see Eq. (4.71)); as a consequence, Eqs. (5.19) for functionals $\Omega[t', t; v(\tau)]$ and $\Omega[t, t; v(\tau)]$ assume the forms

$$\Omega[t', t; v(\tau)] = i \int\limits_{-\infty}^{\infty} d\xi \xi p(\xi) e^{i\xi v(t')} \quad (t' < t),$$

$$\Omega[t, t; v(\tau)] = \frac{\nu}{iv(t)} \int\limits_{-\infty}^{\infty} d\xi \xi p(\xi) \left[e^{i\xi v(t)} - 1 \right] = \nu \int\limits_{-\infty}^{\infty} d\xi p(\xi) \int\limits_{0}^{\xi} d\eta e^{i\eta v(t)},$$

and we obtain the following expression for the correlation of Poisson random process $z(t)$ with a functional of this process

$$\begin{aligned} \left\langle z(t')R[t; z(\tau)] \right\rangle &= \nu \int\limits_{-\infty}^{\infty} d\xi \xi p(\xi) \left\langle R[t; z(\tau) + \xi\delta(\tau - \tau')] \right\rangle \quad (t' \leqslant t), \\ \langle z(t)R[t; z(\tau)] \rangle &= \nu \int\limits_{-\infty}^{\infty} d\xi p(\xi) \int\limits_{0}^{\xi} d\eta \, \langle R[t; z(\tau) + \eta\delta(t - \tau)] \rangle. \end{aligned} \tag{5.31}$$

These expressions also show that statistical averages are discontinuous at $t' = t$. As in the case of the Gaussian process, this discontinuity is dictated entirely by the fact that this process is delta-correlated.

In the general case of delta-correlated process $z(t)$, we can expand the logarithm of characteristic functional in the functional Taylor series

$$\Theta[t; v(\tau)] = \sum_{n=1}^{\infty} \frac{i^n}{n!} \int\limits_{0}^{t} d\tau K_n(\tau) v^n(\tau), \tag{5.32}$$

where cumulant functions assume the form

$$K_n(t_1, \ldots, t_n) = K_n(t_1)\delta(t_1 - t_2) \ldots \delta(t_{n-1} - t_n).$$

As can be seen from Eq. (5.32), a characteristic feature of these processes consists in the validity of the equality

$$\dot{\Theta}[t; v(\tau)] = \dot{\Theta}[t; v(t)] \quad \left(\dot{\Theta}[t; v(\tau)] = \frac{d}{dt}\Theta[t; v(\tau)]\right), \tag{5.33}$$

which is of fundamental significance. This equality shows that, in the case of arbitrary delta-correlated process, quantity $\Theta[t; v(\tau)]$ appears not a functional, but simply a function of time t. In this case, functionals $\Omega[t', t; v(\tau)]$ and $\Omega[t, t; v(\tau)]$ are

$$\Omega[t', t; v(\tau)] = \sum_{n=0}^{\infty} \frac{i^n}{n!} K_{n+1}(t') v^n(t') \quad (t' < t),$$

$$\Omega[t, t; v(\tau)] = \sum_{n=0}^{\infty} \frac{i^n}{(n+1)!} K_{n+1}(t) v^n(t),$$

and formulas for correlation splitting assume the forms

$$\begin{aligned} \left\langle z(t') R[t; z(\tau)] \right\rangle &= \sum_{n=0}^{\infty} \frac{i^n}{n!} K_{n+1}(t') \left\langle \frac{\delta^n R[t; z(\tau)]}{\delta z^n(t')} \right\rangle \quad (t' < t), \\ \left\langle z(t) R[t; z(\tau)] \right\rangle &= \sum_{n=0}^{\infty} \frac{i^n}{(n+1)!} K_{n+1}(t) \left\langle \frac{\delta^n R[t; z(\tau)]}{\delta z^n(t)} \right\rangle. \end{aligned} \tag{5.34}$$

These formulas describe the discontinuity of statistical averages at $t' = t$ in the general case of delta-correlated processes.

Note that, for $t' > t$, delta-correlated processes satisfy the obvious equality

$$\left\langle z(t') R[t; z(\tau)] \right\rangle = \left\langle z(t') \right\rangle \left\langle R[t; z(\tau)] \right\rangle. \tag{5.35}$$

Now, we dwell on the concept of random fields delta-correlated in time.

We will deal with vector field $\boldsymbol{f}(\boldsymbol{x}, t)$, where $\boldsymbol{x}$ describes the spatial coordinates and t is the temporal coordinate. In this case, the logarithm of characteristic functional can be expanded in the Taylor series with coefficients expressed in terms of cumulant functions of random field $\boldsymbol{f}(\boldsymbol{x}, t)$ (see Secttion 4.2). In the special case of cumulant functions

$$\begin{aligned} &K_n^{i_1,\ldots,i_m}(\boldsymbol{x}_1, t_1; \ldots, \boldsymbol{x}_n, t_n) \\ &\qquad = K_n^{i_1,\ldots,i_m}(\boldsymbol{x}_1, \ldots, \boldsymbol{x}_n; t_1)\delta(t_1 - t_2)\ldots\delta(t_{n-1} - t_n), \end{aligned} \tag{5.36}$$

we will call field $\boldsymbol{f}(\boldsymbol{x}, t)$ the random field delta-correlated in time t. In this case, functional $\Theta[t; \boldsymbol{\psi}(\boldsymbol{x}', \tau)]$ assumes the form

$$\Theta[t; \boldsymbol{\psi}(\boldsymbol{x}', \tau)] = \sum_{n=1}^{\infty} \frac{i^n}{n!} \int_0^t d\tau \int \ldots \int d\boldsymbol{x}_1 \ldots d\boldsymbol{x}_n K_n^{i_1,\ldots,i_m}(\boldsymbol{x}_1, \ldots, \boldsymbol{x}_n; \tau)$$
$$\times \psi_{i_1}(\boldsymbol{x}_1, \tau) \ldots \psi_{im}(\boldsymbol{x}_n, \tau), \tag{5.37}$$

An important feature of this functional is the fact that it satisfies the equality similar to Eq. (5.33):

$$\dot{\Theta}[t; \boldsymbol{\psi}(\boldsymbol{x}', \tau)] = \dot{\Theta}[t; \boldsymbol{\psi}(\boldsymbol{x}', t)]. \tag{5.38}$$

5.5.1 Asymptotic Meaning of Delta-Correlated Processes and Fields

The nature knows no delta-correlated processes. All actual processes and fields are characterized by a finite temporal correlation radius, and delta-correlated processes and fields result from asymptotic expansions in terms of their temporal correlation radii.

We illustrate the appearance of delta-correlated processes using the stationary Gaussian process with correlation radius τ_0 as an example. In this case, the logarithm of the characteristic functional is described by the expression

$$\Theta[t; v(\tau)] = -\int_0^t d\tau_1 v(\tau_1) \int_0^{\tau_1} d\tau_2 B\left(\frac{\tau_1 - \tau_2}{\tau_0}\right) v(\tau_2). \tag{5.39}$$

Setting $\tau_1 - \tau_2 = \xi\tau_0$, we transform Eq. (5.39) to the form

$$\Theta[t; v(\tau)] = -\tau_0 \int_0^t d\tau_1 v(\tau_1) \int_0^{\tau_1/\tau_0} d\xi B(\xi) v(\tau_1 - \xi\tau_0).$$

Assume now that $\tau_0 \to 0$. In this case, the leading term of the asymptotic expansion in parameter τ_0 is given by the formula

$$\Theta[t; v(\tau)] = -\tau_0 \int_0^\infty d\xi B(\xi) \int_0^t d\tau_1 v^2(\tau_1)$$

that can be represented in the form

$$\Theta[t; v(\tau)] = -B^{\text{eff}} \int_0^t d\tau_1 v^2(\tau_1), \tag{5.40}$$

where

$$B^{\text{eff}} = \int_0^\infty d\tau B\left(\frac{\tau}{\tau_0}\right) = \frac{1}{2}\int_{-\infty}^{\infty} d\tau B\left(\frac{\tau}{\tau_0}\right). \tag{5.41}$$

Certainly, asymptotic expression (5.40) holds only for the functions $v(t)$ that vary during times about τ_0 only slightly, rather than for arbitrary functions $v(t)$. Indeed, if we specify this function as $v(t) = v\delta(t - t_0)$, asymptotic expression (5.40) appears invalid; in this case, we must replace Eq. (5.39) with the expression

$$\Theta[t; v(\tau)] = -\frac{1}{2}B(0)v^2 \quad (t > t_0)$$

corresponding to the characteristic function of process $z(t)$ at a fixed time $t = t_0$.

Consider now correlation $\langle z(t)R[t; z(\tau)]\rangle$ given, according to the Furutsu–Novikov formula (5.16), page 126, by the expression

$$\langle z(t)R[t; z(\tau)]\rangle = \int_0^t dt_1 B\left(\frac{t - t_1}{\tau_0}\right)\left\langle \frac{\delta}{\delta z(t_1)} R[t; z(\tau)]\right\rangle.$$

The change of integration variable $t - t_1 \to \xi\tau_0$ transforms this expression to the form

$$\langle z(t)R[t; z(\tau)]\rangle = \tau_0 \int_0^{t/\tau_0} d\xi B(\xi)\left\langle \frac{\delta}{\delta z(t - \xi\tau_0)} R[t; z(\tau)]\right\rangle, \tag{5.42}$$

which grades for $\tau_0 \to 0$ into the equality obtained earlier for the Gaussian delta-correlated process

$$\langle z(t)R[t; z(\tau)]\rangle = B^{\text{eff}}\left\langle \frac{\delta}{\delta z(t)} R[t; z(\tau)]\right\rangle$$

provided that the variational derivative in Eq. (5.42) varies only slightly during times of about τ_0.

Thus, the approximation of process $z(t)$ by the delta-correlated one is conditioned by small variations of functionals of this process during times approximating the process temporal correlation radius.

Consider now telegrapher's and generalized telegrapher's processes. In the case of telegrapher's process, the characteristic functional satisfies Eq. (4.79), page 114. The correlation radius of this process is $\tau_0 = 1/2\nu$, and, for $\nu \to \infty$ ($\tau_0 \to 0$), this equation grades for sufficiently smooth functions $v(t)$ into the equation

$$\frac{d}{dt}\Phi[t; v(\tau)] = -\frac{a_0^2}{2\nu}v^2(t)\Phi[t; v(\tau)], \tag{5.43}$$

which corresponds to the Gaussian delta-correlated process. If we additionally assume that $a_0^2 \to \infty$ and

$$\lim_{\nu\to\infty} a_0^2/2\nu = \sigma_0^2,$$

then Eq. (5.43) appears independent of parameter ν. Of course, this fact does not mean that telegrapher's process loses its telegrapher's properties for $\nu \to \infty$. Indeed, for $\nu \to \infty$, the one-time probability distribution of process $z(t)$ will as before correspond to telegrapher's process, i.e., to the process with two possible states. As regards the correlation function and higher-order moment functions, they will possess for $\nu \to \infty$ all properties of delta-functions in view of the fact that

$$\lim_{\nu\to\infty} \left\{2\nu e^{-2\nu|\tau|}\right\} = \begin{cases} 0, & \text{if } \tau \neq 0, \\ \infty, & \text{if } \tau = 0. \end{cases}$$

Such functions should be considered the generalized functions; their delta-functional behavior will manifest itself in the integrals of them. As Eq. (5.43) shows, limit process $\nu \to \infty$ is equivalent for these quantities to the replacement of process $z(t)$ by the Gaussian delta-correlated process. This situation is completely similar to the approximation of the Gaussian random process with a finite correlation radius τ_0 by the delta-correlated process for $\tau_0 \to 0$.

Problems

Problem 5.1

Split the correlator $\langle g(\xi)f(\xi)\rangle_\xi$, where ξ is the random quantity.

Solution

$$\langle g(\xi+\eta_1)f(\xi+\eta_2)\rangle_\xi$$
$$= \exp\left\{\Theta\left(\frac{d}{id\eta_1}+\frac{d}{id\eta_2}\right) - \Theta\left(\frac{d}{id\eta_1}\right) - \Theta\left(\frac{d}{id\eta_2}\right)\right\}\langle g(\xi+\eta_1)\rangle\,\langle f(\xi+\eta_2)\rangle.$$

Problem 5.2

Split the correlator $\left\langle e^{\omega\xi}f(\xi)\right\rangle_\xi$, where parameter ω can assume complex values.

Solution

$$\left\langle e^{\omega\xi}f(\xi+\eta)\right\rangle_\xi = \exp\left\{\Theta\left(\frac{1}{i}\left(\omega+\frac{d}{d\eta}\right)\right) - \Theta\left(\frac{d}{id\eta}\right)\right\}\langle f(\xi+\eta)\rangle_\xi.$$

Problem 5.3

Split the correlator $\left\langle e^{\omega\xi}\right\rangle_\xi$ and $\left\langle \xi e^{\omega\xi}\right\rangle_\xi$ assuming that ξ is the Gaussian random quantity with zero-valued mean.

Solution

$$\left\langle e^{\omega\xi}\right\rangle_\xi = \exp\left\{\frac{\omega^2\sigma^2}{2}\right\}, \quad \left\langle \xi e^{\omega\xi}\right\rangle_\xi = \omega\sigma^2\exp\left\{\frac{\omega^2\sigma^2}{2}\right\}.$$

Problem 5.4

Split the correlator $\left\langle e^{\omega\xi}f(\xi)\right\rangle_\xi$ assuming that ξ is the Gaussian random quantity with zero-valued mean.

Solution

$$\left\langle e^{\omega\xi}f(\xi)\right\rangle_\xi = \exp\left\{\frac{\omega^2\sigma^2}{2}\right\}\left\langle f(z+\omega\sigma^2)\right\rangle.$$

Problem 5.5

Split the correlator $\langle F[z(\tau)]R[z(\tau)]\rangle$.

Solution

$$\begin{aligned}&\langle F[z(\tau)+\eta_1(\tau)]R[z(\tau)+\eta_2(\tau)]\rangle\\ &= \left\langle \exp\left\{\Theta\left[\frac{1}{i}\left(\frac{\delta}{\delta\eta_1(\tau)}+\frac{\delta}{\delta\eta_2(\tau)}\right)\right]-\Theta\left[\frac{1}{i}\frac{\delta}{\delta\eta_1(\tau)}\right]-\Theta\left[\frac{1}{i}\frac{\delta}{\delta\eta_2(\tau)}\right]\right\}\right\rangle\\ &\quad\times\langle F[z(\tau)+\eta_1(\tau)]\rangle\,\langle R[z(\tau)+\eta_2(\tau)]\rangle,\end{aligned}$$

where $\eta_1(\tau)$ and $\eta_2(\tau)$ are arbitrary deterministic functions.

Problem 5.6

Split the correlator $\langle F[z(\tau)]R[z(\tau)]\rangle$ assuming that $z(\tau)$ is the Gaussian random process.

Solution

$$\begin{aligned}&\langle F[z(\tau)+\eta_1(\tau)]R[z(\tau)+\eta_2(\tau)]\rangle\\ &= \exp\left\{\int\limits_{-\infty}^{\infty}\int\limits_{-\infty}^{\infty}d\tau_1 d\tau_2 B(\tau_1,\tau_2)\frac{\delta}{\delta\eta_1(\tau_1)}\frac{\delta}{\delta\eta_2(\tau_2)}\right\}\langle F[z(\tau)+\eta_1(\tau)]\rangle\,\langle R[z(\tau)+\eta_2(\tau)]\rangle,\end{aligned}$$

or

$$\langle F[z(\tau)]R[z(\tau)+\eta(\tau)]\rangle = F\left[z(\tau)+\int\limits_{-\infty}^{\infty} d\tau_1 B(\tau,\tau_1)\frac{\delta}{\delta\eta(\tau_1)}\right]\langle R[z(\tau)+\eta(\tau)]\rangle.$$

Problem 5.7

Split the correlator $\left\langle e^{i\int_{-\infty}^{\infty} d\tau z(\tau)v(\tau)} R[z(\tau)]\right\rangle$ *assuming that* $z(\tau)$ *is the Gaussian random process.*

Solution

$$\left\langle e^{i\int_{-\infty}^{\infty} d\tau z(\tau)v(\tau)} R[z(\tau)]\right\rangle = \Phi[v(\tau)]\left\langle R\left[z(\tau)+i\int\limits_{-\infty}^{\infty} d\tau_1 B(\tau,\tau_1)v(\tau_1)\right]\right\rangle.$$

Problem 5.8

Split the correlator $\left\langle e^{i\int_0^t d\tau z(\tau)v(\tau)} R[t; z(\tau)]\right\rangle$ *assuming that* $z(\tau)$ *is the Gaussian random process.*

Solution

$$\left\langle e^{i\int_0^t d\tau z(\tau)v(\tau)} R[t; z(\tau)]\right\rangle = \Phi[t; v(\tau)]\left\langle R\left[t; z(\tau)+i\int\limits_0^t d\tau_1 B(\tau,\tau_1)v(\tau_1)\right]\right\rangle,$$

where $\Phi[t; v(\tau)]$ is the characteristic functional of the Gaussian random process $z(t)$.

Problem 5.9

Calculate the characteristic functional of process $z(t)=\xi^2(t)$, *where* $\xi(t)$ *is the Gaussian random process with the parameters*

$$\langle\xi(t)\rangle = 0, \quad \langle\xi(t_1)\xi(t_2)\rangle = B(t_1-t_2).$$

Instruction *The characteristic functional*

$$\Phi[t; v(\tau)] = \langle\varphi[t;\xi(\tau)]\rangle, \quad \textit{where} \quad \varphi[t;\xi(\tau)] = \exp\left\{i\int\limits_0^t d\tau v(\tau)\xi^2(\tau)\right\}$$

satisfies the stochastic equation

$$\frac{d}{dt}\Phi[t; v(\tau)] = iv(t)\left\langle \xi^2(t)\varphi[t; \xi(\tau)]\right\rangle.$$

One needs to obtain the integral equation for quantity

$$\Psi(t_1, t) = \langle \xi(t_1)\xi(t)\varphi[t; \xi(\tau)]\rangle,$$

$$\Psi(t_1, t) = B(t_1 - t)\Phi[t; v(\tau)] + 2i \int_0^t d\tau B(t_1 - \tau)v(\tau)\Psi(\tau, t),$$

and represent function $\Psi(t_1, t)$ *in the form*

$$\Psi(t_1, t) = S(t_1, t)\Phi[t; v(\tau)].$$

Solution The characteristic functional $\Phi[t; v(\tau)]$ has the form

$$\Phi[t; v(\tau)] = \exp\left\{ i \int_0^t d\tau v(\tau)S(\tau, \tau) \right\},$$

where

$$S(t, t) = \sum_{n=0}^{\infty} S^{(n)}(t, t),$$

$$S^{(n)}(t, t) = (2i)^n \int_0^t d\tau_1 \ldots \int_0^t d\tau_n\, v(\tau_1)\ldots v(\tau_n)B(t - \tau_1)B(\tau_1 - \tau_2)\ldots B(\tau_n - t).$$

Consequently, the n-th cumulant of process $z(t) = \xi^2(t)$ is expressed in terms of quantity $S^{(n-1)}(t, t)$.

Problem 5.10

Calculate the characteristic functional of process $z(t) = \xi^2(t)$, *where* $\xi(t)$ *is the Gaussian Markovian process with the parameters*

$$\langle \xi(t)\rangle = 0, \quad \langle \xi(t_1)\xi(t_2)\rangle = B(t_1 - t_2) = \sigma^2 \exp\{-\alpha|t_1 - t_2|\}.$$

Solution The characteristic functional $\Phi[t; v(\tau)]$ has the form

$$\Phi[t; v(\tau)] = \exp\left\{ i \int_0^t d\tau v(\tau)S(\tau, \tau) \right\},$$

where function $S(t, t)$ can be described in terms of the initial-value problem

$$\frac{d}{dt}S(t,t) = -2\alpha\left[S(t,t) - \sigma^2\right] + 2iv(t)S^2(t,t), \qquad S(t,t)|_{t=0} = \sigma^2.$$

Problem 5.11

Show that the expression

$$\langle z(t_1)z(t_2)R[z(\tau)]\rangle = \langle z(t_1)z(t_2)\rangle \langle R[z(\tau)]\rangle,$$

holds for arbitrary functional $R[z(\tau)]$ *and* $\tau \le t_2 \le t_1$ *under the condition that* $z(t)$ *is telegrapher's process* [48].

Instruction *Expand functional* $R[z(\tau)]$ *in the Taylor series in* $z(\tau)$ *and use formulas Eq. (4.75).*

Problem 5.12

Show that the expression [48]

$$\begin{aligned}&\langle F[z(\tau_1)]z(t_1)z(t_2)R[z(\tau_2)]\rangle \\ &\quad = \langle F[z(\tau_1)]\rangle \langle z(t_1)z(t_2)\rangle \langle R[z(\tau_2)]\rangle + \langle F[z(\tau_1)]z(t_1)\rangle \langle z(t_2)R[z(\tau_2)]\rangle,\end{aligned}$$

holds for any $\tau_1 \le t_1 \le t_2 \le \tau_2$ *and arbitrary functionals* $F[z(\tau_1)]$ *and* $R[z(\tau_2)]$ *under the condition that* $z(t)$ *is telegrapher's process with random parameter* a *distributed according to the probability density*

$$p(a) = \frac{1}{2}\left[\delta(a - a_0) + \delta(a + a_0)\right].$$

Instruction *In terms of the Taylor expansion in* $z(\tau)$, *functional* $R[z(\tau_2)]$ *can be represented in the form*

$$R[z(\tau_2)] = \sum_{2k} + \sum_{2k+1},$$

where the first sum consists of terms with the even number of co-factors $z(\tau)$ *and the second sum consists of terms with the odd number of such co-factors. The desired result immediately follows from the fact that*

$$\langle R[z(\tau_2)]\rangle = \left\langle \sum_{2k} \right\rangle, \qquad \langle z(t_2)R[z(\tau_2)]\rangle = \left\langle z(t_2)\sum_{2k+1} \right\rangle.$$

Problem 5.13

Show that the equality

$$\left\langle z(t')R[t; z(\tau)]\right\rangle = e^{-2\nu(t'-t)} \langle z(t)R[t; z(\tau)]\rangle \quad (t' \geqslant t)$$

holds for telegrapher's process $z(t)$.

Instruction *Rewrite the correlator $\left\langle z(t')R[t; z(\tau)]\right\rangle$ $(t' \geq t, \tau < t)$ in the form*

$$\left\langle z(t')R[t; z(\tau)]\right\rangle = \frac{1}{a_0^2}\left\langle z(t')z(t)z(t)R[t; z(\tau)]\right\rangle.$$

Lecture 6

General Approaches to Analyzing Stochastic Systems

In this Lecture, we will consider basic methods of determining statistical characteristics of solutions to the stochastic equations.

Consider a linear (differential, integro-differential, or integral) stochastic equation. Averaging of such an equation over an ensemble of realizations of fluctuating parameters will not result generally in a closed equation for the corresponding average value. To obtain a closed equation, we must deal with an additional extended space whose dimension appears to be infinite in most cases. This approach makes it possible to derive the linear equation for average quantity of interest, but this equation will contain variational derivatives.

Consider some special types of dynamic systems.

6.1 Ordinary Differential Equations

Assume dynamics of vector function $\boldsymbol{x}(t)$ is described by the ordinary differential equation

$$\frac{d}{dt}\boldsymbol{x}(t) = \boldsymbol{v}(\boldsymbol{x}, t) + \boldsymbol{f}(\boldsymbol{x}, t), \quad \boldsymbol{x}(t_0) = \boldsymbol{x}_0. \tag{6.1}$$

Here, function $\boldsymbol{v}(\boldsymbol{x}, t)$ is the deterministic function and $\boldsymbol{f}(\boldsymbol{x}, t)$ is the random function.

The solution to Eq. (6.1) is a functional of $\boldsymbol{f}(\boldsymbol{y}, \tau) + \boldsymbol{v}(\boldsymbol{y}, \tau)$ with $\tau \in (t_0, t)$, i.e.,

$$\boldsymbol{x}(t) = \boldsymbol{x}[t; \boldsymbol{f}(\boldsymbol{y}, \tau) + \boldsymbol{v}(\boldsymbol{y}, \tau)].$$

From this fact follows the equality

$$\frac{\delta}{\delta f_j(\boldsymbol{y}, \tau)} F(\boldsymbol{x}(t)) = \frac{\delta}{\delta v_j(\boldsymbol{y}, \tau)} F(\boldsymbol{x}(t)) = \frac{\partial F(\boldsymbol{x}(t))}{\partial x_i} \frac{\delta x_i(t)}{\delta f_j(\boldsymbol{y}, \tau)}$$

valid for arbitrary function $F(\boldsymbol{x})$. In addition, we have

$$\frac{\delta}{\delta f_j(\boldsymbol{y}, t-0)} x_i(t) = \frac{\delta}{\delta v_j(\boldsymbol{y}, t-0)} x_i(t) = \delta_{ij} \delta\left(\boldsymbol{x}(t) - \boldsymbol{y}\right).$$

Lectures on Dynamics of Stochastic Systems. DOI: 10.1016/B978-0-12-384966-3.00006-1

The corresponding Liouville equation for the indicator function

$$\varphi(\boldsymbol{x}, t) = \delta(\boldsymbol{x}(t) - \boldsymbol{x})$$

follows from Eq. (6.1) and has the form

$$\frac{\partial}{\partial t}\varphi(\boldsymbol{x}, t) = -\frac{\partial}{\partial \boldsymbol{x}}\left\{[\boldsymbol{v}(\boldsymbol{x}, t) + \boldsymbol{f}(\boldsymbol{x}, t)]\,\varphi(\boldsymbol{x}, t)\right\}, \quad \varphi(\boldsymbol{x}, t_0) = \delta(\boldsymbol{x} - \boldsymbol{x}_0), \tag{6.2}$$

from which follows the equality

$$\frac{\delta}{\delta \boldsymbol{f}(\boldsymbol{y}, t-0)}\varphi(\boldsymbol{x}, t) = \frac{\delta}{\delta \boldsymbol{v}(\boldsymbol{y}, t-0)}\varphi(\boldsymbol{x}, t) = -\frac{\partial}{\partial \boldsymbol{x}}\left\{\delta(\boldsymbol{x} - \boldsymbol{y})\varphi(\boldsymbol{x}, t)\right\}. \tag{6.3}$$

Using this equality, we can rewrite Eq. (6.2) in the form, which may look at first sight more complicated

$$\left(\frac{\partial}{\partial t} + \frac{\partial}{\partial \boldsymbol{x}}\boldsymbol{v}(\boldsymbol{x}, t)\right)\varphi(\boldsymbol{x}, t) = \int d\boldsymbol{y}\boldsymbol{f}(\boldsymbol{y}, t)\frac{\delta}{\delta \boldsymbol{v}(\boldsymbol{y}, t)}\varphi(\boldsymbol{x}, t). \tag{6.4}$$

Consider now the one-time probability density for solution $\boldsymbol{x}(t)$ of Eq. (6.1)

$$P(\boldsymbol{x}, t) = \langle \varphi(\boldsymbol{x}, t)\rangle = \langle \delta(\boldsymbol{x}(t) - \boldsymbol{x})\rangle.$$

Here, $\boldsymbol{x}(t)$ is the solution of Eq. (6.1) corresponding to a particular realization of random field $\boldsymbol{f}(\boldsymbol{x}, t)$, and angle brackets $\langle \cdots \rangle$ denote averaging over an ensemble of realizations of field $\boldsymbol{f}(\boldsymbol{x}, t)$.

Averaging Eq. (6.4) over an ensemble of realizations of field $\boldsymbol{f}(\boldsymbol{x}, t)$, we obtain the expression

$$\left(\frac{\partial}{\partial t} + \frac{\partial}{\partial \boldsymbol{x}}\boldsymbol{v}(\boldsymbol{x}, t)\right)P(\boldsymbol{x}, t) = \int d\boldsymbol{y}\frac{\delta}{\delta \boldsymbol{v}(\boldsymbol{y}, t)}\langle \boldsymbol{f}(\boldsymbol{y}, t)\varphi(\boldsymbol{x}, t)\rangle. \tag{6.5}$$

Quantity $\langle \boldsymbol{f}(\boldsymbol{y}, t)\varphi(\boldsymbol{x}, t)\rangle$ in the right-hand side of Eq. (6.5) is the correlator of random field $\boldsymbol{f}(\boldsymbol{y}, t)$ with function $\varphi(\boldsymbol{x}, t)$, which is a functional of random field $\boldsymbol{f}(\boldsymbol{y}, \tau)$ and is given either by Eq. (6.2), or by Eq. (6.4).

The characteristic functional

$$\Phi[t, t_0; \boldsymbol{u}(\boldsymbol{y}, \tau)] = \left\langle \exp\left\{i\int_{t_0}^{t} d\tau \int d\boldsymbol{y}\boldsymbol{f}(\boldsymbol{y}, \tau)\boldsymbol{u}(\boldsymbol{y}, \tau)\right\}\right\rangle$$
$$= \exp\left\{\Theta[t, t_0; \boldsymbol{u}(\boldsymbol{y}, \tau)]\right\}$$

exhaustively describes all statistical characteristics of random field $\boldsymbol{f}(\boldsymbol{y}, \tau)$ for $\tau \in (t_0, t)$.

We split correlator $\langle \boldsymbol{f}(\boldsymbol{y},t)\varphi(\boldsymbol{x},t)\rangle$ using the technique of functionals. Introducing functional shift operator with respect to field $\boldsymbol{v}(\boldsymbol{y},\tau)$, we represent functional $\varphi[\boldsymbol{x},t;\boldsymbol{f}(\boldsymbol{y},\tau)+\boldsymbol{v}(\boldsymbol{y},\tau)]$ in the operator form

$$\varphi[\boldsymbol{x},t;\boldsymbol{f}(\boldsymbol{y},\tau)+\boldsymbol{v}(\boldsymbol{y},\tau)]$$
$$=\exp\left[\int\limits_{t_0}^{t} d\tau \int d\boldsymbol{y}\boldsymbol{f}(\boldsymbol{y},\tau)\frac{\delta}{\delta \boldsymbol{v}(\boldsymbol{y},\tau)}\right]\varphi[\boldsymbol{x},t;\boldsymbol{v}(\boldsymbol{y},\tau)].$$

With this representation, the term in the right-hand side of Eq. (6.5) assumes the form

$$\int d\boldsymbol{y}\frac{\delta}{\delta v_j(\boldsymbol{y},t)}\frac{\left\langle f_j(\boldsymbol{y},t)\exp\left\{\int\limits_{t_0}^{t} d\tau \int d\boldsymbol{y}'\boldsymbol{f}(\boldsymbol{y}',\tau)\frac{\delta}{\delta \boldsymbol{v}(\boldsymbol{y}',\tau)}\right\}\right\rangle}{\left\langle \exp\left\{\int\limits_{t_0}^{t} d\tau \int d\boldsymbol{y}'\boldsymbol{f}(\boldsymbol{y}',\tau)\frac{\delta}{\delta \boldsymbol{v}(\boldsymbol{y}',\tau)}\right\}\right\rangle}P(\boldsymbol{x},t)$$
$$=\dot{\Theta}_t\left[t,t_0;\frac{\delta}{i\delta \boldsymbol{v}(\boldsymbol{y},\tau)}\right]P(\boldsymbol{x},t),$$

where we introduced the functional

$$\dot{\Theta}_t[t,t_0;\boldsymbol{u}(\boldsymbol{y},\tau)]=\frac{d}{dt}\ln\Phi[t,t_0;\boldsymbol{u}(\boldsymbol{y},\tau)].$$

Consequently, we can rewrite Eq. (6.5) in the form

$$\left(\frac{\partial}{\partial t}+\frac{\partial}{\partial \boldsymbol{x}}\boldsymbol{v}(\boldsymbol{x},t)\right)P(\boldsymbol{x},t)=\dot{\Theta}_t\left[t,t_0;\frac{\delta}{i\delta \boldsymbol{v}(\boldsymbol{y},\tau)}\right]P(\boldsymbol{x},t). \tag{6.6}$$

Equation (6.6) is the closed equation with variational derivatives in the functional space of all possible functions $\{\boldsymbol{v}(\boldsymbol{y},\tau)\}$. However, for a fixed function $\boldsymbol{v}(\boldsymbol{x},t)$, we arrive at the unclosed equation

$$\left(\frac{\partial}{\partial t}+\frac{\partial}{\partial \boldsymbol{x}}\boldsymbol{v}(\boldsymbol{x},t)\right)P(\boldsymbol{x},t)=\left\langle\dot{\Theta}_t\left[t,t_0;\frac{\delta}{i\delta \boldsymbol{f}(\boldsymbol{y},t)}\right]\varphi[\boldsymbol{x},t;\boldsymbol{f}(\boldsymbol{y},\tau)]\right\rangle,$$
$$P(\boldsymbol{x},t_0)=\delta(\boldsymbol{x}-\boldsymbol{x}_0). \tag{6.7}$$

Equation (6.7) is the exact consequence of the initial dynamic equation (6.1). Statistical characteristics of random field $\boldsymbol{f}(\boldsymbol{x},t)$ appear in this equation only through functional $\dot{\Theta}_t[t,t_0;\boldsymbol{u}(\boldsymbol{y},\tau)]$ whose expansion in the functional Taylor series in powers of $\boldsymbol{u}(\boldsymbol{x},\tau)$ depends on all space-time cumulant functions of random field $\boldsymbol{f}(\boldsymbol{x},t)$.

As we mentioned earlier, Eq. (6.7) is unclosed in the general case with respect to function $P(\boldsymbol{x}, t)$, because quantity

$$\dot{\Theta}_t\left[t, t_0; \frac{\delta}{i\delta \boldsymbol{f}(\boldsymbol{y}, \tau)}\right]\delta(\boldsymbol{x}(t) - \boldsymbol{x})$$

appeared in averaging brackets depends on the solution $\boldsymbol{x}(t)$ (which is a functional of random field $\boldsymbol{f}(\boldsymbol{y}, \tau)$) for all times $t_0 < \tau < t$. However, in some cases, the variational derivative in Eq. (6.7) can be expressed in terms of ordinary differential operators. In such conditions, equations like Eq. (6.7) will be the closed equations in the corresponding probability densities. The corresponding examples will be given below.

Proceeding in a similar way, one can obtain the equation similar to Eq. (6.7) for the m-time probability density that refers to m different instants $t_1 < t_2 < \cdots < t_m$

$$P_m(\boldsymbol{x}_1, t_1; \ldots; \boldsymbol{x}_m, t_m) = \langle \varphi_m(\boldsymbol{x}_1, t_1; \ldots; \boldsymbol{x}_m, t_m) \rangle, \tag{6.8}$$

where the indicator function is defined by the equality

$$\varphi_m(\boldsymbol{x}_1, t_1; \ldots; \boldsymbol{x}_m, t_m) = \delta(\boldsymbol{x}(t_1) - \boldsymbol{x}_1) \ldots \delta(\boldsymbol{x}(t_m) - \boldsymbol{x}_m).$$

Differentiating Eq. (6.8) with respect to time t_m and using then dynamic equation (6.1), one can obtain the equation

$$\begin{aligned}&\left(\frac{\partial}{\partial t_m} + \frac{\partial}{\partial \boldsymbol{x}_m}\boldsymbol{v}(\boldsymbol{x}_m, t_m)\right) P_m(\boldsymbol{x}_1, t_1; \ldots; \boldsymbol{x}_m, t_m)\\ &\quad = \left\langle \dot{\Theta}_{t_m}\left[t_m, t_0; \frac{\delta}{i\delta \boldsymbol{f}(\boldsymbol{y}, \tau)}\right] \varphi_m(\boldsymbol{x}_1, t_1; \ldots; \boldsymbol{x}_m, t_m)\right\rangle.\end{aligned} \tag{6.9}$$

No summation over index m is performed here. The initial value to Eq. (6.9) can be derived from Eq. (6.8). Setting $t_m = t_{m-1}$ in Eq. (6.8), we obtain the equality

$$P_m(\boldsymbol{x}_1, t_1; \ldots; \boldsymbol{x}_m, t_{m-1}) = \delta(\boldsymbol{x}_m - \boldsymbol{x}_{m-1}) P_{m-1}(\boldsymbol{x}_1, t_1; \ldots; \boldsymbol{x}_{m-1}, t_{m-1}),$$

which just determines the initial value for Eq. (6.9).

6.2 Completely Solvable Stochastic Dynamic Systems

Consider now several dynamic systems that allow sufficiently adequate statistical analysis for arbitrary random parameters.

6.2.1 Ordinary Differential Equations

Multiplicative Action

As the first example, we consider the vector stochastic equation with initial value

$$\frac{d}{dt}\boldsymbol{x}(t) = z(t) g(t) \boldsymbol{F}(\boldsymbol{x}), \quad \boldsymbol{x}(0) = \boldsymbol{x}_0, \tag{6.10}$$

where $g(t)$ and $F_i(\boldsymbol{x})$, $i = 1, \ldots, N$, are the deterministic functions and $z(t)$ is the random process whose statistical characteristics are described by the characteristic functional

$$\Phi[t; v(\tau)] = \left\langle \exp\left\{ i \int_0^t d\tau z(\tau) v(\tau) \right\} \right\rangle = e^{\Theta[t; v(\tau)]}.$$

Equation (6.10) has a feature that offers a possibility of determining statistical characteristics of its solutions in the general case of arbitrarily distributed process $z(t)$. The point is that introduction of new 'random' time

$$T = \int_0^t d\tau z(\tau) g(\tau)$$

reduces Eq. (6.10) to the equation looking formally deterministic

$$\frac{d}{dT}\boldsymbol{x}(T) = \boldsymbol{F}(\boldsymbol{x}), \quad \boldsymbol{x}(0) = \boldsymbol{x}_0,$$

so that the solution to Eq. (6.10) has the following structure

$$\boldsymbol{x}(t) = \boldsymbol{x}(T) = \boldsymbol{x}\left(\int_0^t d\tau z(\tau) g(\tau) \right). \tag{6.11}$$

Varying Eq. (6.11) with respect to $z(\tau)$ and using Eq. (6.10), we obtain the equality

$$\frac{\delta}{\delta z(\tau)}\boldsymbol{x}(t) = g(\tau)\frac{d}{dT}\boldsymbol{x}(T) = g(\tau)\boldsymbol{F}(\boldsymbol{x}(t)). \tag{6.12}$$

Thus, the variational derivative of solution $\boldsymbol{x}(t)$ is expressed in terms of the same solution at the same time. This fact makes it possible to immediately write the closed equations for statistical characteristics of problem (6.10).

Let us derive the equation for the one-time probability density

$$P(\boldsymbol{x}, t) = \langle \delta(\boldsymbol{x}(t) - \boldsymbol{x}) \rangle .$$

It has the form

$$\frac{\partial}{\partial t}P(\boldsymbol{x}, t) = \left\langle \dot{\Theta}_t \left[t; \frac{\delta}{i\delta z(\tau)} \right] \delta\left(\boldsymbol{x}(t) - \boldsymbol{x}\right) \right\rangle. \tag{6.13}$$

Consider now the result of operator $\delta/\delta z(\tau)$ applied to the indicator function $\varphi(\boldsymbol{x}, t) = \delta(\boldsymbol{x}(t) - \boldsymbol{x})$. Using formula (6.12), we obtain the expression

$$\frac{\delta}{\delta z(\tau)}\delta\left(\boldsymbol{x}(t) - \boldsymbol{x}\right) = -g(\tau)\frac{\partial}{\partial \boldsymbol{x}}\left\{\boldsymbol{F}(\boldsymbol{x})\varphi(\boldsymbol{x}, t)\right\}.$$

Consequently, we can rewrite Eq. (6.13) in the form of the closed operator equation

$$\frac{\partial}{\partial t}P(\boldsymbol{x},t) = \dot{\Theta}_t\left[t; ig(\tau)\frac{\partial}{\partial \boldsymbol{x}}\boldsymbol{F}(\boldsymbol{x})\right]P(\boldsymbol{x},t), \quad P(\boldsymbol{x},0) = \delta(\boldsymbol{x}-\boldsymbol{x}_0), \tag{6.14}$$

whose particular form depends on the behavior of process $z(t)$.

For the two-time probability density

$$P(\boldsymbol{x},t;\boldsymbol{x}_1,t_1) = \langle \delta(\boldsymbol{x}(t)-\boldsymbol{x})\delta(\boldsymbol{x}(t_1)-\boldsymbol{x}_1)\rangle$$

we obtain similarly the equation (for $t > t_1$)

$$\frac{\partial}{\partial t}P(\boldsymbol{x},t;\boldsymbol{x}_1,t_1) = \dot{\Theta}_t\left[t; ig(\tau)\left\{\frac{\partial}{\partial \boldsymbol{x}}\boldsymbol{F}(\boldsymbol{x}) + \theta(t_1-\tau)\frac{\partial}{\partial \boldsymbol{x}_1}\boldsymbol{F}(\boldsymbol{x}_1)\right\}\right]P(\boldsymbol{x},t;\boldsymbol{x}_1,t_1) \tag{6.15}$$

with the initial value

$$P(\boldsymbol{x},t_1;\boldsymbol{x}_1,t_1) = \delta(\boldsymbol{x}-\boldsymbol{x}_1)P(\boldsymbol{x}_1,t_1),$$

where function $P(\boldsymbol{x}_1,t_1)$ satisfies Eq. (6.14).

One can see from Eq. (6.15) that multidimensional probability density cannot be factorized in terms of the transition probability (see Sect. 4.3), so that process $\boldsymbol{x}(t)$ is not the Markovian process. The particular forms of Eqs. (6.14) and (6.15) are governed by the statistics of process $z(t)$.

If $z(t)$ is the Gaussian process whose mean value and correlation function are

$$\langle z(t)\rangle = 0, \quad B(t,t') = \langle z(t)z(t')\rangle,$$

then functional $\Theta[t; v(\tau)]$ has the form

$$\Theta[t; v(\tau)] = -\frac{1}{2}\int_0^t dt_1 \int_0^t dt_2 B(t_1,t_2)v(t_1)v(t_2)$$

and Eq. (6.14) assumes the following form

$$\frac{\partial}{\partial t}P(\boldsymbol{x},t) = g(t)\int_0^t d\tau B(t,\tau)g(\tau)\frac{\partial}{\partial x_j}F_j(\boldsymbol{x})\frac{\partial}{\partial x_k}F_k(\boldsymbol{x})P(\boldsymbol{x},t) \tag{6.16}$$

and can be considered as the extended *Fokker–Planck equation.*

The class of problems formulated in terms of the system of equations

$$\frac{d}{dt}\boldsymbol{x}(t) = z(t)\boldsymbol{F}(\boldsymbol{x}) - \lambda\boldsymbol{x}(t), \quad \boldsymbol{x}(0) = \boldsymbol{x}_0, \tag{6.17}$$

where $\boldsymbol{F}(\boldsymbol{x})$ are the homogeneous polynomials of power k, can be reduced to problem (6.10). Indeed, introducing new functions

$$\boldsymbol{x}(t) = \widetilde{\boldsymbol{x}}(t)e^{-\lambda t},$$

we arrive at problem (6.10) with function $g(t) = e^{-\lambda(k-1)t}$. In the important special case with $k=2$ and functions $\boldsymbol{F}(\boldsymbol{x})$ such that $\boldsymbol{x}\boldsymbol{F}(\boldsymbol{x}) = 0$, the system of equations (6.10) describes hydrodynamic systems with linear friction (see, e.g., [24]). In this case, the interaction between the components appears to be random.

If $\lambda = 0$, energy conservation holds in hydrodynamic systems for any realization of process $z(t)$. For $t \to \infty$, there is the steady-state probability distribution $P(\boldsymbol{x})$, which is, under the assumption that no additional integrals of motion exist, the uniform distribution over the sphere $x_i^2 = E_0$. If additional integrals of motion exist (as it is the case for finite-dimensional approximation of the two-dimensional motion of liquid), the domain of the steady-state probability distribution will coincide with the phase space region allowed by the integrals of motion.

Note that, in the special case of the Gaussian process $z(t)$ appeared in the one-dimensional linear equation of type Eq. (6.17)

$$\frac{d}{dt}x(t) = -\lambda x(t) + z(t)x(t), \quad x(0) = 1,$$

which determines the simplest logarithmic-normal random process, we obtain, instead of Eq. (6.16), the extended Fokker–Planck equation

$$\left(\frac{\partial}{\partial t} + \lambda\frac{\partial}{\partial x}x\right)P(x,t) = \int_0^t d\tau B(t,\tau)\frac{\partial}{\partial x}x\frac{\partial}{\partial x}xP(x,t),$$
$$P(x,0) = \delta(x-1). \tag{6.18}$$

Additive Action

Consider now the class of linear equations

$$\frac{d}{dt}\boldsymbol{x}(t) = A(t)\boldsymbol{x}(t) + \boldsymbol{f}(t), \quad \boldsymbol{x}(0) = \boldsymbol{x}_0, \tag{6.19}$$

where $A(t)$ is the deterministic matrix and $\boldsymbol{f}(t)$ is the random vector function whose characteristic functional $\Phi[t; \boldsymbol{v}(\tau)]$ is known.

For the probability density of the solution to Eq. (6.19), we have

$$\frac{\partial}{\partial t}P(\boldsymbol{x},t) = -\frac{\partial}{\partial x_i}\left(A_{ik}(t)x_kP(\boldsymbol{x},t)\right) + \left\langle \dot{\Theta}_t\left[t; \frac{\delta}{i\delta\boldsymbol{f}(\tau)}\right]\delta\left(\boldsymbol{x}(t)-\boldsymbol{x}\right)\right\rangle. \tag{6.20}$$

In the problem under consideration, the variational derivative $\dfrac{\delta \boldsymbol{x}(t)}{\delta f(\tau)}$ also satisfies (for $\tau < t$) the linear equation with the initial value

$$\frac{d}{dt}\frac{\delta}{\delta f(\tau)}x_i(t) = A_{ik}(t)\frac{\delta}{\delta f(\tau)}x_k(t), \qquad \left.\frac{\delta}{\delta f_l(\tau)}x_i(t)\right|_{t=\tau} = \delta_{il}. \tag{6.21}$$

Equation (6.21) has no randomness and governs Green's function $G_{il}(t,\tau)$ of homogeneous system (6.19), which means that

$$\frac{\delta}{\delta f_l(\tau)}x_i(t) = G_{il}(t,\tau).$$

As a consequence, we have

$$\frac{\delta}{\delta f_l(\tau)}\delta(\boldsymbol{x}(t)-\boldsymbol{x}) = -\frac{\partial}{\partial x_k}G_{kl}(t,\tau)\delta(\boldsymbol{x}(t)-\boldsymbol{x}),$$

and Eq. (6.20) appears to be converted into the closed equation

$$\frac{\partial}{\partial t}P(\boldsymbol{x},t) = -\frac{\partial}{\partial \boldsymbol{x}_i}\left(A_{ik}(t)x_kP(\boldsymbol{x},t)\right) + \dot{\Theta}_t\left[t;\, iG_{kl}(t,\tau)\frac{\partial}{\partial x_k}\right]P(\boldsymbol{x},t). \tag{6.22}$$

From Eq. (6.22) follows that any moment of quantity $\boldsymbol{x}(t)$ will satisfy a closed linear equation that will include only a finite number of cumulant functions whose order will not exceed the order of the moment of interest.

For the two-time probability density

$$P(\boldsymbol{x},t;\boldsymbol{x}_1,t_1) = \langle\delta(\boldsymbol{x}(t)-\boldsymbol{x})\delta(\boldsymbol{x}(t_1)-\boldsymbol{x}_1)\rangle,$$

we quite similarly obtain the equation

$$\begin{aligned}\frac{\partial}{\partial t}P(\boldsymbol{x},t;\boldsymbol{x}_1,t_1) &= -\frac{\partial}{\partial x_i}\left(A_{ik}(t)x_kP(\boldsymbol{x},t;\boldsymbol{x}_1,t_1)\right)\\ &+ \dot{\Theta}_t\left[t;\, i\{G_{kl}(t,\tau)+G_{kl}(t_1,\tau)\}\frac{\partial}{\partial x_l}\right]P(\boldsymbol{x},t;\boldsymbol{x}_1,t_1) \quad (t>t_1)\end{aligned} \tag{6.23}$$

with the initial value

$$P(\boldsymbol{x},t_1;\boldsymbol{x}_1,t_1) = \delta(\boldsymbol{x}-\boldsymbol{x}_1)P(\boldsymbol{x}_1,t_1),$$

where $P(\boldsymbol{x}_1,t_1)$ is the one-time probability density satisfying Eq. (6.22). From Eq. (6.23) follows that $\boldsymbol{x}(t)$ is not the Markovian process. The particular form of Eqs. (6.22) and (6.23) depends on the structure of functional $\Phi[t;v(\tau)]$, i.e., on the behavior of random function $\boldsymbol{f}(t)$.

For the Gaussian vector process $\boldsymbol{f}(t)$ whose mean value and correlation function are

$$\langle \boldsymbol{f}(t)\rangle = 0, \quad B_{ij}(t,t') = \left\langle f_i(t) f_j(t')\right\rangle,$$

Eq. (6.22) assumes the form of the extended Fokker–Planck equation

$$\frac{\partial}{\partial t}P(\boldsymbol{x},t) = -\frac{\partial}{\partial x_i}\left(A_{ik}(t)x_k P(\boldsymbol{x},t)\right)$$
$$+\int_0^t d\tau B_{jl}(t,\tau)G_{kj}(t,t)G_{ml}(t,\tau)\frac{\partial^2}{\partial x_k \partial x_m}P(\boldsymbol{x},t). \tag{6.24}$$

Consider the dynamics of a particle under random forces in the presence of friction [49] as an example of such a problem.

Inertial Particle Under Random Forces The simplest example of particle diffusion under the action of random external force $\boldsymbol{f}(t)$ and linear friction is described by the linear system of equations (1.75), page 36

$$\begin{aligned} &\frac{d}{dt}\boldsymbol{r}(t) = \boldsymbol{v}(t), \quad \frac{d}{dt}\boldsymbol{v}(t) = -\lambda\left[\boldsymbol{v}(t) - \boldsymbol{f}(t)\right], \\ &\boldsymbol{r}(0) = 0, \quad \boldsymbol{v}(0) = 0. \end{aligned} \tag{6.25}$$

The stochastic solution to Eqs. (6.25) has the form

$$\boldsymbol{v}(t) = \lambda\int_0^t d\tau e^{-\lambda(t-\tau)}\boldsymbol{f}(\tau), \quad \boldsymbol{r}(t) = \int_0^t d\tau\left[1 - e^{-\lambda(t-\tau)}\right]\boldsymbol{f}(\tau). \tag{6.26}$$

In the case of stationary random process $\boldsymbol{f}(t)$ with the correlation tensor

$$\left\langle f_i(t) f_j(t')\right\rangle = B_{ij}(t-t')$$

and temporal correlation radius τ_0 determined from the relationship

$$\int_0^\infty d\tau B_{ii}(\tau) = \tau_0 B_{ii}(0),$$

Eq. (6.25) allows obtaining analytical expressions for correlators between particle velocity components and coordinates

$$\begin{aligned} &\left\langle v_i(t)v_j(t)\right\rangle = \lambda\int_0^t d\tau B_{ij}(\tau)\left[e^{-\lambda\tau} - e^{-\lambda(2t-\tau)}\right], \\ &\frac{1}{2}\frac{d}{dt}\left\langle r_i(t)r_j(t)\right\rangle = \left\langle r_i(t)v_j(t)\right\rangle = \int_0^t d\tau B_{ij}(\tau)\left[1 - e^{-\lambda t}\right]\left[1 - e^{-\lambda(t-\tau)}\right]. \end{aligned} \tag{6.27}$$

In the steady-state regime, when $\lambda t \gg 1$ and $t/\tau_0 \gg 1$, but parameter $\lambda\tau_0$ can be arbitrary, particle velocity is the stationary process with the correlation tensor

$$\langle v_i(t)v_j(t)\rangle = \lambda \int\limits_0^\infty d\tau B_{ij}(\tau)e^{-\lambda\tau}, \tag{6.28}$$

and correlations $\langle r_i(t)v_j(t)\rangle$ and $\langle r_i(t)r_j(t)\rangle$ are as follows

$$\langle r_i(t)v_j(t)\rangle = \int\limits_0^\infty d\tau B_{ij}(\tau), \quad \langle r_i(t)r_j(t)\rangle = 2t \int\limits_0^\infty d\tau B_{ij}(\tau). \tag{6.29}$$

If we additionally assume that $\lambda\tau_0 \gg 1$, the correlation tensor grades into

$$\langle v_i(t)v_j(t)\rangle = B_{ij}(0), \tag{6.30}$$

which is consistent with (6.25), because $\boldsymbol{v}(t) \equiv \boldsymbol{f}(t)$ in this limit.

If the opposite condition $\lambda\tau_0 \ll 1$ holds, then

$$\langle v_i(t)v_j(t)\rangle = \lambda \int\limits_0^\infty d\tau B_{ij}(\tau).$$

This result corresponds to the delta-correlated approximation of random process $\boldsymbol{f}(t)$.

Introduce now the indicator function of the solution to Eq. (6.25)

$$\varphi(\boldsymbol{r}, \boldsymbol{v}, t) = \delta\left(\boldsymbol{r}(t) - \boldsymbol{r}\right)\delta\left(\boldsymbol{v}(t) - \boldsymbol{v}\right),$$

which satisfies the Liouville equation

$$\left(\frac{\partial}{\partial t} + \boldsymbol{v}\frac{\partial}{\partial \boldsymbol{r}} - \lambda\frac{\partial}{\partial \boldsymbol{v}}\boldsymbol{v}\right)\varphi(\boldsymbol{r}, \boldsymbol{v}, t) = -\lambda\boldsymbol{f}(t)\frac{\partial}{\partial \boldsymbol{v}}\varphi(\boldsymbol{r}, \boldsymbol{v}, t),$$
$$\varphi(\boldsymbol{r}, \boldsymbol{v}; 0) = \delta\left(\boldsymbol{r}\right)\delta\left(\boldsymbol{v}\right). \tag{6.31}$$

The mean value of the indicator function $\varphi(\boldsymbol{r}, \boldsymbol{v}, t)$ over an ensemble of realizations of random process $\boldsymbol{f}(t)$ is the joint one-time probability density of particle position and velocity

$$P(\boldsymbol{r}, \boldsymbol{v}, t) = \langle\varphi(\boldsymbol{r}, \boldsymbol{v}; t)\rangle = \langle\delta\left(\boldsymbol{r}(t) - \boldsymbol{r}\right)\delta\left(\boldsymbol{v}(t) - \boldsymbol{v}\right)\rangle_{\boldsymbol{f}}.$$

Averaging Eq. (6.31) over an ensemble of realizations of random process $\boldsymbol{f}(t)$, we obtain the unclosed equation

$$\left(\frac{\partial}{\partial t} + \boldsymbol{v}\frac{\partial}{\partial \boldsymbol{r}} - \lambda\frac{\partial}{\partial \boldsymbol{v}}\boldsymbol{v}\right)P(\boldsymbol{r}, \boldsymbol{v}, t) = -\lambda\frac{\partial}{\partial \boldsymbol{v}}\langle\boldsymbol{f}(t)\varphi(\boldsymbol{r}, \boldsymbol{v}, t)\rangle,$$
$$P(\boldsymbol{r}, \boldsymbol{v}; 0) = \delta\left(\boldsymbol{r}\right)\delta\left(\boldsymbol{v}\right). \tag{6.32}$$

This equation contains correlation $\langle \boldsymbol{f}(t)\varphi(\boldsymbol{r},\boldsymbol{v},t)\rangle$ and is equivalent to the equality

$$\left(\frac{\partial}{\partial t}+\boldsymbol{v}\frac{\partial}{\partial \boldsymbol{r}}-\lambda\frac{\partial}{\partial \boldsymbol{v}}\boldsymbol{v}\right)P(\boldsymbol{r},\boldsymbol{v},t)=\left\langle \dot{\Theta}\left[t;\frac{\delta}{i\delta \boldsymbol{f}(\tau)}\right]\varphi(\boldsymbol{r},\boldsymbol{v},t)\right\rangle, \qquad P(\boldsymbol{r},\boldsymbol{v};0)=\delta(\boldsymbol{r})\,\delta(\boldsymbol{v}), \tag{6.33}$$

where functional $\Theta\,[t;\boldsymbol{v}(\tau)]$ is related to the characteristic functional of random process $\boldsymbol{f}(t)$

$$\Phi\,[t;\boldsymbol{\psi}(\tau)]=\left\langle \exp\left\{i\int_0^t d\tau\ \boldsymbol{\psi}(\tau)\boldsymbol{f}(\tau)\right\}\right\rangle=e^{\Theta[t;\boldsymbol{\psi}(\tau)]}$$

by the formula

$$\dot{\Theta}\,[t;\boldsymbol{\psi}(\tau)]=\frac{d}{dt}\ln\Phi\,[t;\boldsymbol{\psi}(\tau)]=\frac{d}{dt}\Theta\,[t;\boldsymbol{\psi}(\tau)].$$

Functional $\Theta[t;\boldsymbol{\psi}(\tau)]$ can be expanded in the functional power series

$$\Theta[t;\boldsymbol{\psi}(\tau)]=\sum_{n=1}^{\infty}\frac{i^n}{n!}\int_0^t dt_1\ldots\int_0^t dt_n K^{(n)}_{i_1,\ldots,i_n}(t_1,\ldots,t_n)\psi_{i_1}(t_1)\ldots\psi_{i_n}(t_n),$$

where functions

$$K^{(n)}_{i_1,\ldots,i_n}(t_1,\ldots,t_n)=\frac{1}{i^n}\frac{\delta^n}{\delta\psi_{i_1}(t_1)\ldots\delta\psi_{i_n}(t_n)}\Theta[t;\boldsymbol{\psi}(\tau)]\bigg|_{\boldsymbol{\psi}=0}$$

are the n-th order cumulant functions of random process $\boldsymbol{f}(t)$.

Consider the variational derivative

$$\frac{\delta}{\delta f_j(t')}\varphi(\boldsymbol{r},\boldsymbol{v},t)=-\left[\frac{\partial}{\partial r_k}\frac{\delta r_k(t)}{\delta f_j(t')}+\frac{\partial}{\partial v_k}\frac{\delta v_k(t)}{\delta f_j(t')}\right]\varphi(\boldsymbol{r},\boldsymbol{v},t). \tag{6.34}$$

In the context of dynamic problem (6.25), the variational derivatives of functions $\boldsymbol{r}(t)$ and $\boldsymbol{v}(t)$ in Eq. (6.34) can be calculated from Eqs. (6.26) and have the forms

$$\frac{\delta v_k(t)}{\delta f_j(t')}=\lambda\delta_{kj}e^{-\lambda(t-t')},\qquad \frac{\delta r_k(t)}{\delta f_j(t')}=\delta_{kj}\left[1-e^{-\lambda(t-t')}\right]. \tag{6.35}$$

Using Eq. (6.35), we can now rewrite Eq. (6.34) in the form

$$\frac{\delta}{\delta \boldsymbol{f}(t')}\varphi(\boldsymbol{r},\boldsymbol{v};t)=-\left\{\left[1-e^{-\lambda(t-t')}\right]\frac{\partial}{\partial \boldsymbol{r}}+\lambda e^{-\lambda(t-t')}\frac{\partial}{\partial \boldsymbol{v}}\right\}\varphi(\boldsymbol{r},\boldsymbol{v},t),$$

after which Eq. (6.33) assumes the closed form

$$\left(\frac{\partial}{\partial t}+\boldsymbol{v}\frac{\partial}{\partial \boldsymbol{r}}-\lambda\frac{\partial}{\partial \boldsymbol{v}}\boldsymbol{v}\right)P(\boldsymbol{r},\boldsymbol{v},t)$$

$$=\dot{\Theta}\left[t;i\left\{\left[1-e^{-\lambda(t-t')}\right]\frac{\partial}{\partial \boldsymbol{r}}+\lambda e^{-\lambda(t-t')}\frac{\partial}{\partial \boldsymbol{v}}\right\}\right]P(\boldsymbol{r},\boldsymbol{v},t), \tag{6.36}$$

$$P(\boldsymbol{r},\boldsymbol{v};0)=\delta(\boldsymbol{r})\,\delta(\boldsymbol{v}).$$

Note that from Eq. (6.36) follows that equations for the n-th order moment functions include cumulant functions of orders not higher than n.

Assume now that $\boldsymbol{f}(t)$ is the Gaussian stationary process with the zero mean value and correlation tensor

$$B_{ij}(t-t')=\left\langle f_i(t)f_j(t')\right\rangle.$$

In this case, the characteristic functional of process $\boldsymbol{f}(t)$ is

$$\Phi\left[t;\boldsymbol{\psi}(\tau)\right]=\exp\left\{-\frac{1}{2}\int\limits_0^t\int\limits_0^t dt_1dt_2B_{ij}(t_1-t_2)\psi_i(t_1)\psi_j(t_2)\right\},$$

functional $\dot{\Theta}[t;\boldsymbol{\psi}(\tau)]$ is given by the formula

$$\dot{\Theta}\left[t;\boldsymbol{\psi}(\tau)\right]=-\psi_i(t)\int\limits_0^t dt'B_{ij}(t-t')\psi_j(t'),$$

and Eq. (6.36) appears to be an extension of the Fokker–Planck equation

$$\left(\frac{\partial}{\partial t}+\boldsymbol{v}\frac{\partial}{\partial \boldsymbol{r}}-\lambda\frac{\partial}{\partial \boldsymbol{v}}\boldsymbol{v}\right)P(\boldsymbol{r},\boldsymbol{v},t)=\lambda^2\int\limits_0^t d\tau B_{ij}(\tau)e^{-\lambda\tau}\frac{\partial^2}{\partial v_i\partial v_j}P(\boldsymbol{r},\boldsymbol{v},t)$$

$$+\lambda\int\limits_0^t d\tau B_{ij}(\tau)\left[1-e^{-\lambda\tau}\right]\frac{\partial^2}{\partial v_i\partial r_j}P(\boldsymbol{r},\boldsymbol{v},t), \tag{6.37}$$

$$P(\boldsymbol{r},\boldsymbol{v};0)=\delta(\boldsymbol{r})\,\delta(\boldsymbol{v}).$$

Equation (6.37) is the exact equation and remains valid for arbitrary times t. From this equation follows that $\boldsymbol{r}(t)$ and $\boldsymbol{v}(t)$ are the Gaussian functions. For moment functions of processes $\boldsymbol{r}(t)$ and $\boldsymbol{v}(t)$, we obtain in the ordinary way the system of equations

$$\frac{d}{dt}\langle r_i(t)r_j(t)\rangle = 2\langle r_i(t)v_j(t)\rangle,$$

$$\left(\frac{d}{dt}+\lambda\right)\langle r_i(t)v_j(t)\rangle = \langle v_i(t)v_j(t)\rangle + \lambda\int\limits_0^t d\tau B_{ij}(\tau)\left[1-e^{-\lambda\tau}\right], \tag{6.38}$$

$$\left(\frac{d}{dt}+2\lambda\right)\langle v_i(t)v_j(t)\rangle = 2\lambda^2\int\limits_0^t d\tau B_{ij}(\tau)e^{-\lambda\tau}.$$

From system (6.38) follows that steady-state values of all one-time correlators for $\lambda t \gg 1$ and $t/\tau_0 \gg 1$ are given by the expressions

$$\langle v_i(t)v_j(t)\rangle = \lambda\int\limits_0^\infty d\tau B_{ij}(\tau)e^{-\lambda\tau}, \quad \langle r_i(t)v_j(t)\rangle = D_{ij},$$
$$\langle r_i(t)r_j(t)\rangle = 2tD_{ij}, \tag{6.39}$$

where

$$D_{ij} = \int\limits_0^\infty d\tau B_{ij}(\tau) \tag{6.40}$$

is the spatial diffusion tensor, which agrees with expressions (6.28) and (6.29).

Remark

Temporal correlation tensor and temporal correlation radius of process $\boldsymbol{v}(t)$.

We can additionally calculate the temporal correlation radius of velocity $\boldsymbol{v}(t)$, i.e., the scale of correlator $\langle v_i(t)v_j(t_1)\rangle$. Using equalities (6.35), we obtain for $t_1 < t$ the equation

$$\left(\frac{d}{dt}+\lambda\right)\langle v_i(t)v_j(t_1)\rangle = \lambda^2\int\limits_0^{t_1} dt' B_{ij}(t-t')e^{-\lambda(t_1-t')}$$
$$= \lambda^2 e^{\lambda(t-t_1)}\int\limits_{t-t_1}^{t} d\tau B_{ij}(\tau)e^{-\lambda\tau} \tag{6.41}$$

with the initial value

$$\langle v_i(t)v_j(t_1)\rangle|_{t=t_1} = \langle v_i(t_1)v_j(t_1)\rangle. \tag{6.42}$$

In the steady-state regime, i.e., for $\lambda t \gg 1$ and $\lambda t_1 \gg 1$, but at fixed difference $(t - t_1)$, we obtain the equation with initial value $(\tau = t - t_1)$

$$\left(\frac{d}{d\tau} + \lambda\right)\langle v_i(t+\tau)v_j(t)\rangle = \lambda^2 e^{\lambda\tau}\int\limits_{\tau}^{\infty} d\tau_1 B_{ij}(\tau_1)e^{-\lambda\tau_1},$$
$$\langle v_i(t+\tau)v_j(t)\rangle_{\tau=0} = \langle v_i(t)v_j(t)\rangle. \tag{6.43}$$

One can easily write the solution to Eq. (6.43); however, our interest here concerns only the temporal correlation radius τ_v of random process $\boldsymbol{v}(t)$. To obtain this quantity, we integrate Eq. (6.43) with respect to parameter τ over the interval $(0, \infty)$. The result is

$$\lambda\int\limits_0^{\infty} d\tau\,\langle v_i(t+\tau)v_j(t)\rangle = \langle v_i(t)v_j(t)\rangle + \lambda\int\limits_0^{\infty} d\tau_1 B_{ij}(\tau_1)\left[1 - e^{-\lambda\tau_1}\right],$$

and we, using Eq. (6.39), arrive at the expression

$$\tau_v\left\langle \boldsymbol{v}^2(t)\right\rangle = D_{ii} = \tau_0 B_{ii}(0), \tag{6.44}$$

i.e.,

$$\tau_v = \frac{\tau_0 B_{ii}(0)}{\langle \boldsymbol{v}^2(t)\rangle} = \frac{\tau_0 B_{ii}(0)}{\lambda\int_0^{\infty} d\tau B_{ii}(\tau)e^{-\lambda\tau}} = \begin{cases} \tau_0, & \text{for } \lambda\tau_0 \gg 1, \\ 1/\lambda, & \text{for } \lambda\tau_0 \ll 1. \end{cases} \tag{6.45}$$

Integrating Eq. (6.37) over $\boldsymbol{r}$, we obtain the closed equation for the probability density of particle velocity

$$\left(\frac{\partial}{\partial t} - \lambda\frac{\partial}{\partial \boldsymbol{v}}\boldsymbol{v}\right)P(\boldsymbol{v}, t) = \lambda^2\int\limits_0^{t} d\tau B_{ij}(\tau)e^{-\lambda\tau}\frac{\partial^2}{\partial v_i\partial v_j}P(\boldsymbol{r}, \boldsymbol{v}, t),$$
$$P(\boldsymbol{r}, \boldsymbol{v}; 0) = \delta(\boldsymbol{v}).$$

The solution to this equation corresponds to the Gaussian process $\boldsymbol{v}(t)$ with correlation tensor (6.27), which follows from the fact that the second equation of system (6.25) is closed. It can be shown that, if the steady-state probability density exists under the condition $\lambda t \gg 1$, then this probability density satisfies the equation

$$-\frac{\partial}{\partial \boldsymbol{v}}\boldsymbol{v}P(\boldsymbol{v}, t) = \lambda\int\limits_0^{\infty} d\tau B_{ij}(\tau)e^{-\lambda\tau}\frac{\partial^2}{\partial v_i\partial v_j}P(\boldsymbol{v}, t),$$

and the rate of establishing this distribution depends on parameter λ.

The equation for the probability density of particle coordinate $P(\mathbf{r}, t)$ cannot be derived immediately from Eq. (6.37). Indeed, integrating Eq. (6.37) over $\mathbf{v}$, we obtain the equality

$$\frac{\partial}{\partial t}P(\mathbf{r}, t) = -\frac{\partial}{\partial \mathbf{r}} \int \mathbf{v}P(\mathbf{r}, \mathbf{v}, t)d\mathbf{v}, \quad P(\mathbf{r}, 0) = \delta(\mathbf{r}). \tag{6.46}$$

For function $\int v_k P(\mathbf{r}, \mathbf{v}, t)d\mathbf{v}$, we have the equality

$$\left(\frac{\partial}{\partial t} + \lambda\right) \int v_k P(\mathbf{r}, \mathbf{v}, t)d\mathbf{v} = -\frac{\partial}{\partial \mathbf{r}} \int v_k \mathbf{v} P(\mathbf{r}, \mathbf{v}, t)d\mathbf{v}$$
$$-\lambda \int_0^t d\tau B_{kj}(\tau)\left[1 - e^{-\lambda\tau}\right] \frac{\partial}{\partial r_j} P(\mathbf{r}, t), \tag{6.47}$$

and so on, i.e., this approach results in an infinite system of equations.

Random function $\mathbf{r}(t)$ satisfies the first equation of system (6.25) and, if we would know the complete statistics of function $\mathbf{v}(t)$ (i.e., the multi-time statistics), we could calculate all statistical characteristics of function $\mathbf{r}(t)$. Unfortunately, Eq. (6.37) describes only one-time statistical quantities, and only the infinite system of equations similar to Eqs. (6.46), (6.47), and so on appears equivalent to the multi-time statistics of function $\mathbf{v}(t)$. Indeed, function $\mathbf{r}(t)$ can be represented in the form

$$\mathbf{r}(t) = \int_0^t dt_1 \mathbf{v}(t_1),$$

so that the spatial diffusion coefficient in the steady-state regime assumes, in view of Eq. (6.44), the form

$$\frac{1}{2}\frac{d}{dt}\left\langle \mathbf{r}^2(t)\right\rangle = \int_0^\infty d\tau \left\langle \mathbf{v}(t+\tau)\mathbf{v}(t)\right\rangle = \tau_v \left\langle \mathbf{v}^2(t)\right\rangle = D_{ii} = \tau_0 B_{ii}(0), \tag{6.48}$$

from which follows that it depends on the temporal correlation radius τ_v and the variance of random function $\mathbf{v}(t)$.

However, in the case of this simplest problem, we know both variances and all correlations of functions $\mathbf{v}(t)$ and $\mathbf{r}(t)$ (see Eqs. (6.27)) and, consequently, can draw the equation for the probability density of particle coordinate $P(\mathbf{r}, t)$. This equation is the diffusion equation

$$\frac{\partial}{\partial t}P(\mathbf{r}, t) = D_{ij}(t)\frac{\partial^2}{\partial r_i \partial r_j}P(\mathbf{r}, t), \quad P(\mathbf{r}, 0) = \delta(\mathbf{r}),$$

where

$$D_{ij}(t) = \frac{1}{2}\frac{d}{dt}\left\langle r_i(t)r_j(t)\right\rangle = \frac{1}{2}\left\{\left\langle r_i(t)v_j(t)\right\rangle + \left\langle r_j(t)v_i(t)\right\rangle\right\}$$

$$= \int_0^t d\tau B_{ij}(\tau)\left[1 - e^{-\lambda t}\right]\left[1 - e^{-\lambda(t-\tau)}\right].$$

is the diffusion tensor (6.27). Under the condition $\lambda t \gg 1$, we obtain the equation

$$\frac{\partial}{\partial t}P(\boldsymbol{r}, t) = D_{ij}\frac{\partial^2}{\partial r_i \partial r_j}P(\boldsymbol{r}, t), \quad P(\boldsymbol{r}, 0) = \delta(\boldsymbol{r}) \tag{6.49}$$

with the diffusion tensor

$$D_{ij} = \int_0^\infty d\tau B_{ij}(\tau). \tag{6.50}$$

Note that conversion from Eq. (6.37) to the equation in probability density of particle coordinate (6.49) with the diffusion coefficient (6.50) corresponds to the so-called *Kramers problem.*

Under the assumption that $\lambda\tau_0 \ll 1$, where τ_0 is the temporal correlation radius of process $\boldsymbol{f}(t)$, Eq. (6.37) becomes simpler

$$\left(\frac{\partial}{\partial t} + \boldsymbol{v}\frac{\partial}{\partial \boldsymbol{r}} - \lambda\frac{\partial}{\partial \boldsymbol{v}}\boldsymbol{v}\right)P(\boldsymbol{r}, \boldsymbol{v}, t) = \lambda^2 \int_0^t d\tau B_{ij}(\tau)\frac{\partial^2}{\partial v_i \partial v_j}P(\boldsymbol{r}, \boldsymbol{v}, t),$$

$$P(\boldsymbol{r}, \boldsymbol{v}; 0) = \delta(\boldsymbol{r})\,\delta(\boldsymbol{v}),$$

and corresponds to the approximation of random function $\boldsymbol{f}(t)$ by the delta-correlated process. If $t \gg \tau_0$, we can replace the upper limit of the integral with the infinity and proceed to the standard diffusion Fokker–Planck equation

$$\left(\frac{\partial}{\partial t} + \boldsymbol{v}\frac{\partial}{\partial \boldsymbol{r}} - \lambda\frac{\partial}{\partial \boldsymbol{v}}\boldsymbol{v}\right)P(\boldsymbol{r}, \boldsymbol{v}, t) = \lambda^2 D_{ij}\frac{\partial^2}{\partial v_i \partial v_j}P(\boldsymbol{r}, \boldsymbol{v}, t),$$
$$P(\boldsymbol{r}, \boldsymbol{v}; 0) = \delta(\boldsymbol{r})\,\delta(\boldsymbol{v}), \tag{6.51}$$

with diffusion tensor (6.50). In this approximation, the combined random process $\{\boldsymbol{r}(t), \boldsymbol{v}(t)\}$ is the Markovian process.

Under the condition $\lambda t \gg 1$, there are the steady-state equation for the probability density of particle velocity

$$-\lambda\frac{\partial}{\partial \boldsymbol{v}}\boldsymbol{v}P(\boldsymbol{v}) = \lambda^2 D_{ij}\frac{\partial^2}{\partial v_i \partial v_j}P(\boldsymbol{v}),$$

and the nonsteady-state equation for the probability density of particle coordinate

$$\frac{\partial}{\partial t}P(\boldsymbol{r},t) = D_{ij}\frac{\partial^2}{\partial r_i \partial r_j}P(\boldsymbol{r},t), \quad P(\boldsymbol{r},0) = \delta(\boldsymbol{r}).$$

Consider now the limit $\lambda\tau_0 \gg 1$. In this case, we can rewrite Eq. (6.37) in the form

$$\left(\frac{\partial}{\partial t} + \boldsymbol{v}\frac{\partial}{\partial \boldsymbol{r}} - \lambda\frac{\partial}{\partial \boldsymbol{v}}\boldsymbol{v}\right)P(\boldsymbol{r},\boldsymbol{v},t) = \lambda B_{ij}(0)\left[1 - e^{-\lambda t}\right]\frac{\partial^2}{\partial v_i \partial v_j}P(\boldsymbol{r},\boldsymbol{v},t)$$

$$- B_{ij}(0)\left[1 - e^{-\lambda t}\right]\frac{\partial^2}{\partial v_i \partial r_j}P(\boldsymbol{r},\boldsymbol{v},t) + \lambda\int\limits_0^t d\tau B_{ij}(\tau)\frac{\partial^2}{\partial v_i \partial r_j}P(\boldsymbol{r},\boldsymbol{v},t),$$

$$P(\boldsymbol{r},\boldsymbol{v};0) = \delta(\boldsymbol{r})\,\delta(\boldsymbol{v}).$$

Integrating this equation over $\boldsymbol{r}$, we obtain the equation for the probability density of particle velocity

$$\left(\frac{\partial}{\partial t} - \lambda\frac{\partial}{\partial \boldsymbol{v}}\boldsymbol{v}\right)P(\boldsymbol{v},t) = \lambda B_{ij}(0)\left[1 - e^{-\lambda t}\right]\frac{\partial^2}{\partial v_i \partial v_j}P(\boldsymbol{v},t),$$
$$P(\boldsymbol{v};0) = \delta(\boldsymbol{v}),$$

and, under the condition $\lambda t \gg 1$, we arrive at the steady-state Gaussian probability density with variance

$$\left\langle v_i(t) v_j(t)\right\rangle = B_{ij}(0).$$

As regards the probability density of particle position, it satisfies, under the condition $\lambda t \gg 1$, the equation

$$\frac{\partial}{\partial t}P(\boldsymbol{r},t) = D_{ij}\frac{\partial^2}{\partial r_i \partial r_j}P(\boldsymbol{r},t), \quad P(\boldsymbol{r},0) = \delta(\boldsymbol{r}) \tag{6.52}$$

with the same diffusion coefficient as previously. This is a consequence of the fact that Eq. (6.44) is independent of parameter λ. Note that this equation corresponds to the limit process $\lambda \to \infty$ in Eq. (6.25)

$$\frac{d}{dt}\boldsymbol{r}(t) = \boldsymbol{v}(t), \quad \boldsymbol{v}(t) = \boldsymbol{f}(t), \quad \boldsymbol{r}(0) = 0.$$

In the limit $\lambda \to \infty$ (or $\lambda\tau_0 \gg 1$), we have the equality

$$\boldsymbol{v}(t) \approx \boldsymbol{f}(t), \tag{6.53}$$

and all multi-time statistics of random functions $\boldsymbol{v}(t)$ and $\boldsymbol{r}(t)$ will be described in terms of statistical characteristics of process $\boldsymbol{f}(t)$. In particular, the one-time probability density of particle velocity $\boldsymbol{v}(t)$ is the Gaussian probability density with variance $\langle v_i(t)v_j(t)\rangle = B_{ij}(0)$, and the spatial diffusion coefficient is

$$D = \frac{1}{2}\frac{d}{dt}\left\langle r^2(t)\right\rangle = \int\limits_0^\infty d\tau B_{ii}(\tau) = \tau_0 B_{ii}(0).$$

As we have seen earlier, in the case of process $\boldsymbol{f}(t)$ such that it can be correctly described in the delta-correlated approximation (i.e., if $\lambda\tau_0 \ll 1$), the approximate equality (6.53) appears inappropriate to determine statistical characteristics of process $\boldsymbol{v}(t)$. Nevertheless, Eq. (6.52) with the same diffusion tensor remains as before valid for the one-time statistical characteristics of process $\boldsymbol{r}(t)$, which follows from the fact that Eq. (6.48) is valid for any parameter λ and arbitrary probability density of random process $\boldsymbol{f}(t)$.

Above, we considered several types of stochastic ordinary differential equations that allow obtaining closed statistical description in the general form. It is clear that similar situations can appear in dynamic systems formulated in terms of partial differential equations.

6.2.2 Partial Differential Equations

First of all, we note that the first-order partial differential equation

$$\left(\frac{\partial}{\partial t} + z(t)g(t)\frac{\partial}{\partial \boldsymbol{x}}\boldsymbol{F}(\boldsymbol{x})\right)\rho(\boldsymbol{x},t) = 0, \tag{6.54}$$

is equivalent to the system of ordinary differential equations (6.10), page 144 and, consequently, also allows the complete statistical description for arbitrary given random process $z(t)$.

Solution $\rho(\boldsymbol{r},t)$ to Eq. (6.54) is a functional of random process $z(t)$ of the form

$$\rho(\boldsymbol{x},t) = \rho[\boldsymbol{x},t;\ z(\tau)] = \rho(\boldsymbol{x}, T[t;\ z(\tau)]),$$

with functional

$$T[t;\ z(\tau)] = \int\limits_0^t d\tau\, z(\tau)g(\tau), \tag{6.55}$$

so that Eq. (6.54) can be rewritten in the form

$$\left(\frac{\partial}{\partial T} + \frac{\partial}{\partial \boldsymbol{x}}\boldsymbol{F}(\boldsymbol{x})\right)\rho(\boldsymbol{x},T) = 0.$$

The variational derivative is expressed in this case as follows

$$\frac{\delta}{\delta z(\tau)}\rho(\boldsymbol{x},t)=\frac{\delta}{\delta z(\tau)}\rho(\boldsymbol{x},T[t;\,z(\tau)])=\frac{d\rho(\boldsymbol{x},T)}{dT}\frac{\delta T[t;\,z(\tau)]}{\delta z(\tau)}$$
$$=\theta(t-\tau)g(\tau)\frac{d\rho(\boldsymbol{x},T)}{dT}=-\theta(t-\tau)g(\tau)\frac{\partial}{\partial\boldsymbol{x}}\boldsymbol{F}(\boldsymbol{x})\rho(\boldsymbol{x},t).$$

Consider now the class of nonlinear partial differential equations whose parameters are independent of spatial variable $\boldsymbol{x}$,

$$\left(\frac{\partial}{\partial t}+z(t)\frac{\partial}{\partial\boldsymbol{x}}\right)q(\boldsymbol{x},t)=F\left(t,q,\frac{\partial q}{\partial\boldsymbol{x}},\frac{\partial}{\partial\boldsymbol{x}}\otimes\frac{\partial q}{\partial\boldsymbol{x}},\ldots\right),\tag{6.56}$$

where $z(t)$ is the vector random process and F is the deterministic function. Solution to this equation is representable in the form

$$q(\boldsymbol{x},t)=\exp\left\{-\int_0^t d\tau\,z(\tau)\frac{\partial}{\partial\boldsymbol{x}}\right\}Q(\boldsymbol{x},t)=Q\left(\boldsymbol{x}-\int_0^t d\tau\,z(\tau),t\right),\tag{6.57}$$

where we introduced the shift operator. Function $Q(x,t)$ satisfies the equation

$$\exp\left\{-\int_0^t d\tau\,z(\tau)\frac{\partial}{\partial\boldsymbol{x}}\right\}\frac{\partial Q(\boldsymbol{x},t)}{\partial t}$$
$$=F\left(t,\left[\exp\left\{-\int_0^t d\tau\,z(\tau)\frac{\partial}{\partial\boldsymbol{x}}\right\}Q\right],\left[\exp\left\{-\int_0^t d\tau\,z(\tau)\frac{\partial}{\partial\boldsymbol{x}}\right\}\frac{\partial Q}{\partial\boldsymbol{x}}\right],\right.$$
$$\left.\frac{\partial}{\partial\boldsymbol{x}}\otimes\left[\exp\left\{-\int_0^t d\tau\,z(\tau)\frac{\partial}{\partial\boldsymbol{x}}\right\}\frac{\partial Q}{\partial\boldsymbol{x}}\right],\ldots\right).\tag{6.58}$$

The shift operator can be factored out from arguments of function $F\left(t,q,\frac{\partial q}{\partial\boldsymbol{x}},\frac{\partial}{\partial\boldsymbol{x}}\otimes\frac{\partial q}{\partial\boldsymbol{x}},\ldots\right)$, and Eq. (6.58) assumes the form of the deterministic equation

$$\frac{\partial Q(\boldsymbol{x},t)}{\partial t}=F\left(t,Q,\frac{\partial Q}{\partial\boldsymbol{x}},\frac{\partial}{\partial\boldsymbol{x}}\otimes\exp\frac{\partial Q}{\partial\boldsymbol{x}},\ldots\right).\tag{6.59}$$

Thus, the variational derivative can be expressed in the form

$$\frac{\delta q(\boldsymbol{x},t)}{\delta z(\tau)}=\frac{\delta Q\left(\boldsymbol{x}-\int_0^t d\tau\,z(\tau),t\right)}{\delta z(\tau)}=-\theta(t-\tau)\frac{\partial}{\partial\boldsymbol{x}}q(\boldsymbol{x},t),\tag{6.60}$$

i.e., the variational derivatives of the solution to problem (6.56) are expressed in terms of spatial derivatives.

Consider the *Burgers equation* with random drift $z(t)$ as a specific example:

$$\frac{\partial}{\partial t}\boldsymbol{v}(\boldsymbol{x},t)+\left[(\boldsymbol{v}+\boldsymbol{z}(t))\frac{\partial}{\partial \boldsymbol{x}}\right]\boldsymbol{v}(\boldsymbol{x},t)=\nu\frac{\partial^2}{\partial \boldsymbol{x}^2}\boldsymbol{v}(\boldsymbol{x},t). \tag{6.61}$$

In this case, the solution has the form

$$\boldsymbol{v}(\boldsymbol{x},t)=\exp\left\{-\int_0^t d\tau\, \boldsymbol{z}(\tau)\frac{\partial}{\partial \boldsymbol{x}}\right\}\boldsymbol{V}(\boldsymbol{x},t)=\boldsymbol{V}\left(\boldsymbol{x}-\int_0^t d\tau\, \boldsymbol{z}(\tau),t\right),$$

where function $\boldsymbol{V}(\boldsymbol{x},t)$ satisfies the standard Burgers equation

$$\frac{\partial}{\partial t}\boldsymbol{V}(\boldsymbol{x},t)+\left(\boldsymbol{V}(\boldsymbol{x},t)\frac{\partial}{\partial \boldsymbol{x}}\right)\boldsymbol{V}(\boldsymbol{x},t)=\nu\frac{\partial^2}{\partial \boldsymbol{x}^2}\boldsymbol{V}(\boldsymbol{x},t). \tag{6.62}$$

In this example, the variational derivative is given by the expression

$$\frac{\delta}{\delta z_i(\tau)}\boldsymbol{v}(\boldsymbol{x},t)=\frac{\delta}{\delta z_i(\tau)}\boldsymbol{V}\left(\boldsymbol{x}-\int_0^t d\tau\, \boldsymbol{z}(\tau),t\right)=-\theta(t-\tau)\frac{\partial}{\partial x_i}\boldsymbol{v}(\boldsymbol{x},t). \tag{6.63}$$

In the case of such problems, statistical characteristics of the solution can be determined immediately by averaging the corresponding expressions constructed from the solution to the last equation.

Unfortunately, there is only limited number of equations that allow sufficiently complete analysis. In the general case, the analysis of dynamic systems appears possible only on the basis of various asymptotic and approximate techniques. In physics, techniques based on approximating actual random processes and fields by the fields delta-correlated in time are often and successfully used.

6.3 Delta-Correlated Fields and Processes

In the case of random field $\boldsymbol{f}(\boldsymbol{x},t)$ delta-correlated in time, the following equality holds (see Sect. 5.5, page 130)

$$\dot{\Theta}_t[t,t_0;\boldsymbol{v}(\boldsymbol{y},\tau)]\equiv\dot{\Theta}_t[t,t_0;\boldsymbol{v}(\boldsymbol{y},t)],$$

and situation becomes significantly simpler. The fact that field $\boldsymbol{f}(\boldsymbol{x},t)$ is delta-correlated means that

$$\Theta[t,t_0;\boldsymbol{v}(\boldsymbol{y},\tau)]=\sum_{i=1}^{\infty}\frac{i^n}{n!}\int_{t_0}^{t}d\tau\int d\boldsymbol{y}_1\ldots\int d\boldsymbol{y}_n K_n^{i_1,\ldots,i_n}(\boldsymbol{y}_1,\ldots,\boldsymbol{y}_n;\tau)$$
$$\times\, v_{i_1}(\boldsymbol{y}_1,\tau)\ldots v_{i_n}(\boldsymbol{y}_n,\tau),$$

which, in turn, means that field $\boldsymbol{f}(\boldsymbol{x}, t)$ is characterized by cumulant functions of the form

$$K_n^{i_1,\ldots,i_n}(\boldsymbol{y}_1, t_1; \ldots; \boldsymbol{y}_n, t_n) \\ = K_n^{i_1,\ldots,i_n}(\boldsymbol{y}_1, \ldots, \boldsymbol{y}_n; t_1)\delta(t_1 - t_2)\ldots\delta(t_{n-1} - t_n).$$

In view of (6.3), page 142, Eqs. (6.7) and (6.9) appear to be in this case the closed operator equations in functions $P(\boldsymbol{x}, t)$, $p(\boldsymbol{x}, t|\boldsymbol{x}_0, t_0)$, and $P_m(\boldsymbol{x}_1, t_1; \ldots; \boldsymbol{x}_m, t_m)$. Indeed, Eq. (6.7), page 143 is reduced to the equation

$$\left(\frac{\partial}{\partial t} + \frac{\partial}{\partial \boldsymbol{x}}\boldsymbol{v}(\boldsymbol{x}, t)\right) P(\boldsymbol{x},t) = \dot{\Theta}_t\left[t, t_0; i\frac{\partial}{\partial \boldsymbol{x}}\delta(\boldsymbol{y} - \boldsymbol{x})\right] P(\boldsymbol{x}, t), \\ P(\boldsymbol{x}, 0) = \delta(\boldsymbol{x} - \boldsymbol{x}_0), \tag{6.64}$$

whose concrete form is governed by functional $\Theta[t, t_0; \boldsymbol{v}(\boldsymbol{y}, \tau)]$, i.e., by statistical behavior of random field $\boldsymbol{f}(\boldsymbol{x}, t)$. Correspondingly, Eq. (6.9) for the m-time probability density is reduced to the operator equation ($t_1 \le t_2 \le \ldots \le t_m$)

$$\left(\frac{\partial}{\partial t} + \frac{\partial}{\partial \boldsymbol{x}_m}\boldsymbol{v}(\boldsymbol{x}_m, t_m)\right) P_m(\boldsymbol{x}_1, t_1; \ldots; \boldsymbol{x}_m, t_m) \\ = \dot{\Theta}_{t_m}\left[t_m, t_0; i\frac{\partial}{\partial \boldsymbol{x}_m}\delta(\boldsymbol{y} - \boldsymbol{x}_m)\right] P_m(\boldsymbol{x}_1, t_1; \ldots; \boldsymbol{x}_m, t_m), \tag{6.65}$$

$$P_m(\boldsymbol{x}_1, t_1; \ldots; \boldsymbol{x}_m, t_{m-1}) = \delta(\boldsymbol{x}_m - \boldsymbol{x}_{m-1})P_{m-1}(\boldsymbol{x}_1, t_1; \ldots; \boldsymbol{x}_{m-1}, t_{m-1}),$$

We can seek the solution to Eq. (6.65) in the form

$$P_m(\boldsymbol{x}_1, t_1; \ldots; \boldsymbol{x}_m, t_m) \\ = p(\boldsymbol{x}_m, t_m|\boldsymbol{x}_{m-1}, t_{m-1})P_{m-1}(\boldsymbol{x}_1, t_1; \ldots; \boldsymbol{x}_{m-1}, t_{m-1}). \tag{6.66}$$

Because all differential operations in Eq. (6.65) concern only t_m and $\boldsymbol{x}_m$, we can substitute Eq. (6.66) in Eq. (6.65) to obtain the following equation for the transition probability density

$$\left(\frac{\partial}{\partial t} + \frac{\partial}{\partial \boldsymbol{x}}\boldsymbol{v}(\boldsymbol{x}, t)\right) p(\boldsymbol{x}, t|\boldsymbol{x}_0, t_0) = \dot{\Theta}_t\left[t, t_0; i\frac{\partial}{\partial \boldsymbol{x}}\delta(\boldsymbol{y} - \boldsymbol{x})\right] p(\boldsymbol{x}, t|\boldsymbol{x}_0, t_0), \\ p(\boldsymbol{x}, t|\boldsymbol{x}_0, t_0)|_{t \to t_0} = \delta(\boldsymbol{x} - \boldsymbol{x}_0). \tag{6.67}$$

Here, we denoted variables $\boldsymbol{x}_m$ and t_m as $\boldsymbol{x}$ and t and variables $\boldsymbol{x}_{m-1}$ and t_{m-1} as $\boldsymbol{x}_0$ and t_0.

Using formula (6.66) $(m - 1)$ times, we obtain the relationship

$$P_m(\boldsymbol{x}_1, t_1; \ldots; \boldsymbol{x}_m, t_m) \\ = p(\boldsymbol{x}_m, t_m|\boldsymbol{x}_{m-1}, t_{m-1})\ldots p(\boldsymbol{x}_2, t_2|\boldsymbol{x}_1, t_1)P(\boldsymbol{x}_1, t_1), \tag{6.68}$$

where $P(\boldsymbol{x}_1, t_1)$ is the one-time probability density governed by Eq. (6.64). Equality (6.68) expresses the many-time probability density in terms of the product of transition probability densities, which means that random process $\boldsymbol{x}(t)$ is the Markovian process. The transition probability density is defined in this case as follows:

$$p(\boldsymbol{x}, t|\boldsymbol{x}_0, t_0) = \langle \delta(\boldsymbol{x}(t|\boldsymbol{x}_0, t_0) - \boldsymbol{x}) \rangle .$$

Special models of parameter fluctuations can significantly simplify the obtained equations.

For example, in the case of the Gaussian delta-correlated field $\boldsymbol{f}(\boldsymbol{x}, t)$, the correlation tensor has the form ($\langle \boldsymbol{f}(\boldsymbol{x}, t) \rangle = 0$)

$$B_{ij}(\boldsymbol{x}, t; \boldsymbol{x}', t') = 2\delta(t - t')F_{ij}(\boldsymbol{x}, \boldsymbol{x}'; t).$$

Then, functional $\Theta[t, t_0; \boldsymbol{v}(\boldsymbol{y}, \tau)]$ assumes the form

$$\Theta[t, t_0; \boldsymbol{v}(\boldsymbol{y}, \tau)] = -\int\limits_{t_0}^{t} d\tau \int d\boldsymbol{y}_1 \int d\boldsymbol{y}_2 F_{ij}(\boldsymbol{y}_1, \boldsymbol{y}_2; \tau) v_i(\boldsymbol{y}_1, \tau) v_j(\boldsymbol{y}_2, \tau),$$

and Eq. (6.64) reduces to the Fokker–Planck equation

$$\left(\frac{\partial}{\partial t} + \frac{\partial}{\partial x_k} \left[v_k(\boldsymbol{x},t) + A_k(\boldsymbol{x},t) \right] \right) P(\boldsymbol{x}, t) = \frac{\partial^2}{\partial x_k \partial x_l} \left[F_{kl}(\boldsymbol{x}, \boldsymbol{x},t) P(\boldsymbol{x},t) \right], \tag{6.69}$$

where

$$A_k(\boldsymbol{x}, t) = \left. \frac{\partial}{\partial x'_l} F_{kl}(\boldsymbol{x}, \boldsymbol{x}'; t) \right|_{\boldsymbol{x}'=\boldsymbol{x}} .$$

In view of the special role that the Gaussian delta-correlated field $\boldsymbol{f}(\boldsymbol{x}, t)$ plays in physics, we give an alternative and more detailed discussion of this approximation commonly called the *approximation of the Gaussian delta-correlated field* in Lecture 8.

We illustrate the above general theory by the examples of several equations.

6.3.1 One-Dimensional Nonlinear Differential Equation

Consider the one-dimensional stochastic equation

$$\frac{d}{dt} x(t) = f(x, t) + z(t) g(x, t), \quad x(0) = x_0, \tag{6.70}$$

where $f(x, t)$ and $g(x, t)$ are the deterministic functions and $z(t)$ is the random function of time. For indicator function $\varphi(x, t) = \delta(x(t) - x)$, we have the Liouville equation

$$\left(\frac{\partial}{\partial t} + \frac{\partial}{\partial x} f(x, t) \right) \varphi(x, t) = -z(t) \frac{\partial}{\partial x} \{ g(x, t) \varphi(x, t) \},$$

so that the equation for the one-time probability density $P(x, t)$ has the form

$$\left(\frac{\partial}{\partial t} + \frac{\partial}{\partial x} f(x, t)\right) P(x, t) = \left\langle \dot{\Theta}_t \left[t, \frac{\delta}{i\delta z(\tau)} \right] \varphi(x, t) \right\rangle.$$

In the case of the delta-correlated random process $z(t)$, the equality

$$\dot{\Theta}_t[t, v(\tau)] = \dot{\Theta}_t[t, v(t)]$$

holds. Taking into account the equality

$$\frac{\delta}{\delta z(t-0)} \varphi(x, t) = -\frac{\partial}{\partial x} \{g(x, t)\varphi(x, t)\},$$

we obtain the closed operator equation

$$\left(\frac{\partial}{\partial t} + \frac{\partial}{\partial x} f(x, t)\right) P(x, t) = \dot{\Theta}_t \left[t, i\frac{\partial}{\partial x} g(x, t) \right] P(x, t). \tag{6.71}$$

For the Gaussian delta-correlated process, we have

$$\Theta[t, v(\tau)] = -\frac{1}{2} \int\limits_0^t d\tau B(\tau) v^2(\tau), \tag{6.72}$$

and Eq. (6.71) assumes the form of the Fokker–Planck equation

$$\left(\frac{\partial}{\partial t} + \frac{\partial}{\partial x} f(x, t)\right) P(x, t) = \frac{1}{2} B(t) \frac{\partial}{\partial x} g(x, t) \frac{\partial}{\partial x} g(x, t) P(x, t). \tag{6.73}$$

For Poisson's delta-correlated process $z(t)$, we have

$$\Theta[t, v(\tau)] = \nu \int\limits_0^t d\tau \left\{ \int\limits_{-\infty}^{\infty} d\xi p(\xi) e^{i\xi v(\tau)} - 1 \right\}, \tag{6.74}$$

and Eq. (6.71) reduces to the form

$$\left(\frac{\partial}{\partial t} + \frac{\partial}{\partial x} f(x, t)\right) P(x, t) = \nu \left\{ \int\limits_{-\infty}^{\infty} d\xi p(\xi) e^{-\xi \frac{\partial}{\partial x} g(x,t)} - 1 \right\} P(x, t). \tag{6.75}$$

If we set $g(x, t) = 1$, Eq. (6.70) assumes the form

$$\frac{d}{dt} x(t) = f(x, t) + z(t), \quad x(0) = x_0.$$

In this case, the operator in the right-hand side of (6.75) is the shift operator, and Eq. (6.75) assumes the form of the Kolmogorov–Feller equation

$$\left(\frac{\partial}{\partial t}+\frac{\partial}{\partial x}f(x,t)\right)P(x,t)=\nu\int\limits_{-\infty}^{\infty}d\xi p(\xi)P(x-\xi,t)-\nu P(x,t).$$

Define now $g(x,t)=x$, so that Eq. (6.70) reduces to the form

$$\frac{d}{dt}x(t)=f(x,t)+z(t)x(t),\quad x(0)=x_0.$$

In this case, Eq. (6.75) assumes the form

$$\left(\frac{\partial}{\partial t}+\frac{\partial}{\partial x}f(x,t)\right)P(x,t)=\nu\left\{\int\limits_{-\infty}^{\infty}d\xi p(\xi)e^{-\xi\frac{\partial}{\partial x}x}-1\right\}P(x,t). \tag{6.76}$$

To determine the action of the operator in the right-hand side of Eq. (6.76), we expand it in series in ξ

$$\left\{e^{-\xi\frac{\partial}{\partial x}x}-1\right\}P(x,t)=\sum_{n=1}^{\infty}\frac{(-\xi)^n}{n!}\left(\frac{\partial}{\partial x}x\right)^n P(x,t)$$

and consider the action of every term.

Representing x in the form $x=e^{\varphi}$, we can transform this formula as follows (the fact that x is the alternating quantity is insignificant here)

$$\begin{aligned}&\sum_{n=1}^{\infty}\frac{(-\xi)^n}{n!}e^{-\varphi}\frac{\partial^n}{\partial\varphi^n}e^{\varphi}P(e^{\varphi},t)\\&\quad=e^{-\varphi}\left\{e^{-\xi\frac{\partial}{\partial\varphi}}-1\right\}e^{\varphi}P(e^{\varphi},t)=e^{-\xi}P(e^{\varphi-\xi},t)-P(e^{\varphi},t).\end{aligned}$$

Reverting to variable x, we can represent Eq. (6.76) in the final form of the integro-differential equation similar to the Kolmogorov–Feller equation

$$\left(\frac{\partial}{\partial t}+\frac{\partial}{\partial x}f(x,t)\right)P(x,t)=\nu\int\limits_{-\infty}^{\infty}d\xi p(\xi)e^{-\xi}P(xe^{-\xi},t)-\nu P(x,t).$$

In Lecture 4, page 89, we mentioned that formula

$$x(t)=\int\limits_0^t d\tau g(t-\tau)z(\tau)$$

relates Poisson's process $x(t)$ with arbitrary impulse function $g(t)$ to Poisson's delta-correlated random process $z(t)$. Let $g(t) = e^{-\lambda t}$. In this case, process $x(t)$ satisfies the stochastic differential equation

$$\frac{d}{dt}x(t) = -\lambda x(t) + z(t)$$

and, consequently, both transition probability density and one-time probability density of this process satisfy, according to Eq. (6.75), the equations

$$\frac{\partial}{\partial t}p(x,t|x_0,t_0) = \widehat{L}(x)p(x,t|x_0,t_0), \qquad \frac{\partial}{\partial t}P(x,t) = \widehat{L}(x)P(x,t),$$

where operator

$$\widehat{L}(x) = \lambda\frac{\partial}{\partial x}x + \nu\left\{\int\limits_{-\infty}^{\infty} d\xi p(\xi)e^{-\xi\frac{\partial}{\partial x}} - 1\right\}. \tag{6.77}$$

6.3.2 Linear Operator Equation

Consider now the linear operator equation

$$\frac{d}{dt}\boldsymbol{x}(t) = \widehat{A}(t)\boldsymbol{x}(t) + z(t)\widehat{B}(t)\boldsymbol{x}(t), \quad \boldsymbol{x}(0) = \boldsymbol{x}_0, \tag{6.78}$$

where $\widehat{A}(t)$ and $\widehat{B}(t)$ are the deterministic operators (e.g., differential operators with respect to auxiliary variables or regular matrixes). We will assume that function $z(t)$ is the random delta-correlated function.

Averaging system (6.78), we obtain, according to general formulas,

$$\frac{d}{dt}\langle\boldsymbol{x}(t)\rangle = \widehat{A}(t)\langle\boldsymbol{x}(t)\rangle + \left\langle\dot{\Theta}_t\left[t, \frac{\delta}{i\delta z(t)}\right]\boldsymbol{x}(t)\right\rangle. \tag{6.79}$$

Then, taking into account the equality

$$\frac{\delta}{\delta z(t-0)}\boldsymbol{x}(t) = \widehat{B}(t)\boldsymbol{x}(t)$$

that follows immediately from Eq. (6.78), we can rewrite Eq. (6.79) in the form

$$\frac{d}{dt}\langle\boldsymbol{x}(t)\rangle = \widehat{A}(t)\langle\boldsymbol{x}(t)\rangle + \dot{\Theta}_t\left[t, -i\widehat{B}\right]\langle\boldsymbol{x}(t)\rangle. \tag{6.80}$$

Thus, in the case of linear system (6.78), equations for average values also are the linear equations.

We can expand the logarithm of the characteristic functional $\Theta[t; v(\tau)]$ of delta-correlated processes in the functional Fourier series

$$\Theta[t; v(\tau)] = \sum_{n=1}^{\infty} \frac{i^n}{n!} \int_0^t d\tau K_n(\tau) v^n(\tau), \tag{6.81}$$

where $K_n(t)$ determine the cumulant functions of process $z(t)$. Substituting Eq. (6.81) in Eq. (6.80), we obtain the equation

$$\frac{d}{dt} \langle \boldsymbol{x}(t) \rangle = \widehat{A}(t) \langle \boldsymbol{x}(t) \rangle + \sum_{n=1}^{\infty} \frac{1}{n!} K_n(t) \left[\widehat{B}(t)\right]^n \langle \boldsymbol{x}(t) \rangle. \tag{6.82}$$

If there exists power l such that $\widehat{B}^l(t) = 0$, then Eq. (6.82) assumes the form

$$\frac{d}{dt} \langle \boldsymbol{x}(t) \rangle = \widehat{A}(t) \langle \boldsymbol{x}(t) \rangle + \sum_{n=1}^{l-1} \frac{1}{n!} K_n(t) \left[\widehat{B}(t)\right]^n \langle \boldsymbol{x}(t) \rangle. \tag{6.83}$$

In this case, the equation for average value depends only on a finite number of cumulants of process $z(t)$. This means that there is no necessity in knowledge of probability distribution of function $z(t)$ in the context of the equation for average value; sufficient information includes only certain cumulants of process and knowledge of the fact that process $z(t)$ can be considered as the delta-correlated random process. Statistical description of an oscillator with fluctuating frequency is a good example of such system in physics.

Problems

Problem 6.1

Starting from the backward Liouville equation (3.8), page 71 describing the behavior of dynamic system (6.1), page 141 as a function of initial values t_0 and $\boldsymbol{x}_0$, derive the backward equation for probability density $P(\boldsymbol{x}, t|\boldsymbol{x}_0, t_0) = \langle \delta(\boldsymbol{x}(t|\boldsymbol{x}_0, t_0) - \boldsymbol{x}) \rangle$.

Solution

$$\left(\frac{\partial}{\partial t_0} + \boldsymbol{v}(\boldsymbol{x}_0, t_0) \frac{\partial}{\partial \boldsymbol{x}_0}\right) P(\boldsymbol{x}, t|\boldsymbol{x}_0, t_0) = \left\langle \dot{\Theta}_{t_0} \left[t, t_0; \frac{\delta}{i\delta \boldsymbol{f}(\boldsymbol{y}, \tau)}\right] \delta\left(\boldsymbol{x}(t|\boldsymbol{x}_0, t_0) - \boldsymbol{x}\right) \right\rangle,$$

where

$$\dot{\Theta}_{t_0}[t, t_0; \boldsymbol{u}(\boldsymbol{y}, \tau)] = \frac{d}{dt_0} \ln \Phi[t, t_0; \boldsymbol{u}(\boldsymbol{y}, \tau)]$$

and $\Phi[t, t_0; \boldsymbol{u}(\boldsymbol{y}, \tau)]$ is the characteristic functional of field $\boldsymbol{f}(\boldsymbol{x}_0, t_0)$.

Problem 6.2

Derive the equation for probability density of the solution to the one-dimensional problem on passive tracer diffusion

$$\frac{\partial}{\partial t}\rho(x,t) + v(t)\frac{\partial}{\partial x}f(x)\rho(x,t) = 0,$$

where $v(t)$ *is the Gaussian stationary random process with the parameters*

$$\langle v(t)\rangle = 0, \quad \langle v(t)v(t')\rangle = B(t-t') \quad \left(B_v(0) = \left\langle v^2(t)\right\rangle\right)$$

and $f(x)$ *is the deterministic function.*

Instruction *Average the Liouville equation (3.19), page 74 that can be rewritten in the form*

$$\frac{\partial}{\partial t}\varphi(x,t;\rho) = -v(t)\left\{\frac{\partial}{\partial x}f(x) - \frac{\partial f(x)}{\partial x}\left(1+\frac{\partial}{\partial \rho}\rho\right)\right\}\varphi(x,t;\rho),$$
$$\varphi(x,0;\rho) = \delta(\rho_0(x)-\rho)$$

over an ensemble of realizations of random process $v(t)$. *Use the Furutsu–Novikov formula (5.16), page 126 for functional* $\varphi[x,t;\rho;v(\tau)]$ *and take into account Eq. (6.12), page 145 that assumes in this problem the form*

$$\frac{\delta}{\delta v(t')}\varphi\left[x,t;\rho;v(\tau)\right] = \frac{\partial \varphi(x,T(t),\rho)}{\partial T}\theta(t-t')$$
$$= -\theta(t-t')\left\{\frac{\partial}{\partial x}f(x) - \frac{\partial f(x)}{\partial x}\left(1+\frac{\partial}{\partial \rho}\rho\right)\right\}\varphi(x,t;\rho),$$

where $T[t,v(\tau)] = \int_0^t d\tau\, v(\tau)$, *is the new "random" time, and* $\theta(t)$ *is the Heaviside step function.*

Solution

$$\frac{\partial}{\partial t}P(x,t;\rho) = \int\limits_0^t d\tau B(\tau)\left\{\frac{\partial}{\partial x}f(x) - \frac{df(x)}{dx}\left(1+\frac{\partial}{\partial \rho}\rho\right)\right\}$$
$$\times\left\{f(x)\frac{\partial}{\partial x} - \frac{df(x)}{dx}\frac{\partial}{\partial \rho}\rho\right\}P(x,t;\rho),$$

$$P(0,x;\rho) = \delta(\rho_0(x)-\rho),$$

which can be rewritten in the form

$$\frac{\partial}{\partial t}P(x,t;\rho)=\int\limits_0^t dt'\,B(t')\left\{\frac{\partial}{\partial x}f^2(x)\frac{\partial}{\partial x}-\frac{\partial}{\partial x}f(x)\frac{df(x)}{dx}\frac{\partial}{\partial \rho}\rho\right.$$

$$\left.-\frac{df(x)}{dx}f(x)\frac{\partial}{\partial x}\left(1+\frac{\partial}{\partial \rho}\rho\right)+\left[\frac{df(x)}{dx}\right]^2\frac{\partial}{\partial \rho}\rho\left(1+\frac{\partial}{\partial \rho}\rho\right)\right\}P(x,t;\rho),$$

$$P(x,0;\rho)=\delta(\rho_0(x)-\rho).$$

Remark Note that, if function $f(x)$ varies along x with characteristic scale k^{-1} and is a periodic function ('fast variation'), then additional averaging over x results in an equation for 'slow' spatial variations

$$\frac{\partial}{\partial t}P(x,t;\rho)=\int\limits_0^t dt'\,B(t')\left\{\overline{f^2(x)}\frac{\partial^2}{\partial x^2}+\overline{\left[\frac{df(x)}{dx}\right]^2}\frac{\partial^2}{\partial \rho^2}\rho^2\right\}P(x,t;\rho),$$

$$P(x,0;\rho)=\delta(\rho_0(x)-\rho).$$

If, moreover, the initial condition is independent of x (a homogeneous initial state), the average probability density will be also independent of x and satisfy the equation

$$\frac{\partial}{\partial t}P(t;\rho)=\overline{\left[\frac{df(x)}{dx}\right]^2}\int\limits_0^t dt'\,B(t')\frac{\partial^2}{\partial \rho^2}\rho^2P(t;\rho),\quad P(0;\rho)=\delta(\rho_0-\rho).$$

The solution to this equation is the probability density of the lognormal random process.

For $t\gg\tau_0$ this equation grades into the equation

$$\frac{\partial}{\partial t}P(t;\rho)=D_\rho\frac{\partial^2}{\partial \rho^2}\rho^2P(t;\rho),\quad P(0;\rho)=\delta(\rho_0-\rho),\tag{6.84}$$

where D_ρ is the diffusion coefficient in ρ-space,

$$D_\rho=\overline{\left[\frac{df(x)}{dx}\right]^2}\int\limits_0^\infty dt'\,B(t')=\sigma^2\tau_0\overline{\left[\frac{df(x)}{dx}\right]^2},$$

and τ_0 is the temporal correlation radius of random process $v(t)$. Equation (6.84) can be rewritten in the form

$$\frac{\partial}{\partial t}P(t;\rho)=D_\rho\frac{\partial}{\partial \rho}\rho P(t;\rho)+D_\rho\frac{\partial}{\partial \rho}\rho\frac{\partial}{\partial \rho}\rho P(t;\rho),\quad P(0;\rho)=\delta(\rho_0-\rho).\tag{6.85}$$

Features of a logarithmically normal random process will be considered in detail later, in Sect. 8.3.3, page 200.

Problem 6.3

Derive the equation for the average of the field satisfying the Burgers equation with random shear

$$\frac{\partial}{\partial t}q(x,t)+(q+z(t))\,\frac{\partial}{\partial x}q(x,t)=\nu\frac{\partial^2}{\partial x^2}q(x,t), \tag{6.86}$$

where $z(t)$ is the stationary Gaussian process with correlation function

$$B(t-t')=\left\langle z(t)z(t')\right\rangle.$$

Instruction *Averaging Eq. (6.86), use the Furutsu–Novikov formula*

$$\langle z(t)q(x,t)\rangle=\int\limits_0^t d\tau\, B(t-\tau)\left\langle\frac{\delta}{\delta z(\tau)}q(x,t)\right\rangle,$$

and the expression

$$\begin{aligned}&\langle q[z(\tau)+\eta_1(\tau)]q[z(\tau)+\eta_2(\tau)]\rangle\\&\quad=\exp\left\{\int\limits_0^t d\tau_1\int\limits_0^t d\tau_2\, B(\tau_1-\tau_2)\frac{\delta^2}{\delta\eta_1(\tau_1)\delta\eta_2(\tau_2)}\right\}\\&\qquad\times\langle q[z(\tau)+\eta_1(\tau)]q[z(\tau)+\eta_2(\tau)]\rangle\end{aligned}$$

to split correlations (see Lecture 5 and Problem 5.5, page 136) and rewrite these formulas, in view of Eq. (6.63), in the form

$$\langle z(t)q(x,t)\rangle=-\int\limits_0^t d\tau\, B(\tau)\frac{\partial}{\partial x}\langle q(x,t)\rangle,$$

$$\begin{aligned}\left\langle q^2(x,t)\right\rangle&=\exp\left\{2\int\limits_0^t d\tau(t-\tau)B(\tau)\frac{\partial^2}{\partial\eta_1\partial\eta_2}\right\}\langle q(x,t+\eta_1)\rangle\,\langle q(x,t+\eta_2)\rangle\,|_{\eta=0}\\&=\sum_{n=0}^{\infty}\frac{2^n}{n!}\left[\int\limits_0^t d\tau\,(t-\tau)\,B(\tau)\right]^n\left[\frac{\partial^n}{\partial x^n}\langle q(x,t)\rangle\right]^2.\end{aligned}$$

Solution

$$\frac{\partial}{\partial t}\langle q(x,t)\rangle+\frac{1}{2}\frac{\partial}{\partial x}\sum_{n=0}^{\infty}\frac{2^n}{n!}\left[\int_0^t d\tau(t-\tau)B(\tau)\right]^n\left[\frac{\partial^n}{\partial x^n}\langle q(x,t)\rangle\right]^2$$
$$=\left(\nu+\int_0^t d\tau B(\tau)\right)\frac{\partial^2}{\partial x^2}\langle q(x,t)\rangle\,. \tag{6.87}$$

Problem 6.4

Derive Eq. (6.87) for average field starting from the one-dimensional Burger's equation with random drift (6.86) and Eq. (6.57), where $z(t)$ is the stationary Gaussian process with correlation function

$$B(t-t')=\left\langle z(t)z(t')\right\rangle.$$

Instruction *Use equality*

$$\langle q(x,t)\rangle=\exp\left\{\frac{1}{2}\int_0^t d\tau_1\int_0^t d\tau_2\,B(\tau_1-\tau_2)\frac{\partial^2}{\partial x^2}\right\}Q(x,t)$$
$$=\exp\left\{\int_0^t d\tau\,(t-\tau)B(\tau)\frac{\partial^2}{\partial x^2}\right\}Q(x,t).$$

Problem 6.5

Under the assumption that random process $z(t)$ is delta-correlated, derive the equation for probability density of the solution to stochastic equation (1.30), page 19 describing the stochastic parametric resonance.

Instruction *Rewrite Eq. (1.30) in the form of the system of equations*

$$\frac{d}{dt}x(t)=y(t),\quad \frac{d}{dt}y(t)=-\omega_0^2[1+z(t)]x(t),$$
$$x(0)=x_0,\quad y(0)=y_0. \tag{6.88}$$

Solution

$$\frac{\partial}{\partial t}P(x,y,t)=\left(-y\frac{\partial}{\partial x}+\omega_0^2 x\frac{\partial}{\partial y}\right)P(x,y,t)+\dot{\Theta}_t\left[t;-i\omega_0^2 x\frac{\partial}{\partial y}\right]P(x,y,t),$$
$$P(x,y,0)=\delta(x-x_0)\delta(y-y_0), \tag{6.89}$$

where $\dot{\Theta}_t[t; v(\tau)] = \frac{d}{dt}\Theta[t; v(\tau)]$, $\Theta[t; v(\tau)] = \ln\Phi[t; v(\tau)]$, and $\Phi[t; v(\tau)]$ is the characteristic functional of random process $z(t)$.

Problem 6.6

Starting from system of equations (6.88), derive the equation for the average of quantity

$$A_k(t) = x^k(t)y^{N-k}(t) \quad (k = 0, \ldots, N)$$

under the assumption that $z(t)$ is the delta-correlated random process.

Solution

$$\frac{d}{dt}\langle A_k(t)\rangle = k\langle A_{k-1}(t)\rangle - \omega_0^2(N-k)\langle A_{k+1}(t)\rangle + \sum_{n=1}^{N}\frac{1}{n!}K_n\left[B^n\right]_{kl}\langle A_l(t)\rangle \quad (k = 0, \ldots, N),$$

where K_n are the cumulants of random process $z(t)$ and matrix

$$B_{ij} = -\omega_0^2(N-i)\delta_{i,j-1}.$$

The square of this matrix has the form $\widehat{B}^2_{ij} = -\omega_0^4(N-i)(N-j+i)\delta_{i,j-2}$ and so forth for higher powers, so that $\widehat{B}^{N+1} \equiv 0$.

Problem 6.7

Derive the Fokker–Planck equation for system of equations (6.88) in the case of the Gaussian delta-correlated process $z(t)$ with the parameters

$$\langle z(t)\rangle = 0, \quad \langle z(t)z(t')\rangle = 2\sigma^2\tau_0\delta(t-t').$$

Solution

$$\frac{\partial}{\partial t}P(x,y,t) = \left(-y\frac{\partial}{\partial x} + \omega_0^2 x\frac{\partial}{\partial y}\right)P(x,y,t) + D\omega_0^2x^2\frac{\partial^2}{\partial y^2}P(x,y,t), \tag{6.90}$$
$$P(x,y,0) = \delta(x-x_0)\,\delta(y-y_0),$$

where $D = \sigma^2\tau_0\omega_0^2$ is the coefficient of diffusion in space $\{x, y/\omega_0\}$.

Problem 6.8

Obtain the solutions for the first- and second-order moments, which follow from Eq. (6.90) for $D/\omega_0 \ll 1$ and $x(0) = 0$, $y(0) = \omega_0$.

Solution

$$\langle x(t)\rangle = \sin\omega_0 t, \qquad \langle y(t)\rangle = \omega_0\cos\omega_0 t,$$

which coincides with the solution to system (6.88) for absent fluctuations, and

$$\begin{aligned}
\left\langle x^2(t)\right\rangle &= \frac{1}{2}\left\{e^{Dt} - e^{-Dt/2}\left[\cos(2\omega_0 t) + \frac{3D}{4\omega_0}\sin(2\omega_0 t)\right]\right\},\\
\langle x(t)y(t)\rangle &= \frac{\omega_0}{4}\left\{2e^{-Dt/2}\sin(2\omega_0 t) + \frac{D}{\omega_0}\left[e^{Dt} - e^{-Dt/2}\cos(2\omega_0 t)\right]\right\},\\
\left\langle y^2(t)\right\rangle &= \frac{\omega_0^2}{2}\left\{e^{Dt} + e^{-Dt/2}\left[\cos(2\omega_0 t) - \frac{D}{4\omega_0}\sin(2\omega_0 t)\right]\right\}
\end{aligned} \tag{6.91}$$

Remark Solution (6.19) includes terms increasing with time, which corresponds to statistical parametric build-up of oscillations in a dynamic system (6.88) because of fluctuations of frequency. Solutions to a statistical problem (6.88) have two characteristic temporal scales $t_1 \sim 1/\omega_0$ and $t_2 \sim 1/D$. The first temporal scale corresponds to the period of oscillations in system (6.88) without fluctuations (fast processes), and the second scale characterizes slow variations caused in statistical characteristics by fluctuations (slow processes). The ratio of these scales is small:

$$t_1/t_2 = D/\omega_0 \ll 1.$$

We can explicitly obtain slow variations of statistical characteristics of processes $x(t)$ and $y(t)$ by excluding fast motions by averaging the corresponding quantities over the period $T = 2\pi/\omega_0$. Denoting such averaging with the overbar, we have

$$\overline{\left\langle x^2(t)\right\rangle} = \frac{1}{2}e^{Dt}, \quad \overline{\langle x(t)y(t)\rangle} = 0, \quad \overline{\left\langle y^2(t)\right\rangle} = \frac{\omega_0^2}{2}e^{Dt}.$$

Problem 6.9

Starting from (6.90), derive the third-order equation for average potential energy of an oscillator $\langle U(t)\rangle = \left\langle x^2(t)\right\rangle$.

Solution

$$\frac{d^3}{dt^3}\langle U(t)\rangle + 4\omega_0^2\frac{d}{dt}\langle U(t)\rangle - 4D\omega_0^2\langle U(t)\rangle = 0, \tag{6.92}$$

Problem 6.10

Derive the Fokker–Planck equation for the stochastic oscillator with linear friction

$$\frac{d}{dt}x(t) = y(t), \quad \frac{d}{dt}y(t) = -2\gamma y(t) - \omega_0^2[1 + z(t)]x(t),$$
$$x(0) = x_0, \quad y(0) = y_0, \tag{6.93}$$

and the condition of stochastic excitation of second moments.

Solution

$$\frac{\partial}{\partial t}P(x, y, t) = \left(2\gamma\frac{\partial}{\partial y}y - y\frac{\partial}{\partial x} + \omega_0^2 x\frac{\partial}{\partial y}\right)P(x, y, t) + D\omega_0^2 x^2\frac{\partial^2}{\partial y^2}P(x, y, t),$$
$$P(x, y, 0) = \delta(x - x_0)\delta(y - y_0),$$

where $D = \sigma^2\tau_0\omega_0^2$ is the diffusion coefficient. If we substitute in the system of equations for second moments the solutions proportional to $\exp\{\lambda t\}$, we obtain the characteristic equation in λ of the form

$$\lambda^3 + 6\gamma\lambda^2 + 4(\omega_0^2 + 2\gamma^2)\lambda + 4\omega_0^2(2\gamma - D) = 0.$$

Under the condition

$$2\gamma < D, \tag{6.94}$$

second moments are the functions exponentially increasing with time, which means the statistical parametric excitation of second moments.

Problem 6.11

Obtain steady-state values of second moments formed under the action of random forces on the stochastic parametric oscillator with friction

$$\frac{d}{dt}x(t) = y(t), \quad x(0) = x_0,$$
$$\frac{d}{dt}y(t) = -2\gamma y(t) - \omega_0^2[1 + z(t)]x(t) + f(t), \quad y(0) = y_0. \tag{6.95}$$

Here, $f(t)$ is the Gaussian delta-correlated process statistically independent of process $z(t)$ and having the parameters

$$\langle f(t)\rangle = 0, \quad \langle f(t)f(t')\rangle = 2\sigma_f^2\tau_f\delta(t-t'),$$

where σ_f^2 is the variance and τ_f is the temporal correlation radius.

Solution In the case of stochastic system of equations (6.95), the one-time probability density satisfies the Fokker–Planck equation

$$\frac{\partial}{\partial t}P(x,y,t) = \left(2\gamma\frac{\partial}{\partial y}y - y\frac{\partial}{\partial x} + \omega_0^2 x\frac{\partial}{\partial y}\right)P(x,y,t) + D\omega_0^2 x^2\frac{\partial^2}{\partial y^2}P(x,y,t) + \sigma_f^2\tau_f\frac{\partial^2}{\partial y^2}P(x,y,t),$$

$$P(x,y,0) = \delta(x-x_0)\,\delta(y-y_0).$$

Consequently, the steady-state solution for second moments exists for $t \to \infty$ under the condition (6.94) and has the form

$$\langle x(t)\rangle = 0, \quad \langle y(t)\rangle = 0,$$

$$\langle x(t)y(t)\rangle = 0, \quad \left\langle x^2(t)\right\rangle = \frac{\sigma_f^2\tau_f}{\omega_0^2(D-2\gamma)}, \quad \left\langle y^2(t)\right\rangle = \frac{\sigma_f^2\tau_f}{D-2\gamma}.$$

Problem 6.12

Starting from (6.89), derive the Kolmogorov–Feller equation in the case of the Poisson delta-correlated random process $z(t)$.

Solution

$$\frac{\partial}{\partial t}P(x,y,t) = \left(-y\frac{\partial}{\partial x} + \omega_0^2 x\frac{\partial}{\partial y}\right)P(x,y,t) + \nu\int_{-\infty}^{\infty} d\xi p(\xi)P(t;x,y+\xi\omega_0^2 x) - \nu P(x,y,t). \tag{6.96}$$

Remark For sufficiently small ξ, Eq. (6.96) grades into the Fokker–Planck equation (6.90) with the diffusion coefficient

$$D = \frac{1}{2}\nu\left\langle\xi^2\right\rangle\omega_0^2 x.$$

Problem 6.13

Derive the equation for the characteristic functional of the solution to linear parabolic equation (1.89), page 39 by averaging Eq. (3.48), page 82.

Solution Characteristic functional of the problem solution

$$\Phi[x; v, v^*] = \left\langle\varphi[x; v, v^*]\right\rangle,$$

where

$$\varphi[x; v, v^*] = \exp\left\{i\int d\boldsymbol{R}'\left[u(x,\boldsymbol{R}')v(\boldsymbol{R}')+u^*(x,\boldsymbol{R}')v^*(\boldsymbol{R}')\right]\right\},$$

satisfies the equation

$$\begin{aligned}\frac{\partial}{\partial x}\Phi[x; v, v^*] &= \left\langle\dot{\Theta}_x\left[x; \frac{\delta}{i\delta\varepsilon(\xi,\boldsymbol{R}')}\right]\varphi[x; v, v^*]\right\rangle\\ &+ \frac{i}{2k}\left\{\int d\boldsymbol{R}'\left[v(\boldsymbol{R}')\Delta_{\boldsymbol{R}'}\frac{\delta}{\delta v(\boldsymbol{R}')} - v^*(\boldsymbol{R}')\Delta_{\boldsymbol{R}'}\frac{\delta}{\delta v^*(\boldsymbol{R}')}\right]\right\}\\ &\times\Phi[x; v, v^*],\end{aligned} \tag{6.97}$$

where

$$\dot{\Theta}_x\left[x; \psi(\xi,\boldsymbol{R}')\right] = \frac{d}{dx}\ln\left\langle\exp\left\{i\int\limits_0^x d\xi\int d\boldsymbol{R}'\varepsilon(\xi,\boldsymbol{R}')\psi(\xi,\boldsymbol{R}')\right\}\right\rangle$$

is the derivative of the characteristic functional logarithm of field $\varepsilon(x,\boldsymbol{R})$.

Problem 6.14

Starting from Eq. (6.97), derive the equation for the characteristic functional of the solution to linear parabolic equation (1.89), page 39 assuming that $\varepsilon(x,\boldsymbol{R})$ is the homogeneous random field delta-correlated in x.

Solution Characteristic functional $\Phi[x; v, v^*]$ of the problem solution satisfies the equation

$$\frac{\partial}{\partial x}\Phi[x; v, v^*] = \dot{\Theta}_x\left[x, \frac{k}{2}\widehat{M}(\mathbf{R}')\right]\Phi[x; v, v^*]$$
$$+\frac{i}{2k}\left\{\int d\mathbf{R}'\left[v(\mathbf{R}')\Delta_{\mathbf{R}'}\frac{\delta}{\delta v(\mathbf{R}')} - v^*(\mathbf{R}')\Delta_{\mathbf{R}'}\frac{\delta}{\delta v^*(\mathbf{R}')}\right]\right\}$$
$$\times\,\Phi[x; v, v^*] \qquad (6.98)$$

with the Hermitian operator

$$\widehat{M}(\mathbf{R}') = v(\mathbf{R}')\frac{\delta}{\delta v(\mathbf{R}')} - v^*(\mathbf{R}')\frac{\delta}{\delta v^*(\mathbf{R}')}.$$

Here, functional

$$\dot{\Theta}_x\left[x; \psi(\xi, \mathbf{R}')\right] = \frac{d}{dx}\ln\left\langle \exp\left\{i\int\limits_0^x d\xi\int d\mathbf{R}'\varepsilon(\xi, \mathbf{R}')\psi(\xi, \mathbf{R}')\right\}\right\rangle$$

is the derivative of the logarithm of the characteristic functional of filed $\varepsilon(x, \mathbf{R})$.

Remark Equation (6.98) yields the equations for the moment functions of field $u(x, \mathbf{R})$,

$$M_{m,n}(x; \mathbf{R}_1, \ldots, \mathbf{R}_m; \mathbf{R}'_1, \ldots, \mathbf{R}'_n)$$
$$= \left\langle u(x, \mathbf{R}_1)\ldots u(x, \mathbf{R}_m)u^*(x, \mathbf{R}'_1)\ldots u^*(x, \mathbf{R}'_n)\right\rangle,$$

(for $m = n$, these functions are usually called the *coherence functions* of order $2n$)

$$\frac{\partial}{\partial x}M_{m,n} = \frac{i}{2k}\left(\sum_{p=1}^{m}\Delta_{\mathbf{R}_p} - \sum_{q=1}^{n}\Delta_{\mathbf{R}'_q}\right)M_{m,n}$$
$$+\,\dot{\Theta}_x\left[x, \frac{1}{k}\left(\sum_{p=1}^{m}\delta(\mathbf{R}' - \mathbf{R}_p) - \sum_{q=1}^{n}\delta(\mathbf{R}' - \mathbf{R}'_q)\right)\right]M_{m,n}. \qquad (6.99)$$

Problem 6.15

Assuming that $\varepsilon(x, \boldsymbol{R})$ is the homogeneous Gaussian random process delta-correlated in x with the correlation function

$$B_{\varepsilon}(x, \boldsymbol{R}) = A(\boldsymbol{R})\delta(x), \qquad A(\boldsymbol{R}) = \int_{-\infty}^{\infty} dx B_{\varepsilon}(x, \boldsymbol{R})$$

and starting from the solution to the previous problem, derive the equations for the characteristic functional of wave field $u(x, \boldsymbol{R})$, which is the solution to linear parabolic equation (1.89), page 39, and moment functions of this field [50].

Solution

$$\frac{\partial}{\partial x}\Phi[x; v, v^*] = -\frac{k^2}{8}\int d\boldsymbol{R}' \int d\boldsymbol{R} A(\boldsymbol{R}' - \boldsymbol{R})\widehat{M}(\boldsymbol{R}')\widehat{M}(\boldsymbol{R})\Phi[x; v, v^*]$$

$$+ \frac{i}{2k}\int d\boldsymbol{R}' \left[v(\boldsymbol{R}')\Delta_{\boldsymbol{R}'}\frac{\delta}{\delta v(\boldsymbol{R}')} - v^*(\boldsymbol{R}')\Delta_{\boldsymbol{R}'}\frac{\delta}{\delta v^*(\boldsymbol{R}')}\right]\Phi[x; v, v^*],$$

$$\partial x M_{m,n} = \frac{i}{2k}\left(\sum_{p=1}^{m}\Delta_{\boldsymbol{R}_p} - \sum_{q=1}^{n}\Delta_{\boldsymbol{R}'_q}\right) M_{m,n}$$

$$- \frac{k^2}{8} Q(\boldsymbol{R}_1, \ldots, \boldsymbol{R}_m; \boldsymbol{R}'_1, \ldots, \boldsymbol{R}'_m) M_{m,n},$$

where

$$Q(\boldsymbol{R}_1, \ldots, \boldsymbol{R}_m; \boldsymbol{R}'_1, \ldots, \boldsymbol{R}'_m) =$$

$$= \sum_{i=1}^{m}\sum_{j=1}^{m} A(\boldsymbol{R}_i - \boldsymbol{R}_j) - 2\sum_{i=1}^{m}\sum_{j=1}^{n} A(\boldsymbol{R}_i - \boldsymbol{R}'_j) + \sum_{i=1}^{n}\sum_{j=1}^{n} A(\boldsymbol{R}'_i - \boldsymbol{R}'_j).$$

Problem 6.16

Derive the equation for the one-time characteristic functional of the solution to stochastic nonlinear integro-differential equation (1.102), page 44.

Solution Characteristic functional of the velocity field

$$\Phi[t; z(\boldsymbol{k}')] = \Phi[t; z] = \langle \varphi[t; z(\boldsymbol{k}')] \rangle,$$

where

$$\varphi[t; z(\boldsymbol{k}')] = \exp\left\{i \int d\boldsymbol{k}' \widehat{\boldsymbol{u}}(\boldsymbol{k}', t) z(\boldsymbol{k}')\right\},$$

satisfies the unclosed variational-derivative equation

$$\frac{\partial}{\partial t}\Phi[t; z] = \int d\boldsymbol{k} z_i(\boldsymbol{k}) \left\{-\frac{1}{2}\int d\boldsymbol{k}_1 \int d\boldsymbol{k}_2 \Lambda_{i,\alpha\beta}(\boldsymbol{k}_1, \boldsymbol{k}_2, \boldsymbol{k}) \frac{\delta^2}{\delta z_\alpha(\boldsymbol{k}_1)\delta z_\beta(\boldsymbol{k}_2)} - \nu k^2 \frac{\delta}{\delta z_i(\boldsymbol{k})}\Phi[t; z]\right\} + \left\langle \dot{\Theta}_t \left[t; \frac{\delta}{i\delta \boldsymbol{f}(\boldsymbol{\kappa}, \tau)}\right] \varphi[t; z]\right\rangle,$$

where

$$\dot{\Theta}_t [t; \boldsymbol{\psi}(\boldsymbol{\kappa}, \tau)] = \frac{d}{dt} \ln \left\langle \exp\left\{i \int_0^t d\tau \int d\boldsymbol{\kappa} \boldsymbol{f}(\boldsymbol{\kappa}, \tau) \boldsymbol{\psi}(\boldsymbol{\kappa}, \tau)\right\}\right\rangle$$

is the derivative of the characteristic functional logarithm of external forces $\widehat{\boldsymbol{f}}(\boldsymbol{k}, t)$.

Problem 6.17

Assuming that $\boldsymbol{f}(\boldsymbol{x}, t)$ is the homogeneous stationary random field delta-correlated in time, derive the equation for the one-time characteristic functional of the solution to stochastic nonlinear integro-differential equation (1.102), page 44. Consider the case of the Gaussian field $\boldsymbol{f}(\boldsymbol{x}, t)$ [1].

Solution

$$\frac{\partial}{\partial t}\Phi[t; z] = \int d\boldsymbol{k} z_i(\boldsymbol{k}) \left\{-\frac{1}{2}\int d\boldsymbol{k}_1 \int d\boldsymbol{k}_2 \Lambda_{l,\alpha\beta}(\boldsymbol{k}_1, \boldsymbol{k}_2, \boldsymbol{k}) \frac{\delta^2}{\delta z_\alpha(\boldsymbol{k}_1)\delta z_\beta(\boldsymbol{k}_2)} - \nu k^2 \frac{\delta}{\delta z_i(\boldsymbol{k})}\right\} \Phi[t; z] + \dot{\Theta}_t [t; z(\boldsymbol{k})] \Phi[t; z], \tag{6.100}$$

where

$$\dot{\Theta}_t [t; \boldsymbol{\psi}(\boldsymbol{\kappa}, \tau)] = \frac{d}{dt} \ln \left\langle \exp\left\{i \int_0^t d\tau \int d\boldsymbol{\kappa} \widehat{\boldsymbol{f}}(\boldsymbol{\kappa}, \tau) \boldsymbol{\psi}(\boldsymbol{\kappa}, \tau)\right\}\right\rangle$$

is the derivative of the characteristic functional logarithm of external forces $\widehat{\boldsymbol{f}}(\boldsymbol{k}, t)$.

If we assume now that $\boldsymbol{f}(\boldsymbol{x}, t)$ is the Gaussian random field homogeneous and isotropic in space and stationary in time with the correlation tensor

$$B_{ij}(\boldsymbol{x}_1 - \boldsymbol{x}_2, t_1 - t_2) = \left\langle f_i(\boldsymbol{x}_1, t_1) f_j(\boldsymbol{x}_2, t_2) \right\rangle,$$

then the field $\widehat{\boldsymbol{f}}(\boldsymbol{k}, t)$ will also be the Gaussian stationary random field with the correlation tensor

$$\left\langle \widehat{f}_i(\boldsymbol{k}, t+\tau) \widehat{f}_j(\boldsymbol{k}', t) \right\rangle = \frac{1}{2} F_{ij}(\boldsymbol{k}, \tau) \delta(\boldsymbol{k} + \boldsymbol{k}),$$

where $F_{ij}(\boldsymbol{k}, \tau)$ is the external force spatial spectrum given by the formula

$$F_{ij}(\boldsymbol{k}, \tau) = 2(2\pi)^3 \int d\boldsymbol{x} B_{ij}(\boldsymbol{x}, \tau) e^{-i\boldsymbol{k}\boldsymbol{x}}.$$

In view of the fact that forces are spatially isotropic, we have

$$F_{ij}(\boldsymbol{k}, \tau) = F(k, \tau) \Delta_{ij}(\boldsymbol{k}).$$

As long as field $\widehat{\boldsymbol{f}}(\boldsymbol{k}, t)$ is delta-correlated in time, we have that $F(k, \tau) = F(k)\delta(\tau)$, so that functional $\Theta\,[t; \boldsymbol{\psi}(\boldsymbol{\kappa}, \tau)]$ is given by the formula

$$\Theta\,[t; \boldsymbol{\psi}(\boldsymbol{\kappa}, \tau)] = -\frac{1}{4} \int\limits_0^t d\tau \int d\boldsymbol{\kappa}\, F(\kappa) \Delta_{ij}(\boldsymbol{\kappa}) \psi_i(\boldsymbol{\kappa}, \tau) \psi_j(-\boldsymbol{\kappa}, \tau),$$

and Eq. (6.100) assumes the closed form

$$\frac{\partial}{\partial t} \Phi[t; z] = -\frac{1}{4} \int d\boldsymbol{k} F(k) \Delta_{ij}(\boldsymbol{k}) z_i(\boldsymbol{k}) z_j(-\boldsymbol{k}) \Phi[t; z]$$

$$- \int d\boldsymbol{k} z_i(\boldsymbol{k}) \left\{ \frac{1}{2} \int d\boldsymbol{k}_1 \int d\boldsymbol{k}_2 \Lambda_{i,\alpha\beta}(\boldsymbol{k}_1, \boldsymbol{k}_2, \boldsymbol{k}) \frac{\delta^2}{\delta z_\alpha(\boldsymbol{k}_1) \delta z_\beta(\boldsymbol{k}_2)} \Phi[t; z] \right.$$

$$\left. + \nu k^2 \frac{\delta}{\delta z_i(\boldsymbol{k})} \int d\boldsymbol{k} z_i(\boldsymbol{k}) \frac{\delta}{\delta z_i(\boldsymbol{k})} \right\} \Phi[t; z].$$

Problem 6.18

Average the linear stochastic integral equation

$$S(\boldsymbol{r},\boldsymbol{r}') = S_0(\boldsymbol{r},\boldsymbol{r}') + \int d\boldsymbol{r}_1 \int d\boldsymbol{r}_2 \int d\boldsymbol{r}_3 S_0(\boldsymbol{r},\boldsymbol{r}_1)\Lambda(\boldsymbol{r}_1,\boldsymbol{r}_2,\boldsymbol{r}_3)\left[\eta(\boldsymbol{r}_2)+f(\boldsymbol{r}_2)\right]S(\boldsymbol{r}_3,\boldsymbol{r}'),$$

Solution

$$G[\boldsymbol{r},\boldsymbol{r}';\eta(\boldsymbol{r})] = S_0(\boldsymbol{r},\boldsymbol{r}') + \int d\boldsymbol{r}_1 \int d\boldsymbol{r}_2 \int d\boldsymbol{r}_3 S_0(\boldsymbol{r},\boldsymbol{r}_1)\Lambda(\boldsymbol{r}_1,\boldsymbol{r}_2,\boldsymbol{r}_3)\eta(\boldsymbol{r}_2)G[\boldsymbol{r}_3,\boldsymbol{r}';\eta(\boldsymbol{r})] + \int d\boldsymbol{r}_1 \int d\boldsymbol{r}_2 \int d\boldsymbol{r}_3 S_0(\boldsymbol{r},\boldsymbol{r}_1)\Lambda(\boldsymbol{r}_1,\boldsymbol{r}_2,\boldsymbol{r}_3)\Omega_{\boldsymbol{r}_2}\left[\frac{\delta}{i\delta\eta(\boldsymbol{r})}\right]G[\boldsymbol{r}_3,\boldsymbol{r}';\eta(\boldsymbol{r})].$$

Here,

$$G[\boldsymbol{r},\boldsymbol{r}';\eta(\boldsymbol{r})] = \left\langle S[\boldsymbol{r},\boldsymbol{r}';f(\boldsymbol{r})+\eta(\boldsymbol{r})]\right\rangle$$

and functional

$$\Omega_{\boldsymbol{r}}\left[v(\boldsymbol{r})\right] = \frac{\delta}{i\delta v(\boldsymbol{r})}\Theta\left[v(\boldsymbol{r})\right], \quad \Theta\left[v(\boldsymbol{r})\right] = \ln\left\langle \exp\left\{i\int d\boldsymbol{r} f(\boldsymbol{r})v(\boldsymbol{r})\right\}\right\rangle.$$

Problem 6.19

Assuming that random external force $\boldsymbol{f}(\boldsymbol{r},t)$ is the homogeneous stationary Gaussian field, derive the equation for the characteristic functional of space-time harmonics of turbulent velocity field (1.103), page 45 [51].

Solution Characteristic functional of the velocity field

$$\Phi[z] = \left\langle \exp\left\{i\int d^4\boldsymbol{K}'\widehat{\boldsymbol{u}}\left(\boldsymbol{K}'\right)z\left(\boldsymbol{K}'\right)\right\}\right\rangle$$

and functional

$$G_{ij}\left[\boldsymbol{K},\boldsymbol{K}';z\right] = \left\langle \frac{\delta u_i(\boldsymbol{K})}{\delta\widehat{f}_j(\boldsymbol{K}')}\varphi[z]\right\rangle$$

satisfy the closed system of functional equations

$$(i\omega+\nu\boldsymbol{k}^2)\frac{\delta}{\delta z_i(\boldsymbol{K})}\Phi[z]=-\frac{1}{2}F_{ij}(\boldsymbol{K})\int d^4\boldsymbol{K}_1 z_\alpha(\boldsymbol{K}_1)\,G_{\alpha j}[\boldsymbol{K}_1,-\boldsymbol{K};z]$$
$$-\frac{1}{2}\int d^4\boldsymbol{K}_1\int d^4\boldsymbol{K}_2\Lambda_i^{\alpha\beta}(\boldsymbol{K}_1,\boldsymbol{K}_2,\boldsymbol{K})\frac{\delta^2\Phi[z]}{\delta z_\alpha(\boldsymbol{K}_1)\,\delta z_\beta(\boldsymbol{K}_2)},$$
$$(i\omega+\nu\boldsymbol{k}^2)G_{ij}\left[\boldsymbol{K},\boldsymbol{K}';z\right]$$
$$+\int d^4\boldsymbol{K}_1\int d^4\boldsymbol{K}_2\Lambda_i^{\alpha\beta}(\boldsymbol{K}_1,\boldsymbol{K}_2,\boldsymbol{K})\frac{\delta}{\delta z_\alpha(\boldsymbol{K}_1)}G_{\beta j}\left[\boldsymbol{K}_2,\boldsymbol{K}';z\right]$$
$$=\delta_{ij}\delta^4\left(\boldsymbol{K}-\boldsymbol{K}'\right)\Phi[z],$$

where $\left\langle\widehat{f}_i(\boldsymbol{K}_1)\widehat{f}_j(\boldsymbol{K}_2)\right\rangle=\frac{1}{2}\delta^4(\boldsymbol{K}_1+\boldsymbol{K}_2)\,F_{ij}(\boldsymbol{K}_1)$, and $F_{ij}(\boldsymbol{K})$ is the space-time spectrum of external forces.

Remark Expansion of functionals $\Phi[z]$ and $G_{ij}\left[\boldsymbol{K},\boldsymbol{K}';z\right]$ in the functional Taylor series in $z(\boldsymbol{K})$ determines velocity field cumulants and correlators of Green's function analog $S_{ij}\left(\boldsymbol{K},\boldsymbol{K}'\right)=\delta u_i(\boldsymbol{K})/\delta f_j\left(\boldsymbol{K}'\right)$ with the velocity field.

Lecture 7

Stochastic Equations with the Markovian Fluctuations of Parameters

In the preceding Lecture, we dealt with the statistical description of dynamic systems in terms of general methods that assumed knowledge of the characteristic functional of fluctuating parameters. However, this functional is unknown in most cases, and we are forced to resort either to assumptions on the model of parameter fluctuations, or to asymptotic approximations.

The methods based on approximating fluctuating parameters with the Markovian random processes and fields with a finite temporal correlation radius are widely used. Such approximations can be found, for example, as solutions to the dynamic equations with delta-correlated fluctuations of parameters. Consider such methods in greater detail by the examples of the Markovian random processes.

Consider stochastic equations of the form

$$\frac{d}{dt}\boldsymbol{x}(t) = f\left(t, \boldsymbol{x}, z(t)\right), \quad \boldsymbol{x}(0) = \boldsymbol{x}_0, \tag{7.1}$$

where $f\left(t, \boldsymbol{x}, z(t)\right)$ is the deterministic function and

$$z(t) = \{z_1(t), \ldots, z_n(t)\}$$

is the Markovian vector process.

Our task consists in the determination of statistical characteristics of the solution to Eq. (7.1) from known statistical characteristics of process $z(t)$.

In the general case of arbitrary Markovian process $z(t)$, we cannot judge about process $\boldsymbol{x}(t)$. We can only assert that the joint process $\{\boldsymbol{x}(t), z(t)\}$ is the Markovian process with joint probability density $P(\boldsymbol{x}, z, t)$.

If we could solve the equation for $P(\boldsymbol{x}, z, t)$, then we could integrate the solution over z to obtain the probability density of the solution to Eq. (7.1), i.e., function $P(\boldsymbol{x}, t)$. In this case, process $\boldsymbol{x}(t)$ by itself would not be the Markovian process.

There are several types of process $z(t)$ that allow us to obtain equations for density $P(\boldsymbol{x}, t)$ without solving the equation for $P(\boldsymbol{x}, z, t)$. Among these processes, we mention first of all the telegrapher's and generalized telegrapher's processes, Markovian

Lectures on Dynamics of Stochastic Systems. DOI: 10.1016/B978-0-12-384966-3.00007-6

processes with finite number of states, and Gaussian Markovian processes. Below, we discuss the telegrapher's and Gaussian Markovian processes in more detail as examples of processes widely used in different applications.

7.1 Telegrapher's Processes

Recall that telegrapher's random process $z(t)$ (the two-state, or binary process) is defined by the equality

$$z(t) = a(-1)^{n(0,t)},$$

where random quantity a assumes values $a = \pm a_0$ with probabilities $1/2$ and $n(t_1, t_2)$, $t_1 < t_2$ is Poisson's integer-valued process with mean value $\overline{n(t_1, t_2)} = \nu|t_1 - t_2|$.

Telegrapher's process $z(t)$ is stationary in time and its correlation function

$$\langle z(t)z(t')\rangle = a_0^2 e^{-2\nu|t-t'|}$$

has the temporal correlation radius $\tau_0 = 1/(2\nu)$.

For splitting the correlation between telegrapher's process $z(t)$ and arbitrary functional $R[t; z(\tau)]$, where $\tau \leq t$, we obtained the relationship (5.25), (page 129)

$$\langle z(t)R[t; z(\tau)]\rangle = a_0^2 \int_0^t dt_1 e^{-2\nu(t-t_1)} \left\langle \frac{\delta}{\delta z(t_1)} \widetilde{R}[t, t_1; z(\tau)] \right\rangle, \tag{7.2}$$

where functional $\widetilde{R}[t, t_1; z(\tau)]$ is given by the formula

$$\widetilde{R}[t, t_1; z(\tau)] = R[t; z(\tau)\theta(t_1 - \tau + 0)] \quad (t_1 < t). \tag{7.3}$$

Formula (7.2) is appropriate for analyzing stochastic equations linear in process $z(t)$. Functional $R[t; z(\tau)]$ is the solution to a system of differential equations of the first order in time with initial values at $t = 0$. Functional $\widetilde{R}[t, t_1; z(\tau)]$ will also satisfy the same system of equations with product $z(t)\theta(t_1 - t)$ instead of $z(t)$. Consequently, we obtain that functional $\widetilde{R}[t, t_1; z(\tau)] = R[t; 0]$ for all times $t > t_1$; moreover, it satisfies the same system of equations for absent fluctuations (i.e., at $z(t) = 0$) with the initial value $\widetilde{R}[t_1, t_1; z(\tau)] = R[t_1; z(\tau)]$.

Another formula convenient in the context of stochastic equations linear in random telegrapher's process $z(t)$ concerns the differentiation of a correlator of this process with arbitrary functional $R[t; z(\tau)](\tau \leq t)$ (5.26)

$$\frac{d}{dt}\langle z(t)R[t; z(\tau)]\rangle = -2\nu \langle z(t)R[t; z(\tau)]\rangle + \left\langle z(t)\frac{d}{dt}R[t; z(\tau)]\right\rangle. \tag{7.4}$$

Formula (7.4) determines the rule of factoring the differentiation operation out of averaging brackets

$$\left\langle z(t)\frac{d^n}{dt^n}R[t;z(\tau)]\right\rangle = \left(\frac{d}{dt}+2\nu\right)^n \langle z(t)R[t;z(\tau)]\rangle . \tag{7.5}$$

We consider some special examples to show the usability of these formulas. It is evident that both methods give the same result. However, the method based on the differentiation formula appears more practicable.

The first example concerns the system of linear operator equations

$$\frac{d}{dt}\boldsymbol{x}(t) = \widehat{A}(t)\boldsymbol{x}(t) + z(t)\widehat{B}(t)\boldsymbol{x}(t), \quad \boldsymbol{x}(0) = \boldsymbol{x}_0, \tag{7.6}$$

where $\widehat{A}(t)$ and $\widehat{B}(t)$ are certain differential operators (they may include differential operators with respect to auxiliary variables). If operators $\widehat{A}(t)$ and $\widehat{B}(t)$ are matrices, then Eqs. (7.6) describe the linear dynamic system.

Average Eqs. (7.6) over an ensemble of random functions $z(t)$. The result will be the vector equation

$$\frac{d}{dt}\langle\boldsymbol{x}(t)\rangle = \widehat{A}(t)\langle\boldsymbol{x}(t)\rangle + \widehat{B}(t)\boldsymbol{\psi}(t), \quad \langle\boldsymbol{x}(0)\rangle = \boldsymbol{x}_0, \tag{7.7}$$

where we introduced new functions

$$\boldsymbol{\psi}(t) = \langle z(t)\boldsymbol{x}(t)\rangle .$$

We can use formula Eqs. (7.4) for these functions; as a result, we obtain the equality

$$\frac{d}{dt}\boldsymbol{\psi}(t) = -2\nu\boldsymbol{\psi}(t) + \left\langle z(t)\frac{d}{dt}\boldsymbol{x}(t)\right\rangle . \tag{7.8}$$

Substituting now derivative $d\boldsymbol{x}/dt$ (7.6) in Eq. (7.8), we obtain the equation for the vector function $\boldsymbol{\psi}(t)$

$$\left(\frac{d}{dt}+2\nu\right)\boldsymbol{\psi}(t) = \widehat{A}(t)\boldsymbol{\psi}(t) + \widehat{B}(t)\left\langle z^2(t)\boldsymbol{x}(t)\right\rangle . \tag{7.9}$$

Because $z^2(t) \equiv a_0^2$ for telegrapher's process, we obtain finally the closed system of linear equations for vectors $\langle\boldsymbol{x}(t)\rangle$ and $\boldsymbol{\psi}(t)$

$$\begin{aligned} \frac{d}{dt}\langle\boldsymbol{x}(t)\rangle &= \widehat{A}(t)\langle\boldsymbol{x}(t)\rangle + \widehat{B}(t)\boldsymbol{\psi}(t), \quad \langle\boldsymbol{x}(0)\rangle = \boldsymbol{x}_0, \\ \left(\frac{d}{dt}+2\nu\right)\boldsymbol{\psi}(t) &= \widehat{A}(t)\boldsymbol{\psi}(t) + a_0^2\widehat{B}(t)\langle\boldsymbol{x}(t)\rangle , \quad \boldsymbol{\psi}(0) = (0). \end{aligned} \tag{7.10}$$

If operators $\widehat{A}(t)$ and $\widehat{B}(t)$ are the time-independent matrices A and B, we can solve system (7.10) using the Laplace transform. After the Laplace transform, system (7.10) grades into the algebraic system of equations

$$\begin{aligned}(pE - A)\,\langle \boldsymbol{x}\rangle_p - B\boldsymbol{\psi}_p &= \boldsymbol{x}_0,\\ [(p+2\nu)\,E - A]\boldsymbol{\psi}_p - a_0^2 B\,\langle \boldsymbol{x}\rangle_p &= 0,\end{aligned} \tag{7.11}$$

where E is the unit matrix. From this system, we obtain solution $\langle \boldsymbol{x}\rangle_p$ in the form

$$\langle \boldsymbol{x}\rangle_p = \left[(pE - A) - a_0^2 B \frac{1}{(p+2\nu)E - A} B\right]^{-1} \boldsymbol{x}_0. \tag{7.12}$$

Assume now operators $\widehat{A}(t)$ and $\widehat{B}(t)$ in Eq. (7.6) be the matrices. If only one component is of interest, we can obtain for it the operator equation

$$\widehat{L}\left(\frac{d}{dt}\right)x(t) + \sum_{i+j=0}^{n-1} b_{ij}(t)\frac{d^i}{dt^i}z(t)\frac{d^j}{dt^j}x(t) = f(t), \tag{7.13}$$

where

$$\widehat{L}\left(\frac{d}{dt}\right) = \sum_{i=0}^{n} a_i(t)\frac{d^i}{dt^i}$$

and n is the order of matrices A and B in Eq. (7.6). The initial conditions for x are included in function $f(t)$ through the corresponding derivatives of delta function. Note that function $f(t)$ can depend additionally on derivatives of random process $z(t)$ at $t = 0$, i.e., $f(t)$ is also the random function statistically related to process $z(t)$.

Averaging Eq. (7.13) over an ensemble of realizations of process $z(t)$ with the use of formula (7.5), we obtain the equation

$$\widehat{L}\left(\frac{d}{dt}\right)\langle x(t)\rangle + M\left[\frac{d}{dt}, \frac{d}{dt} + 2\nu\right]\langle z(t)x(t)\rangle = \langle f(t)\rangle, \tag{7.14}$$

where

$$M[p,q] = \sum_{i+j=0}^{n-1} b_{ij}(t)p^i q^j.$$

However, Eq. (7.14) is unclosed because of the presence of correlation $\langle z(t)x(t)\rangle$. Multiplying Eq. (7.13) by $z(t)$ and averaging the result, we obtain the equation for correlation $\langle z(t)x(t)\rangle$

$$\widehat{L}\left(\frac{d}{dt} + 2\nu\right)\langle z(t)x(t)\rangle + a_0^2 M\left[\frac{d}{dt} + 2\nu, \frac{d}{dt}\right]\langle x(t)\rangle = \langle z(t)f(t)\rangle. \tag{7.15}$$

System of equations (7.14) and Eq. (7.15) is the closed system. If functions $a_i(t)$ and $b_{ij}(t)$ are independent of time, this system can be solved using Laplace transform. The solution is as follows:

$$\langle x\rangle_p = \frac{\widehat{L}(p+2\nu)\,\langle f\rangle_p - M[p,p+2\nu]\,\langle zf\rangle_p}{\widehat{L}(p)\widehat{L}(p+2\nu) - a_0^2 M[p+2\nu,p]M[p,p+2\nu]}. \tag{7.16}$$

7.2 Gaussian Markovian Processes

Here, we consider several examples associated with the Gaussian Markovian processes.

Define random process $z(t)$ by the formula

$$z(t) = z_1(t) + \cdots + z_N(t), \tag{7.17}$$

where $z_i(t)$ are statistically independent telegrapher's processes with correlation functions

$$\left\langle z_i(t)z_j(t')\right\rangle = \delta_{ij}\left\langle z^2\right\rangle e^{-\alpha|t-t'|} \quad (\alpha = 2\nu).$$

If we set $\left\langle z^2\right\rangle = \sigma^2/N$, then this process grades for $N \to \infty$ into the Gaussian Markovian process with correlation function

$$\left\langle z_i(t)z_j(t')\right\rangle = \left\langle z^2\right\rangle e^{-\alpha|t-t'|}.$$

Thus, process $z(t)$ (7.17) approximates the Gaussian Markovian process in terms of the Markovian process with a finite number of states.

It is evident that the differentiation formula and the rule of factoring the derivative out of averaging brackets for process $z(t)$ assume the forms

$$\begin{aligned}
&\left(\frac{d}{dt} + \alpha k\right)\langle z_1(t)\cdots z_k(t)R[t; z(\tau)]\rangle = \left\langle z_1(t)\cdots z_k(t)\frac{d}{dt}R[t; z(\tau)]\right\rangle,\\
&\left\langle z_1(t)\cdots z_k(t)\frac{d^n}{dt^n}R[t; z(\tau)]\right\rangle = \left(\frac{d}{dt} + \alpha k\right)^n \langle z_1(t)\cdots z_k(t)R[t; z(\tau)]\rangle,
\end{aligned} \tag{7.18}$$

where $R[t; z(\tau)]$ is an arbitrary functional of process $z(t)$ $(\tau \le t)$.

Consider again Eq. (7.6), which we represent here in the form

$$\frac{d}{dt}\boldsymbol{x}(t) = \widehat{A}(t)\boldsymbol{x}(t) + [z_1(t) + \cdots + z_N(t)]\widehat{B}(t)\boldsymbol{x}(t), \quad \boldsymbol{x}(0) = \boldsymbol{x}_0, \tag{7.19}$$

and introduce vector-function

$$\boldsymbol{X}_k(t) = \langle z_1(t)\ldots z_k(t)\boldsymbol{x}(t)\rangle, \quad k = 1,\ldots,N; \quad \boldsymbol{X}_0(t) = \langle \boldsymbol{x}(t)\rangle. \tag{7.20}$$

Using formula (7.18) for differentiating correlations (7.20) and Eq. (7.19), we obtain the recurrence equation for $\boldsymbol{x}_k(t)$ $(k = 0, 1, \ldots, N)$

$$\frac{d}{dt}\boldsymbol{X}_k(t) = \widehat{A}(t)\boldsymbol{X}_k(t) + \left\langle z_1(t)\cdots z_k(t)[z_1(t) + \cdots + z_N(t)]\widehat{B}(t)\boldsymbol{x}(t)\right\rangle$$

$$= \widehat{A}(t)\boldsymbol{X}_k(t) + k\left\langle z^2\right\rangle\widehat{B}(t)\boldsymbol{X}_{k-1}(t) + (N-k)\widehat{B}(t)\boldsymbol{X}_{k+1}(t), \tag{7.21}$$

with the initial condition

$$\boldsymbol{X}_k(0) = \boldsymbol{x}_0\delta_{k,0}.$$

Thus, the mean value of the solution to system Eq. (7.19) satisfies the closed system of $(N+1)$ vector equations. If operators $\widehat{A}(t)$, $\widehat{B}(t)$ are time-independent matrices, system (7.21) can be easily solved using the Laplace transform. It is clear that such a solution will have the form of a finite segment of continued fraction. If we set $\left\langle z^2\right\rangle = \sigma^2/N$ and proceed to the limit $N \to \infty$, then random process (7.17) will grade, as was mentioned earlier, into the Gaussian Markovian process and solution to system (7.21) will assume the form of the infinite continued fraction.

Problems

Problem 7.1

Assuming that $z(t)$ is telegrapher's random process, find average potential energy of the stochastic oscillator described by the equation

$$\frac{d^3}{dt^3}U(t) + 4\omega_0^2\frac{d}{dt}U(t) + 2\omega_0^2\left(z(t)\frac{d}{dt}U(t) + \frac{d}{dt}z(t)U(t)\right) = 0,$$

$$x(0) = 0, \quad \left.\frac{d}{dt}U(t)\right|_{t=0} = 0, \quad \left.\frac{d^2}{dt^2}U(t)\right|_{t=0} = 2y_0^2.$$

Solution

$$\frac{d^3}{dt^3}\langle U(t)\rangle + 4\omega_0^2\frac{d}{dt}\langle U(t)\rangle + 4\omega_0^2\left(\frac{d}{dt} + 2\nu\right)\langle z(t)U(t)\rangle = 0,$$

$$\left(\frac{d}{dt} + 2\nu\right)^3\langle z(t)U(t)\rangle + 4\omega_0^2\left(\frac{d}{dt} + 2\nu\right)\langle z(t)U(t)\rangle$$
$$+ 4\omega_0^2 a_0^2\left(\frac{d}{dt} + 2\nu\right)\langle U(t)\rangle = 0. \tag{7.22}$$

System of equations (7.22) can be solved using Laplace transform, and we obtain the solution in the form

$$\langle U\rangle_p = 2y_0^2 \frac{L(p+2\nu)}{L(p)L(p+2\nu) - a_0^2 M^2(p)},$$
$$L(p) = p\left(p^2 + 4\omega_0^2\right), \quad M(p) = 4\omega_0^2\left(p^2 + \nu\right). \tag{7.23}$$

Remark In the limiting case of great parameters ν and a_0^2, but finite ratio $a_0^2/2\nu = \sigma^2\tau_0$, we obtain from the second equation of system (7.22)

$$\langle z(t)U(t)\rangle = -\frac{\omega_0^2\sigma^2\tau_0}{\nu}\langle U(t)\rangle .$$

Consequently, average potential energy $\langle U(t)\rangle$ satisfies in this limiting case the closed third-order equation

$$\frac{d^3}{dt^3}\langle U(t)\rangle + 4\omega_0^2\frac{d}{dt}\langle U(t)\rangle - 4\omega_0^4\sigma^2\tau_0\langle U(t)\rangle = 0,$$

which coincides with Eq. (6.92), page 173, and corresponds to the Gaussian delta-correlated process $z(t)$.

Problem 7.2

Assuming that $z(t)$ is telegrapher's random process, obtain the steady-state probability distribution of the solution to the stochastic equation

$$\frac{d}{dt}x(t) = f(x) + z(t)g(x), \quad x(0) = x_0.$$

Solution Under the condition that $|f(x)| < a_0|g(x)|$, we have

$$P(x) = \frac{C|g(x)|}{a_0^2 g^2(x) - f^2(x)} \exp\left\{\frac{2\nu}{a_0^2}\int dx \frac{f(x)}{a_0^2 g^2(x) - f^2(x)}\right\}, \tag{7.24}$$

where the positive constant C is determined from the normalization condition.

Remark In the limit $\nu \to \infty$ and $a_0^2 \to \infty$ under the condition that $a_0^2\tau_0 = \text{const}$ $(\tau_0 = 1/(2\nu))$, probability distribution (7.24) grades into the expression

$$P(x) = \frac{C}{|g(x)|} \exp\left\{\frac{2\nu}{a_0^2}\int dx \frac{f(x)}{g^2(x)}\right\}$$

corresponding to the Gaussian delta-correlated process $z(t)$, i.e., the Gaussian process with correlation function $\langle z(t)z(t')\rangle = 2a_0^2\tau_0\delta(t-t')$.

Problem 7.3

Assuming that $z(t)$ is telegrapher's random process, obtain the steady-state probability distribution of the solution to the stochastic equation

$$\frac{d}{dt}x(t) = -x + z(t), \quad x(0) = x_0. \tag{7.25}$$

Solution

$$P(x) = \frac{1}{B(\nu, 1/2)}(1-x^2)^{\nu-1} \quad (|x| < 1), \tag{7.26}$$

where $B(\nu, 1/2)$ is the beta-function. This distribution has essentially different behaviors for $\nu > 1$, $\nu = 1$ and $\nu < 1$. One can easily see that this system resides mainly near state $x = 0$ if $\nu > 1$, and near states $x = \pm 1$ if $\nu < 1$. In the case $\nu = 1$, we obtain the uniform probability distribution on segment $[-1, 1]$ (Fig. 7.1).

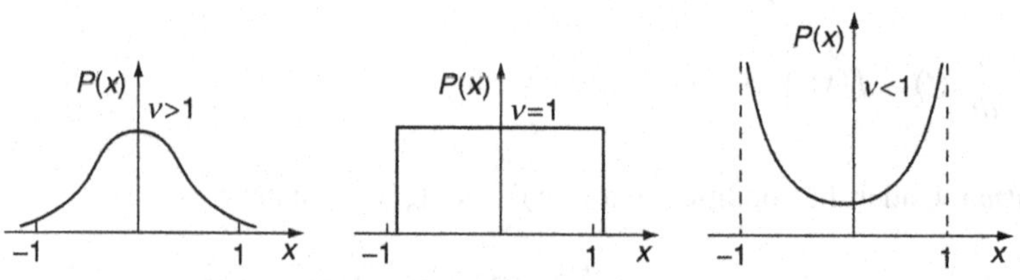

Figure 7.1 Steady-state probability distribution (7.26) of the solution to the Eq. (7.25) versus parameter ν.

Lecture 8

Approximations of Gaussian Random Field Delta-Correlated in Time

In the foregoing Lectures, we considered in detail the general methods of analyzing stochastic equations. Here, we give an alternative and more detailed consideration of the approximation of the Gaussian random delta-correlated (in time) field in the context of stochastic equations and discuss the physical meaning of this widely used approximation.

8.1 The Fokker–Planck Equation

Let vector function $\boldsymbol{x}(t) = \{x_1(t), x_2(t), \ldots, x_n(t)\}$ satisfies the dynamic equation

$$\frac{d}{dt}\boldsymbol{x}(t) = \boldsymbol{v}(\boldsymbol{x}, t) + \boldsymbol{f}(\boldsymbol{x}, t), \quad \boldsymbol{x}(t_0) = \boldsymbol{x}_0, \tag{8.1}$$

where $v_i(\boldsymbol{x}, t)$ $(i = 1, \ldots, n)$ are the deterministic functions and $f_i(\boldsymbol{x}, t)$ are the random functions of $(n+1)$ variable that have the following properties:

(a) $f_i(\boldsymbol{x}, t)$ is the Gaussian random field in the $(n+1)$-dimensional space $(\boldsymbol{x}, t)$;
(b) $\langle f_i(\boldsymbol{x}, t)\rangle = 0$.

For definiteness, we assume that t is the temporal variable and $\boldsymbol{x}$ is the spatial variable.

Statistical characteristics of field $f_i(\boldsymbol{x}, t)$ are completely described by correlation tensor

$$B_{ij}(\boldsymbol{x}, t; \boldsymbol{x}', t') = \left\langle f_i(\boldsymbol{x}, t) f_j(\boldsymbol{x}', t')\right\rangle.$$

Because Eq. (8.1) is the first-order equation with the initial value, its solution satisfies the dynamic causality condition

$$\frac{\delta}{\delta f_j(\boldsymbol{x}', t')} x_i(t) = 0 \quad \text{for} \quad t' < t_0 \quad \text{and} t' > t, \tag{8.2}$$

Lectures on Dynamics of Stochastic Systems. DOI: 10.1016/B978-0-12-384966-3.00008-8

which means that solution $\boldsymbol{x}(t)$ depends only on values of function $f_j(\boldsymbol{x}, t')$ for times t' preceding time t, i.e., $t_0 \le t' \le t$. In addition, we have the following equality for the variational derivative

$$\frac{\delta}{\delta f_j(\boldsymbol{x}', t-0)} x_i(t) = \delta_{ij}\delta(\boldsymbol{x}(t) - \boldsymbol{x}'). \tag{8.3}$$

Nevertheless, the statistical relationship between $\boldsymbol{x}(t)$ and function values $f_j(\boldsymbol{x}, t'')$ for consequent times $t'' > t$ can exist, because such function values $f_j(\boldsymbol{x}, t'')$ are correlated with values $f_j(\boldsymbol{x}, t')$ for $t' \le t$. It is obvious that the correlation between function $\boldsymbol{x}(t)$ and consequent values $f_j(\boldsymbol{x}, t'')$ is appreciable only for $t'' - t \le \tau_0$, where τ_0 is the correlation radius of field $\boldsymbol{f}(\boldsymbol{x}, t)$ with respect to variable t.

For many actual physical processes, characteristic temporal scale T of function $\boldsymbol{x}(t)$ significantly exceeds correlation radius τ_0 ($T \gg \tau_0$); in this case, the problem has small parameter τ_0/T that can be used to construct an approximate solution.

In the first approximation with respect to this small parameter, one can consider the asymptotic solution for $\tau_0 \to 0$. In this case values of function $\boldsymbol{x}(t')$ for $t' < t$ will be independent of values $\boldsymbol{f}(\boldsymbol{x}, t'')$ for $t'' > t$ not only functionally, but also statistically. This approximation is equivalent to the replacement of correlation tensor B_{ij} with the effective tensor

$$B_{ij}^{\text{eff}}(\boldsymbol{x}, t; \boldsymbol{x}', t') = 2\delta(t-t')F_{ij}(\boldsymbol{x}, \boldsymbol{x}'; t). \tag{8.4}$$

Here, quantity $F_{ij}(\boldsymbol{x}, \boldsymbol{x}', t)$ is determined from the condition that integrals of $B_{ij}(\boldsymbol{x}, t; \boldsymbol{x}', t')$ and $B_{ij}^{\text{eff}}(\boldsymbol{x}, t; \boldsymbol{x}', t')$ over t' coincide

$$F_{ij}(\boldsymbol{x}, \boldsymbol{x}', t) = \frac{1}{2}\int_{-\infty}^{\infty} dt' B_{ij}(\boldsymbol{x}, t; \boldsymbol{x}', t'),$$

which just corresponds to the passage to the Gaussian random field delta-correlated in time t.

Introduce the indicator function

$$\varphi(\boldsymbol{x}, t) = \delta(\boldsymbol{x}(t) - \boldsymbol{x}), \tag{8.5}$$

where $\boldsymbol{x}(t)$ is the solution to Eq. (8.1), which satisfies the Liouville equation

$$\left(\frac{\partial}{\partial t} + \frac{\partial}{\partial \boldsymbol{x}}\boldsymbol{v}(\boldsymbol{x}, t)\right)\varphi(\boldsymbol{x}, t) = -\frac{\partial}{\partial \boldsymbol{x}}\boldsymbol{f}(\boldsymbol{x}, t)\varphi(\boldsymbol{x}, t) \tag{8.6}$$

and the equality

$$\frac{\delta}{\delta f_j(\boldsymbol{x}', t-0)}\varphi(\boldsymbol{x}, t) = -\frac{\partial}{\partial x_j}\left\{\delta(\boldsymbol{x}-\boldsymbol{x}')\varphi(\boldsymbol{x}, t)\right\}. \tag{8.7}$$

The equation for the probability density of the solution to Eq. (8.1)

$$P(\boldsymbol{x}, t) = \langle \varphi(\boldsymbol{x}, t) \rangle = \langle \delta(\boldsymbol{x}(t) - \boldsymbol{x}) \rangle$$

can be obtained by averaging Eq. (8.6) over an ensemble of realizations of field $\boldsymbol{f}(\boldsymbol{x}, t)$,

$$\left(\frac{\partial}{\partial t} + \frac{\partial}{\partial \boldsymbol{x}} \boldsymbol{v}(\boldsymbol{x}, t) \right) P(\boldsymbol{x}, t) = -\frac{\partial}{\partial \boldsymbol{x}} \langle \boldsymbol{f}(\boldsymbol{x}, t) \varphi(\boldsymbol{x}, t) \rangle. \tag{8.8}$$

We rewrite Eq. (8.8) in the form

$$\begin{aligned} &\left(\frac{\partial}{\partial t} + \frac{\partial}{\partial \boldsymbol{x}} \boldsymbol{v}(\boldsymbol{x}, t) \right) P(\boldsymbol{x}, t) \\ &\quad = -\frac{\partial}{\partial x_i} \int d\boldsymbol{x}' \int_{t_0}^{t} dt' B_{ij}(\boldsymbol{x}, t; \boldsymbol{x}', t') \left\langle \frac{\delta \varphi(\boldsymbol{x}, t)}{\delta f_j(\boldsymbol{x}', t')} \right\rangle, \end{aligned} \tag{8.9}$$

where we used the Furutsu–Novikov formula (5.13), page 126,

$$\langle f_k(\boldsymbol{x}, t) R[t; \boldsymbol{f}(\boldsymbol{y}, \tau)] \rangle = \int d\boldsymbol{x}' \int dt' B_{kl}(\boldsymbol{x}, t; \boldsymbol{x}', t') \left\langle \frac{\delta R[t; \boldsymbol{f}(\boldsymbol{y}, \tau)]}{\delta f_l(\boldsymbol{x}', t')} \right\rangle \tag{8.10}$$

for the correlator of the Gaussian random field $\boldsymbol{f}(\boldsymbol{x}, t)$ with arbitrary functional $R[t; \boldsymbol{f}(\boldsymbol{y}, \tau)]$ of it and the dynamic causality condition (8.2).

Equation (8.9) shows that the one-time probability density of solution $\boldsymbol{x}(t)$ at instant t is governed by functional dependence of solution $\boldsymbol{x}(t)$ on field $\boldsymbol{f}(\boldsymbol{x}', t)$ for all times in the interval (t_0, t).

In the general case, there is no closed equation for the probability density $P(\boldsymbol{x}, t)$. However, if we use approximation (8.4) for the correlation function of field $\boldsymbol{f}(\boldsymbol{x}, t)$, there appear terms related to variational derivatives $\delta\varphi[\boldsymbol{x}, t; \boldsymbol{f}(\boldsymbol{y}, \tau)]/\delta f_j(\boldsymbol{x}', t')$ at coincident temporal arguments $t' = t - 0$,

$$\left(\frac{\partial}{\partial t} + \frac{\partial}{\partial \boldsymbol{x}} \boldsymbol{v}(\boldsymbol{x}, t) \right) P(\boldsymbol{x}, t) = -\frac{\partial}{\partial x_i} \int d\boldsymbol{x}' F_{ij}(\boldsymbol{x}, \boldsymbol{x}', t) \left\langle \frac{\delta \varphi(\boldsymbol{x}, t)}{\delta f_j(\boldsymbol{x}', t - 0)} \right\rangle.$$

According to Eq. (8.7), these variational derivatives can be expressed immediately in terms of quantity $\varphi[\boldsymbol{x}, t; \boldsymbol{f}(\boldsymbol{y}, \tau)]$. Thus, we obtain the closed Fokker-Planck equation

$$\left(\frac{\partial}{\partial t} + \frac{\partial}{\partial x_k} [v_k(\boldsymbol{x}, t) + A_k(\boldsymbol{x}, t)] \right) P(\boldsymbol{x}, t) = \frac{\partial^2}{\partial x_k \partial x_l} [F_{kl}(\boldsymbol{x}, \boldsymbol{x}, t) P(\boldsymbol{x}, t)], \tag{8.11}$$

where

$$A_k(\boldsymbol{x}, t) = \left. \frac{\partial}{\partial x_l'} F_{kl}(\boldsymbol{x}, \boldsymbol{x}', t) \right|_{\boldsymbol{x}' = \boldsymbol{x}}.$$

Equation (8.11) should be solved with the initial condition

$$P(\boldsymbol{x}, t_0) = \delta(\boldsymbol{x} - \boldsymbol{x}_0),$$

or with a more general initial condition $P(\boldsymbol{x}, t_0) = W_0(\boldsymbol{x})$ if the initial conditions are also random, but statistically independent of field $\boldsymbol{f}(\boldsymbol{x}, t)$.

The Fokker–Planck equation (8.11) is a partial differential equation and its further analysis essentially depends on boundary conditions with respect to $\boldsymbol{x}$ whose form can vary depending on the problem at hand.

Consider the quantities appeared in Eq. (8.11). In this equation, the terms containing $A_k(\boldsymbol{x}, t)$ and $F_{kl}(\boldsymbol{x}, \boldsymbol{x}', t)$ are stipulated by fluctuations of field $\boldsymbol{f}(\boldsymbol{x}, t)$. If field $\boldsymbol{f}(\boldsymbol{x}, t)$ is stationary in time, quantities $A_k(\boldsymbol{x})$ and $F_{kl}(\boldsymbol{x}, \boldsymbol{x}')$ are independent of time. If field $f(\boldsymbol{x}, t)$ is additionally homogeneous and isotropic in all spatial coordinates, then

$$F_{kl}(\boldsymbol{x}, \boldsymbol{x}, t) = \text{const},$$

which corresponds to the constant tensor of diffusion coefficients, and $A_k(\boldsymbol{x}, t) = 0$ (note however that quantities $F_{kl}(\boldsymbol{x}, \boldsymbol{x}', t)$ and $A_k(\boldsymbol{x}, t)$ can depend on $\boldsymbol{x}$ because of the use of a curvilinear coordinate systems).

8.2 Transition Probability Distributions

Turn back to dynamic system (8.1) and consider the m-time probability density

$$P_m(\boldsymbol{x}_1, t_1; \ldots; \boldsymbol{x}_m, t_m) = \langle \delta(\boldsymbol{x}(t_1) - \boldsymbol{x}_1) \cdots \delta(\boldsymbol{x}(t_m) - \boldsymbol{x}_m) \rangle \tag{8.12}$$

for m different instants $t_1 < t_2 < \cdots < t_m$. Differentiating Eq. (8.12) with respect to time t_m and using then dynamic equation (8.1), dynamic causality condition (8.2), definition of function $F_{kl}(\boldsymbol{x}, \boldsymbol{x}', t)$, and the Furutsu–Novikov formula (8.10), one can obtain the equation similar to the Fokker–Planck equation (8.11),

$$\begin{aligned} &\frac{\partial}{\partial t_m} P_m(\boldsymbol{x}_1, t_1; \ldots; \boldsymbol{x}_m, t_m) \\ &\quad + \sum_{k=1}^{n} \frac{\partial}{\partial x_{mk}} [v_k(\boldsymbol{x}_m, t_m) + A_k(\boldsymbol{x}_m, t_m)] P_m(\boldsymbol{x}_1, t_1; \ldots; \boldsymbol{x}_m, t_m) \\ &= \sum_{k=1}^{n} \sum_{l=1}^{n} \frac{\partial^2}{\partial x_{mk} \partial x_{ml}} [F_{kl}(\boldsymbol{x}_m, \boldsymbol{x}_m, t_m) P_m(\boldsymbol{x}_1, t_1; \ldots; \boldsymbol{x}_m, t_m))]. \end{aligned} \tag{8.13}$$

No summation over index m is performed here. The initial value to Eq. (8.13) can be determined from Eq. (8.12). Setting $t_m = t_{m-1}$ in (8.12), we obtain

$$\begin{aligned} &P_m(\boldsymbol{x}_1, t_1; \ldots; \boldsymbol{x}_m, t_{m-1}) \\ &\quad = \delta(\boldsymbol{x}_m - \boldsymbol{x}_{m-1}) P_{m-1}(\boldsymbol{x}_1, t_1; \ldots; \boldsymbol{x}_{m-1}, t_{m-1}). \end{aligned} \tag{8.14}$$

Equation (8.13) assumes the solution in the form

$$
\begin{aligned}
&P_m(\boldsymbol{x}_1, t_1; \ldots; \boldsymbol{x}_m, t_m) \\
&\quad = p(\boldsymbol{x}_m, t_m|\boldsymbol{x}_{m-1}, t_{m-1})P_{m-1}(\boldsymbol{x}_1, t_1; \ldots; \boldsymbol{x}_{m-1}, t_{m-1}).
\end{aligned} \tag{8.15}
$$

Because all differential operations in Eq. (8.13) concern only t_m and $\boldsymbol{x}_m$, we can find the equation for the *transitional probability density* by substituting Eq. (8.15) in Eq. (8.13) and (8.14):

$$
\begin{aligned}
&\left(\frac{\partial}{\partial t} + \frac{\partial}{\partial x_k}\left[v_k(\boldsymbol{x}, t) + A_k(\boldsymbol{x}, t)\right]\right) p(\boldsymbol{x}, t|\boldsymbol{x}_0, t_0) \\
&\quad = \frac{\partial^2}{\partial x_k \partial x_l}\left[F_{kl}(\boldsymbol{x}, \boldsymbol{x}, t)p(\boldsymbol{x}, t|\boldsymbol{x}_0, t_0)\right]
\end{aligned} \tag{8.16}
$$

with initial condition

$$
p(\boldsymbol{x}, t|\boldsymbol{x}_0, t_0)|_{t\to t_0} = \delta(\boldsymbol{x} - \boldsymbol{x}_0),
$$

where

$$
p(\boldsymbol{x}, t|\boldsymbol{x}_0, t_0) = \langle \delta(\boldsymbol{x}(t) - \boldsymbol{x})|\boldsymbol{x}(t_0) = \boldsymbol{x}_0\rangle.
$$

In Eq. (8.16) we denoted variables $\boldsymbol{x}_m$ and t_m as $\boldsymbol{x}$ and t, and variables $\boldsymbol{x}_{m-1}$ and t_{m-1} as $\boldsymbol{x}_0$ and t_0.

Using formula (8.15) $(m - 1)$ times, we obtain the relationship

$$
\begin{aligned}
&P_m(\boldsymbol{x}_1, t_1; \ldots; \boldsymbol{x}_m, t_m) \\
&\quad = p(\boldsymbol{x}_m, t_m|\boldsymbol{x}_{m-1}, t_{m-1}) \cdots p(\boldsymbol{x}_2, t_2|\boldsymbol{x}_1, t_1)P(\boldsymbol{x}_1, t_1),
\end{aligned} \tag{8.17}
$$

where $P(\boldsymbol{x}_1, t_1)$ is the one-time probability density governed by Eq. (8.11). Equality (8.17) expresses the multi-time probability density as the product of transitional probability densities, which means that random process $\boldsymbol{x}(t)$ is the Markovian process.

Equation (8.11) is usually called the *forward Fokker–Planck equation*. The *backward Fokker–Planck equation* (it describes the transitional probability density as a function of the initial parameters t_0 and $\boldsymbol{x}_0$) can also be easily derived.

Indeed, in Lecture 3 (page 69), we obtained the backward Liouville equation (3.8), page 71, for indicator function

$$
\left(\frac{\partial}{\partial t_0} + \boldsymbol{v}(\boldsymbol{x}_0, t_0)\frac{\partial}{\partial \boldsymbol{x}_0}\right)\varphi(\boldsymbol{x}, t|\boldsymbol{x}_0, t_0) = -\boldsymbol{f}(\boldsymbol{x}_0, t_0)\frac{\partial}{\partial \boldsymbol{x}_0}\varphi(\boldsymbol{x}, t|\boldsymbol{x}_0, t_0), \tag{8.18}
$$

with the initial value $\varphi(\boldsymbol{x}, t|\boldsymbol{x}_0, t) = \delta(\boldsymbol{x} - \boldsymbol{x}_0)$. This equation describes the dynamic system evolution in terms of initial parameters t_0 and $\boldsymbol{x}_0$. From Eq. (8.18) follows the equality similar to Eq. (8.7),

$$
\frac{\delta}{\delta f_j(\boldsymbol{x}', t_0 + 0)}\varphi(\boldsymbol{x}, t|\boldsymbol{x}_0, t_0) = \delta(\boldsymbol{x} - \boldsymbol{x}')\frac{\partial}{\partial x_{0j}}\varphi(\boldsymbol{x}, t|\boldsymbol{x}_0, t_0). \tag{8.19}
$$

Averaging now the backward Liouville equation (8.18) over an ensemble of realizations of random field $\boldsymbol{f}(\boldsymbol{x},t)$ with the effective correlation tensor (8.4), using the Furutsu–Novikov formula (8.10), and relationship (8.19) for the variational derivative, we obtain the backward Fokker–Planck equation

$$\left(\frac{\partial}{\partial t_0} + [v_k(\boldsymbol{x}_0,t_0) + A_k(\boldsymbol{x}_0,t_0)]\frac{\partial}{\partial x_{0k}}\right) p(\boldsymbol{x},t|\boldsymbol{x}_0,t_0)$$
$$= -F_{kl}(\boldsymbol{x}_0,\boldsymbol{x}_0,t_0)\frac{\partial^2}{\partial x_{0k}\partial x_{0l}} p(\boldsymbol{x},t|\boldsymbol{x}_0,t_0), \tag{8.20}$$
$$p(\boldsymbol{x},t|\boldsymbol{x}_0,t) = \delta(\boldsymbol{x}-\boldsymbol{x}_0).$$

The forward and backward Fokker–Planck equations are equivalent. The forward equation is more convenient for analyzing the temporal behavior of statistical characteristics of the solution to Eq. (8.1). The backward equation appears to be more convenient for studying statistical characteristics related to initial values, such as the time during which process $\boldsymbol{x}(t)$ resides in certain spatial region and the time at which the process arrives at region's boundary. In this case the probability of the fact that random process $\boldsymbol{x}(t)$ rests in spatial region V is given by the integral

$$G(t;\boldsymbol{x}_0,t_0) = \int\limits_V d\boldsymbol{x} p(\boldsymbol{x},t|\boldsymbol{x}_0,t_0),$$

which, according to Eq. (8.20), satisfies the closed equation

$$\left(\frac{\partial}{\partial t_0} + [v_k(\boldsymbol{x}_0,t_0) + A_k(\boldsymbol{x}_0,t_0)]\frac{\partial}{\partial x_{0k}}\right) G(t;\boldsymbol{x}_0,t_0)$$
$$= -F_{kl}(\boldsymbol{x}_0,\boldsymbol{x}_0,t_0)\frac{\partial^2}{\partial x_{0k}\partial x_{0l}} G(t;\boldsymbol{x}_0,t_0), \tag{8.21}$$
$$G(t;\boldsymbol{x}_0,t_0) = \begin{cases} 1 & (\boldsymbol{x}_0 \in V), \\ 0 & (\boldsymbol{x}_0 \notin V). \end{cases}$$

For Eq. (8.21), we must formulate additional boundary conditions, which depend on characteristics of both region V and its boundaries.

8.3 The Simplest Markovian Random Processes

There are only few Fokker–Planck equations that allow an exact solution. First of all, among them are the Fokker–Planck equations corresponding to the stochastic equations that are themselves solvable in the analytic form. Such problems often allow determination of not only the one-point and transitional probability densities, but also the characteristic functional and other statistical characteristics important for practice.

The simplest special case of Eq. (8.11) is the equation that defines the *Wiener random process*. In view of the significant role that such processes plays in physics (for example, they describe the *Brownian motion of particles*), we consider the Wiener process in detail.

8.3.1 Wiener Random Process

The Wiener random process is defined as the solution to the stochastic equation

$$\frac{d}{dt}w(t) = z(t), \quad w(0) = 0,$$

where $z(t)$ is the Gaussian process delta-correlated in time and described by the parameters

$$\langle z(t)\rangle = 0, \quad \left\langle z(t)z(t')\right\rangle = 2\sigma^2\tau_0\delta(t-t').$$

The solution to this equation $w(t) = \int_0^t d\tau z(\tau)$ is the continuous Gaussian nonstationary random process with the parameters

$$\langle w(t)\rangle = 0, \quad \left\langle w(t)w(t')\right\rangle = 2\sigma^2\tau_0 \min(t, t').$$

8.3.2 Wiener Random Process with Shear

Consider a more general process that includes additionally the drift dependent on parameter α

$$w(t;\alpha) = -\alpha t + w(t), \quad \alpha > 0.$$

Process $w(t;\alpha)$ is the Markovian process, and its probability density

$$P(w,t;\alpha) = \langle\delta(w(t;\alpha) - w)\rangle$$

satisfies the Fokker–Planck equation

$$\left(\frac{\partial}{\partial t} - \alpha\frac{\partial}{\partial w}\right)P(w,t;\alpha) = D\frac{\partial^2}{\partial w^2}P(w,t;\alpha), \quad P(w,0;\alpha) = \delta(w), \tag{8.22}$$

where $D = \sigma^2\tau_0$ is the diffusion coefficient. The solution to this equation has the form of the Gaussian distribution

$$P(w,t;\alpha) = \frac{1}{2\sqrt{\pi Dt}}\exp\left\{-\frac{(w+\alpha t)^2}{4Dt}\right\}. \tag{8.23}$$

The corresponding integral distribution function defined as the probability of the event that $w(t;\alpha) < w$ is given by the formula

$$F(w,t;\alpha) = \int_{-\infty}^{w} dwP(w,t;\alpha) = \Pr\left(\frac{w}{\sqrt{2Dt}} + \alpha\sqrt{\frac{t}{2D}}\right), \tag{8.24}$$

where function $\Pr(z)$ is the *probability integral* (4.20), page 94. In this case, the typical realization curve of the Wiener random process with shear is the linear function of time in accordance with Eqs. (4.60), page 108

$$w^*(t;\alpha) = -\alpha t.$$

In addition to the initial value, supplement Eq. (8.22) with the boundary condition

$$P(w,t;\alpha)|_{w=h} = 0, \quad (t > 0). \tag{8.25}$$

This condition breaks down realizations of process $w(t;\alpha)$ at the instant they reach boundary h. For $w < h$, the solution to the boundary-value problem (8.22), (8.25) (we denote it as $P(w,t;\alpha,h)$) describes the probability distribution of those realizations of process $w(t;\alpha)$ that survived instant t, i.e., never reached boundary h during the whole temporal interval. Correspondingly, the norm of the probability density appears not unity, but the probability of the event that $t < t^*$, where t^* is the instant at which process $w(t;\alpha)$ reaches boundary h for the first time

$$\int_{-\infty}^{h} dwP(w,t;\alpha,h) = P(t < t^*). \tag{8.26}$$

Introduce the integral distribution function and probability density of random instant at which the process reaches boundary h

$$F(t;\alpha,h) = P(t^* < t) = 1 - P(t < t^*) = 1 - \int_{-\infty}^{h} dwP(w,t;\alpha,h),$$
$$P(t;\alpha,h) = \frac{\partial}{\partial t}F(t;\alpha,h) = -\frac{\partial}{\partial w}P(w,t;\alpha,h)|_{w=h}. \tag{8.27}$$

If $\alpha > 0$, process $w(t;\alpha)$ moves on average out of boundary h; as a result, probability $P(t < t^*)$ (8.26) tends for $t \to \infty$ to the probability of the event that process $w(t;\alpha)$ never reaches boundary h. In other words, limit

$$\lim_{t\to\infty} \int_{-\infty}^{h} dwP(w,t;\alpha,h) = P\left(w_{\max}(\alpha) < h\right) \tag{8.28}$$

is equal to the probability of the event that the process absolute maximum

$$w_{\max}(\alpha) = \max_{t\in(0,\infty)} w(t;\alpha)$$

is less than h. Thus, from Eq. (8.28) follows that the integral distribution function of the absolute maximum $w_{\max}(\alpha)$ is given by the formula

$$F(h;\alpha) = P\left(w_{\max}(\alpha) < h\right) = \lim_{t\to\infty} \int_{-\infty}^{h} dwP(w,t;\alpha,h). \tag{8.29}$$

After we solve boundary-value problem (8.22), (8.25) by using, for example, the reflection method, we obtain

$$P(w,t;\alpha,h) = \frac{1}{2\sqrt{\pi Dt}}\left\{\exp\left[-\frac{(w+\alpha t)^2}{4Dt}\right] - \exp\left[-\frac{h\alpha}{D} - \frac{(w-2h+\alpha t)^2}{4Dt}\right]\right\}. \tag{8.30}$$

Substituting this expression in Eq. (8.27), we obtain the probability density of instant t^* at which process $w(t;\alpha)$ reaches boundary h for the first time

$$P(t;\alpha,h) = \frac{1}{2Dt\sqrt{\pi Dt}} \exp\left\{-\frac{(h+\alpha t)^2}{4Dt}\right\}.$$

Finally, integrating Eq. (8.30) over w and setting $t \to \infty$, we obtain, in accordance with Eq. (8.29), the integral distribution function of absolute maximum $w_{\max}(\alpha)$ of process $w(t;\alpha)$ in the form [1, 2]

$$F(h;\alpha) = P\left(w_{\max}(\alpha) < h\right) = 1 - \exp\left\{-\frac{h\alpha}{D}\right\}. \tag{8.31}$$

Consequently, the absolute maximum of the Wiener process has the exponential probability density

$$P(h;\alpha) = \left\langle \delta\left(w_{\max}(\alpha) - h\right)\right\rangle = \frac{\alpha}{D} \exp\left\{-\frac{h\alpha}{D}\right\}.$$

The Wiener random process offers the possibility of constructing other processes convenient for modeling different physical phenomena. In the case of positive quantities, the simplest approximation is the logarithmic-normal (lognormal) process. We will consider this process in greater detail.

8.3.3 Logarithmic-Normal Random Process

We define the lognormal process (logarithmic-normal process) by the formula

$$y(t;\alpha) = e^{w(t;\alpha)} = \exp\left\{-\alpha t + \int_0^t d\tau z(\tau)\right\}, \tag{8.32}$$

where $z(t)$ is the Gaussian white noise process with the parameters

$$\langle z(t)\rangle = 0, \quad \langle z(t)z(t')\rangle = 2\sigma^2\tau_0\delta(t-t').$$

The lognormal process satisfies the stochastic equation

$$\frac{d}{dt}y(t;\alpha) = \{-\alpha + z(t)\}\, y(t;\alpha), \qquad y(0;\alpha) = 1.$$

The one-time probability density of the lognormal process is given by the formula

$$\begin{aligned} P(y,t;\alpha) &= \langle\delta\left(y(t;\alpha) - y\right)\rangle = \left\langle\delta\left(e^{w(t;\alpha)} - y\right)\right\rangle \\ &= \frac{1}{y}\langle\delta\left(w(t;\alpha) - \ln y\right)\rangle = \frac{1}{y}P(w,t;\alpha)|_{w=\ln y}, \end{aligned}$$

where $P(w,t;\alpha)$ is the one-time probability density of the Wiener process with a drift, which is given by Eq. (8.23), so that

$$P(y,t;\alpha) = \frac{1}{2y\sqrt{\pi Dt}}\exp\left\{-\frac{(\ln y + \alpha t)^2}{4Dt}\right\} = \frac{1}{2y\sqrt{\pi Dt}}\exp\left\{-\frac{\ln^2\left(ye^{\alpha t}\right)}{4Dt}\right\}, \tag{8.33}$$

where $D = \sigma^2\tau_0$.

Note that the one-time probability density of random process $\widetilde{y}(t;\alpha) = 1/y(t;\alpha)$ is also lognormal and is given by the formula

$$P(\widetilde{y},t;\alpha) = \frac{1}{2\widetilde{y}\sqrt{\pi Dt}}\exp\left\{-\frac{\ln^2\left(\widetilde{y}e^{-\alpha t}\right)}{4Dt}\right\}, \tag{8.34}$$

which coincides with Eq. (8.33) with parameter α of opposite sign. Correspondingly, the integral distribution functions are given, in accordance with Eq. (8.24), by the expressions

$$F(y,t;\alpha) = P\left(y(t;\alpha) < y\right) = \Pr\left(\frac{1}{\sqrt{2Dt}}\ln\left(ye^{\pm\alpha t}\right)\right), \tag{8.35}$$

where $\Pr(z)$ is the probability integral (4.20), page 94.

Figure 8.1 shows the curves of the lognormal probability densities (8.33) and (8.34) for $\alpha/D = 1$ and dimensionless times $\tau = Dt = 0.1$ and 1. Figure 8.2 shows these probability densities at $\tau = 1$ in logarithmic scale.

Structurally, these probability distributions are absolutely different. The only common feature of these distributions consists in the existence of long flat *tails* that appear in distributions at $\tau = 1$; these tails increase the role of high peaks of processes $y(t; \alpha)$ and $\widetilde{y}(t; \alpha)$ in the formation of the one-time statistics.

Having only the one-point statistical characteristics of processes $y(t; \alpha)$ and $\widetilde{y}(t; \alpha)$, one can obtain a deeper insight into the behavior of realizations of these processes on the whole interval of times $(0, \infty)$ [2, 5]. In particular,

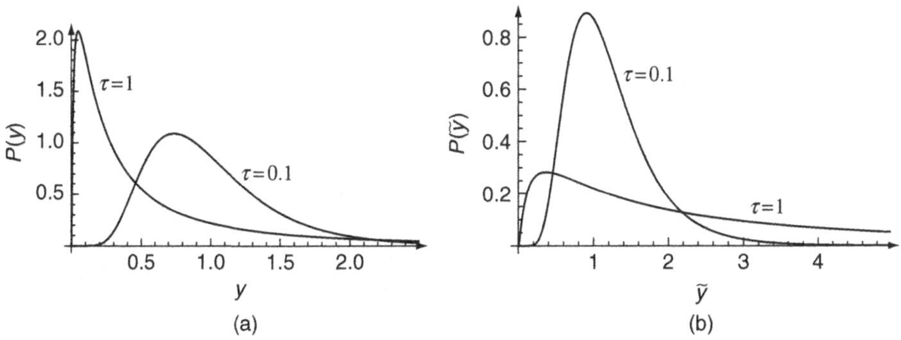

Figure 8.1 Logarithmic-normal probability densities (8.33) (a) and (8.34) (b) for $\alpha/D = 1$ and dimensionless times $\tau = 0.1$ and 1.

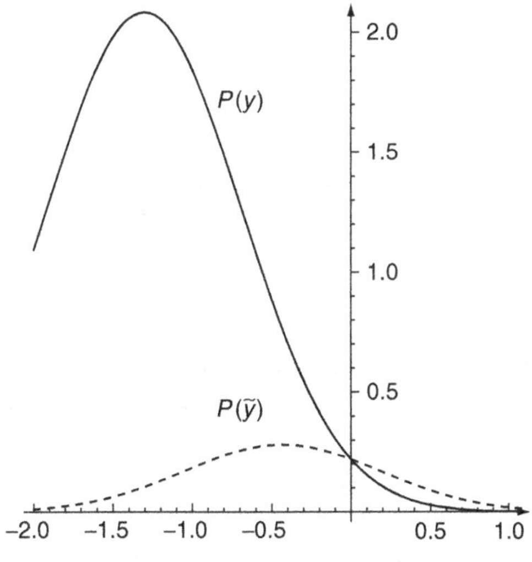

Figure 8.2 Probability densities of processes $y(t; \alpha)$ (solid line) and $\widetilde{y}(t; \alpha)$ (dashed line) at $\tau = 1$ in common logarithmic scale.

(1) The lognormal process $y(t;\alpha)$ is the Markovian process and its one-time probability density (8.33) satisfies the Fokker–Planck equation

$$\left(\frac{\partial}{\partial t} - \alpha\frac{\partial}{\partial y}y\right)P(y,t;\alpha) = D\frac{\partial}{\partial y}y\frac{\partial}{\partial y}yP(y,t;\alpha), \quad P(y,0;\alpha) = \delta(y-1). \tag{8.36}$$

From Eq. (8.36), one can easily derive the equations for the moment functions of processes $y(t;\alpha)$ and $\widetilde{y}(t;\alpha)$; solutions to these equations are given by the formulas $(n = 1, 2, \ldots)$

$$\left\langle y^n(t;\alpha)\right\rangle = e^{n(n-\alpha/D)Dt}, \quad \left\langle \widetilde{y}^n(t;\alpha)\right\rangle = \left\langle \frac{1}{y^n(t;\alpha)}\right\rangle = e^{n(n+\alpha/D)Dt}, \tag{8.37}$$

from which follows that moments exponentially grow with time.

From Eq. (8.36), one can easily obtain the equality

$$\langle \ln y(t)\rangle = -\alpha t. \tag{8.38}$$

Consequently, parameter α appeared in Eq. (8.38) can be rewritten in the form

$$-\alpha = \frac{1}{t}\langle \ln y(t)\rangle \quad \text{or} \quad \alpha = \frac{1}{t}\langle \ln \widetilde{y}(t)\rangle. \tag{8.39}$$

Note that many investigators give great attention to the approach based on the Lyapunov analysis of stability of solutions to deterministic ordinary differential equations

$$\frac{d}{dt}\boldsymbol{x}(t) = A(t)\boldsymbol{x}(t).$$

This approach deals with the upper limit of problem solution

$$\lambda_{\boldsymbol{x}(t)} = \overline{\lim_{t\to+\infty}}\,\frac{1}{t}\ln|\boldsymbol{x}(t)|$$

called the characteristic index of the solution. In the context of this approach applied to stochastic dynamic systems, these investigators often use statistical analysis at the last stage to interpret and simplify the obtained results; in particular, they calculate statistical averages such as

$$\left\langle \lambda_{\boldsymbol{x}(t)}\right\rangle = \overline{\lim_{t\to+\infty}}\,\frac{1}{t}\langle \ln|\boldsymbol{x}(t)|\rangle. \tag{8.40}$$

Parameter α (8.38) is the *Lyapunov exponent* of the lognormal random process $y(t)$.

(2) From the integral distribution functions, one can calculate the typical realization curves of lognormal processes $y(t;\alpha)$ and $\widetilde{y}(t;\alpha)$

$$y^*(t) = e^{\langle \ln y(t)\rangle} = e^{-\alpha t}, \quad \widetilde{y}^*(t) = e^{\langle \ln \widetilde{y}(t)\rangle} = e^{\alpha t}, \tag{8.41}$$

which, in correspondence with Eqs. (4.60), page 108, are the exponentially decaying curve in the case of process $y(t;\alpha)$ and the exponentially increasing curve in the case of process $\widetilde{y}(t;\alpha)$.

Consequently, the exponential increase of moments of random processes $y(t;\alpha)$ and $\widetilde{y}(t;\alpha)$ are caused by deviations of these processes from the typical realization curves $y^*(t;\alpha)$ and $\widetilde{y}^*(t;\alpha)$ towards both large and small values of y and $\widetilde{y}$.

As it follows from Eq. (8.37) at $\alpha/D = 1$, the average value of process $y(t;D)$ is independent of time and is equal to unity. Despite this fact, according to Eq. (8.35), the probability of the event that $y < 1$ for $Dt \gg 1$ rapidly approaches the unity by the law

$$P\left(y(t;D) < 1\right) = \Pr\left(\sqrt{\frac{Dt}{2}}\right) = 1 - \frac{1}{\sqrt{\pi Dt}} e^{-Dt/4},$$

i.e., the curves of process realizations run mainly below the level of the process average $\langle y(t;D)\rangle = 1$, which means that large peaks of the process govern the behavior of statistical moments of process $y(t;D)$.

Here, we have a clear contradiction between the behavior of statistical characteristics of process $y(t;\alpha)$ and the behavior of process realizations.

(3) The behavior of realizations of process $y(t;\alpha)$ on the whole temporal interval can also be evaluated with the use of the p-majorant curves $M_p(t,\alpha)$ whose definition is as follows [2, 5]. We call the majorant curve the curve $M_p(t,\alpha)$ for which inequality $y(t;\alpha) < M_p(t,\alpha)$ is satisfied for all times t with probability p, i.e.,

$$P\left\{y(t;\alpha) < M_p(t,\alpha) \text{ for all } t \in (0,\infty)\right\} = p.$$

The above statistics (8.31) of the absolute maximum of the Wiener process with a drift $w(t;\alpha)$ make it possible to outline a wide enough class of the majorant curves. Indeed, let p be the probability of the event that the absolute maximum $w_{\max}(\beta)$ of the auxiliary process $w(t;\beta)$ with arbitrary parameter β in the interval $0 < \beta < \alpha$ satisfies inequality $w(t;\beta) < h = \ln A$. It is clear that the whole realization of process $y(t;\alpha)$ will run in this case below the majorant curve

$$M_p(t,\alpha,\beta) = Ae^{(\beta-\alpha)t} \tag{8.42}$$

with the same probability p. As may be seen from Eq. (8.31), the probability of the event that process $y(t;\alpha)$ never exceeds the majorant curve (8.42) depends on this curve parameters according to the formula

$$p = 1 - A^{-\beta/D}.$$

This means that we derived the one-parameter class of exponentially decaying majorant curves

$$M_p(t,\alpha,\beta) = \frac{1}{(1-p)^{D/\beta}} e^{(\beta-\alpha)t}. \tag{8.43}$$

Notice the remarkable fact that, despite statistical average $\langle y(t;D)\rangle = 1$ remains constant and higher-order moments of process $y(t;D)$ are exponentially increasing functions, one can always select an exponentially decreasing majorant curve (8.43) such that realizations of process $y(t;D)$ will run below it with arbitrary predetermined probability $p < 1$.

In particular, inequality ($\tau = Dt$)

$$y(t; D) < M_{1/2}(t, D, D/2) = M(\tau) = 4e^{-\tau/2} \tag{8.44}$$

is satisfied with probability $p = 1/2$ for any instant t from interval $(0, \infty)$.

Figure 8.3 schematically shows the behaviors of a realization of process $y(t; D)$ and the majorant curve (8.44). This schematic is an additional fact in favor of our conclusion that the exponential growth of moments of process $y(t; D)$ with time is the purely statistical effect caused by averaging over the whole ensemble of realizations.

Note that the area below the exponentially decaying majorant curves has a finite value. Consequently, high peaks of process $y(t; \alpha)$, which are the reason of the exponential growth of higher moments, only insignificantly contribute to the area below realizations; this area appears finite for almost all realizations, which means that the peaks of the lognormal process $y(t; \alpha)$ are sufficiently narrow.

(4) In this connection, it is of interest to investigate immediately the statistics of random area below realizations of process $y(t; \alpha)$

$$S_n(t; \alpha) = \int_0^t d\tau y^n(\tau; \alpha). \tag{8.45}$$

This function satisfies the system of stochastic equations

$$\begin{aligned} &\frac{d}{dt} S_n(t; \alpha) = y^n(t; \alpha), \quad S_n(0; \alpha) = 0, \\ &\frac{d}{dt} y(t; \alpha) = \{-\alpha + z(t)\} y(t; \alpha), \quad y(0; \alpha) = 1, \end{aligned} \tag{8.46}$$

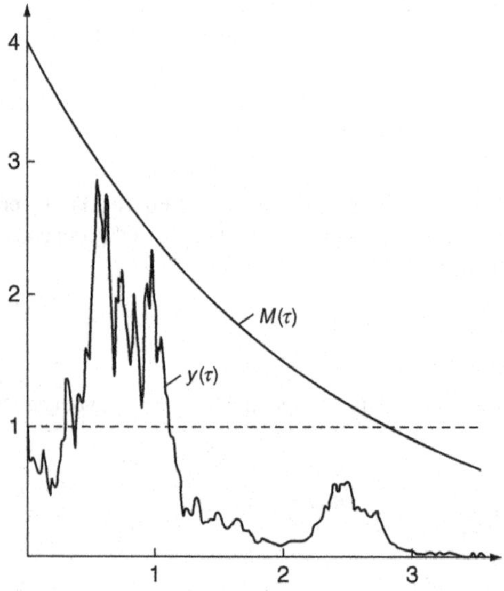

Figure 8.3 Schematic behaviors of a realization of process $y(t; D)$ and majorant curve $M(\tau)$ (8.44).

so that the two-component process $\{y(t;\alpha), S_n(t;\alpha)\}$ is the Markovian process whose one-time probability density

$$P(S_n, y, t;\alpha) = \langle \delta\left(S_n(t;\alpha) - S_n\right)\delta\left(y(t;\alpha) - y\right)\rangle$$

and transition probability density satisfy the Fokker–Planck equation

$$\begin{aligned}\left(\frac{\partial}{\partial t} + y^n\frac{\partial}{\partial S_n} - \alpha\frac{\partial}{\partial y}y\right)P(S_n, y, t;\alpha) &= D\frac{\partial}{\partial y}y\frac{\partial}{\partial y}yP(S_n, y, t;\alpha), \\ P(S_n, y, 0;\alpha) &= \delta(S_n)\delta(y-1).\end{aligned} \tag{8.47}$$

Unfortunately, Eq. (8.47) cannot be solved analytically, which prevents us from studying the statistics of process $S_n(t;\alpha)$ exhaustively. However, for the one-time statistical averages of process $S_n(t;\alpha)$, i.e., averages at a fixed instant, the corresponding statistics can be studied in sufficient detail.

With this goal in view, we rewrite Eq. (8.45) in the form

$$\begin{aligned}S_n(t;\alpha) &= \int_0^t d\tau \exp\left\{-n\alpha\tau + n\int_0^\tau d\tau_1 z(\tau_1)\right\} \\ &= \int_0^t d\tau \exp\left\{-n\alpha(t-\tau) + n\int_0^{t-\tau} d\tau_1 z(t-\tau-\tau_1)\right\},\end{aligned}$$

from which it follows that quantity $S_n(t;\alpha)$ in the context of the one-time statistics is statistically equivalent to the quantity (see section 4.2.1, page 95)

$$S_n(t;\alpha) = \int_0^t d\tau e^{-n\alpha(t-\tau)+n\int_0^{t-\tau} d\tau_1 z(\tau+\tau_1)}. \tag{8.48}$$

Differentiating now Eq. (8.48) with respect to time, we obtain the statistically equivalent stochastic equation

$$\frac{d}{dt}S_n(t;\alpha) = 1 - n\{\alpha - z(t)\}S_n(t;\alpha), \quad S_n(0;\alpha) = 0,$$

whose one-time statistical characteristics are described by the one-time probability density $P(S_n, t;\alpha) = \langle \delta\left(S_n(t;\alpha) - S_n\right)\rangle$ that satisfies the Fokker–Planck equation

$$\left(\frac{\partial}{\partial t} + \frac{\partial}{\partial S_n} - n\alpha\frac{\partial}{\partial S_n}S_n\right)P(S_n, t;\alpha) = n^2D\frac{\partial}{\partial S_n}S_n\frac{\partial}{\partial S_n}S_nP(S_n, y, t;\alpha). \tag{8.49}$$

As may be seen from Eq. (8.49), random integrals

$$S_n(\alpha) = \int_0^\infty d\tau y^n(\tau;\alpha)$$

are distributed according to the steady-state probability density

$$P(S_n;\alpha)=\frac{1}{\left(n^2D\right)^{\alpha/nD}\Gamma\left(\frac{\alpha}{D}\right)S_n^{1+\alpha/D}}\exp\left\{-\frac{1}{n^2DS_n}\right\},$$

where $\Gamma(z)$ is the *gamma function*. In the special case $n=1$, quantity

$$S(\alpha)=S_1(\alpha)=\int\limits_0^\infty d\tau y(\tau;\alpha)$$

has the following probability density

$$P(S;\alpha)=\frac{1}{D^{\alpha/D}\Gamma\left(\frac{\alpha}{D}\right)S^{1+\alpha/D}}\exp\left\{-\frac{1}{DS}\right\}. \tag{8.50}$$

If we set now $\alpha=D$, then the steady-state probability density and the corresponding integral distribution function will have the form

$$P(S;D)=\frac{1}{DS^2}\exp\left\{-\frac{1}{DS}\right\},\quad F(S;D)=\exp\left\{-\frac{1}{DS}\right\}. \tag{8.51}$$

The time-dependent behavior of the probability density of random process

$$\widetilde{S}(t,\alpha)=\int\limits_t^\infty d\tau y(\tau;\alpha) \tag{8.52}$$

gives an additional information about the behavior of realizations of process $y(t;\alpha)$ with time t. The integral in the right-hand side of Eq. (8.52) can be represented in the form

$$\widetilde{S}(t,\alpha)=y(t;\alpha)\int\limits_0^\infty d\tau\exp\left\{-\alpha\tau+\int\limits_0^\tau d\tau_1 z(\tau_1+t)\right\}. \tag{8.53}$$

In Eq. (8.53), random process $y(t;\alpha)$ is statistically independent of the integral factor, because they depend functionally on process $z(\tau)$ for nonoverlapping intervals of times τ; in addition, the integral factor by itself appears statistically equivalent to random quantity $S(\alpha)$. Consequently, the one-time probability density $P(\widetilde{S},t;\alpha)=\left\langle\delta\left(\widetilde{S}(t;\alpha)-\widetilde{S}\right)\right\rangle$ of random process $\widetilde{S}(t,\alpha)$ is described by the expression

$$P(\widetilde{S},t;\alpha)=\int\limits_0^\infty\int\limits_0^\infty dydS\delta(yS-\widetilde{S})P(y,t;\alpha)=\int\limits_0^\infty\frac{dy}{y}P(y,t;\alpha)P\left(\frac{\widetilde{S}}{y};\alpha\right), \tag{8.54}$$

where $P(y,t;\alpha)$ is the one-time probability density of lognormal process $y(t;\alpha)$ (8.33) and $P\left(\frac{\widetilde{S}}{y};\alpha\right)$ is the probability density (8.50) of random area.

The corresponding integral distribution function

$$F(\widetilde{S}, t; \alpha) = P\left(\widetilde{S}(t; \alpha) < \widetilde{S}\right) = \int_0^{\widetilde{S}} d\widetilde{S} P(\widetilde{S}, t; \alpha)$$

is given by the integral

$$F(\widetilde{S}, t; \alpha) = \int_0^{\infty} dy P(y, t; \alpha) F\left(\frac{\widetilde{S}}{y}; \alpha\right),$$

where $F(S; \alpha)$ is the integral distribution function of random area $S(t; \alpha)$. In the special case $\alpha = D$, we obtain, according to Eq. (8.33) and (8.51), the expression

$$F(\widetilde{S}, t; D) = \frac{1}{2\sqrt{\pi D t}} \int_0^{\infty} \frac{dy}{y} \exp\left\{-\frac{\ln^2\left(ye^{Dt}\right)}{4Dt} - \frac{y}{D\widetilde{S}}\right\}$$

from which follows that the probability of the event that inequality $\widetilde{S}(t; D) < \widetilde{S}$ is satisfied monotonously tends to unity with increasing Dt for any predetermined value of $D\widetilde{S}$. This is an additional evidence in favor of the fact that every separate realization of the lognormal process tends to zero with increasing Dt, though moment functions of process $y(t; \alpha)$ show the exponential growth caused by large spikes.

(5) Now, we dwell on positive random fields $E(\boldsymbol{r}, t)$ closely related to lognormal processes whose one-point probability density is described by the equation

$$\left(\frac{\partial}{\partial t} - D_0 \frac{\partial^2}{\partial \boldsymbol{r}^2}\right) P(\boldsymbol{r}, t; E) = \left\{\alpha \frac{\partial}{\partial E} E + D \frac{\partial}{\partial E} E \frac{\partial}{\partial E} E\right\} P(\boldsymbol{r}, t; E),$$
$$P(\boldsymbol{r}, 0; E) = \delta\left(E - E_0(\boldsymbol{r})\right), \tag{8.55}$$

where D_0 is the coefficient of diffusion in $\boldsymbol{r}$-space, and coefficients α and D characterize diffusion in E-space. Parameter α can be both positive and negative. In the context of the one-point characteristics, the change of sign of parameter α means transition from random field $E(\boldsymbol{r}, t)$ to random field $\widetilde{E}(\boldsymbol{r}, t) = 1/E(\boldsymbol{r}, t)$. For definiteness, we will term field $E(\boldsymbol{r}, t)$ energy.

The solution to this equation has the form

$$P(\boldsymbol{r}, t; E) = \exp\left\{D_0 t \frac{\partial^2}{\partial \boldsymbol{r}^2}\right\} \widetilde{P}(\boldsymbol{r}, t; E),$$

where function $\widetilde{P}(\boldsymbol{r}, t; E)$ satisfies the equation (cf. with Eq. (8.36))

$$\frac{\partial}{\partial t} \widetilde{P}(\boldsymbol{r}, t; E) = \left\{\alpha \frac{\partial}{\partial E} E + D \frac{\partial}{\partial E} E \frac{\partial}{\partial E} E\right\} \widetilde{P}(\boldsymbol{r}, t; E),$$

$$\widetilde{P}(\boldsymbol{r}, 0; E) = \delta\left(E - E_0(\boldsymbol{r})\right).$$

As may be seen, dependence of function $\widetilde{P}(\boldsymbol{r}, t; E)$ on parameter $\boldsymbol{r}$ is determined solely by initial value $E_0(\boldsymbol{r})$, i.e.,

$$\widetilde{P}(\boldsymbol{r}, t; E) \equiv \widetilde{P}(t; E|E_0(\boldsymbol{r}))$$

and, hence, function $\widetilde{P}(t; E|E_0(\boldsymbol{r}))$ is the lognormal probability density of random process $E(t, |E_0(\boldsymbol{r}))$ dependent on $\boldsymbol{r}$ parametrically,

$$\widetilde{P}(t; E|E_0(\boldsymbol{r})) = \frac{1}{2E\sqrt{\pi Dt}} \exp\left\{-\frac{\ln^2\left[Ee^{\alpha t}/E_0(\boldsymbol{r})\right]}{4Dt}\right\}.$$

Thus, the solution to Eq. (8.55) has the form

$$P(\boldsymbol{r}, t; E) = \frac{1}{2E\sqrt{\pi Dt}} \exp\left\{D_0 t \frac{\partial^2}{\partial \boldsymbol{r}^2}\right\} \exp\left\{-\frac{\ln^2\left[Ee^{\alpha t}/E_0(\boldsymbol{r})\right]}{4Dt}\right\}. \tag{8.56}$$

A consequence of equation (8.55) or expression (8.56) are relationships for spatial integrals of moment functions (for all $n > 0$ and $n < 0$)

$$\int d\boldsymbol{r} \left\langle E^n(\boldsymbol{r}, t)\right\rangle = e^{n(nD-\alpha)t} \int d\boldsymbol{r} E_0^n(\boldsymbol{r}),$$

which are independent of the coefficient of diffusion in $\boldsymbol{r}$-space (coefficient D_0), and, in particular, the expression for average total energy in the whole space,

$$\int d\boldsymbol{r} \left\langle E(\boldsymbol{r}, t)\right\rangle = e^{\gamma t} \int d\boldsymbol{r} E_0(\boldsymbol{r}),$$

where $\gamma = D - \alpha$. If the initial energy distribution is homogeneous, $E_0(\boldsymbol{r}) = E_0$, probability density (8.56) is independent of $\boldsymbol{r}$ and is given by Eq. (8.33),

$$P(t; E) = \frac{1}{2E\sqrt{\pi Dt}} \exp\left\{-\frac{\ln^2\left[Ee^{\alpha t}/E_0\right]}{4Dt}\right\}. \tag{8.57}$$

Thus, in this case, the one-point statistical characteristics of energy $E(\boldsymbol{r}, t)$ are statistically equivalent to statistical characteristics of random process $E(t) = y(t; \alpha)$ (8.32). A characteristic feature of distribution (8.57) consists in a long flat *tail* appeared for $Dt \gg 1$, which is indicative of an increased role of great peaks of process $E(t; \alpha)$ in the formation of the one-point statistics. For this distribution, all moments of energy are exponential functions and, in particular,

$$\langle E(t)\rangle = E_0 e^{\gamma t}, \quad \gamma = D - \alpha$$

at $n = 1$, and

$$\langle \ln \left(E(t)/E_0\right)\rangle = -\alpha t,$$

so that the typical realization curve determining the behavior of energy at arbitrary spatial point in specific realizations of random process $E(t)$ is the exponential quantity

$$E^*(t) = E_0 e^{-\alpha t}$$

increasing or decreasing with time. For $\alpha < 0$, the typical realization curve exponentially increases with time, which is evidence of general increase of energy at every spatial point. Otherwise, for $\alpha > 0$, the typical realization curve exponentially decreases at every spatial point, which is indicative of cluster structure of energy field.

Figure 8.4 schematically shows random realizations of energy field for different signs of parameter α.

The knowledge of the one-point probability density of energy (8.56) yields general knowledge of the spatial structure of energy. In particular, functionals of the energy field such as total average volume (in the three-dimensional case) or area (in the two-dimensional case) of the region in which $E(\boldsymbol{r}, t) > E$ and total average energy contained in this region are given by the equalities

$$\langle V(t, E)\rangle = \int\limits_E^\infty d\widetilde{E} \int d\boldsymbol{r}\, P(\boldsymbol{r}, t; \widetilde{E}),$$

$$\langle \mathcal{E}(t, E)\rangle = \int\limits_E^\infty \widetilde{E} d\widetilde{E} \int d\boldsymbol{r}\, P(\boldsymbol{r}, t; \widetilde{E}).$$

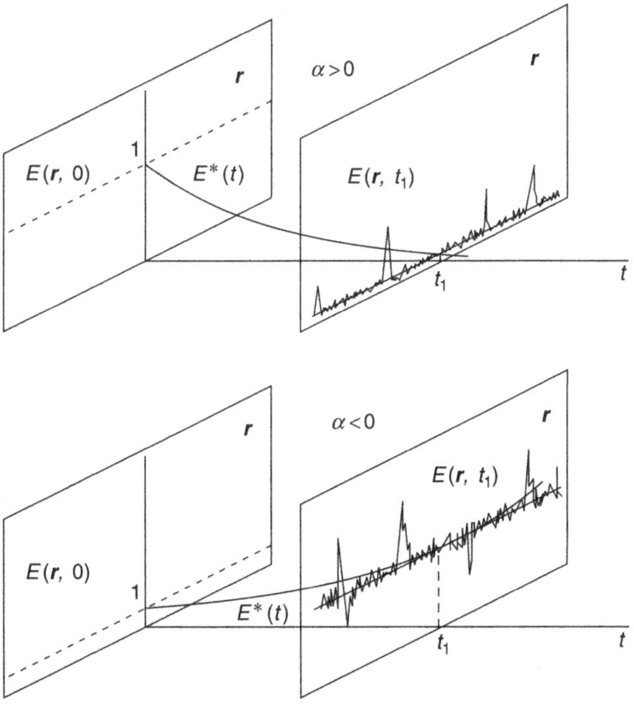

Figure 8.4 Schematic behavior of random realization of energy for $\alpha > 0$ and $\alpha < 0$.

The values of these functionals are independent of the coefficient of diffusion in $\boldsymbol{r}$-space (coefficient D_0), and, using probability density (8.56), we obtain the expressions

$$\langle V(t,E)\rangle = \int d\boldsymbol{r}\,\Pr\left(\frac{1}{\sqrt{2Dt}}\ln\left(\frac{E_0(\boldsymbol{r})}{E}e^{-\alpha t}\right)\right),$$

$$\langle \mathcal{E}(t,E)\rangle = e^{\gamma t}\int d\boldsymbol{r}\,E_0(\boldsymbol{r})\Pr\left(\frac{1}{\sqrt{2Dt}}\ln\left(\frac{E_0(\boldsymbol{r})}{E}e^{(2D-\alpha)t}\right)\right), \tag{8.58}$$

where probability integral $\Pr(z)$ is defined by Eq. (4.20), page 94. Asymptotic expressions of function $\Pr(z)$ for $z\to\infty$ and $z\to-\infty$ (4.23), page 95, offer a possibility of studying temporal evolution of these functionals. Namely, average volume asymptotically decays (for $\alpha>0$) with time ($t\to\infty$) according to the law

$$\langle V(t,E)\rangle \approx \frac{1}{\alpha}\sqrt{\frac{D}{\pi E^{\alpha/D}t}}e^{-\alpha^2 t/4D}\int d\boldsymbol{r}\,\sqrt{E_0^{\alpha/D}(\boldsymbol{r})}.$$

On the contrary, for $\alpha<0$, average volume occupies the whole space for $t\to\infty$.

Asymptotic behavior of total energy for $t\to\infty$ has the form (in the most interesting case of $\gamma>0$, $\alpha<D$)

$$\langle \mathcal{E}(t,E)\rangle$$
$$\approx e^{\gamma t}\int d\boldsymbol{r}E_0(\boldsymbol{r})\left[1-\frac{1}{(2D-\alpha)}\sqrt{\frac{D}{\pi t}\left(\frac{E}{E_0(\boldsymbol{r})}\right)^{(2D-\alpha)/D}}e^{-(2D-\alpha)^2 t/4}\right],$$

which means that 100% of total average energy is contained in clusters for $\alpha>0$.

In the case of homogeneous initial conditions, the corresponding expressions without integration over $\boldsymbol{r}$ present specific values of volume occupied by large peaks and their total energy per unit volume:

$$\langle \mathfrak{V}(t,E)\rangle = \Pr\left(\frac{1}{\sqrt{2Dt}}\ln\left(\frac{E_0(\boldsymbol{r})}{E}e^{-\alpha t}\right)\right),$$

$$\langle \mathfrak{E}(t,E)\rangle = E_0 e^{\gamma t}\Pr\left(\frac{1}{\sqrt{2Dt}}\ln\left(\frac{E_0(\boldsymbol{r})}{E}e^{(2D-\alpha)t}\right)\right), \tag{8.59}$$

where $\gamma=D-\alpha$.

If we take section at level $E>E_0$, then the initial values of these quantities will be equal to zero at the initial instant, $\langle\mathfrak{V}(0,E)\rangle=0$ and $\langle\mathfrak{E}(0,E)\rangle=0$. Then, spatial disturbances of energy field appear with time and, for $t\to\infty$, they are given by the following asymptotic expressions:

$$\langle \mathfrak{V}(t,E)\rangle \approx \begin{cases} \dfrac{1}{\alpha}\sqrt{\dfrac{D}{\pi t}\left(\dfrac{E_0}{E}\right)^{\alpha/D}}e^{-\alpha^2 t/4D} & (\alpha>0), \\[2ex] 1-\dfrac{1}{|\alpha|}\sqrt{\dfrac{D}{\pi t}\left(\dfrac{E}{E_0}\right)^{|\alpha|/D}}e^{-\alpha^2 t/4D} & (\alpha<0), \end{cases}$$

and, for $D > \alpha$,

$$\langle \mathfrak{E}(t, E)) \rangle \approx E_0 e^{\gamma t} \left[1 - \frac{1}{(2D - \alpha)} \sqrt{\frac{D}{\pi t} \left(\frac{E}{E_0} \right)^{(2D-\alpha)/D}} e^{-(2D-\alpha)^2 t/4D} \right].$$

Thus, for $\alpha > 0$, specific total volume tends to zero and specific total energy increases with time as average energy does, which is evidence of clustering the field of energy in this case.

For $\alpha < 0$, no clustering occurs and specific volume occupies the whole space in which specific energy increases with time as average energy does.

8.4 Applicability Range of the Fokker–Planck Equation

To estimate the applicability range of the Fokker–Planck equation, we must include into consideration the finite-valued correlation radius τ_0 of field $\boldsymbol{f}(\boldsymbol{x}, t)$ with respect to time. Thus, smallness of parameter τ_0/T is the necessary but generally not sufficient condition in order that one can describe the statistical characteristics of the solution to Eq. (8.1), using the approximation of the delta-correlated random field of which a consequence is the Fokker–Planck equation. Every particular problem requires more detailed investigation. Below, we give a more physical method called the *diffusion approximation.* This method also leads to the Markovian property of the solution to Eq. (8.1); however, it considers to some extent the finite value of the temporal correlation radius.

Here, we emphasize that the approximation of the delta-correlated random field does not reduce to the formal replacement of random field $\boldsymbol{f}(\boldsymbol{x}, t)$ in Eq. (8.1) with the random field with correlation function (8.4). This approximation corresponds to the construction of an asymptotic expansion in temporal correlation radius τ_0 of filed $\boldsymbol{f}(\boldsymbol{x}, t)$ for $\tau_0 \to 0$. It is in such a limited process that exact average quantities like

$$\left\langle \boldsymbol{f}(\boldsymbol{x}, t) R[t; \boldsymbol{f}(\boldsymbol{x}', \tau)] \right\rangle$$

grade into the expressions obtained by the formal replacement of the correlation tensor of field $\boldsymbol{f}(\boldsymbol{x}, t)$ with the effective tensor (8.4).

8.4.1 Langevin Equation

We illustrate the above speculation with the example of the *Langevin equation* that allows an exhaustive statistical analysis. This equation has the form

$$\frac{d}{dt} x(t) = -\lambda x(t) + f(t), \quad x(t_0) = 0 \tag{8.60}$$

and assumes that the sufficiently fine smooth function $f(t)$ is the stationary Gaussian process with zero-valued mean and correlation function

$$\left\langle f(t) f(t') \right\rangle = B_f(t - t').$$

For any individual realization of random force $f(t)$, the solution to Eq. (8.60) has the form

$$x(t) = \int_{t_0}^{t} d\tau f(\tau) e^{-\lambda(t-\tau)}.$$

Consequently, this solution $x(t)$ is also the Gaussian process with the parameters

$$\langle x(t)\rangle = 0, \quad \langle x(t)x(t')\rangle = \int_{t_0}^{t} d\tau_1 \int_{t_0}^{t'} d\tau_2 B_f(\tau_1 - \tau_2) e^{-\lambda(t+t'-\tau_1-\tau_2)}.$$

In addition, we have, for example,

$$\langle f(t)x(t)\rangle = \int_{0}^{t-t_0} d\tau B_f(\tau) e^{-\lambda\tau}.$$

Note that the one-point probability density $P(x, t) = \langle \delta(x(t) - x)\rangle$ for Eq. (8.60) satisfies the equation

$$\left(\frac{\partial}{\partial t} - \lambda \frac{\partial}{\partial x} x\right) P(x, t) = \int_{0}^{t-t_0} d\tau B_f(\tau) e^{-\lambda\tau} \frac{\partial^2}{\partial x^2} P(x, t), \quad P(x, t_0) = \delta(x),$$

which rigorously follows from Eq. (6.24), page 149. As a consequence, we obtain

$$\frac{d}{dt}\langle x^2(t)\rangle = -2\lambda\langle x^2(t)\rangle + 2\int_{0}^{t-t_0} d\tau B_f(\tau) e^{-\lambda\tau}.$$

For $t_0 \to -\infty$, process $x(t)$ grades into the stationary Gaussian process with the following one-time statistical parameters ($\langle x(t)\rangle = 0$)

$$\sigma_x^2 = \langle x^2(t)\rangle = \frac{1}{\lambda}\int_{0}^{\infty} d\tau B_f(\tau) e^{-\lambda\tau}, \quad \langle f(t)x(t)\rangle = \int_{0}^{\infty} d\tau B_f(\tau) e^{-\lambda\tau}.$$

In particular, for exponential correlation function $B_f(t)$,

$$B_f(t) = \sigma_f^2 e^{-|\tau|/\tau_0},$$

we obtain the expressions

$$\langle x(t)\rangle = 0, \quad \left\langle x^2(t)\right\rangle = \frac{\sigma_f^2 \tau_0}{\lambda(1+\lambda\tau_0)}, \quad \langle f(t)x(t)\rangle = \frac{\sigma_f^2 \tau_0}{1+\lambda\tau_0}, \tag{8.61}$$

which grade into the asymptotic expressions

$$\langle x^2(t)\rangle = \frac{\sigma_f^2\tau_0}{\lambda}, \quad \langle f(t)x(t)\rangle = \sigma_f^2\tau_0 \tag{8.62}$$

for $\tau_0 \to 0$.

Multiply now Eq. (8.60) by $x(t)$. Assuming that function $x(t)$ is sufficiently fine function, we obtain the equality

$$x(t)\frac{d}{dt}x(t) = \frac{1}{2}\frac{d}{dt}x^2(t) = -\lambda x^2(t) + f(t)x(t).$$

Averaging this equation over an ensemble of realizations of function $f(t)$, we obtain the equation

$$\frac{1}{2}\frac{d}{dt}\langle x^2(t)\rangle = -\lambda\langle x^2(t)\rangle + \langle f(t)x(t)\rangle, \tag{8.63}$$

whose steady-state solution (it corresponds to the limit process $t_0 \to -\infty$ and $\tau_0 \to 0$)

$$\langle x^2(t)\rangle = \frac{1}{\lambda}\langle f(t)x(t)\rangle$$

coincides with Eq. (8.61) and (8.62).

Taking into account the fact that $\delta x(t)/\delta f(t-0) = 1$, we obtain the same result for correlation $\langle f(t)x(t)\rangle$ by using the formula

$$\langle f(t)x(t)\rangle = \int_{-\infty}^{t} d\tau B_f(t-\tau)\left\langle \frac{\delta}{\delta f(\tau)}x(t)\right\rangle \tag{8.64}$$

with the effective correlation function

$$B_f^{\text{eff}}(t) = 2\sigma_f^2\tau_0\delta(t).$$

Earlier, we mentioned that statistical characteristics of solutions to dynamic problems in the approximation of the delta-correlated random process (field) coincide with the statistical characteristics of the Markovian processes. However, one should clearly understand that this is the case only for statistical averages and equations for these averages. In particular, realizations of process $x(t)$ satisfying the Langevin equation (8.60) drastically differ from realizations of the corresponding Markovian process. The latter satisfies Eq. (8.60) in which function $f(t)$ in the right-hand side is the ideal white noise with the correlation function $B_f(t) = 2\sigma_f^2\tau_0\delta(t)$; moreover, this equation must be treated in the sense of generalized functions, because the Markovian processes are not differentiable in the ordinary sense. At the same time, process $x(t)$ – whose statistical characteristics coincide with the characteristics of the Markovian process – behaves as a sufficiently fine function and is differentiable in the ordinary sense. For example,

$$x(t)\frac{d}{dt}x(t) = \frac{1}{2}\frac{d}{dt}x^2(t),$$

and we have for $t_0 \to -\infty$ in particular

$$\left\langle x(t)\frac{d}{dt}x(t)\right\rangle = 0. \tag{8.65}$$

On the other hand, in the case of the ideal Markovian process $x(t)$ satisfying (in the sense of generalized functions) the Langevin equation (8.60) with the white noise in the right-hand side, Eq. (8.65) makes no sense at all, and the meaning of the relationship

$$\left\langle x(t)\frac{d}{dt}x(t)\right\rangle = -\lambda\left\langle x^2(t)\right\rangle + \langle f(t)x(t)\rangle \tag{8.66}$$

depends on the definition of averages. Indeed, if we will treat Eq. (8.66) as the limit of the equality

$$\left\langle x(t+\Delta)\frac{d}{dt}x(t)\right\rangle = -\lambda\,\langle x(t)x(t+\Delta)\rangle + \langle f(t)x(t+\Delta)\rangle \tag{8.67}$$

for $\Delta \to 0$, the result will be essentially different depending on whether we use limit processes $\Delta \to +0$, or $\Delta \to -0$. For limit process $\Delta \to +0$, we have

$$\lim_{\Delta\to+0}\langle f(t)x(t+\Delta)\rangle = 2\sigma_f^2\tau_0,$$

and, taking into account Eq. (8.64), we can rewrite Eq. (8.67) in the form

$$\left\langle x(t+0)\frac{d}{dt}x(t)\right\rangle = \sigma_f^2\tau_0. \tag{8.68}$$

On the contrary, for limit process $\Delta \to -0$, we have

$$\langle f(t)x(t-0)\rangle = 0$$

because of the dynamic causality condition, and Eq. (8.67) assumes the form

$$\left\langle x(t-0)\frac{d}{dt}x(t)\right\rangle = -\sigma_f^2\tau_0. \tag{8.69}$$

Comparing Eq. (8.65) with Eq. (8.68) and (8.69), we see that, for the ideal Markovian process described by the solution to the Langevin equation with the white noise in the right-hand side and commonly called the *Ohrnstein–Ulenbeck process*, we have

$$\left\langle x(t+0)\frac{d}{dt}x(t)\right\rangle \neq \left\langle x(t-0)\frac{d}{dt}x(t)\right\rangle \neq \frac{1}{2}\frac{d}{dt}\left\langle x^2(t)\right\rangle.$$

Note that equalities (8.68) and (8.69) can also be obtained from the correlation function

$$\langle x(t)x(t+\tau)\rangle = \frac{\sigma_f^2\tau_0}{\lambda}e^{-\lambda|\tau|}$$

of process $x(t)$.

To conclude with a discussion of the approximation of the delta-correlated random process (field), we emphasize that, in all further examples, we will treat the statement 'dynamic system (equation) with the delta-correlated parameter fluctuations' as the asymptotic limit in which these parameters have temporal correlation radii small in comparison with all characteristic temporal scales of the problem under consideration.

8.5 Causal Integral Equations

In problems discussed earlier, we succeeded in deriving the closed statistical description in the approximation of the delta-correlated random field due to the fact that each of these problems corresponded to a system of the first-order (in temporal coordinate) differential equations with given initial conditions at $t = 0$. Such systems possess the dynamic causality property, which means that the solution at instant t depends only on system parameter fluctuations for preceding times and is independent of fluctuations for consequent times.

However, problems described in terms of integral equations that generally cannot be reduced to a system of differential equations can also possess the causality property. In short, we illustrate this fact by the simplest example of the one-dimensional causal equation $(t > t')$

$$G(t;t') = g(t;t') + \Lambda\int\limits_{t'}^{t} d\tau g(t;\tau)z(\tau)G(\tau;t'), \tag{8.70}$$

where function $g(t;t')$ is Green's function for the problem with absent parameter fluctuations, i.e., for $z(t) = 0$, and we assume that $z(t)$ is the Gaussian delta-correlated random function with the parameters

$$\langle z(t)\rangle = 0, \quad \left\langle z(t)z(t')\right\rangle = 2D\delta(t-t') \quad (D = \sigma_z^2\tau_0).$$

Averaging then Eq. (8.70) over an ensemble of realizations of random function $z(t)$, we obtain the equation

$$\langle G(t;t')\rangle = g(t;t') + \Lambda\int\limits_{t'}^{t} d\tau g(t;\tau)\langle z(\tau)G(\tau;t')\rangle. \tag{8.71}$$

Taking into account equality

$$\frac{\delta}{\delta z(t)}G(t;t') = g(t;t)\Lambda G(t;t') \tag{8.72}$$

following from Eq. (8.70), we can rewrite the correlator in the right-hand side of Eq. (8.71) in the form

$$\langle z(\tau)G(\tau;t')\rangle = D\left\langle\frac{\delta}{\delta z(\tau)}G(\tau;t')\right\rangle = \Lambda Dg(\tau;\tau)\langle G(\tau;t')\rangle.$$

As a consequence, Eq. (8.71) grades into the closed integral equation for average Green's function

$$\langle G(t;t')\rangle = g(t;t') + \Lambda^2 D\int\limits_{t'}^{t} d\tau g(t;\tau)g(\tau;\tau)\langle G(\tau;t')\rangle, \tag{8.73}$$

which has the form of the *Dyson equation* (in the terminology of the quantum field theory)

$$\langle G(t;t')\rangle = g(t;t') + \Lambda\int\limits_{t'}^{t} d\tau g(t;\tau)\int\limits_{t'}^{\tau} d\tau' Q(\tau;\tau')\langle G(\tau';t')\rangle,$$

or

$$\langle G(t;t')\rangle = g(t;t') + \Lambda\int\limits_{t'}^{t} d\tau\,\langle G(t;\tau)\rangle\int\limits_{t'}^{\tau} d\tau' Q(\tau;\tau')g(\tau';t'), \tag{8.74}$$

with the *mass function*

$$Q(\tau;\tau') = \Lambda^2 Dg(\tau;\tau)\delta(\tau-\tau').$$

Derive now the equation for the correlation function

$$\Gamma(t,t';t_1,t_1') = \langle G(t;t')G^*(t_1;t_1')\rangle \quad (t>t',\quad t_1>t_1'),$$

where $G^*(t;t')$ is complex conjugated Green's function. With this goal in view, we multiply Eq. (8.70) by $G^*(t_1;t_1')$ and average the result over an ensemble of realizations of random function $z(t)$. The result is the equation that can be symbolically represented as

$$\Gamma = g\left\langle G^*\right\rangle + \Lambda g\left\langle zGG^*\right\rangle. \tag{8.75}$$

Taking into account the Dyson equation (8.74)

$$\langle G\rangle = \{1 + \langle G\rangle\, Q\}g,$$

we apply operator $\{1 + \langle G\rangle\, Q\}$ to Eq. (8.75). As a result, we obtain the symbolic-form equation

$$\Gamma = \langle G\rangle\left\langle G^*\right\rangle + \langle G\rangle\, \Lambda\left\{\left\langle zGG^*\right\rangle - Q\Gamma\right\},$$

which can be represented in common variables as

$$\begin{aligned}\Gamma(t,t';t_1,t_1') &= \left\langle G(t;t')\right\rangle\left\langle G^*(t_1;t_1')\right\rangle\\ &+ \Lambda D\int_0^t d\tau\, \langle G(t;\tau)\rangle\left[\left\langle\frac{\delta G(\tau;t')}{\delta z(\tau)}G^*(t_1;t_1') + 2G(\tau;t')\frac{\delta G^*(t_1;t_1')}{\delta z(\tau)}\right\rangle\right]\\ &- \Lambda^2 D\int_0^t d\tau\, \langle G(t;\tau)\rangle\, g(\tau;\tau)\Gamma(\tau,t';t_1,t_1').\end{aligned} \tag{8.76}$$

Deriving Eq. (8.76), we used additionally Eq. (5.30), page 130, for splitting correlations between the Gaussian delta-correlated process $z(t)$ and functionals of this process

$$\left\langle z(t')R[t;z(\tau)]\right\rangle >= \begin{cases} D\left\langle\dfrac{\delta}{\delta z(t)}R[t;z(\tau)]\right\rangle & (t'=t,\quad \tau<t),\\[2ex] 2D\left\langle\dfrac{\delta}{\delta z(t')}R[t;z(\tau)]\right\rangle & (t'<t,\quad \tau<t).\end{cases}$$

Taking into account Eq. (8.72) and equality

$$\frac{\delta}{\delta z(\tau)}G^*(t_1;t_1') = \Lambda G^*(t_1;\tau)G^*(\tau;t_1'),$$

following from Eq. (8.70) we can rewrite Eq. (8.76) as

$$\begin{aligned}\Gamma(t,t';t_1,t_1') &= \left\langle G(t;t')\right\rangle\left\langle G^*(t_1;t_1')\right\rangle\\ &\qquad + 2|\Lambda|^2 D\int_0^t d\tau\, \langle G(t;\tau)\rangle\left\langle G^*(t_1;\tau)G(\tau;t')G^*(\tau;t_1')\right\rangle.\end{aligned} \tag{8.77}$$

Now, we take into account the fact that function $G^*(t_1;\tau)$ functionally depends on random process $z(\widetilde{\tau})$ for $\widetilde{\tau} \geq \tau$ while functions $G(\tau;t')$ and $G^*(\tau;t_1')$ depend on it for $\widetilde{\tau} \leq \tau$. Consequently, these functions are statistically independent in the case of the delta-correlated process $z(\widetilde{\tau})$, and we can rewrite Eq. (8.77) in the form of the closed equation ($t_1 \geq t$)

$$\Gamma(t,t';t_1,t_1') = \langle G(t;t')\rangle \langle G^*(t_1;t_1')\rangle + 2|\Lambda|^2 D \int_0^t d\tau \, \langle G(t;\tau)\rangle \langle G^*(t_1;\tau)\rangle \Gamma(\tau;t';\tau;t_1'). \tag{8.78}$$

8.6 Diffusion Approximation

Applicability of the approximation of the delta-correlated random field $\boldsymbol{f}(\boldsymbol{x},t)$ (i.e., applicability of the Fokker–Planck equation) is restricted by the smallness of the temporal correlation radius τ_0 of random field $\boldsymbol{f}(\boldsymbol{x},t)$ with respect to all temporal scales of the problem under consideration. The effect of the finite-valued temporal correlation radius of random field $\boldsymbol{f}(\boldsymbol{x},t)$ can be considered within the framework of the diffusion approximation. The diffusion approximation appears to be more obvious and physical than the formal mathematical derivation of the approximation of the delta-correlated random field. This approximation also holds for sufficiently weak parameter fluctuations of the stochastic dynamic system and allows us to describe new physical effects caused by the finite-valued temporal correlation radius of random parameters, rather than only obtaining the applicability range of the delta-correlated approximation. The diffusion approximation assumes that the effect of random actions is insignificant during temporal scales about τ_0, i.e., the system behaves during these time intervals as the free system.

Again, assume vector function $\boldsymbol{x}(t)$ satisfies the dynamic equation (8.1), page 191,

$$\frac{d}{dt}\boldsymbol{x}(t) = \boldsymbol{v}(\boldsymbol{x},t) + \boldsymbol{f}(\boldsymbol{x},t), \quad \boldsymbol{x}(t_0) = \boldsymbol{x}_0, \tag{8.79}$$

where $\boldsymbol{v}(\boldsymbol{x},t)$ is the deterministic vector function and $\boldsymbol{f}(\boldsymbol{x},t)$ is the random statistically homogeneous and stationary Gaussian vector field with the statistical characteristics

$$\langle \boldsymbol{f}(\boldsymbol{x},t)\rangle = 0, \quad B_{ij}(\boldsymbol{x},t;\boldsymbol{x}',t') = B_{ij}(\boldsymbol{x}-\boldsymbol{x}',t-t') = \langle f_i(\boldsymbol{x},t) f_j(\boldsymbol{x}',t')\rangle.$$

Introduce the indicator function

$$\varphi(\boldsymbol{x},t) = \delta(\boldsymbol{x}(t)-\boldsymbol{x}), \tag{8.80}$$

where $\boldsymbol{x}(t)$ is the solution to Eq. (8.79) satisfying the Liouville equation (8.6)

$$\left(\frac{\partial}{\partial t} + \frac{\partial}{\partial \boldsymbol{x}}\boldsymbol{v}(\boldsymbol{x},t)\right)\varphi(\boldsymbol{x},t) = -\frac{\partial}{\partial \boldsymbol{x}}\boldsymbol{f}(\boldsymbol{x},t)\varphi(\boldsymbol{x},t). \tag{8.81}$$

As earlier, we obtain the equation for the probability density of the solution to Eq. (8.79)

$$P(\boldsymbol{x}, t) = \langle \varphi(\boldsymbol{x}, t) \rangle = \langle \delta(\boldsymbol{x}(t) - \boldsymbol{x}) \rangle$$

by averaging Eq. (8.81) over an ensemble of realizations of field $\boldsymbol{f}(\boldsymbol{x}, t)$

$$\left(\frac{\partial}{\partial t} + \frac{\partial}{\partial \boldsymbol{x}} \boldsymbol{v}(\boldsymbol{x}, t) \right) P(\boldsymbol{x}, t) = -\frac{\partial}{\partial \boldsymbol{x}} \langle \boldsymbol{f}(\boldsymbol{x}, t) \varphi(\boldsymbol{x}, t) \rangle,$$
$$P(\boldsymbol{x}, t_0) = \delta(\boldsymbol{x} - \boldsymbol{x}_0). \tag{8.82}$$

Using the Furutsu–Novikov formula (8.10), page 193,

$$\langle f_k(\boldsymbol{x}, t) R[t; \boldsymbol{f}(\boldsymbol{y}, \tau)] \rangle = \int d\boldsymbol{x}' \int dt' B_{kl}(\boldsymbol{x}, t; \boldsymbol{x}', t') \left\langle \frac{\delta}{\delta f_l(\boldsymbol{x}', t')} R[t; \boldsymbol{f}(\boldsymbol{y}, \tau)] \right\rangle$$

valid for the correlation between the Gaussian random field $\boldsymbol{f}(\boldsymbol{x}, t)$ and arbitrary functional $R[t; \boldsymbol{f}(\boldsymbol{y}, \tau)]$ of this field, we can rewrite Eq. (8.82) in the form

$$\left(\frac{\partial}{\partial t} + \frac{\partial}{\partial \boldsymbol{x}} \boldsymbol{v}(\boldsymbol{x}, t) \right) P(\boldsymbol{x}, t) = -\frac{\partial}{\partial x_i} \int d\boldsymbol{x}' \int_{t_0}^{t} dt' B_{ij}(\boldsymbol{x}, t; \boldsymbol{x}', t') \left\langle \frac{\delta}{\delta f_j(\boldsymbol{x}', t')} \varphi(\boldsymbol{x}, t) \right\rangle. \tag{8.83}$$

In the diffusion approximation, Eq. (8.83) is the exact equation, and the variational derivative and indicator function satisfy, within temporal intervals of about temporal correlation radius τ_0 of random field $\boldsymbol{f}(\boldsymbol{x}, t)$, the system of dynamic equations

$$\frac{\partial}{\partial t} \frac{\delta \varphi(\boldsymbol{x}, t)}{\delta f_i(\boldsymbol{x}', t')} = -\frac{\partial}{\partial \boldsymbol{x}} \left\{ \boldsymbol{v}(\boldsymbol{x}, t) \frac{\delta \varphi(\boldsymbol{x}, t)}{\delta f_i(\boldsymbol{x}', t')} \right\},$$
$$\left. \frac{\delta \varphi(\boldsymbol{x}, t)}{\delta f_i(\boldsymbol{x}', t')} \right|_{t=t'} = -\frac{\partial}{\partial x_i} \left\{ \delta(\boldsymbol{x} - \boldsymbol{x}') \varphi(\boldsymbol{x}, t') \right\}, \tag{8.84}$$
$$\frac{\partial}{\partial t} \varphi(\boldsymbol{x}, t) = -\frac{\partial}{\partial \boldsymbol{x}} \{ \boldsymbol{v}(\boldsymbol{x}, t) \varphi(\boldsymbol{x}, t) \}, \quad \varphi(\boldsymbol{x}, t)|_{t=t'} = \varphi(\boldsymbol{x}, t').$$

The solution to problem (8.83), (8.84) holds for all times t. In this case, the solution $\boldsymbol{x}(t)$ to problem (8.79) cannot be considered as the Markovian vector random process because its multi-time probability density cannot be factorized in terms of the transition probability density. However, in asymptotic limit $t \gg \tau_0$, the diffusion-approximation solution to the initial dynamic system (8.79) will be the Markovian random process, and the corresponding conditions of applicability are formulated as smallness of all statistical effects within temporal intervals of about temporal correlation radius τ_0.

Problems

Problem 8.1

Assuming that $\varepsilon(x, \boldsymbol{R})$, is the homogeneous isotropic delta-correlated Gaussian field with zero-valued mean and correlation function

$$B_\varepsilon(x - x', \boldsymbol{R} - \boldsymbol{R}') = \langle \varepsilon(x, \boldsymbol{R})\varepsilon(x', \boldsymbol{R}') \rangle = A(\boldsymbol{R} - \boldsymbol{R}')\delta(x - x'),$$

derive the Fokker–Planck equation for the diffusion of rays satisfying system of equations (1.92), page 42,

$$\frac{d}{dx}\boldsymbol{R}(x) = \boldsymbol{p}(x), \quad \frac{d}{dx}\boldsymbol{p}(x) = \frac{1}{2}\nabla_{\boldsymbol{R}}\varepsilon(x, \boldsymbol{R}). \tag{8.85}$$

Solution Function $P(x; \boldsymbol{R}, \boldsymbol{p}) = \langle \delta(\boldsymbol{R}(x) - \boldsymbol{R})\delta(\boldsymbol{p}(x) - \boldsymbol{p}) \rangle$ satisfies the Fokker–Planck equation

$$\left(\frac{\partial}{\partial x} + \boldsymbol{p}\frac{\partial}{\partial \boldsymbol{R}}\right) P(x; \boldsymbol{R}, \boldsymbol{p}) = D\Delta_{\boldsymbol{R}} P(x; \boldsymbol{R}, \boldsymbol{p})$$

with the diffusion coefficient $D = -\frac{1}{8}\ \Delta_{\boldsymbol{R}} A(\boldsymbol{R})|_{\boldsymbol{R}=0} = \pi^2 \int_0^\infty d\kappa \kappa^3 \Phi_\varepsilon(0, \kappa)$.

Here,

$$\Phi_\varepsilon(q, \boldsymbol{\kappa}) = \frac{1}{(2\pi)^3} \int_{-\infty}^{\infty} dx \int d\boldsymbol{R} B_\varepsilon(x, \boldsymbol{R}) e^{-i(qx + \boldsymbol{\kappa}\boldsymbol{R})},$$

$$A(\boldsymbol{R}) = 2\pi \int d\boldsymbol{\kappa} \Phi_\varepsilon(0, \boldsymbol{\kappa}) e^{i\boldsymbol{\kappa}\boldsymbol{R}},$$

is the three-dimensional spectral density of random field $\varepsilon(x, \boldsymbol{R})$.

Solution corresponding to the initial condition $P(0; \boldsymbol{R}, \boldsymbol{p}) = \delta(\boldsymbol{R})\delta(\boldsymbol{p})$ is the Gaussian probability density with the parameters

$$\langle R_j(x) R_k(x) \rangle = \frac{2}{3} D\delta_{jk} x^3, \quad \langle R_j(x) p_k(x) \rangle = D\delta_{jk} x^2,$$

$$\langle p_j(x) p_k(x) \rangle = 2D\delta_{jk} x.$$

Problem 8.2

Assuming that $\varepsilon(x, \boldsymbol{R})$ is the homogeneous isotropic delta-correlated Gaussian field with zero-valued mean and correlation function

$$B_\varepsilon(x - x', \boldsymbol{R} - \boldsymbol{R}') = \langle \varepsilon(x, \boldsymbol{R})\varepsilon(x', \boldsymbol{R}') \rangle = A(\boldsymbol{R} - \boldsymbol{R}')\delta(x - x'),$$

derive, starting from Eqs. (8.85), the longitudinal correlation function of ray displacement.

Solution

$$\left\langle \boldsymbol{R}(x)\boldsymbol{R}(x')\right\rangle = 2D(x')^2\left(x - \frac{1}{3}x'\right).$$

Problem 8.3

Consider the diffusion of two rays described by the system of equations

$$\frac{d}{dx}\boldsymbol{R}_\nu(x) = \boldsymbol{p}_\nu(x), \qquad \frac{d}{dx}\boldsymbol{p}_\nu(x) = \frac{1}{2}\nabla_{\boldsymbol{R}_\nu}\varepsilon(x, \boldsymbol{R}_\nu),$$

where index $\nu = 1, 2$ *denotes the number of the corresponding ray.*

Solution The joint probability density

$$P(x; \boldsymbol{R}_1, \boldsymbol{p}_1, \boldsymbol{R}_2, \boldsymbol{p}_2) = \left\langle \delta(\boldsymbol{R}_1(x) - \boldsymbol{R}_1)\delta(\boldsymbol{p}_1(x) - \boldsymbol{p}_1)\delta(\boldsymbol{R}_2(x) - \boldsymbol{R}_2)\delta(\boldsymbol{p}_2(x) - \boldsymbol{p}_2)\right\rangle$$

satisfies the Fokker–Planck equation

$$\left(\frac{\partial}{\partial x} + \boldsymbol{p}_1\frac{\partial}{\partial \boldsymbol{R}_1} + \boldsymbol{p}_2\frac{\partial}{\partial \boldsymbol{R}_2}\right)P(x; \boldsymbol{R}_1, \boldsymbol{p}_1, \boldsymbol{R}_2, \boldsymbol{p})$$
$$= \widehat{L}\left(\frac{\partial}{\partial \boldsymbol{p}_1}, \frac{\partial}{\partial \boldsymbol{p}_2}\right)P(x; \boldsymbol{R}_1, \boldsymbol{p}_1, \boldsymbol{R}_2, \boldsymbol{p}),$$

where operator

$$\widehat{L}\left(\frac{\partial}{\partial \boldsymbol{p}_1}, \frac{\partial}{\partial \boldsymbol{p}_2}\right) = \frac{\pi}{4}\int d\boldsymbol{\kappa}\,\Phi_\varepsilon(0, \boldsymbol{\kappa})\left[\left(\boldsymbol{\kappa}\frac{\partial}{\partial \boldsymbol{p}_1}\right)^2 + \left(\boldsymbol{\kappa}\frac{\partial}{\partial \boldsymbol{p}_2}\right)^2 \right.$$
$$\left. + 2\cos\left[\boldsymbol{\kappa}(\boldsymbol{R}_1 - \boldsymbol{R}_2)\right]\left(\boldsymbol{\kappa}\frac{\partial}{\partial \boldsymbol{p}_1}\right)\left(\boldsymbol{\kappa}\frac{\partial}{\partial \boldsymbol{p}_1}\right)\right].$$

Problem 8.4

Consider the relative diffusion of two rays.

Solution Probability density of relative diffusion of two rays

$$P(x; \boldsymbol{R}, \boldsymbol{p}) = \left\langle \delta(\boldsymbol{R}_1(x) - \boldsymbol{R}_2(x) - \boldsymbol{R})\delta(\boldsymbol{p}_1(x) - \boldsymbol{p}_2(x) - \boldsymbol{p})\right\rangle,$$

satisfies the Fokker–Planck equation

$$\left(\frac{\partial}{\partial x}+\boldsymbol{p}\frac{\partial}{\partial \boldsymbol{R}}\right)P(x;\boldsymbol{R},\boldsymbol{p})=D_{\alpha\beta}(\boldsymbol{R})\frac{\partial^2}{\partial p_\alpha \partial p_\beta}P(x;\boldsymbol{R},\boldsymbol{p}), \tag{8.86}$$

where

$$D_{\alpha\beta}(\boldsymbol{R})=2\pi\int d\boldsymbol{\kappa}\,[1-\cos(\boldsymbol{\kappa}\boldsymbol{R})]\,\kappa_\alpha\kappa_\beta\Phi_\varepsilon(0,\boldsymbol{\kappa}).$$

Remark If l_0 is the correlation radius of random field $\varepsilon(x,\boldsymbol{R})$, than under the condition $R \gg l_0$ we have

$$D_{\alpha\beta}(\boldsymbol{R})=2D\delta_{\alpha\beta},$$

i.e., relative diffusion of rays is characterized by the diffusion coefficient exceeding the diffusion coefficient of a single ray by a factor of 2, which corresponds to statistically independent rays. In this case, the joint probability density of relative diffusion is Gaussian.

In the general case, Eq. (8.86) can hardly be solved in the analytic form. The only clear point is that the solution corresponding to variable diffusion coefficient is not the Gaussian distribution.

Asymptotic case $R \ll l_0$ can be analyzed in more detail. In this case, we can expand function $\{1-\cos(\boldsymbol{\kappa}\boldsymbol{R})\}$ in the Taylor series to obtain the diffusion matrix in the form

$$D_{\alpha\beta}(\boldsymbol{R})=\pi B\left(\boldsymbol{R}^2\delta_{\alpha\beta}+2R_\alpha R_\beta\right), \tag{8.87}$$

where

$$B=\frac{\pi}{4}\int\limits_0^\infty d\kappa\kappa^5\Phi_\varepsilon(0,\kappa).$$

From Eq. (8.86) with coefficients (8.87) follow three equations for moment functions

$$\frac{d}{dx}\left\langle\boldsymbol{p}^2(x)\right\rangle=8\pi B\left\langle\boldsymbol{R}^2(x)\right\rangle,\quad \frac{d}{dx}\left\langle\boldsymbol{R}^2(x)\right\rangle=2\left\langle\boldsymbol{R}(x)\boldsymbol{p}(x)\right\rangle,$$

$$\frac{d}{dx}\left\langle\boldsymbol{R}(x)\boldsymbol{p}(x)\right\rangle=\left\langle\boldsymbol{p}^2(x)\right\rangle.$$

These equations can be easily solved. From the solution follows that, for x from the interval in which $\alpha x \gg 1$ ($\alpha = (16\pi B)^{1/3}$), but $R_0^2 e^{\alpha x} \ll l_0^2$ (such an interval always exists for sufficiently small distances R_0 between the rays under consideration), quantities $\langle \boldsymbol{R}^2(x) \rangle$, $\langle \boldsymbol{R}(x)\boldsymbol{p}(x) \rangle$, and $\langle \boldsymbol{p}^2(x) \rangle$ are exponentially increasing functions of x.

Problem 8.5

Derive an equation for the probability density of magnetic field with homogeneous initial conditions with the use of the simplest model of velocity field (1.3), (1.4), page 4,

$$\boldsymbol{u}(\boldsymbol{r}, t) = \boldsymbol{v}(t) f(kx),$$

where $\boldsymbol{v}(t)$ is the stationary Gaussian vector random process with the correlation tensor

$$\langle v_i(t) v_j(t') \rangle = 2\sigma^2 \delta_{ij} \tau_0 \delta(t - t'),$$

where σ^2 is the variance of every velocity component, τ_0 is the temporal correlation radius of these components, and $f(kx)$ is a deterministic function.

Solution In the dimensionless variables

$$t \to k^2 \sigma^2 \tau_0 t, \quad x \to kx, \quad \langle v_i(t) v_j(t') \rangle \to 2\delta_{ij}\delta(t - t'),$$

the indicator function of the transverse portion of magnetic field is described by Eq. (3.27), page 76. Averaging this equation over an ensemble of realizations of vector function $\boldsymbol{v}(t)$, we obtain the equation

$$\begin{aligned} \frac{\partial}{\partial t} P(x, t; \boldsymbol{H}) &= \frac{\partial}{\partial x} \left(f^2(x) \frac{\partial}{\partial x} - f(x) f'(x) \frac{\partial}{\partial \boldsymbol{H}} \boldsymbol{H} \right) P(x, t; \boldsymbol{H}) \\ &+ \left(1 + \frac{\partial}{\partial \boldsymbol{H}} \boldsymbol{H} \right) \left(-f(x) f'(x) \frac{\partial}{\partial x} + \left[f'(x) \right]^2 \frac{\partial}{\partial \boldsymbol{H}} \boldsymbol{H} \right) P(x, t; \boldsymbol{H}) \\ &+ \left[f'(x) \right]^2 E_{x0} \frac{\partial}{\partial H_i} \frac{\partial}{\partial H_i} P(x, t; \boldsymbol{H}). \end{aligned}$$

Remark If function $f(x)$ varies along x with characteristic scale k^{-1} and is a periodic function ('fast variation'), then additional averaging over x results in an equation for 'slow' spatial variations

$$\frac{\partial}{\partial t}P(x,t;\boldsymbol{H}) = \frac{\partial}{\partial x}\left(\overline{f^2(x)}\frac{\partial}{\partial x}\right)P(x,t;\boldsymbol{H})$$

$$+\overline{[f'(x)]^2}\left(1+\frac{\partial}{\partial \boldsymbol{H}}\boldsymbol{H}\right)\frac{\partial}{\partial \boldsymbol{H}}\boldsymbol{H}P(x,t;\boldsymbol{H}) + \overline{[f'(x)]^2}E_{x0}\frac{\partial^2}{\partial \boldsymbol{H}^2}P(x,t;\boldsymbol{H}).$$

Taking into account the fact that function $P(x,t;\boldsymbol{H})$ is independent of x in view of the initial conditions, we obtain the equation

$$\frac{\partial}{\partial t}P(t;\boldsymbol{H}) = D\left(\frac{\partial}{\partial \boldsymbol{H}}\boldsymbol{H} + \frac{\partial}{\partial \boldsymbol{H}}\boldsymbol{H}\frac{\partial}{\partial \boldsymbol{H}}\boldsymbol{H}\right)P(t;\boldsymbol{H}) + DE_{x0}\frac{\partial^2}{\partial \boldsymbol{H}^2}P(t;\boldsymbol{H}),$$

where $D = \overline{[f'(x)]^2}$. Then, introducing the normalized time ($t \to Dk^2\sigma^2\tau_0 t$), we obtain the equation in the final form

$$\frac{\partial}{\partial t}P(t;\boldsymbol{H}) = \left(\frac{\partial}{\partial \boldsymbol{H}}\boldsymbol{H} + \frac{\partial}{\partial \boldsymbol{H}}\boldsymbol{H}\frac{\partial}{\partial \boldsymbol{H}}\boldsymbol{H}\right)P(t;\boldsymbol{H}) + E_{x0}\frac{\partial^2}{\partial \boldsymbol{H}^2}P(t;\boldsymbol{H}). \quad (8.88)$$

Problem 8.6

Starting from Eq. (8.88), derive an equation for the probability density of generated energy of the transverse portion of magnetic field with the use of the simplest model of velocity field (1.3), (1.4), page 4.

Solution Multiplying Eq. (8.88) by $\delta(\boldsymbol{H}^2 - E)$ and integrating over $\boldsymbol{H}$, we obtain the equation

$$\frac{\partial}{\partial t}P(t;E) = \left(2\frac{\partial}{\partial E}E + 4\frac{\partial}{\partial E}E\frac{\partial}{\partial E}E\right)P(t;E) + 4E_{x0}\frac{\partial}{\partial E}E\frac{\partial}{\partial E}P(t;E). \quad (8.89)$$

Introduce dimensionless energy by normalizing it by E_{x0}. The result will be the equation

$$\frac{\partial}{\partial t}P(t;E) = \left(2\frac{\partial}{\partial E}E + 4\frac{\partial}{\partial E}E\frac{\partial}{\partial E}E\right)P(t;E) + 4\frac{\partial}{\partial E}E\frac{\partial}{\partial E}P(t;E), \quad (8.90)$$

with the initial condition

$$P(0;E) = \delta(E-\beta),$$

where $\beta = E_{\perp 0}/E_{x0}$.

Remark It should be remembered that problem (8.90) is the boundary-value problem, and boundary conditions are the requirements of vanishing the probability density at $E = 0$ and $E = \infty$ for arbitrary time $t > 0$. In addition, Eq. (8.90) generally has a steady-state solution independent of parameter β and characterized by infinite moments,

$$P(\infty; E) = \frac{1}{2\,(E+1)^{3/2}}. \tag{8.91}$$

It should be kept in mind that this distribution is a generalized function and is valid on interval $(0, \infty)$. As for point $E = 0$, the probability density must vanish at this point.

Note that, at $\beta = 1$ and $E_{x0} = 0$, probability density of energy normalized by $E_{\perp 0}$ satisfies Eq. (8.90) without the last term and, consequently, the corresponding probability density is lognormal,

$$P(t; E) = \frac{1}{4\sqrt{\pi t}E} \exp\left\{-\frac{1}{16t} \ln^2\left(Ee^{2t}\right)\right\}. \tag{8.92}$$

This structure of probability density will evidently be valid only for very small time intervals during which the generating term is still insignificant. The solid lines in Fig. 8.5 show the solutions to Eq. (8.90) obtained numerically for parameter $\beta = 1$ at (a) $t = 0.05$ and (b) $t = 0.25$. The dashed lines in Fig. 8.5 show the lognormal distribution (8.92). These curves show that the term responsible for energy generation in Eq. (8.90) prevails even for small times.

Figure 8.6 shows the solution to 8.90) (the solid lines) for $\beta = 0$ (i.e., the probability density of the generated magnetic energy) at $t = 10$ and the asymptotic curve (8.91) (the dashed line). These curves indicate that only the tail of the solution to Eq. (8.90) verges towards the asymptotic solution (8.91) (see Fig. 8.6a), and this process is very slow (see Fig. 8.6b). The curves intersect because of normalization of the probability densities under consideration. It seems that probability density $P(t; E)$ will approach the tail of the steady-state probability density with further increasing time.

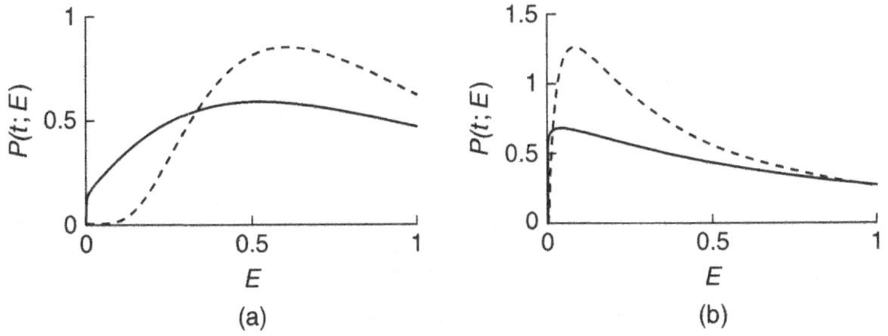

Figure 8.5 Probability distributions at $\beta = 1$ at times (a) $t = 0.05$ and (b) $t = 0.25$.

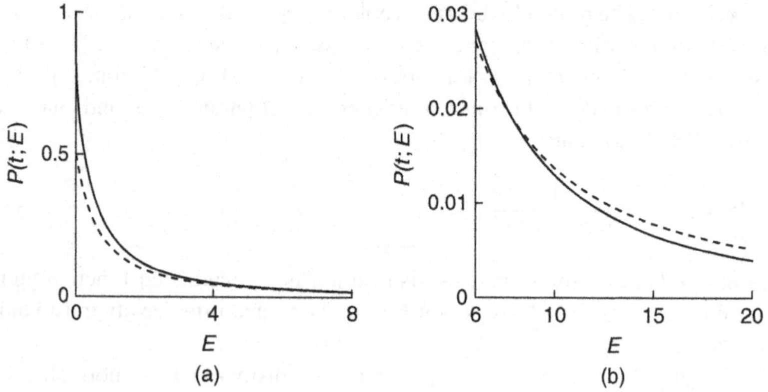

Figure 8.6 Probability distributions at $\beta = 0$ at time $t = 10$ for energy in intervals (a) (0, 8) and (b) (6, 20).

Problem 8.7

Find the relationship between fields $q(\mathbf{r}, t)$ and $p(\mathbf{r}, t)$ whose moment functions are given by the expressions

$$\langle q^n(\mathbf{r}, t)\rangle = \exp\{n(n + 0,5)t\}, \quad \langle p^n(\mathbf{r}, t)\rangle = \exp\{n(n - 0,5)t\}.$$

Which field possesses the property of clustering?

Solution $q(\mathbf{r}, t) = 1/p(\mathbf{r}, t)$ and clustering occurs for field $p(\mathbf{r}, t)$.

Problem 8.8

Show that integral equation (8.70) is equivalent to the variational derivative equation

$$\frac{\delta}{\delta z(\tau)} G(t, t') = G(t, \tau)\Lambda G(\tau, t') \quad (t' \le \tau \le t)$$

with the initial condition

$$G(t, t')\big|_{z(\tau)=0} = g(t, t').$$

Problem 8.9

Construct the diffusion approximation for the problem on the dynamics of a particle with linear friction under random forces, which is described by the stochastic system (1.12), page 8,

$$\frac{d}{dt}\boldsymbol{r}(t) = \boldsymbol{v}(t), \quad \frac{d}{dt}\boldsymbol{v}(t) = -\lambda \boldsymbol{v}(t) + \boldsymbol{f}(\boldsymbol{r}, t),$$
$$\boldsymbol{r}(0) = \boldsymbol{r}_0, \quad \boldsymbol{v}(0) = \boldsymbol{v}_0.$$

Solution The one-time probability density of particle position and velocity

$$P(\boldsymbol{r}, \boldsymbol{v}, t) = \langle \delta(\boldsymbol{r}(t) - \boldsymbol{r})\delta(\boldsymbol{v}(t) - \boldsymbol{v})\rangle$$

satisfies the Fokker–Planck equation

$$\left(\frac{\partial}{\partial t} + \boldsymbol{v}\frac{\partial}{\partial \boldsymbol{r}} - \lambda \frac{\partial}{\partial \boldsymbol{v}}\boldsymbol{v}\right) P(\boldsymbol{r}, \boldsymbol{v}, t) = \frac{\partial}{\partial v_i}\left\{D_{ij}^{(1)}(\boldsymbol{v})\frac{\partial}{\partial v_j} + D_{ij}^{(2)}(\boldsymbol{v})\frac{\partial}{\partial r_j}\right\} P(\boldsymbol{r}, \boldsymbol{v}, t). \tag{8.93}$$

with the diffusion coefficients

$$D_{ij}^{(1)}(\boldsymbol{v}) = \int\limits_0^\infty d\tau e^{-\lambda\tau} B_{ij}\left(\frac{1}{\lambda}\left[e^{\lambda\tau} - 1\right]\boldsymbol{v}, \tau\right),$$
$$D_{ij}^{(2)}(\boldsymbol{v}) = \frac{1}{\lambda}\int\limits_0^\infty d\tau \left[1 - e^{-\lambda\tau}\right] B_{ij}\left(\frac{1}{\lambda}\left[e^{\lambda\tau} - 1\right]\boldsymbol{v}, \tau\right). \tag{8.94}$$

Problem 8.10

In the diffusion approximation, find the steady-state probability distribution of the velocity of a particle described by dynamic system (1.12), page 8, in the one-dimensional case assuming that the force correlation function has the form

$$B_f(x, t) = \sigma_f^2 \exp\left\{-\frac{|x|}{l_0} - \frac{|t|}{\tau_0}\right\}.$$

Solution From Eq. (8.93) follows that steady-state probability density of particle's velocity (i.e., probability density in limit $t \to \infty$) satisfies the equation

$$-\lambda v P(v) = D(v)\frac{\partial}{\partial v}P(v), \tag{8.95}$$

where, in accordance with Eq. (8.94),

$$D(v) = \int_0^{\infty} d\tau e^{-\lambda\tau} B\left(\frac{1}{\lambda}\left[e^{\lambda\tau} - 1\right] v, \tau\right).$$

For a sufficiently small friction ($\lambda\tau_0 \ll 1$), the solution to Eq. (8.95) has the form

$$P(v) = C\exp\left\{-\frac{\lambda v^2}{2\sigma_f^2\tau_0}\left[1 + \frac{2}{3}\frac{|v|\tau_0}{l_0}\right]\right\}. \tag{8.96}$$

For small particle velocity $|v|\tau_0 \ll l_0$, probability distribution (8.96) grades into the Gaussian distribution corresponding to the approximation of the delta-correlated (in time) random field $f(x, t)$. However, in the opposite limiting case $|v|\tau_0 \gg l_0$, probability distribution (8.96) decreases significantly faster than in the case of the approximation of the delta-correlated (in time) random field $f(x, t)$, namely,

$$P(v) = C\exp\left\{-\frac{\lambda v^2 |v|}{3\sigma_f^2 l_0}\right\}, \tag{8.97}$$

which corresponds to the diffusion coefficient decreasing according to the law $D(v) \sim 1/|v|$ for great particle velocities. Physically, it means that the effect of random force $f(x, t)$ on faster particles is significantly smaller than on slower ones.

Lecture 9

Methods for Solving and Analyzing the Fokker–Planck Equation

The Fokker–Planck equations for the one-point probability density (8.11), page 193, and for the transitional probability density (8.16), page 195, are the partial differential equations of parabolic type, so that we can use methods from the theory of mathematical physics equations to solve them. In this context, the basic methods are those such as the method of separation of variables, Fourier transformation with respect to spatial coordinates, and other integral transformations.

9.1 Integral Transformations

Integral transformations are very practicable for solving the Fokker–Planck equation. Indeed, earlier we mentioned the convenience of the Fourier transformation in Eq. (8.11), page 193, if the tensor of diffusion coefficients $F_{kl}(\boldsymbol{x}, \boldsymbol{x}; t)$ is independent of $\boldsymbol{x}$. Different integral transformations related to eigenfunctions of the diffusion operator

$$\widehat{L} = \frac{\partial^2}{\partial x_k \partial x_l} F_{kl}(\boldsymbol{x}, \boldsymbol{x}; t)$$

can be used in other situations.

For example, in the case of the Legendre operator

$$\widehat{L} = \frac{\partial}{\partial x}(x^2 - 1)\frac{\partial}{\partial x},$$

it is quite natural to use the integral transformation related to the *Legendre functions*. This transformation is called the *Meler–Fock* transform and is defined by the formula

$$F(\mu) = \int\limits_1^\infty dx f(x) P_{-1/2+i\mu}(x) \quad (\mu > 0), \tag{9.1}$$

Lectures on Dynamics of Stochastic Systems. DOI: 10.1016/B978-0-12-384966-3.00009-X

where $P_{-1/2+i\mu}(x)$ is the *complex index Legendre function of the first kind*, which satisfies the equation

$$\frac{d}{dx}(x^2-1)\frac{d}{dx}P_{-1/2+i\mu}(x) = -\left(\mu^2+\frac{1}{4}\right)P_{-1/2+i\mu}(x). \tag{9.2}$$

The inversion of the transform (9.1) has the form

$$f(x) = \int_0^\infty d\mu\,\mu\tanh(\pi\mu)F(\mu)P_{-1/2+i\mu}(x) \quad (1 \le x \le \infty), \tag{9.3}$$

where $F(\mu)$ is given by formula (9.1).

Another integral transformation called the *Kontorovich–Lebedev* transform, is related to the diffusion operator

$$\hat{L} = \frac{\partial}{\partial x}x^2\frac{\partial}{\partial x}$$

and has the form

$$F(\tau) = \int_0^\infty dx f(x)K_{i\tau}(x) \quad (\tau > 0), \tag{9.4}$$

where $K_{i\tau}(x)$ is the *imaginary index McDonald's function of the first kind*, which satisfies the equations

$$\begin{gathered}\left(x^2\frac{d^2}{dx^2}+x\frac{d}{dx}-x^2+\tau^2\right)K_{i\tau}(x)=0,\\ \left(\frac{d}{dx}x^2\frac{d}{dx}-x\frac{d}{dx}\right)K_{i\tau}(x)=\left(x^2-\tau^2\right)K_{i\tau}(x).\end{gathered} \tag{9.5}$$

The corresponding inversion has the form

$$f(x) = \frac{2}{\pi^2 x}\int_0^\infty d\tau\sinh(\pi\tau)F(\tau)K_{i\tau}(x). \tag{9.6}$$

9.2 Steady-State Solutions of the Fokker–Planck Equation

In previous sections, we discussed the general methods for solving the Fokker–Planck equation for both transition and one-point probability densities. However, the problem of the one-point probability density can have peculiarities related to the possible existence of the steady-state solution; in a number of cases, such a solution can be obtained immediately. The steady-state solution, if it exists, is independent of initial values and is the solution of the Fokker–Planck equation in the limit $t \to \infty$.

There are two classes of problems for which the steady-state solution of the Fokker–Planck equation can be easily found. These classes deal with one-dimensional differential equations and with the Hamiltonian systems of equations. We will consider them in greater detail.

9.2.1 One-Dimensional Nonlinear Differential Equation

The one-dimensional nonlinear systems are described by the stochastic equation

$$\frac{d}{dt}x(t) = f(x) + z(t)g(x), \quad x(0) = x_0, \tag{9.7}$$

where $z(t)$ is, as earlier, the Gaussian delta-correlated process with the parameters

$$\langle z(t)\rangle = 0, \quad \langle z(t)z(t')\rangle = 2D\delta(t - t') \quad (D = \sigma_z^2\tau_0).$$

The corresponding Fokker–Planck equation has the form

$$\left(\frac{\partial}{\partial t} + \frac{\partial}{\partial x}f(x)\right)P(x, t) = D\frac{\partial}{\partial x}g(x)\frac{\partial}{\partial x}g(x)P(x, t). \tag{9.8}$$

The steady-state probability distribution $P(x)$, if it exists, satisfies the equation

$$f(x)P(x) = Dg(x)\frac{d}{dx}g(x)P(x) \tag{9.9}$$

(we assume that $P(x)$ is distributed over the whole space, i.e., for $-\infty < x < \infty$) whose solution is as follows

$$P(x) = \frac{C}{|g(x)|}\exp\left\{\frac{1}{D}\int dx\frac{f(x)}{g^2(x)}\right\}, \tag{9.10}$$

where constant C is determined from the normalization condition

$$\int_{-\infty}^{\infty} dxP(x) = 1.$$

In the special case of the Langevin equation (8.60), page 211,

$$f(x) = -\lambda x, \; g(x) = 1,$$

Eq. (9.10) grades into the Gaussian probability distribution

$$P(x) = \sqrt{\frac{\lambda}{2\pi D}} \exp\left\{-\frac{\lambda}{2D}x^2\right\}. \tag{9.11}$$

9.2.2 Hamiltonian Systems

Another type of dynamic system that allows us to obtain the steady-state probability distribution for N particles is described by the Hamiltonian system of equations with linear friction and external random forces

$$\begin{aligned} \frac{d}{dt}\boldsymbol{r}_i(t) &= \frac{\partial}{\partial \boldsymbol{p}_i} H(\{\boldsymbol{r}_i\}, \{\boldsymbol{p}_i\}), \\ \frac{d}{dt}\boldsymbol{p}_i(t) &= -\frac{\partial}{\partial \boldsymbol{r}_i} H(\{\boldsymbol{r}_i\}, \{\boldsymbol{p}_i\}) - \lambda \boldsymbol{p}_i + \boldsymbol{f}_i(t), \end{aligned} \tag{9.12}$$

where $i = 1, 2, \ldots, N$;

$$H(\{\boldsymbol{r}_i\}, \{\boldsymbol{p}_i\}) = \sum_{i=1}^{N} \frac{\boldsymbol{p}_i^2(t)}{2} + U(\boldsymbol{r}_1, \ldots, \boldsymbol{r}_N)$$

is the Hamiltonian function; $\{\boldsymbol{r}_i\}$, $\{\boldsymbol{p}_i\}$ stand for the totalities of all quantities $\boldsymbol{r}(t)$ and $\boldsymbol{p}(t)$, i.e.,

$$\{\boldsymbol{r}_i\} = \{\boldsymbol{r}_1, \ldots, \boldsymbol{r}_N\}, \quad \{\boldsymbol{p}_i\} = \{\boldsymbol{p}_1, \ldots, \boldsymbol{p}_N\};$$

λ is a constant coefficient (friction), and random forces $\boldsymbol{f}_i(t)$ are the Gaussian delta-correlated random vector functions with the correlation tensor

$$\left\langle f_i^{\alpha}(t) f_j^{\beta}(t') \right\rangle = 2D\delta_{ij}\delta_{\alpha\beta}\delta(t - t'), \quad D = \sigma_f^2 \tau_0. \tag{9.13}$$

Here, α and β are the vector indices.

The system of equations (9.12) describes the Brownian motion of a system of N interacting particles. The corresponding indicator function $\varphi(\{\boldsymbol{r}_i\}, \{\boldsymbol{p}_i\}, t)$ satisfies the equation

$$\frac{\partial}{\partial t}\varphi(\{\boldsymbol{r}_i\}, \{\boldsymbol{p}_i\}, t) + \sum_{k=1}^{N} \{H, \varphi\}_{(k)} - \lambda \sum_{k=1}^{N} \frac{\partial}{\partial \boldsymbol{p}_k}\{\boldsymbol{p}_k \varphi\} = \sum_{k=1}^{N} \frac{\partial}{\partial \boldsymbol{p}_k}\{\boldsymbol{f}_k(t)\varphi\},$$

which is a generalization of Eq. (3.58), page 85, and, consequently, the Fokker–Planck equation for the joint probability density of the solution to system (9.12) has the form

$$\begin{aligned} &\frac{\partial}{\partial t} P(\{\boldsymbol{r}_i\}, \{\boldsymbol{p}_i\}, t) + \sum_{k=1}^{N} \{H, P(\{\boldsymbol{r}_i\}, \{\boldsymbol{p}_i\}, t)\}_{(k)} \\ &\quad - \lambda \sum_{k=1}^{N} \frac{\partial}{\partial \boldsymbol{p}_k} \left\{\boldsymbol{p}_k P(\{\boldsymbol{r}_i\}, \{\boldsymbol{p}_i\}, t)\right\} = D \sum_{k=1}^{N} \frac{\partial^2}{\partial \boldsymbol{p}_k^2} P(\{\boldsymbol{r}_i\}, \{\boldsymbol{p}_i\}, t), \end{aligned} \tag{9.14}$$

where

$$\{\varphi, \psi\}_{(k)} = \frac{\partial \varphi}{\partial \boldsymbol{p}_k} \frac{\partial \psi}{\partial \boldsymbol{r}_k} - \frac{\partial \psi}{\partial \boldsymbol{p}_k} \frac{\partial \varphi}{\partial \boldsymbol{r}_k}$$

is the Poisson bracket for the kth particle.

One can easily check that the steady-state solution to Eq. (9.14) is the *canonical Gibbs distribution*

$$P(\{\boldsymbol{r}_i\}, \{\boldsymbol{p}_i\}) = C \exp\left\{-\frac{\lambda}{D} H(\{\boldsymbol{r}_i\}, \{\boldsymbol{p}_i\})\right\}. \tag{9.15}$$

The specificity of this distribution consists in the Gaussian behavior with respect to momenta and statistical independence of particle coordinates and momenta.

Integrating Eq. (9.15) over all $\boldsymbol{r}$, we can obtain the *Maxwell distribution* that describes velocity fluctuations of the Brownian particles. The case $U(\boldsymbol{r}_1, \ldots, \boldsymbol{r}_N) = 0$ corresponds to the Brownian motion of a system of free particles.

If we integrate probability distribution (9.15) over momenta (velocities), we obtain the *Boltzmann distribution* of particle coordinates

$$P(\{\boldsymbol{r}_i\}) = C \exp\left\{-\frac{\lambda}{D} U(\{\boldsymbol{r}_i\})\right\}. \tag{9.16}$$

In the case of sufficiently strong friction, the equilibrium distribution (9.15) is formed in two stages. First, the Gaussian momentum distribution (the Maxwell distribution) is formed relatively quickly and then, the spatial distribution (the Boltzmann distribution) is formed at a much slower rate. The latter stage is described by the Fokker–Planck equation

$$\frac{\partial}{\partial t} P(\{\boldsymbol{r}_i\}, t) = \frac{1}{\lambda} \sum_{k=1}^{N} \frac{\partial}{\partial \boldsymbol{r}_k} \left(\frac{\partial U(\{\boldsymbol{r}_i\})}{\partial \boldsymbol{r}_k} P(\{\boldsymbol{r}_i\}, t) \right) + \frac{D}{\lambda^2} \sum_{k=1}^{N} \frac{\partial^2}{\partial \boldsymbol{r}_k^2} P(\{\boldsymbol{r}_i\}, t), \tag{9.17}$$

which is usually called the *Einstein–Smolukhovsky equation*. Derivation of Eq. (9.17) from the Fokker–Planck equation (9.14) is called *Kramers problem* (see, e.g., the corresponding discussion in Sect. 6.2.1, page 144), where dynamics of particles under a random force is considered as an example). Note that Eq. (9.17) statistically corresponds to the stochastic equation

$$\frac{d}{dt} \boldsymbol{r}_i(t) = -\frac{1}{\lambda} \frac{\partial}{\partial \boldsymbol{r}_i} U(\{\boldsymbol{r}_i\}) + \frac{1}{\lambda} \boldsymbol{f}_i(t), \tag{9.18}$$

which, nevertheless, cannot be considered as the limit of Eq. (9.12) for $\lambda \to \infty$.

In the one-dimensional case, Eqs. (9.12) are simplified and assume the form of the system of two equations

$$\frac{d}{dt} x(t) = y(t), \quad \frac{d}{dt} y(t) = -\frac{\partial}{\partial x} U(x) - \lambda y(t) + f(t). \tag{9.19}$$

The corresponding steady-state probability distribution has the form

$$P(x,y) = C\exp\left\{-\frac{\lambda}{D}H(x,y)\right\}, \quad H(x,y) = \frac{y^2}{2} + U(x). \tag{9.20}$$

9.2.3 Systems of Hydrodynamic Type

In Sect. 1.1.3, page 10, we considered the general dynamics of simplest hydrodynamic-type systems (HTS). Now, we consider these systems in terms of the statistical description.

Hydrodynamic-type systems with linear friction are described by the dynamic equations

$$\frac{d}{dt}v_i(t) = F_i(\boldsymbol{v}) - \lambda^{(i)}v_i(t) \qquad (i = 1, \ldots, N), \tag{9.21}$$

where $\lambda^{(i)}$ is the friction coefficient of the i-th component of the N-dimensional vector $\boldsymbol{v}$ [1] and $F_i(\boldsymbol{v})$ is the function quadratic in $\boldsymbol{v}$ and having the following properties:

(*a*) $v_iF_i(\boldsymbol{v}) = 0$,

energy conservation holds at $\lambda^{(i)} = 0$: $\frac{d}{dt}E(t) = 0$, $E(t) = \frac{\boldsymbol{v}_i^2(t)}{2}$;

(*b*) $\frac{\partial}{\partial v_i}F_i(\boldsymbol{v}) = 0$,

conditions of the Liouville theorem are satisfied at $\lambda^{(i)} = 0$, and this equality is the equation of incompressibility in the phase space.

Equilibrium Thermal Fluctuations in the Hydrodynamic-Type Systems

Here, we dwell on a class of phenomena closely related to the Brownian motion; namely, we dwell on equilibrium thermal fluctuations in solids.

Microscopic equations describe the behavior of physical systems only in terms of spatial scales large in comparison with the molecule free path in the medium and temporal scales great in comparison with the time intervals between molecule collisions. This means that macroscopic equations adequately describe the behavior of systems only on average. However, in view of molecule thermal motion, macroscopic variables are in general terms stochastic variables, and a complete macroscopic theory must describe not only system behavior on average, but also the fluctuations about the average.

[1] In the general case, the dissipative term in Eq. (9.21) has the form $\lambda_{ik}v_k$. However, we can always choose the coordinate system in which two positively defined quadratic forms – energy $E = v_i^2/2$ and dissipation $\lambda_{ik}v_iv_k$ – have the diagonal representation. The form of Eq. (9.21) assumes the use of namely such a coordinate system.

Such a description can be performed in terms of macroscopic variables by supplementing the corresponding macroscopic equations with the 'external forces' specified as the Gaussian random fields delta-correlated in time (this approach is closely related to the *fluctuation – dissipation theorem*, or the *Callen–Welton theorem*). See [53, 54] for the corresponding correlation theories of equilibrium thermal fluctuations in electrodynamics, hydrodynamics, and viscoelastic media.

In HTS, equilibrium thermal fluctuations are described by Eq. (9.21) supplemented with external forces $f_i(t)$

$$\frac{d}{dt}v_i(t) = F_i(\boldsymbol{v}) - \lambda^{(i)}v_i(t) + f_i(t) \qquad (i = 1, \ldots, N). \tag{9.22}$$

External forces are assumed to be the Gaussian random functions delta-correlated in time with the correlation tensor of the form

$$\left\langle f_i(t)f_j(t')\right\rangle = 2\delta_{ij}\sigma_{(i)}^2\delta(t-t') \qquad (\langle \boldsymbol{f}(t)\rangle = 0). \tag{9.23}$$

Note that one can consider Eqs. (9.22) as the system describing the Brownian motion in HTS. Such a treatment assumes that coefficients $\lambda^{(i)}$ are certain effective friction coefficients. For example, at $N = 3$, system (9.22) describes (in the velocity space) the rotary Brownian motion of a solid in a medium, and quantities $\lambda^{(i)}v_i(t)$ play the role of the corresponding resistance forces. The probability density of solution $\boldsymbol{v}(t)$ to Eqs. (9.22), i.e., function $P(\boldsymbol{v}, t) = \langle \delta(\boldsymbol{v}(t) - \boldsymbol{v})\rangle$, satisfies the Fokker–Planck equation

$$\frac{\partial}{\partial t}P(\boldsymbol{v}, t) = -\frac{\partial}{\partial v_i}\{F_i(\boldsymbol{v})P(\boldsymbol{v}, t)\} + \lambda^{(i)}\frac{\partial}{\partial v_i}\{v_iP(\boldsymbol{v}, t)\} + \sigma_{(i)}^2\frac{\partial^2}{\partial v_i^2}P(\boldsymbol{v}, t). \tag{9.24}$$

The steady-state, initial data-independent solution to Eq. (9.24) must behave like the Maxwell distribution that corresponds to the uniform distribution of energy over the degrees of freedom

$$P(\boldsymbol{v}) = C\exp\left\{-\frac{v_i^2}{2kT}\right\}, \tag{9.25}$$

where k is the Boltzmann constant and T is the equilibrium temperature in the system.

Substituting Eq. (9.25) in Eq. (9.24), we obtain the expression for $\sigma_{(i)}^2$

$$\sigma_{(i)}^2 = \lambda^{(i)}kT \tag{9.26}$$

called the *Einstein formula*. Here, we used that, in view of conditions (a) and (b),

$$\frac{\partial}{\partial v_i}\{F_i(\boldsymbol{v})P(\boldsymbol{v})\} = 0.$$

Thus, Eqs. (9.23) and (9.26) completely determine statistics of external forces, and the nonlinear term in Eq. (9.22) plays no role for the one-time fluctuations of $\mathbf{v}(t)$. Namely this feature constitutes the subject matter of the fluctuation – dissipation theorem (the Callen–Welton theorem) in the context of HTS.

Note that Eqs. (9.22) with the correlation relationships (9.23) describe not only equilibrium thermal fluctuations in HTS, but also the interaction of small-scale motions (such as microturbulence, for example) with the motions of larger scales. If such an HTS allows the description in terms of phenomenological equations (9.22) with $\lambda^{(i)} = \lambda = \text{const}$, $\sigma^2_{(i)} = \sigma^2 = \text{const}$, the steady-state probability distribution of $\mathbf{v}$ will have the form similar to distribution (9.25):

$$P(\mathbf{v}) = C \exp\left\{-\frac{\lambda}{2\sigma^2} v_i^2\right\}. \tag{9.27}$$

Noises in Hydrodynamic-Type Systems Under the Regular Force

We illustrate statistical description of HTS by the example of the simplest system with three degrees of freedom (S_3), which is described by system of equations (1.23), page 14:

$$\begin{aligned}
\frac{d}{dt}v_0(t) &= v_2^2(t) - v_1^2(t) - v_0(t) + R + f_0(t),\\
\frac{d}{dt}v_1(t) &= v_0(t)v_1(t) - v_1(t) + f_1(t),\\
\frac{d}{dt}v_2(t) &= -v_0(t)v_2(t) - v_2(t) + f_2(t).
\end{aligned} \tag{9.28}$$

We assume that $f_i(t)$ are the Gaussian random forces delta-correlated in time with the correlation tensor

$$\langle f_i(t) f_j(t')\rangle = 2\delta_{ij}\sigma^2\delta(\tau).$$

In the absence of external forces, this system is equivalent to the Euler equations of the dynamics of a gyroscope with isotropic friction, which is excited by a constant moment of force relative to the unstable axis. The steady-state solution to this system depends on parameter R (an analog to the Reynolds number), and the critical value of this parameter is $R_{\text{cr}} = 1$.

For $R < 1$, it has the stable steady-state solution

$$v_{1\,\text{s-s}} = v_{\text{s-s}} = 0, \quad v_{0\,\text{s-s}} = R. \tag{9.29}$$

For $R > 1$, this solution becomes unstable with respect to small disturbances of parameters, and new steady-state regime is formed:

$$v_{0\,\text{s-s}} = 1, \quad v_{2\,\text{s-s}} = 0, \quad v_{1\,\text{s-s}} = \pm\sqrt{R-1}. \tag{9.30}$$

Here, we have an element of randomness because quantity $v_{1\,\text{s-s}}$ can be either positive or negative, depending on the amplitude of small disturbances.

As was mentioned earlier, the steady-state probability distribution at $R = 0$ has the form

$$P(v) = C \exp\left\{-\frac{v_i^2}{2\sigma^2}\right\}. \tag{9.31}$$

Assume now $R \neq 0$ and $R < 1$. In this case, we can easily obtain that fluctuations of components relative to their steady-state values ($\widetilde{v}_i = v_i - v_{i\,\text{s-s}}$) satisfy the system of equations

$$\begin{aligned}
\frac{d}{dt}\widetilde{v}_0(t) &= \widetilde{v}_2^2(t) - \widetilde{v}_1^2(t) - \widetilde{v}_0(t) + f_0(t),\\
\frac{d}{dt}\widetilde{v}_1(t) &= \widetilde{v}_0(t)\widetilde{v}_1(t) - (1-R)\widetilde{v}_1(t) + f_1(t),\\
\frac{d}{dt}\widetilde{v}_2(t) &= -\widetilde{v}_0(t)\widetilde{v}_2(t) - (1+R)\widetilde{v}_2(t) + f_2(t).
\end{aligned} \tag{9.32}$$

Statistical characteristics of the solution to system (9.32) can be determined using the perturbation theory with respect to small parameter σ^2. The second moments of components $\widetilde{v}_i$ will then be described by the linearized system of equations (9.32); the second moment being determined, mean values can be obtained by averaging system (9.32) directly. The corresponding steady-state variances of fluctuations $\widetilde{v}_i$ have the form

$$\left\langle \widetilde{v}_0^2(t)\right\rangle = \sigma^2, \quad \left\langle \widetilde{v}_1^2(t)\right\rangle = \frac{\sigma^2}{1-R}, \quad \left\langle \widetilde{v}_2^2(t)\right\rangle = \frac{\sigma^2}{1+R}. \tag{9.33}$$

Note that expressions (9.33) hold for $R \ll 1$. With increasing R, both intensity and temporal correlation radius of component $\widetilde{v}_1$ increase, while the intensity of component $\widetilde{v}_2$ decreases. In this process, maximum fluctuations of quantity $\widetilde{v}_1$ occur when the dynamic system gets over the critical regime.

Consider now the case of $R > 1$. The steady-state probability distribution of component v_1 has two maxima near $v_1 = \pm\sqrt{R-1}$ (they correspond to the stable steady-state positions) and a minimum $v_1 = 0$ that corresponds to the unstable position. This probability distribution is formed by ensemble averaging over realizations of random forces $f_i(t)$. In a single realization, the system arrives with a probability of 1/2 at one of the stable positions corresponding to distribution maxima. In this case, averaging over time or ensemble averaging over the force realizations that bring the system to this state will form the probability distribution near the maximum, and we can determine statistical characteristics of the solution using the perturbation theory in parameter σ^2.

Assume the system arrives at the state corresponding to the stable steady-state position $v_{2\,\text{s-s}} = 0$, $v_{1\,\text{s-s}} = \sqrt{R-1}$. Then, fluctuations relative to this state will satisfy the system of equations

$$\frac{d}{dt}\widetilde{v}_0(t) = \widetilde{v}_2^2(t) - \widetilde{v}_1^2(t) - 2\sqrt{R-1}\widetilde{v}_1(t) - \widetilde{v}_0(t) + f_0(t),$$

$$\frac{d}{dt}\widetilde{v}_1(t) = \widetilde{v}_0(t)\widetilde{v}_1(t) + \sqrt{R-1}\widetilde{v}_0(t) + f_1(t), \tag{9.34}$$

$$\frac{d}{dt}\widetilde{v}_2(t) = -\widetilde{v}_0(t)\widetilde{v}_2(t) - 2\widetilde{v}_2(t) + f_2(t).$$

For $R \gg 1$, the second moments of fluctuations $\widetilde{v}_i$ are obtained from the linearized system (9.34) and mean values are then determined by direct averaging of Eqs. (9.34). Using this procedure, we obtain that, for $R \gg 1$, the steady-state variances and correlation coefficients of triplet components have the form

$$\begin{aligned} &\langle \widetilde{v}_0(t)\widetilde{v}_2(t)\rangle = \langle \widetilde{v}_1(t)\widetilde{v}_2(t)\rangle = 0, \quad \langle \widetilde{v}_0(t)\widetilde{v}_1(t)\rangle = -\frac{\sigma^2}{\sqrt{R}}, \\ &\left\langle \widetilde{v}_0^2(t)\right\rangle = 3\sigma^2, \quad \left\langle \widetilde{v}_1^2(t)\right\rangle = \frac{3}{2}\sigma^2, \quad \left\langle \widetilde{v}_2^2(t)\right\rangle = \frac{\sigma^2}{2}. \end{aligned} \tag{9.35}$$

As we mentioned earlier, these statistical characteristics correspond either to averaging over time, or to ensemble averaging over realizations of forces $f_i(t)$ that bring the system to the steady-state position. It should be noted that if the system has arrived at one of the most probable states under the action of certain realization of forces $f_i(t)$, it will be transferred into another most probable state after the lapse of certain time T (the greater T, the greater R and the less σ^2) in view of availability of sufficiently great values of $f_i(t)$. This process simulated for a realization of random forces is shown in Fig. 1.5 on page 14.

Consider now fluctuations of components $v_i(t)$ at critical regime (i.e., at $R = 1$). Equations for the fluctuations relative to the state $v_{1\,\mathrm{CT}} = v_{2\,\mathrm{CT}} = 0$, $v_{0\,\mathrm{CT}} = R$ can be obtained by setting $R = 1$ either in Eq. (9.32), or in Eq. (9.34). In this case, fluctuations of component $\widetilde{v}_2(t)$ can be described in terms of the linearized equation ($\sigma^2 \ll 1$), and the corresponding steady-state variance is given by the formula

$$\left\langle \widetilde{v}_2^2(t)\right\rangle = \frac{\sigma^2}{2}.$$

Components $\widetilde{v}_0(t)$ and $\widetilde{v}_1(t)$ are described by the nonlinear system of equations

$$\begin{aligned} &\frac{d}{dt}\widetilde{v}_0(t) = -\widetilde{v}_1^2(t) - \widetilde{v}_0(t) + f_0(t), \\ &\frac{d}{dt}\widetilde{v}_1(t) = \widetilde{v}_0(t)\widetilde{v}_1(t) + f_1(t), \end{aligned} \tag{9.36}$$

and no linearization is allowed here.

Averaging system (9.36), we obtain the following relationships for steady-state values of fluctuations

$$\langle \widetilde{v}_0(t)\widetilde{v}_1(t)\rangle = 0, \quad \langle \widetilde{v}_0(t)\rangle = -\left\langle \widetilde{v}_1^2(t)\right\rangle .$$

As we noted earlier, intensity of fluctuations $\widetilde{v}_1$ increases in the critical regime and, consequently, quantity $\langle \widetilde{v}_0(t)\rangle$ increases. At the same time, the variance of fluctuation $\widetilde{v}_0(t)$ remains nearly intact. Indeed, as follows from Eqs. (9.36), the variance of steady-state fluctuation $\widetilde{v}_0(t)$ is given by the formula

$$\left\langle \widetilde{v}_0^2(t)\right\rangle = 2\sigma^2.$$

It becomes clear that fluctuations $\widetilde{v}_0(t)$ and $\widetilde{v}_1(t)$ can be estimated from the simplified system

$$\frac{d}{dt}\widetilde{v}_0(t) = -\widetilde{v}_1^2(t) - \widetilde{v}_0(t) + f_0(t),$$

$$\frac{d}{dt}\widetilde{v}_1(t) = \langle \widetilde{v}_0(t)\rangle\, \widetilde{v}_1(t) + f_1(t)$$

obtained by the replacement of quantity $\widetilde{v}_0(t)$ in the second equation of system (9.36) by $\langle \widetilde{v}_0(t)\rangle$.

Using this system, we obtain that steady-state fluctuations $\widetilde{v}_0(t)$ and $\widetilde{v}_1(t)$ satisfy the expressions

$$\left\langle \widetilde{v}_1^2(t)\right\rangle = -\langle \widetilde{v}_0(t)\rangle = \sigma.$$

In a similar way, we obtain the estimate of the temporal correlation radius of quantity $\widetilde{v}_1(t)$:

$$\tau \sim 1/\sigma.$$

Earlier, we considered the behavior of a triplet under the assumption that random forces act on all variables. Let us see what will be changed if only one unstable component $v_0(t)$ is affected by the random force (we assume that the force acting on component $v_0(t)$ has regular (R) and random components). In this case, the dynamic system assumes the form ($R > 1$)

$$\begin{aligned} \frac{d}{dt}v_0(t) &= -v_1^2(t) - v_0(t) + R + f(t), \\ \frac{d}{dt}v_1(t) &= v_0(t)v_1(t) - v_1(t). \end{aligned} \tag{9.37}$$

In system (9.37), we omitted the terms related to component $v_2(t)$, because one can easily see that it will not be excited in the problem at hand.

Represent component $v_0(t)$ as $v_0(t) = 1 + \widetilde{v}_0(t)$. The system of equations (9.37) assumes then the form

$$\frac{d}{dt}\widetilde{v}_0(t) = -v_1^2(t) - \widetilde{v}_0(t) + (R-1) + f(t),$$
$$\frac{d}{dt}v_1(t) = \widetilde{v}_0(t)v_1(t), \tag{9.38}$$

from which follows that temporal evolution of component $v_1(t)$ depends on its initial value.

If $v_1(0) > 0$, then $v_1(t) > 0$ too. In this case, we can represent $v_1(t)$ in the form

$$v_1(t) = e^{\varphi(t)}$$

and rewrite the system of equations (9.38) in the Hamiltonian form (9.19)

$$\frac{d}{dt}\widetilde{v}_0(t) = -\frac{\partial U(\varphi)}{\partial \varphi} - \widetilde{v}_0(t) + f(t), \quad \frac{d}{dt}\varphi(t) = \widetilde{v}_0(t), \tag{9.39}$$

where

$$U(\varphi) = \frac{1}{2}e^{2\varphi} - (R-1)\,\varphi.$$

Here, variable $\varphi(t)$ plays the role of particle's coordinate and variable $v_0(t)$ plays the role of particle's velocity.

The solid line in Fig. 9.1 shows the behavior of function $U(\varphi)$. At point $\varphi_0 = \ln\sqrt{R-1}$, this function has a minimum

$$U(\varphi_0) = \frac{1}{2}(R-1)\,[1 - \ln(R-1)]$$

corresponding to the stable equilibrium state $v_1 = \sqrt{R-1}$. Thus, the steady-state probability distribution of $\varphi(t)$ and $\widetilde{v}_0(t)$ is similar to the Gibbs distribution (9.20)

$$P(\widetilde{v}_0, \varphi) = C\exp\left\{-\frac{1}{D}H(\widetilde{v}_0, \varphi)\right\}, \quad H(\widetilde{v}_0, \varphi) = \frac{\widetilde{v}_0^2}{2} + U(\varphi). \tag{9.40}$$

From Eq. (9.40) follows that, for $R > 1$, the steady-state probability distribution is composed of two independent steady-state distributions, of which the distribution of component $v_0(t)$ of system (9.37) is the Gaussian distribution

$$P(v_0) = \frac{1}{\sqrt{2\pi D}}\exp\left\{-\frac{(v_0-1)^2}{2D}\right\}, \tag{9.41}$$

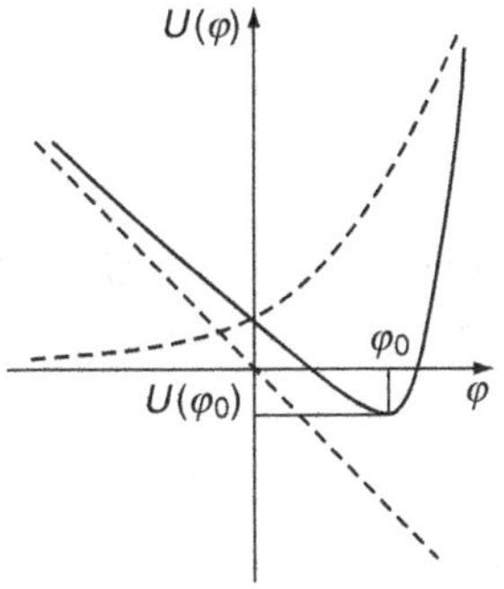

Figure 9.1 Potential function $U(\varphi)$. The dashed lines show curve $U(\varphi) = \frac{1}{2}\exp\{2\varphi\}$ and straight line $U(\varphi) = -(R-1)\varphi$.

and the distribution of quantity $\varphi(t)$ is the non-Gaussian distribution. If we turn back to variable $v_1(t)$, we obtain the corresponding steady-state probability distribution in the form

$$P(v_1) = \text{const}\, v_1^{\frac{R-1}{D}-1} \exp\left\{-\frac{v_1^2}{2D}\right\}. \tag{9.42}$$

As may be seen from Eq. (9.42), no steady-state probability distribution exists for component $v_1(t)$ in critical regime($R = 1$), which contrasts with the above case when random forces acted on all components of the triplet.

Earlier, we mentioned that two stages are characteristic of the formation of distribution (9.40). First, the Maxwellian distribution with respect to v_0 (9.41) is rapidly formed and the distribution with respect to φ is then formed at a much slower rate. The second stage is described by the Einstein–Smolukhovsky equation (9.17) that, in terms of variable v_1, assumes the form

$$\frac{\partial}{\partial t}P(v_1,t) = \frac{\partial}{\partial v_1}\left\{v_1\left[v_1^2 - (R-1)\right]P(v_1,t)\right\} + \sigma^2\frac{\partial}{\partial v_1}\left\{v_1\frac{\partial}{\partial v_1}\left[v_1 P(v_1,t)\right]\right\}. \tag{9.43}$$

This equation is the Fokker–Planck equation for the stochastic dynamic equation

$$\frac{d}{dt}v_1(t) = -v_1(t)\left[v_1^2(t) - (R-1)\right] + v_1(t)f(t).$$

At critical regime $R = 1$, this equation assumes the form

$$\frac{d}{dt}v_1(t) = -v_1^3(t) + v_1(t)f(t),$$

from which follows that, despite strong nonlinear friction, its solution is stochastically unstable due to the specific form of the random term. Stochastic stabilization occurs, as we saw earlier, due to inclusion of random forces into the equations for other components.

9.3 Boundary-Value Problems for the Fokker–Planck Equation (Hopping Phenomenon)

The Fokker–Planck equations are the partial differential equations and they generally require boundary conditions whose particular form depends on the problem under consideration. One can proceed from both forward and backward Fokker–Planck equations, which are equivalent. Consider one typical example.

Consider the nonlinear oscillator with friction described by the equation

$$\frac{d^2}{dt^2}x(t) + \lambda\frac{d}{dt}x(t) + \omega_0^2 x(t) + \beta x^3(t) = f(t) \quad (\beta,\ \lambda > 0) \tag{9.44}$$

and assume that random force $f(t)$ is the delta-correlated random function with the parameters

$$\langle f(t)\rangle = 0, \quad \left\langle f(t)f(t')\right\rangle = 2D\delta(t-t') \quad (D = \sigma_f^2\tau_0).$$

At $\lambda = 0$ and $f(t) = 0$, this equation is called the *Duffing equation.*

We can rewrite Eq. (9.44) in the standard form of the Hamiltonian system in functions $x(t)$ and $v(t) = \dfrac{d}{dt}x(t)$,

$$\frac{d}{dt}x(t) = \frac{\partial}{\partial v}H(x,v), \quad \frac{d}{dt}v(t) = -\frac{\partial}{\partial x}H(x,v) - \lambda v + f(t),$$

where

$$H(x,v) = \frac{v^2}{2} + U(x), \quad U(x) = \frac{\omega_0^2 x^2}{2} + \beta\frac{x^4}{4}$$

is the Hamiltonian.

According to Eq. (9.20), the steady-state solution to the corresponding Fokker–Planck equation has the form

$$P(x,v) = C\exp\left\{-\frac{\lambda}{D}H(x,v)\right\}. \tag{9.45}$$

It is clear that this distribution is the product of two independent distributions, of which one – the steady-state probability distribution of quantity $v(t)$ – is the Gaussian distribution and the other – the steady-state probability distribution of quantity $x(t)$ – is the non-Gaussian distribution. Integrating Eq. (9.45) over v, we obtain the steady-state probability distribution of $x(t)$

$$P(x) = C\exp\left\{-\frac{\lambda}{D}\left(\frac{\omega_0^2 x^2}{2} + \beta\frac{x^4}{4}\right)\right\}.$$

This distribution is maximum at the stable equilibrium point $x = 0$.

Consider now the equation

$$\frac{d^2}{dt^2}x(t) + \lambda\frac{d}{dt}x(t) - \omega_0^2 x(t) + \beta x^3(t) = f(t) \quad (\beta,\ \lambda > 0). \tag{9.46}$$

In this case again, the steady-state probability distribution has the form (9.45), where now

$$H(x, v) = \frac{v^2}{2} + U(x), \quad U(x) = -\frac{\omega_0^2 x^2}{2} + \beta\frac{x^4}{4}.$$

The steady-state probability distribution of $x(t)$ assumes now the form

$$P(x) = C\exp\left\{-\frac{\lambda}{D}\left(-\frac{\omega_0^2 x^2}{2} + \beta\frac{x^4}{4}\right)\right\} \tag{9.47}$$

and has maxima at points $x = \pm\sqrt{\omega_0^2/\beta}$ and a minimum at point $x = 0$; the maxima correspond to the stable equilibrium points of problem (9.46) for $f(t) = 0$ and the minimum, to the instable equilibrium point. Figure 9.2 shows the behavior of the probability distribution (9.47).

As we mentioned earlier, the formation of distribution (9.47) is described by the Einstein–Smolukhovsky equation (9.17), which has in this case the form

$$\frac{\partial}{\partial t}P(x, t) = \frac{1}{\lambda}\frac{\partial}{\partial x}\left(\frac{\partial U(x)}{\partial x}P(x, t)\right) + \frac{1}{\lambda^2}\frac{\partial^2}{\partial x^2}P(x, t). \tag{9.48}$$

This equation is statistically equivalent to the dynamic equation

$$\frac{d}{dt}x(t) = -\frac{1}{\lambda}\frac{\partial U(x)}{\partial x} + \frac{1}{\lambda}f(t). \tag{9.49}$$

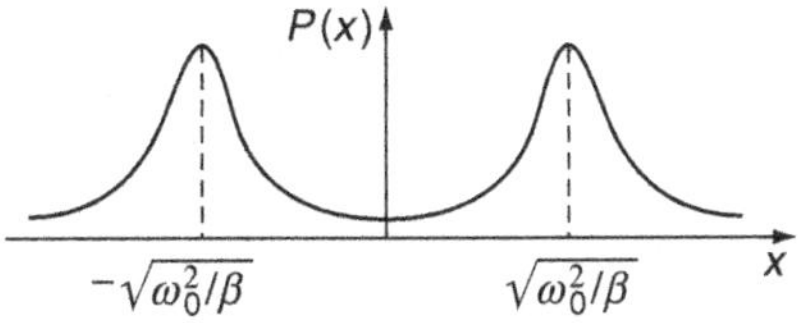

Figure 9.2 Probability distribution (9.47).

Probability distribution (9.47) corresponds to averaging over an ensemble of realizations of random process $f(t)$. If we deal with a single realization, the system arrives at one of states corresponding to the distribution maxima with a probability of $1/2$. In this case, averaging over time will form the probability distribution around the maximum position. However, after a lapse of certain time T (the longer, the smaller D), the system will be transferred in the vicinity of the other maximum due to the fact that function $f(t)$ can assume sufficiently large values. For this reason, temporal averaging will form probability distribution (9.48) only if averaging time $t \gg T$.

Introducing dimensionless coordinate $x \to \sqrt{\dfrac{\omega_0^2}{\beta}}x$ and time $t \to \dfrac{\lambda}{\omega_0^2}t$, we can rewrite Eq. (9.48) in the form

$$\frac{\partial}{\partial t}P(x,t) = \frac{\partial}{\partial x}\left(\frac{\partial U(x)}{\partial x}P(x,t)\right) + \mu\frac{\partial^2}{\partial x^2}P(x,t), \tag{9.50}$$

where

$$\mu = \frac{\beta D}{\lambda\omega_0^4}, \quad U(x) = -\frac{x^2}{2} + \frac{x^4}{4}.$$

In this case, the equivalent stochastic equation (9.49) assumes the form of Eq. (1.15), page 10

$$\frac{d}{dt}x(t) = -\frac{\partial U(x)}{\partial x} + f(t). \tag{9.51}$$

Estimate the time required for the system to switch from a most probable state $x = -1$ to the other $x = 1$.

Assume the system described by stochastic equation (9.51) was at a point from the interval (a, b) at instant t_0. The corresponding probability for the system to leave this interval

$$G(t; x_0, t_0) = 1 - \int_a^b dx p(x, t|x_0, t_0)$$

satisfies Eq. (8.21) following from the backward Fokker–Planck equation (8.20), page 196, i.e., the equation

$$\frac{\partial}{\partial t_0}G(t; x_0, t_0) = \frac{\partial U(x_0)}{\partial x_0}\frac{\partial}{\partial x_0}G(t; x_0, t_0) - \mu\frac{\partial^2}{\partial x_0^2}G(t; x_0, t_0)$$

with the boundary conditions

$$G(t; x_0, t) = 0, \quad G(t; a, t_0) = G(t; b, t_0) = 1.$$

Taking into account the fact that $G(t; x_0, t_0) = G(t - t_0; x_0)$ in our problem, we can denote $(t - t_0) = \tau$ and rewrite the boundary-value problem in the form

$$\frac{\partial}{\partial \tau} G(\tau; x_0) = \frac{\partial U(x_0)}{\partial x_0} \frac{\partial}{\partial x_0} G(\tau; x_0) - \mu \frac{\partial^2}{\partial x_0^2} G(\tau; x_0),$$
$$G(0; x_0) = 0, \; G(\tau; a,) = G(\tau; b) = 1 \quad \left(\lim_{\tau \to \infty} G(\tau; x_0) = 0 \right). \tag{9.52}$$

From Eq. (9.52), one can easily see that average time required for the system to leave interval (a, b)

$$T(x_0) = \int_0^\infty d\tau \, \tau \frac{\partial G(\tau; x_0)}{\partial \tau}$$

satisfies the boundary-value problem

$$\mu \frac{d^2 T(x_0)}{dx_0^2} - \frac{dU(x_0)}{dx_0} \frac{dT(x_0)}{dx_0} = -1, \quad T(a) = T(b) = 0. \tag{9.53}$$

Equation (9.53) can be easily solved, and we obtain that the average time required for the system under random force to switch its state from $x_0 = -1$ to $x_0 = 1$ (this time is usually called the *Kramers time*) is given by the expression

$$T = \frac{1}{\mu} \int_{-1}^{1} d\xi \int_{-\infty}^{\xi} d\eta \exp\left\{ \frac{1}{\mu} \left[U(\xi) - U(\eta) \right] \right\} = \frac{C(\mu)}{\mu} \int_0^1 d\xi \exp\left\{ \frac{1}{\mu} U(\xi) \right\}, \tag{9.54}$$

where $C(\mu) = \int_{-\infty}^{\infty} d\xi \exp\left\{ \frac{1}{\mu} U(\xi) \right\}$. For $\mu \ll 1$, we obtain

$$T \approx \sqrt{2}\pi \, exp \left\{ \frac{1}{4\mu} \right\},$$

i.e., the average switching time increases exponentially with decreasing the intensity of fluctuations of the force.

9.4 Method of Fast Oscillation Averaging

If fluctuations of dynamic system parameters are sufficiently small, we can analyze the Fokker–Planck equation using different asymptotic and approximate techniques.

Here, we consider in greater detail the method that is most often used in the statistical analysis.

Assume the stochastic system is described by the dynamic equations

$$
\begin{aligned}
\frac{d}{dt}x(t) &= A(x,\widetilde{\phi}) + z(t)B(x,\widetilde{\phi}),\\
\frac{d}{dt}\phi(t) &= C(x,\widetilde{\phi}) + z(t)D(x,\widetilde{\phi}),
\end{aligned}
\tag{9.55}
$$

where

$$
\widetilde{\phi}(t) = \omega_0 t + \phi(t),
$$

functions $A(x,\widetilde{\phi})$, $B(x,\widetilde{\phi})$, $C(x,\widetilde{\phi})$, and $D(x,\widetilde{\phi})$ are the periodic functions of variable $\widetilde{\phi}$, and $z(t)$ is the Gaussian delta-correlated process with the parameters

$$
\langle z(t)\rangle = 0, \quad \langle z(t)z(t')\rangle = 2D\delta(t-t'), \quad D = \sigma^2\tau_0.
$$

Variables $x(t)$ and $\phi(t)$ can mean the vector module and phase, respectively. The Fokker–Planck equation corresponding to system of equations (9.55) has the form

$$
\begin{aligned}
\frac{\partial}{\partial t}P(x,\phi,t) &= -\frac{\partial}{\partial x}A(x,\widetilde{\phi})P(x,\phi,t) - \frac{\partial}{\partial \phi}C(x,\widetilde{\phi})P(x,\phi,t)\\
&\quad + D\left[\frac{\partial}{\partial x}B(x,\widetilde{\phi}) + \frac{\partial}{\partial \phi}D(x,\widetilde{\phi})\right]^2 P(x,\phi,t).
\end{aligned}
\tag{9.56}
$$

Commonly, Eq. (9.56) is very complicated for immediately analyzing the joint probability density. We rewrite this equation in the form

$$
\begin{aligned}
\frac{\partial}{\partial t}P(x,\phi,t) &= -\frac{\partial}{\partial x}A(x,\widetilde{\phi})P(x,\phi,t) - \frac{\partial}{\partial \phi}C(x,\widetilde{\phi})P(x,\phi,t)\\
&\quad - D\frac{\partial}{\partial x}\left(\frac{\partial B^2(x,\widetilde{\phi})}{2\partial x} + \frac{\partial B(x,\widetilde{\phi})}{\partial \widetilde{\phi}}D(x,\widetilde{\phi})\right)P(x,\phi,t)\\
&\quad - D\frac{\partial}{\partial \phi}\left(\frac{\partial D(x,\widetilde{\phi})}{\partial x}B(x,\widetilde{\phi}) + \frac{\partial D^2(x,\widetilde{\phi})}{2\partial \widetilde{\phi}}\right)P(x,\phi,t)\\
&\quad + D\left\{\frac{\partial^2}{\partial x^2}B^2(x,\widetilde{\phi}) + 2\frac{\partial^2}{\partial x\partial \phi}B(x,\widetilde{\phi})D(x,\widetilde{\phi}) + \frac{\partial^2}{\partial \phi^2}D^2(x,\widetilde{\phi})\right\}P(x,\phi,t).
\end{aligned}
\tag{9.57}
$$

Now, we assume that functions $A(x,\widetilde{\phi})$ and $C(x,\widetilde{\phi})$ are sufficiently small and fluctuation intensity of process $z(t)$ is also small. In this case, statistical characteristics of system of equations (9.55) only slightly vary during times $\sim 1/\omega_0$. To study these small variations (accumulated effects), we can average Eq. (9.57) over the period of all oscillating functions. Assuming that function $P(x, \phi, t)$ remains intact under averaging, we obtain the equation

$$\frac{\partial}{\partial t}\overline{P(x,\phi,t)} = -\frac{\partial}{\partial x}\overline{A(x,\widetilde{\phi})}\,\overline{P(x,\phi,t)} - \frac{\partial}{\partial \phi}\overline{C(x,\widetilde{\phi})}\,\overline{P(x,\phi,t)}$$
$$-D\left\{\frac{\partial}{\partial x}\left(\overline{\frac{\partial B^2(x,\widetilde{\phi})}{2\partial x}} + \overline{\frac{\partial B(x,\widetilde{\phi})}{\partial \widetilde{\phi}}D(x,\widetilde{\phi})}\right) + \frac{\partial}{\partial \phi}\overline{\frac{\partial D(x,\widetilde{\phi})}{\partial x}B(x,\widetilde{\phi})}\right\}\overline{P(x,\phi,t)}$$
$$+D\left\{\frac{\partial^2}{\partial x^2}\overline{B^2(x,\widetilde{\phi})} + 2\frac{\partial^2}{\partial x\partial \phi}\overline{B(x,\widetilde{\phi})D(x,\widetilde{\phi})} + \frac{\partial^2}{\partial \phi^2}\overline{D^2(x,\widetilde{\phi})}\right\}\overline{P(x,\phi,t)}, \quad (9.58)$$

where the overbar denotes quantities averaged over the oscillation period.

Integrating Eq. (9.58) over ϕ, we obtain the Fokker–Planck equation for function $\overline{P(x,t)}$

$$\frac{\partial}{\partial t}\overline{P(x,t)} = -\frac{\partial}{\partial x}\overline{A(x,\widetilde{\phi})}\,\overline{P(x,t)} + D\frac{\partial}{\partial x}\left(\overline{\frac{\partial B^2(x,\widetilde{\phi})}{2\partial x}} + \overline{\frac{\partial B(x,\widetilde{\phi})}{\partial \widetilde{\phi}}D(x,\widetilde{\phi})}\right)\overline{P(x,t)}$$
$$+D\frac{\partial}{\partial x}\overline{B^2(x,\widetilde{\phi})}\frac{\partial}{\partial x}\overline{P(x,t)}. \quad (9.59)$$

Note that quantity $x(t)$ appears to be the one-dimensional Markovian random process in this approximation.

If we assume that $\overline{B(x,\widetilde{\phi})D(x,\widetilde{\phi})} = \overline{\frac{\partial D(x,\widetilde{\phi})}{\partial x}B(x,\widetilde{\phi})} = 0$, and $\overline{C(x,\widetilde{\phi})} = \text{const}$, $\overline{D^2(x,\widetilde{\phi})} = \text{const}$ in Eq. (9.58), then processes $x(t)$ and $\phi(t)$ become statistically independent, and process $\phi(t)$ becomes the Markovian Gaussian process whose variance is the linear increasing function of time t. This means that probability distribution of quantity $\phi(t)$ on segment $[0, 2\pi]$ becomes uniform for large t (at $\overline{C(x,\widetilde{\phi})} = 0$).

Problems

Problem 9.1

Solve the Fokker–Planck equation

$$\frac{\partial}{\partial t}p(x,t|x_0,t_0) = D\frac{\partial}{\partial x}(x^2-1)\frac{\partial}{\partial x}p(x,t|x_0,t_0) \quad (x \geq 1),$$
$$p(x,t_0|x_0,t_0) = \delta(x-x_0). \quad (9.60)$$

Instruction *Use the integral Meler–Fock transform.*

Solution

$$p(x,t|x_0,t_0) = \int_0^\infty d\mu\, \mu \tanh(\pi\mu) \exp\left\{-D\left(\mu^2+\frac{1}{4}\right)(t-t_0)\right\}$$
$$P_{-1/2+i\mu}(x)P_{-1/2+i\mu}(x_0).$$

If $x_0 = 1$ at the initial instant $t_0 = 0$, then

$$P(x,t) = \int_0^\infty d\mu\, \mu \tanh(\pi\mu) \exp\left\{-D\left(\mu^2+\frac{1}{4}\right)(t-t_0)\right\} P_{-1/2+i\mu}(x).$$

Problem 9.2

In the context of singular stochastic problem (1.29), page 18, for $\lambda = 1$

$$\frac{d}{dt}x(t) = -x^2(t) + f(t), \quad x(0) = x_0, \tag{9.61}$$

where $f(t)$ *is the Gaussian delta-correlated process with the parameters*

$$\langle f(t)\rangle = 0, \quad \langle f(t)f(t')\rangle = 2D\delta(t-t') \quad (D = \sigma_f^2\tau_0),$$

estimate average time required for the system to switch from state x_0 *to state* $(-\infty)$ *and average time between two singularities.*

Solution

$$\langle T(x_0)\rangle = \int_{-\infty}^{x_0} d\xi \int_{\xi}^{\infty} d\eta \exp\left\{\frac{1}{3}\left(\xi^3-\eta^3\right)\right\},$$

$$\langle T(\infty)\rangle = \sqrt{\pi}\frac{12^{1/6}}{3}\Gamma\left(\frac{1}{6}\right) \approx 4.976.$$

Problem 9.3

Find the steady-state probability distribution for stochastic problem (9.61).

Instruction *Solve the Fokker–Planck equation under the condition that function* $x(t)$ *is discontinuous and defined for all times* t *in such a way that its value of*

$-\infty$ at instant $t \to t_0 - 0$ is immediately followed by its value of ∞ at instant $t \to t_0 + 0$. This behavior corresponds to the boundary condition of the Fokker–Planck equation formulated as continuity of probability flux density.

Solution

$$P(x) = J \int_{-\infty}^{x} d\xi \exp\left\{\frac{1}{3}\left(\xi^3 - x^3\right)\right\}, \tag{9.62}$$

where

$$J = \frac{1}{\langle T(\infty)\rangle}$$

is the steady-state probability flux density.

Remark From Eq. (9.62) follows the asymptotic formula

$$P(x) \approx \frac{1}{\langle T(\infty)\rangle\, x^2} \tag{9.63}$$

for great x.

Problem 9.4

Show that asymptotic formula (9.63) is formed by discontinuities of function $x(t)$.

Instruction *Near discontinuous points t_k, represent the solution $x(t)$ in the form*

$$x(t) = \frac{1}{t - t_k}.$$

Problem 9.5

Derive the Fokker–Planck equation for slowly varying amplitude and phase of the solution to the problem

$$\frac{d}{dt}x(t) = y(t), \quad \frac{d}{dt}y(t) = -2\gamma y(t) - \omega_0^2[1 + z(t)]x(t), \tag{9.64}$$

where $z(t)$ is the Gaussian process with the parameters

$$\langle z(t)\rangle = 0, \quad \left\langle z(t)z(t')\right\rangle = 2\sigma^2\tau_0\delta(t - t').$$

Instruction *Replace functions $x(t)$ and $y(t)$ with new variables (oscillation amplitude and phase) according to the equalities*

$$x(t) = A(t)\sin(\omega_0 t + \phi(t)), \quad y(t) = \omega_0 A(t)\cos(\omega_0 t + \phi(t))$$

and represent amplitude $A(t)$ in the form $A(t) = e^{u(t)}$.

Solution

$$\frac{\partial}{\partial t}\overline{P(u,\phi,t)} = \gamma\frac{\partial}{\partial u}\overline{P(u,\phi,t)} - \frac{D}{4}\frac{\partial}{\partial u}\overline{P(u,\phi,t)} + \frac{D}{8}\frac{\partial^2}{\partial u^2}\overline{P(t;u,\phi)} + \frac{3D}{8}\frac{\partial^2}{\partial \phi^2}\overline{P(u,\phi,t)}, \tag{9.65}$$

where $D = \sigma^2\tau_0\omega_0^2$. From Eq. (9.65) follows that statistical characteristics of oscillation amplitude and phase (averaged over the oscillation period) are statistically independent and have the Gaussian probability densities

$$\overline{P(u,t)} = \frac{1}{\sqrt{2\pi\sigma_u^2(t)}}\exp\left\{-\frac{(u-\langle u(t)\rangle)^2}{2\sigma_u^2(t)}\right\},$$
$$\overline{P(\phi,t)} = \frac{1}{\sqrt{2\pi\sigma_\phi^2(t)}}\exp\left\{-\frac{(\phi-\phi_0)^2}{2\sigma_\phi^2(t)}\right\}, \tag{9.66}$$

where

$$\langle u(t)\rangle = u_0 - \gamma t + \frac{D}{4}t, \quad \sigma_u^2(t) = \frac{D}{4}t,$$

$$\langle \phi(t)\rangle = \phi_0, \quad \sigma_\phi^2(t) = \frac{3D}{4}t.$$

Problem 9.6

Using probability distributions (9.66), calculate quantities $\langle x(t)\rangle$ and $\langle x^2(t)\rangle$. Derive the condition of stochastic excitation of the second moment.

Solution Average value

$$\langle x(t)\rangle = e^{-\gamma t}\sin(\omega_0 t)$$

coincides with the problem solution in the absence of fluctuations.

Quantity

$$\left\langle x^2(t)\right\rangle = \frac{1}{2}e^{(D-2\gamma)t}\left\{1 - e^{-3Dt/2}\cos(2\omega_0 t)\right\}, \tag{9.67}$$

coincides in the absence of absorption with Eq. (6.91) to terms about $D/\omega_0 \ll 1$.

Problem 9.7

Find the condition of stochastic parametric excitation of function $\langle A^n(t)\rangle$.

Solution We have

$$\langle A^n(t)\rangle = \left\langle e^{nu(t)}\right\rangle = A_0^n \exp\left\{-n\gamma t + \frac{1}{8}n(n+2)Dt\right\},$$

and, consequently, the stochastic dynamic system is parametrically excited beginning from moment function of order n if the condition

$$8\gamma < (n+2)D$$

is satisfied.

Remark The typical realization curve of random amplitude has the form

$$A^*(t) = A_0 \exp\left\{-\left(\gamma - \frac{D}{4}\right)t\right\}$$

and exponentially decays in time if absorption is sufficiently small, namely, if

$$1 < 4\frac{\gamma}{D} < 1 + \frac{1}{2}n,$$

which always holds for sufficiently great n, whereas all moment functions of random amplitude $A(t)$ of order n and higher exponentially grow in time. This means that statistics of random amplitude $A(t)$ are formed mainly by large spikes over the exponentially decaying typical realization curve. This fact follows from the lognormal property of random amplitude $A(t)$.

Lecture 10

Some Other Approximate Approaches to the Problems of Statistical Hydrodynamics

In previous Lectures, we analyzed statistics of solutions to the nonlinear equations of hydrodynamics using the rigorous approach based on deriving and investigating the exact variational differential equations for characteristic functionals of nonlinear random fields. However, this approach encounters severe difficulties caused by the lack of development of the theory of variational differential equations. For this reason, many researchers prefer to proceed from more habitual partial differential equations for different moment functions of fields of interest. The nonlinearity of the input dynamic equations governing random fields results in the appearance of higher moment functions of fields of interest in the equations governing any moment function. As a result, even determination of average field or correlation functions requires, in the strict sense, solving an infinite system of linked equations.

Thus, the main problem of this approach consists in cutting the mentioned system of equations on the basis of one or another physical hypothesis. The most known example of such a hypothesis is the *Millionshchikov hypothesis* according to which the higher moment functions of even orders are expressed in terms of the lower ones by the laws of the Gaussian statistics. The disadvantage of such approaches consists in the fact that the validity of the hypothesis usually cannot be proved; moreover, cutting the system of equations may often yield physically contradictory results, namely, energy spectra of turbulence can appear negative for certain wave numbers. Nevertheless, these approximate approaches provide a deeper insight into physical mechanisms of forming the statistics of strongly nonlinear random fields and make it possible to derive quantitative expressions for fields' correlation functions and spectra. It seems that the Millionshchikov hypothesis provides correct spectra of the developed turbulence in viscous interval [35].

We emphasize additionally that the mentioned approximate equations reveal many nontrivial effects inherent in nonlinear random fields and have no analogs in the behavior of deterministic fields and waves. Here, we illustrate these methods of analysis by the example of an interesting physical effect, namely, that average flows of

Lectures on Dynamics of Stochastic Systems. DOI: 10.1016/B978-0-12-384966-3.00010-6

noncompressible liquids superimposed on the background of developed turbulent pulsations acquire quasi-elastic wave properties. This effect was first mentioned by Moffatt [57] who studied the reaction of turbulence on variations of the transverse gradient of average velocity. Moffatt noted that turbulent medium behaves in some sense like an elastic medium; namely, variations of average velocity profile of a plane-parallel flow satisfy the wave equation.

10.1 Quasi-Elastic Properties of Isotropic and Stationary Noncompressible Turbulent Media

Let $\boldsymbol{u}^{\mathrm{T}}(\boldsymbol{r},t)$ be the random velocity field of developed turbulence stationary in time and homogeneous and isotropic in space (we assume that $\langle \boldsymbol{u}^{\mathrm{T}}(\boldsymbol{r},t)\rangle = 0$). In actuality, regular average flows are imposed on turbulent pulsations of liquid. We denote $\boldsymbol{U}(\boldsymbol{r},t)$ the velocity field of these regular flows. Because the turbulent pulsations and average flows interact nonlinearly, we can represent the total velocity field in the form

$$\boldsymbol{u}(\boldsymbol{r},t) = \boldsymbol{U}(\boldsymbol{r},t) + \boldsymbol{u}^{\mathrm{T}}(\boldsymbol{r},t) + \boldsymbol{u}'(\boldsymbol{r},t), \tag{10.1}$$

Here, $\boldsymbol{u}^{\mathrm{T}}(\boldsymbol{r},t)$ is the unperturbed turbulent field in the absence of average flows and $\boldsymbol{u}'(\boldsymbol{r},t)$ is the turbulent field disturbance caused by the interaction with the regular flow (we assume that $\langle \boldsymbol{u}'(\boldsymbol{r},t)\rangle = 0$).

Investigation of nonlinear interactions between regular flows and turbulent pulsations (and the analysis of turbulent pulsations $\boldsymbol{u}^{\mathrm{T}}(\boldsymbol{r},t)$ by themselves, though) is very difficult in the general case. However, in the case of weak average flows, i.e., under the condition that

$$\boldsymbol{U}^2(\boldsymbol{r},t) \ll 2T(t),$$

where $T(t) = \dfrac{1}{2}\left\langle\left[\boldsymbol{u}^{\mathrm{T}}(\boldsymbol{r},t)\right]^2\right\rangle$ is the average density of turbulent energy pulsation, we can discuss the effect of turbulence on the average flow evolution in sufficient detail by considering the linear approximation in small disturbances of fields $\boldsymbol{U}(\boldsymbol{r},t)$ and $\boldsymbol{u}'(\boldsymbol{r},t)$ under the assumption that the spectrum of turbulent pulsations $\boldsymbol{u}^{\mathrm{T}}(\boldsymbol{r},t)$ is known.

The total velocity field (10.1) and its turbulent component $\boldsymbol{u}^{\mathrm{T}}(\boldsymbol{r},t)$ satisfy the Navier–Stokes equation (1.95), page 43. Hence, substituting Eq. (10.1) in Eq. (1.95) and linearizing the equations in regular (average) field $\boldsymbol{U}(\boldsymbol{r},t)$ and fluctuation component $\boldsymbol{u}'(\boldsymbol{r},t)$, we arrive at the approximate system of equations

$$\frac{\partial}{\partial t}U_i(\boldsymbol{r},t) + \frac{\partial}{\partial r_k}T_{ik}(\boldsymbol{r},t) = -\frac{\partial}{\partial r_i}P(\boldsymbol{r},t), \tag{10.2}$$

$$\begin{aligned}\frac{\partial}{\partial t}u'_i(\boldsymbol{r},t) &+ \frac{\partial}{\partial r_k}\left[u_i(\boldsymbol{r},t)u_k(\boldsymbol{r},t) - \langle u_i(\boldsymbol{r},t)u_k(\boldsymbol{r},t)\rangle\right]\\ &+ u_l^{\mathrm{T}}(\boldsymbol{r},t)\frac{\partial U_i(\boldsymbol{r},t)}{\partial r_l} + U_l(\boldsymbol{r},t)\frac{\partial u_i^{\mathrm{T}}(\boldsymbol{r},t)}{\partial r_l} = -\frac{\partial}{\partial r_i}p'(\boldsymbol{r},t),\end{aligned} \tag{10.3}$$

where $T_{ik}(\boldsymbol{r},t) = \langle u_i(\boldsymbol{r},t)u_k(\boldsymbol{r},t)\rangle$ is the *Reynolds stress tensor* and $P(\boldsymbol{r},t)$ and $p'(\boldsymbol{r},t)$ are the average and perturbed turbulent pressure components, respectively. Here and below, we neglect the effect of viscosity on dynamics and statistics of regular component $\boldsymbol{U}(\boldsymbol{r},t)$ and perturbed field $\boldsymbol{u}'(\boldsymbol{r},t)$. In addition, we will bear in mind that fields $\boldsymbol{U}(\boldsymbol{r},t)$ and $\boldsymbol{u}'(\boldsymbol{r},t)$ satisfy the incompressibility condition (1.95), page 43,

$$\operatorname{div}\boldsymbol{U}(\boldsymbol{r},t) = 0, \quad \operatorname{div}\boldsymbol{u}'(\boldsymbol{r},t) = 0. \tag{10.4}$$

Being combined with Eq. (1.95), page 43, for turbulent pulsations $\boldsymbol{u}^{\mathrm{T}}(\boldsymbol{r},t)$, Eq. (10.3) yields the following equation for the Reynolds stress tensor $T_{ik}(\boldsymbol{r},t)$

$$\begin{aligned}
&\frac{\partial}{\partial t}T_{ik}(\boldsymbol{r},t) + \frac{\partial}{\partial r_l}\langle u_i(\boldsymbol{r},t)u_k(\boldsymbol{r},t)u_l(\boldsymbol{r},t)\rangle \\
&\quad + \left\langle u_k^{\mathrm{T}}(\boldsymbol{r},t)u_l^{\mathrm{T}}(\boldsymbol{r},t)\right\rangle \frac{\partial U_i(\boldsymbol{r},t)}{\partial r_l} + \left\langle u_i^{\mathrm{T}}(\boldsymbol{r},t)u_l^{\mathrm{T}}(\boldsymbol{r},t)\right\rangle \frac{\partial U_k(\boldsymbol{r},t)}{\partial r_l} \\
&\quad + U_l(\boldsymbol{r},t)\frac{\partial\left\langle u_i^{\mathrm{T}}(\boldsymbol{r},t)u_k^{\mathrm{T}}(\boldsymbol{r},t)\right\rangle}{\partial r_l} \\
&\quad - \left\langle\left[u_k^{\mathrm{T}}(\boldsymbol{r},t)\frac{\partial}{\partial r_i}p'(\boldsymbol{r},t) + u_i^{\mathrm{T}}(\boldsymbol{r},t)\frac{\partial}{\partial r_k}p'(\boldsymbol{r},t)\right]\right\rangle.
\end{aligned} \tag{10.5}$$

Here,

$$u_i(\boldsymbol{r},t)u_k(\boldsymbol{r},t)u_l(\boldsymbol{r},t) = u_i^{\mathrm{T}}(\boldsymbol{r},t)u_k^{\mathrm{T}}(\boldsymbol{r},t)u_l'(\boldsymbol{r},t) + \cdots .$$

Pressure pulsations p' in this equations can be expressed in terms of perturbed velocities $\boldsymbol{U}(\boldsymbol{r},t)$ and $\boldsymbol{u}'(\boldsymbol{r},t)$,

$$\begin{aligned}
&p'(\boldsymbol{r},t) \\
&\quad = -\Delta^{-1}(\boldsymbol{r},\boldsymbol{r}')\left\{\frac{\partial^2}{\partial r'_m\partial r'_n}\left[u_m(\boldsymbol{r}',t)u_n(\boldsymbol{r}',t) - \left\langle u_m(\boldsymbol{r}',t)u_n(\boldsymbol{r}',t)\right\rangle\right]\right. \\
&\qquad\qquad \left. + 2\frac{\partial u_l^{\mathrm{T}}(\boldsymbol{r}',t)}{\partial r'_m}\frac{\partial U_m(\boldsymbol{r}',t)}{\partial r'_l}\right\},
\end{aligned}$$

where $\Delta^{-1}(\boldsymbol{r},\boldsymbol{r}')$ is the integral operator inverse to the Laplace operator. Equations (10.2), (10.5), and the expression for pressure $p'(\boldsymbol{r},t)$ form the system of equations in average field $\boldsymbol{U}(\boldsymbol{r},t)$ and stress tensor $T_{ik}(\boldsymbol{r},t)$. These equations are not closed because they depend on higher-order velocity correlators like the triple correlator $\left\langle u_i^{\mathrm{T}}(\boldsymbol{r},t)u_k^{\mathrm{T}}(\boldsymbol{r},t)u_l'(\boldsymbol{r},t)\right\rangle$. Assuming that such correlators only slightly affect the dynamics of perturbations and taking into account conditions (10.4) and statistical homogeneity of field $\boldsymbol{u}^{\mathrm{T}}(\boldsymbol{r},t)$, we reduce the system of equations (10.2), (10.5) to the form ($\tau_i(\boldsymbol{r},t) = \frac{\partial}{\partial r_k}T_{ik}(\boldsymbol{r},t)$)

$$\frac{\partial}{\partial t}U_i(\boldsymbol{r},t) + \tau_i(\boldsymbol{r},t) = -\frac{\partial}{\partial r_i}P(\boldsymbol{r},t), \tag{10.6}$$

$$\frac{\partial}{\partial t}\tau_i(\boldsymbol{r},t) + \left\langle u_k^{\mathrm{T}}(\boldsymbol{r},t)u_l^{\mathrm{T}}(\boldsymbol{r},t)\right\rangle \frac{\partial^2 U_i(\boldsymbol{r},t)}{\partial r_l \partial r_k}$$
$$= 2\frac{\partial}{\partial r_k}\left[\left\langle u_k^{\mathrm{T}}(\boldsymbol{r},t)\frac{\partial}{\partial r_i}\Delta^{-1}(\boldsymbol{r},\boldsymbol{r}')\left(\frac{\partial u_l^{\mathrm{T}}(\boldsymbol{r}',t)}{\partial r'_m}\frac{\partial U_m(\boldsymbol{r}',t)}{\partial r'_l}\right)\right\rangle\right.$$
$$\left.+\left\langle u_i^{\mathrm{T}}(\boldsymbol{r},t)\frac{\partial}{\partial r_k}\Delta^{-1}(\boldsymbol{r},\boldsymbol{r}')\left(\frac{\partial u_l^{\mathrm{T}}(\boldsymbol{r}',t)}{\partial r'_m}\frac{\partial U_m(\boldsymbol{r}',t)}{\partial r'_l}\right)\right\rangle\right]. \tag{10.7}$$

This system of equations is closed in $\boldsymbol{U}(\boldsymbol{r},t)$. The coefficients in the left-hand side of Eq. (10.7) and the integral operator in the right-hand side can be expressed in terms of the correlation tensor of vortex field of unperturbed turbulence $\boldsymbol{u}^{\mathrm{T}}(\boldsymbol{r},t)$

$$\left\langle u_i^{\mathrm{T}}(\boldsymbol{r},t)u_j^{\mathrm{T}}(\boldsymbol{r}',t)\right\rangle = \int d\boldsymbol{q}\Phi_{ij}(\boldsymbol{q})e^{i\boldsymbol{q}(\boldsymbol{r}-\boldsymbol{r}')}. \tag{10.8}$$

In view of the fact that field $\boldsymbol{u}^{\mathrm{T}}(\boldsymbol{r},t)$ is the solenoidal field, we have

$$\Phi_{ij}(\boldsymbol{q}) = \Delta_{ij}(\boldsymbol{q})F(q),$$

where

$$\Delta_{ij}(\boldsymbol{q}) = \delta_{ij} - \frac{q_i q_j}{\boldsymbol{q}^2}.$$

We will assume the energy spectrum of turbulent pulsations $F(q)$ known. From Eq. (10.8) follows in particular that the energy of turbulent pulsations is expressed through the energy spectrum by the relationship

$$T = \frac{1}{2}\left\langle\left[\boldsymbol{u}^{\mathrm{T}}(\boldsymbol{r},t)\right]^2\right\rangle = 4\pi\int\limits_0^\infty dq\, q^2 F(q),$$

and coefficients in the left-hand side of Eq. (10.7) are given by the formula

$$\left\langle u_k^{\mathrm{T}}(\boldsymbol{r},t)u_l^{\mathrm{T}}(\boldsymbol{r},t)\right\rangle = \frac{2}{3}T\delta_{kl}. \tag{10.9}$$

Finally, using Eq. (10.8), the right-hand side of Eq. (10.7) can be represented in the form

$$-\int d\boldsymbol{q}e^{i\boldsymbol{q}\boldsymbol{r}}U_m(\boldsymbol{q},t)g_{im}(\boldsymbol{q}), \tag{10.10}$$

where $\boldsymbol{U}(\boldsymbol{q},t)$ is the spatial Fourier transform of average velocity field $\boldsymbol{U}(\boldsymbol{r},t)$ and tensor $g_{im}(\boldsymbol{q})$ is given by the formula

$$g_{im}(\boldsymbol{p}) = 2\int d\boldsymbol{q}\frac{q_m p_l p_k}{(\boldsymbol{q}+\boldsymbol{p})^2}\left[(q_i+p_i)\Delta_{kl}(\boldsymbol{q}) + (q_k+p_k)\Delta_{il}(\boldsymbol{q})\right]F(q). \tag{10.11}$$

From the obvious fact that tensor $g_{im}(\boldsymbol{p})$ is invariant relative rotations in space, it follows that it must have the form

$$g_{im}(\boldsymbol{p}) = A(p)\delta_{im} + B(p)\frac{p_i p_m}{\boldsymbol{p}^2}. \tag{10.12}$$

Moreover, the term proportional to $B(p)$ disappears in Eq. (10.12) in view of Eq. (10.4) (it expresses the property of incompressibility of average field $\boldsymbol{U}(\boldsymbol{r}, t)$) and the identity $\boldsymbol{pU}(\boldsymbol{p}, t) = 0$ following from Eq. (10.4), so that the right-hand side (10.10) of Eq. (10.7) assumes the form

$$-\int d\boldsymbol{q} e^{i\boldsymbol{qr}} U_i(\boldsymbol{q}, t) A(q).$$

Substituting this expression in the right-hand side of Eq. (10.7) and taking into account Eq. (10.9), we rewrite Eq. (10.6), (10.7) in the form

$$\begin{aligned} &\frac{\partial}{\partial t} U_i(\boldsymbol{r}, t) + \tau_i(\boldsymbol{r}, t) = -\frac{\partial}{\partial r_i} P(\boldsymbol{r}, t), \\ &\frac{\partial}{\partial t} \tau_i(\boldsymbol{r}, t) + \frac{2}{3} T \Delta U_i(\boldsymbol{r}, t) = -A(-i\nabla) U_i(\boldsymbol{r}, t). \end{aligned} \tag{10.13}$$

The kernel of the integral operator appeared in the second equation can be obtained from comparison of Eq. (10.11) with (10.12). The result is as follows

$$A(p) = \int d\boldsymbol{q} \frac{F(q)}{(q^2 + p^2)} \left[p^2 - \frac{(\boldsymbol{qp})^2}{q^2} \right] \left[q^2 - (\boldsymbol{qp}) - 2\frac{(\boldsymbol{qp})^2}{p^2} \right]. \tag{10.14}$$

We can perform integration in Eq. (10.14) over angular coordinate to obtain the expression

$$A(p) = 2\pi p^2 \int_0^\infty dq\, F(q) \left\{ \frac{2q^2}{3p^2} - \frac{q^4 - p^4}{4p^4} + \frac{(q^2 - p^2)^3}{8qp^5} \ln \left| \frac{q+p}{q-p} \right| \right\}. \tag{10.15}$$

We note additionally that Eq. (10.13) with allowance for Eq. (10.4) yields the identity $\Delta P(\boldsymbol{r}, t) = 0$, so that $P(\boldsymbol{r}, t) = P_0 = \text{const}$. Then, eliminating quantity $\tau_i(\boldsymbol{r}, t)$ from system of equations (10.11) and (10.4), we arrive at a single equation in the vector field of average velocity of flow $\boldsymbol{U}(\boldsymbol{r}, t)$,

$$\frac{\partial^2}{\partial t^2} \boldsymbol{U}(\boldsymbol{r}, t) - \left[\frac{2}{3} T\Delta + A(-i\nabla) \right] \boldsymbol{U}(\boldsymbol{r}, t) = 0. \tag{10.16}$$

The corresponding dispersion equation

$$\omega^2(p) = \frac{2}{3} T p^2 - A(p)$$

can be rewritten in the form

$$\omega^2(p) = 2\pi p^2 \int_0^\infty dq\, F(q) f\left(\frac{q}{p}\right),$$

where

$$f(x) = \frac{2-x^2}{3} + \frac{x^4-1}{8} - \frac{\left(x^2-1\right)^3}{16x} \ln\left|\frac{1+x}{1-x}\right|.$$

Function $f(x)$ has the following asymptotics

$$f(x) = \begin{cases} 2/3, & x \ll 1, \\ 4/15 & x \gg 1. \end{cases}$$

Moreover, the inequalities

$$\frac{2}{3} \geq f(x) \geq \frac{4}{15}$$

hold for arbitrary x.

Thus, the time-dependent development of disturbances in average flow is governed by the hyperbolic equation (10.16). As a result, the turbulent medium possesses certain quasi-elastic properties; namely, disturbances diffuse in the turbulent medium as transverse waves showing dispersion property. The phase and group velocities of these waves vary in the limits

$$\sqrt{\frac{2}{3}T} \geq c \geq \sqrt{\frac{4}{15}T}.$$

10.2 Sound Radiation by Vortex Motions

In the previous section, we considered the physical effect immediately related to statistical averaging of the nonlinear system of hydrodynamical equations in the case of noncompressible flow. Here, we consider an effect related to weakly compressible media; namely, we consider the problem on sound radiation by a weakly compressible medium. This problem corresponds to the inclusion of random fields in the linearized equations of hydrodynamics. Note that, within the framework of linear equations, parametric action of turbulent medium yields the equations of acoustics with random refractive index.

Turbulent motion of a flow in certain finite spatial region excites acoustic waves outside this region. In the case of a weakly compressible flow, this sound field is as if it were generated by the static distribution of acoustic quadrupoles whose instantaneous power per unit volume is given by the relationship

$$T_{ij}(\boldsymbol{r},t) = \rho_0 v_i^{\mathrm{T}}(\boldsymbol{r},t) v_j^{\mathrm{T}}(\boldsymbol{r},t),$$

where ρ_0 is the average density and $v_i^{\mathrm{T}}(\boldsymbol{r},t)$ are the components of the velocity of flow in turbulent region V inside which the turbulence is assumed homogeneous and isotropic in space and stationary in time (we use the coordinate system in which the whole of the flow is quiescent). Turbulent motions cause the fluctuating waves of density $\rho(\boldsymbol{r},t)$ that satisfy the wave equation

$$\left(\frac{\partial^2}{\partial t^2} - c_0^2 \Delta\right) \rho(\boldsymbol{r},t) = \frac{\partial^2}{\partial r_i \partial r_j} T_{ij}(\boldsymbol{r},t), \tag{10.17}$$

where c_0 is the sound velocity in the homogeneous portion of the medium, i.e., outside the region of turbulent motions.

The solution to this equation has the form of the retarded solution

$$\rho(\boldsymbol{r},t) = \frac{1}{4\pi c_0^2} \frac{\partial^2}{\partial r_i \partial r_j} \int\limits_V d\boldsymbol{y} T_{ij}\left(\boldsymbol{y}, t - \frac{|\boldsymbol{r}-\boldsymbol{y}|}{c_0}\right).$$

For distances significantly exceeding linear sizes of turbulent region V, this solution can be reduced to the asymptotic expression

$$\rho(\boldsymbol{r},t) = \frac{1}{4\pi c_0^4} \frac{r_i r_j}{x^3} \int\limits_V d\boldsymbol{y} \ddot{T}_{ij}\left(\boldsymbol{y}, t - \frac{|\boldsymbol{r}-\boldsymbol{y}|}{c_0}\right), \tag{10.18}$$

where

$$\ddot{T}_{ij}(\boldsymbol{y},t) = \frac{\partial^2}{\partial t^2} T_{ij}(\boldsymbol{y},t).$$

Correspondingly, average energy flux density of acoustic waves excited by turbulent motions

$$q(\boldsymbol{r},t) = \frac{c_0^3 \left\langle \rho^2(\boldsymbol{r},t) \right\rangle}{\rho_0}$$

is given by the expression

$$q(\boldsymbol{r},t) = \frac{1}{16\pi^2 c_0^5} \frac{r_i r_j r_k r_l}{r^6} \times \int\limits_V d\boldsymbol{y} \int\limits_V d\boldsymbol{z} \left\langle \ddot{T}_{ij}\left(\boldsymbol{y}, t - \frac{|\boldsymbol{r}-\boldsymbol{y}|}{c_0}\right) \ddot{T}_{kl}\left(\boldsymbol{z}, t - \frac{|\boldsymbol{r}-\boldsymbol{z}|}{c_0}\right) \right\rangle. \tag{10.19}$$

Further calculations require the knowledge of the space–time correlators of the velocity field. The current theory of strong turbulence is incapable of yielding the corresponding relationships. For this reason, investigators limit themselves to various plausible hypotheses that allow complete calculations to be performed (see, e.g., [58–60]). In particular, it was shown that, with allowance for incompressibility of flow in volume V and the use of the Kolmogorov–Obukhov hypothesis and a number of simplifying hypotheses for splitting space-time correlators, from Eq. (10.19) follows that both average energy flux density and acoustic power are proportional to $\sim M^5$, where

$$M = \frac{\sqrt{\langle v^2(\boldsymbol{r}, t)\rangle}}{c_0}$$

is the *Mach number* (significantly smaller than unity).

We note that this result can be explained purely hydrodynamically, by analyzing vortex interactions in weakly compressible medium [61]. The simplest sound-radiating vortex systems are the pair of vortex lines (radiating cylindrical waves) and the pair of vortex rings (radiating spherical waves).

10.2.1 Sound Radiation by Vortex Lines

Consider two parallel vortex lines separated by distance $2h$ and characterized by equal intensities

$$\kappa = \frac{1}{2}\pi\xi\sigma,$$

where ξ is the vorticity (the size of the vortex uniformly distributed over the area of infinitely small section σ), so that the circulation about each of vortex line is

$$\Gamma = 2\pi\kappa.$$

We will call these vortex lines simply vortices. In a noncompressible flow, these vortices revolve with angular velocity

$$\omega = \frac{\kappa}{2h^2}$$

around the center of the line connecting these vortices (see Fig. 10.1a and, e.g., [62]).

Select the coordinate system with the origin at a fixed point and z-axis along the vortex line. In this coordinate system, velocity potential $\varphi_0(\boldsymbol{r}, t)$ ($\boldsymbol{v}_0(\boldsymbol{r}, t) = -\mathrm{Re}\nabla\varphi_0(\boldsymbol{r}, t)$) and squared velocity $\boldsymbol{v}_0^2(\boldsymbol{r}, t)$ assume the forms

$$\varphi_0(\boldsymbol{r}, t) = ik\ln\left[r^2e^{2i\theta} - h^2e^{2i\omega t}\right],$$
$$v_0^2(\boldsymbol{r}, t) = \frac{4\kappa^2 r^2}{r^4 + h^4 - 2r^2h^2\cos 2(\omega t - \theta)}. \qquad (10.20)$$

Here $re^{i\theta}$ is the radius-vector of the observation point.

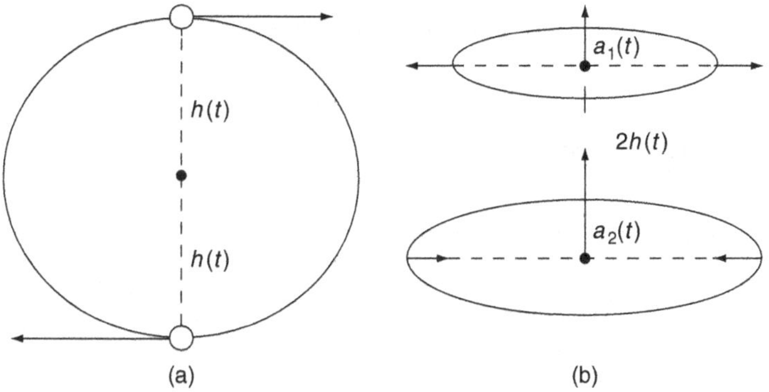

Figure 10.1 Schematics of dynamics of (*a*) vortex lines and (*b*) vortex rings.

According to the Bernoulli equation, local velocity pulsations described by Eq. (10.20) must produce the corresponding pressure pulsations; in the case of weakly compressible medium, namely, under the condition that $M \ll 1$, where $M = \frac{\kappa}{2hc_0}$ is the Mach number and c_0 is the sound velocity, these pressure pulsations will propagate at large distances as sound waves.

Encircle the origin with a circle of radius R such that $h \ll R \ll \lambda$, where λ is the sound wavelength. This is possible because

$$\frac{\lambda}{h} = \frac{\pi}{M} \gg 1 \quad \text{for} \quad M \ll 1.$$

In region $r < R$, dynamics of flow approximately coincides with the dynamics of noncompressible flow. In other words, the dynamics of flow is described by Eqs. (10.20) in this region.

In region $r > R$, equations of motion coincide with the standard equations of acoustics (see, e.g., [63])

$$\left(\frac{\partial^2}{\partial t^2} - c_0^2 \Delta\right) \varphi(\boldsymbol{r}, t) = 0, \quad p'(\boldsymbol{r}, t) = \rho_0 \frac{\partial}{\partial t} \varphi(\boldsymbol{r}, t). \tag{10.21}$$

Here, $\varphi(\boldsymbol{r}, t)$ is the potential of velocity in the acoustic wave, $p'(\boldsymbol{r}, t)$ is the pressure in the wave, and ρ_0 is the density of the medium.

Taking into account the fact that velocity $\boldsymbol{v}_0(\boldsymbol{r}, t)$ depends on t and θ only in combination $2(\omega t - \theta)$, we will seek the solution to Eq. (10.21) in the form

$$\varphi(\boldsymbol{r}, t) = f(r) e^{2i(\omega t - \theta)}, \tag{10.22}$$

Substituting Eq. (10.22) in Eq. (10.21) and solving the resulting equation in function $f(r)$ with allowance for the radiation condition, we obtain

$$\varphi(\boldsymbol{r}, t) = A H_2^2 (2\omega r / c_0)\, e^{2i(\omega t - \theta)}, \tag{10.23}$$

where $H_2^2(z)$ is the Hankel function of the second kind and A is a constant.

For $r \gg c_0/2\omega$ we have the standard divergent cylindrical wave with wavelength $\lambda = \pi c_0/\omega$

$$\varphi(\boldsymbol{r}, t) = A\sqrt{\frac{c}{\pi\omega r}} \exp\{2i(\omega t - \omega r/c_0 - \theta - 3\pi/8)\}.$$

In the opposite case $r \ll \lambda$, we obtain

$$\varphi(\boldsymbol{r}, t) = i\frac{A}{\pi}\left(\frac{c}{\omega r}\right)^2 e^{2i(\omega t-\theta)}.$$

Potential $\varphi(\boldsymbol{r}, t)$ in region $h \ll r \ll \lambda$ must coincide with the oscillating portion of potential $\varphi_0(\boldsymbol{r}, t)$ (see, e.g., [63]), i.e., with

$$\varphi_0^{(1)}(\boldsymbol{r}, t) = -i\kappa\frac{h^2}{r^2}e^{2i(\omega t-\theta)}.$$

This condition yields the following expression for constant A

$$A = -\pi\kappa M^2 = -\frac{\pi\kappa^3}{4h^2c_0^2}.$$

Consequently, potential $\varphi(\boldsymbol{r}, t)$ in the wave zone assumes the form

$$\varphi(\boldsymbol{r}, t) = -\kappa M^{3/2}\sqrt{\frac{\pi h}{r}} \exp\{2i(\omega t - \omega r/c_0 - \theta - 3\pi/8)\}. \tag{10.24}$$

The pressure in sound waves can be determined from potential (10.24) using Eq. (10.21). The result is as follows

$$p'(\boldsymbol{r}, t) = -2\kappa\omega\rho_0 M^{3/2}\sqrt{\frac{\pi h}{r}} \exp\{2i(\omega t - \omega r/c_0 - \theta - 3\pi/8)\}.$$

The sound intensity (energy radiated per unit time) can be obtained by integrating along the circle of radius $R \gg \lambda$

$$I = \frac{c_0}{\rho_0}\oint dl\left\langle\left[p'(r,t)\right]^2\right\rangle = 2\pi^2\rho_0 M^4\frac{\kappa^3}{h^2}. \tag{10.25}$$

The radiated energy must coincide with the interaction energy of vortices located in region $r < R$. Total energy in region $r < R$ is

$$E = \frac{\rho_0}{2}\int dS v_0^2(\boldsymbol{r}, t). \tag{10.26}$$

Substituting Eq. (10.20) in (10.26) and discarding infinite terms corresponding to the energy of motion of vortices themselves (we assume vortices the point vortices), we obtain the interaction energy in the form

$$E_1 = 4\pi\kappa^2\rho_0 \ln(R/h). \tag{10.27}$$

The interaction energy can vary only at the expense of varying the distance between vortices ($h = h(t)$), because the circulation remains intact due to the fact that we consider nonviscous medium.

Differentiating Eq. (10.27) with respect to time, we obtain energy variation rate

$$I(t) = -4\pi\rho_0 \frac{\kappa^2}{h(t)} \frac{d}{dt} h(t), \tag{10.28}$$

which is just transferred into the energy of acoustic waves. Using Eq. (10.25) and (10.28), we obtain that the distance between vortices satisfies the equation

$$\frac{d}{dt} h(t) = \frac{\pi\kappa M^4}{2h(t)}. \tag{10.29}$$

Integrating Eq. (10.29) with allowance for the fact that

$$M = M(t) = \frac{\kappa}{2h(t)c_0},$$

we obtain

$$h(t) = h_0 \left[1 + 6\pi M_0^4 \omega_0 t\right]^{1/6}.$$

Thus, the intensity of radiated sound is proportional to M^4, $I \sim M^4$. It is obvious that, in the case of statistically distributed system of vortex line pairs, this estimate will remain valid for certain portion of the plane.

10.2.2 Sound Radiation by Vortex Rings

In a noncompressible flow, a vortex ring of intensity κ causes the flow to move with a velocity

$$\boldsymbol{v}(\boldsymbol{r}, t) = \frac{\kappa}{2} a \int\limits_0^{2\pi} d\phi \frac{\boldsymbol{s} \times \boldsymbol{r}}{r^3},$$

which follows from the the *Biot–Savart law* (see, e.g. [62]). Here, $\boldsymbol{s}$ is the unit vector tangent to the vortex ring (it is directed along the vortex vector), a is the radius of the ring, and $\boldsymbol{r}$ is the vector specifying observation point position relative to points lying on the ring.

In the cylindrical coordinate system with origin at the center of ring and z-axis directed along the ring axis, we have

$$v_R = \frac{\kappa}{2}a\int\limits_0^{2\pi} d\phi \frac{\cos\phi}{r^3}, \quad v_\theta = 0, \quad v_z = \frac{\kappa}{2}a\int\limits_0^{2\pi} d\phi \frac{a - R\cos\phi}{r^3}, \tag{10.30}$$

where

$$r = \left(R^2 + z^2 + a^2 - 2Ra\cos\phi\right)^{1/2}.$$

Here (R, θ, z) are the coordinates of the radius-vector of the observation point.

We now have two vortex rings of equal intensities and equal radii a_0 at distance $2h_0$. In this case, the front ring will increase in size, while the rear ring will decrease and pursue the front ring (Fig. 10.1b). At a certain instant, it will penetrate through the front ring, and the rings switch places. This phenomenon is called the *game of vortex rings*. In a weakly compressible flow, these motions of rings produce local regions of compression and rarefaction; these regions propagate in the medium and, for large distances assume the form of spherical acoustic waves. To determine the structure of radiated sound, we must know relative motions of rings in a noncompressible flow.

Assume rings have radii $a_1(t)$ and $a_2(t)$ and are separated by distance $2h(t)$ at certain instant t. The rates of variations of ring radii are equal to radial velocities that rings induce at each other, and the rate of variation of the distance between the rings is equal to the difference of the z-components of velocities induced by the rings. Consequently, we have

$$\begin{aligned}
\frac{d}{dt}a_1(t) &= 2\kappa a_2(t)h(t)\int\limits_0^{\pi} d\phi \frac{\cos\phi}{|a_1(t) - a_2(t)|^3}, \\
\frac{d}{dt}a_2(t) &= -2\kappa a_1(t)h(t)\int\limits_0^{\pi} d\phi \frac{\cos\phi}{|a_1(t) - a_2(t)|^3}, \\
\frac{d}{dt}h(t) &= -\frac{\kappa}{2}\left[a_1^2(t) - a_2^2(t)\right]\int\limits_0^{\pi} d\phi \frac{1}{|a_1(t) - a_2(t)|^3},
\end{aligned} \tag{10.31}$$

where

$$|a_1(t) - a_2(t)| = \left[a_1^2(t) + a_2^2(t) + 4h^2(t) - 2a_1(t)a_2(t)\cos\phi\right]^{1/2}.$$

Equations (10.31) should be solved with the initial conditions at $t = 0$

$$a_1(0) = a_2(0) = a_0, \quad h(0) = h_0.$$

The first pair of equations immediately yields the relationship between $a_1(t)$ and $a_2(t)$; namely,

$$a_1^2(t) + a_2^2(t) = 2a_0^2. \tag{10.32}$$

This relationship shows that the moment of inertia of rings relative to the z-axis is conserved.

The integrals in the right-hand side of Eq. (10.31) can be expressed in terms of elliptic functions.

If rings are far from each other ($\gamma = h_0/a_0 \gg 1$), they interact only slightly, and we can assume that, in the first approximation, they move independently with the velocities determined by the ring areas. In the opposite limiting case $\gamma \ll 1$ (namely this case will be considered in what follows), the rings interact actively. The integrals in the right-hand side of Eq.(10.31) are mainly contributed by the neighborhood of point $\phi = 0$. For this reason, we can replace $\cos\phi$ in the numerators of the first pair of equations by unity. Thus we obtain the second integral of motion

$$4h^2(t) + [a_1(t) - a_2(t)]^2 = 4h_0^2. \tag{10.33}$$

Integral (10.33) means that the distance between points on different rings at the same polar angle is the conserved quantity.

In view of the existence of integrals (10.32) and (10.33), we can reduce system (10.31) to a single equation in variable $\theta(t)$ determined by the equalities

$$a_1(t) = \sqrt{2}a_0 \cos\left(\frac{\pi}{4} - \gamma \sin\theta(t)\right),\ a_2(t) = \sqrt{2}a_0 \sin\left(\frac{\pi}{4} - \gamma \sin\theta(t)\right),$$
$$h(t) = h_0 \cos\theta(t). \tag{10.34}$$

With this definition, conservation laws (10.32) and (10.33) are satisfied automatically (in the first order with respect to γ). Substituting Eq. (10.34) in Eq. (10.31), expanding the result in the series, and calculating the integral, we obtain

$$\theta(t) = \frac{\kappa}{2h_0^2}t.$$

Consequently, the ring radii and the distance between rings are given by the expression

$$a_1(t) = a_0\left(1 + \gamma \sin(\omega t)\right),\ a_2(t) = a_0\left(1 - \gamma \sin(\omega t)\right),$$
$$h(t) = h_0 \cos(\omega t), \tag{10.35}$$

where

$$\omega = \frac{\kappa}{2h_0^2}, \tag{10.36}$$

as in the above case of vortex lines.

We note that the infinitely thin rings move with an infinite velocity. However, actual vortex rings move with a finite velocity significantly smaller than the sound velocity, but the dynamics of relative motion of rings only slightly differs from that we just obtained. The fact that angular velocity (10.36) coincides with the corresponding angular velocity in the case of vortex lines shows that the points lying on rings at equal polar angles revolve relative the center point of the line connecting them at the rotational speed coinciding with the rotational speed of vortex lines located at the same distance and having the same intensity as the rings.

To study the structure of sound radiated by the system of rings, we need to know the velocity field for large distances from the system. We associate the coordinate system with the point located at the center of the segment connecting the centers of the rings. The velocity of flow outside the rings is given by the formula

$$\boldsymbol{v}(\boldsymbol{r},t)=\frac{\kappa}{2}a_1(t)\int\limits_0^{2\pi}d\phi\frac{\boldsymbol{s}_1\times\boldsymbol{r}_1}{r_1^3}+\frac{\kappa}{2}a_2(t)\int\limits_0^{2\pi}d\phi\frac{\boldsymbol{s}_2\times\boldsymbol{r}_2}{r_2^3}. \tag{10.37}$$

Here, $\boldsymbol{s}_1$ and $\boldsymbol{s}_2$ are the unit vectors tangent to the vortex rings and $\boldsymbol{r}_1$ and $\boldsymbol{r}_2$ are the vectors specifying observation point position relative to points lying on the rings. For large distances from the rings, the oscillating parts of the velocity have the form (in the cylindrical coordinates)

$$\begin{aligned} v_R^{(1)}(t)&=\frac{3\kappa}{4}\frac{R\left(4z^2-R^2\right)}{(R^2+z^2)^{3/2}}h\left(a_1^2-a_2^2\right),\quad v_\theta^{(1)}(t)=0,\\ v_z^{(1)}(t)&=\frac{3\kappa}{4}\frac{z\left(2z^2-3R^2\right)}{(R^2+z^2)^{3/2}}h\left(a_1^2-a_2^2\right). \end{aligned} \tag{10.38}$$

Introducing potential by the formula $\boldsymbol{v}^{(1)}=-\nabla\varphi^{(1)}$, we obtain

$$\varphi^{(1)}(t)=-\frac{\kappa}{4}\frac{R^2-2z^2}{(R^2+z^2)^{5/2}}h\left(a_1^2-a_2^2\right). \tag{10.39}$$

Substituting Eq. (10.35) in Eq. (10.39) and changing to spherical coordinates, we can rewrite Eq. (10.39) in the complex form

$$\varphi^{(1)}(t)=i\frac{\kappa}{2}a_0^3\gamma^2\frac{1-3\cos^2\theta}{r^3}e^{2i\omega t}. \tag{10.40}$$

Now, we will proceed similar to the case of two vortex lines. We encircle the origin by a sphere of radius L such that $a_0\ll L\ll\lambda$, where λ is the wavelength of radiated sound waves. We will use the fact that the motion of flow inside the sphere approximately coincides with the motion of noncompressible flow. Outside the sphere, the equations of motion will have the form of Eq. (10.21) with the only difference that now Δ is the spatial Laplace operator.

Represent potential φ in the form

$$\varphi(r,\theta) = f(r,\theta)e^{2i\omega t}.$$

Substituting this expression in Eq. (10.21) and taking into account that we must obtain divergent spherical waves for $r \gg \lambda$, we obtain $f(r,\theta)$ in the form

$$f(r,\theta) = \sum_{n=0}^{\infty} \frac{A_n}{\sqrt{r}} H^{(2)}_{n+1/2}\left(\frac{2\omega r}{c_0}\right) P_n(\cos\theta),$$

where $H^{(2)}_{n+1/2}(z)$ is the Hankel function of the second kind and $P_n(z)$ is the Legendre polynomial. Comparing potential $\varphi(r,\theta)$ for $r \ll \lambda$ with $\varphi^{(1)}(r,\theta)$, we obtain that $n = 2$ and

$$A_2 = \frac{2\sqrt{\pi}}{3} \frac{\kappa\gamma^2 a_0^3 \omega^{5/2}}{c_0^{5/2}}.$$

Consequently, potential and pressure in the wave zone have the forms

$$\varphi(r,\theta) = \frac{2\kappa}{3}\left(\frac{\kappa}{2h_0c_0}\right)^2 \frac{a_0}{r}\left(1 - 3\cos^2\theta\right)\exp\left\{2i\left(\omega t - \frac{\omega r}{c_0} - \frac{3\pi}{4}\right)\right\},$$

$$p'(r,\theta) = \frac{4i\kappa}{3}\left(\frac{\kappa}{2h_0c_0}\right)^2 \frac{a_0\omega}{r}\rho_0\left(1 - 3\cos^2\theta\right)\exp\left\{2i\left(\omega t - \frac{\omega r}{c_0} - \frac{3\pi}{4}\right)\right\},$$

so that the angular distribution of energy radiated per unit time is

$$I(\theta) = \frac{8\pi}{9}\frac{\kappa^3 a_0^2}{h_0^3}\left(\frac{\kappa}{2h_0c_0}\right)^5 \rho_0\left(1 - 3\cos^2\theta\right)^2.$$

Vortex rings radiate energy as a quadrupole; the main portion of energy is radiated in a cone about z-axis with corner angle 106° in both positive and negative z-directions. Integrating over θ, we obtain that total energy radiated per unit time is given by the expression

$$I = \frac{64\pi}{45}\frac{\kappa^3 a_0^2}{h_0^3}\left(\frac{\kappa}{2h_0c_0}\right)^5 \rho_0.$$

Thus, intensity of sound radiated by a pair of vortex rings is proportional to M^5. In the case of statistically distributed system of pairs of vortex rings, this proportionality will remain valid for certain portion of space, which agrees with estimates obtained in [58–60].

Part III

Examples of Coherent Phenomena in Stochastic Dynamic Systems

Lecture 11

Passive Tracer Clustering and Diffusion in Random Hydrodynamic and Magnetohydrodynamic Flows

11.1 General Remarks

Diffusion of such passive fields as scalar density (particle concentration) field and magnetic field is an important problem of the theory of turbulence in magnetohydrodynamics. The basic stochastic equations for the scalar density field $\rho(\boldsymbol{r}, t)$ and the magnetic field $\boldsymbol{H}(\boldsymbol{r}, t)$ are the continuity equation

$$\left(\frac{\partial}{\partial t} + \frac{\partial}{\partial \boldsymbol{r}} \boldsymbol{U}(\boldsymbol{r}, t)\right) \rho(\boldsymbol{r}, t) = \mu \Delta \rho(\boldsymbol{r}, t), \qquad \rho(\boldsymbol{r}, 0) = \rho_0(\boldsymbol{r}), \tag{11.1}$$

and the induction equation (1.59), page 30, subject to the dissipative effects

$$\left(\frac{\partial}{\partial t} + \frac{\partial}{\partial \boldsymbol{r}} \boldsymbol{U}(\boldsymbol{r}, t)\right) \boldsymbol{H}(\boldsymbol{r}, t) = \left(\boldsymbol{H}(\boldsymbol{r}, t) \cdot \frac{\partial}{\partial \boldsymbol{r}}\right) \boldsymbol{U}(\boldsymbol{r}, t) + \mu_H \Delta \boldsymbol{H}(\boldsymbol{r}, t), \tag{11.2}$$

where μ_ρ is the dynamic molecular diffusion coefficient, and $\mu_H = c^2/4\pi\sigma$ is the dynamic diffusion coefficient associated with the conductivity of the flow σ.

In these equations, $\boldsymbol{U}(\boldsymbol{r}, t) = \boldsymbol{u}_0(\boldsymbol{r}, t) + \boldsymbol{u}(\boldsymbol{r}, t)$; $\boldsymbol{u}_0(\boldsymbol{r}, t)$ is the deterministic component of the velocity field (mean flow), and $\boldsymbol{u}(\boldsymbol{r}, t)$ is the random component. In the general case, random field $\boldsymbol{u}(\boldsymbol{r}, t)$ can be composed of both solenoidal (for which $\operatorname{div} \boldsymbol{u}(\boldsymbol{r}, t) = 0$) and potential (for which $\operatorname{div} \boldsymbol{u}(\boldsymbol{r}, t) \neq 0$) components. Field $\boldsymbol{H}(\boldsymbol{r}, t)$ is here nondivergent, i.e., $\operatorname{div} \boldsymbol{H}(\boldsymbol{r}, t) = 0$.

Dynamic systems (11.1) and (11.2) are conservative and both the total scalar mass

$$M = M(t) = \int d\boldsymbol{r} \rho(\boldsymbol{r}, t) = \int d\boldsymbol{r} \rho_0(\boldsymbol{r}) = \text{const}$$

Lectures on Dynamics of Stochastic Systems. DOI: 10.1016/B978-0-12-384966-3.00011-8

and the magnetic flux

$$\int d\mathbf{r}\, \mathbf{H}(\mathbf{r}, t) = \int d\mathbf{r}\, \mathbf{H}_0(\mathbf{r}) = \text{const}$$

remain constant during the evolution. For homogeneous initial conditions, the following equalities are a corollary of the conservatism of dynamic systems (11.1) and (11.2):

$$\langle \rho(\mathbf{r}, t)\rangle = \rho_0, \quad \langle \mathbf{H}(\mathbf{r}, t)\rangle = \mathbf{H}_0,$$

where $\langle \cdots \rangle$ denotes averaging over an ensemble of realizations of random field $\{u(\mathbf{r}, t)\}$. If initial conditions are homogeneous, a statistical average of a quantity depending, for example, on magnetic field $\int d\mathbf{r}\, \langle f(\mathbf{H}(\mathbf{r}, t))\rangle$ changes over in the general case to $\langle f(\mathbf{H}(\mathbf{r}, t))\rangle$,

$$\int d\mathbf{r}\, \langle f(\mathbf{H}(\mathbf{r}, t))\rangle \Leftrightarrow \langle f(\mathbf{H}(\mathbf{r}, t))\rangle,$$

which means that quantity $\langle f(\mathbf{H}(\mathbf{r}, t))\rangle$ is a specific quantity per unit volume and is, hence, an integral quantity. For example, quantity $\langle \mathbf{H}^2(\mathbf{r}, t)\rangle$ is the average specific energy per unit volume in the case of homogeneous initial conditions, but it is the average density of energy in the case of inhomogeneous initial conditions. Integral quantities characterize dynamic systems at large, they allow separating the processes of field generation, which offers a possibility of discarding dynamic details related to advection of these quantities by random velocity field [64].

Our interest consists in the temporal evolution of different one-point statistical characteristics of density field $\rho(\mathbf{r}, t)$, its gradient $\mathbf{p}(\mathbf{r}, t) = \nabla\rho(\mathbf{r}, t)$, magnetic field $\mathbf{H}(\mathbf{r}, t)$, and its different spatial derivatives from given initial distributions $\rho_0(\mathbf{r})$ and $\mathbf{H}_0(\mathbf{r})$ (and, in particular, homogeneous ones, $\rho_0(\mathbf{r}) = \rho_0$ and $\mathbf{H}_0(\mathbf{r}) = \mathbf{H}_0$). Such characteristics are probability densities of these fields by themselves, isotropization (decrease of anisotropy) of vector correlations, helicity of these vector fields, and their dissipation.

The effect of dynamic diffusion can be neglected during the initial stages of the development of diffusion. In this case, Eqs. (11.1) and (11.2) become simpler and assume the form

$$\left(\frac{\partial}{\partial t} + \mathbf{U}(\mathbf{r}, t)\frac{\partial}{\partial \mathbf{r}}\right)\rho(\mathbf{r}, t) + \frac{\partial \mathbf{U}(\mathbf{r}, t)}{\partial \mathbf{r}}\rho(\mathbf{r}, t) = 0, \tag{11.3}$$

$$\left(\frac{\partial}{\partial t} + \mathbf{U}(\mathbf{r}, t)\frac{\partial}{\partial \mathbf{r}}\right)\mathbf{H}(\mathbf{r}, t) + \frac{\partial \mathbf{U}(\mathbf{r}, t)}{\partial \mathbf{r}}\mathbf{H}(\mathbf{r}, t) = \left(\mathbf{H}(\mathbf{r}, t)\cdot\frac{\partial}{\partial \mathbf{r}}\right)\mathbf{U}(\mathbf{r}, t). \tag{11.4}$$

In these conditions, such nonstationary and stochastic phenomena occur as mixing, generation and rapid increase (in time) of smaller-scale disturbances (e.g., stochastic (turbulent) dynamo), and, in certain situations, clustering in the phase or physical space. Clustering of a field (such as tracer density, magnetic field energy, etc.) is the appearance of compact regions, where the content of this field is increased, which are surrounded by regions with decreased content of this field.

For example, in the case of homogeneous initial conditions, the density field remains constant in time if liquid is noncompressible (nondivergent velocity field) and, consequently, the gradient of density is identically equal to zero at any spatial point. On the contrary, in compressible flows (divergent velocity field), density field is always clustered, which naturally results in great gradients of the density field. We have seen a similar pattern in the example of the simplest problem whose solution is given in Fig. 1.12, page 27. The characteristic time of the formation of cluster structure of the density field is determined by the relationship $D^{\mathrm{p}} t \sim 1$, where quantity D^{p} given by Eq. (4.59), page 108, depends on the potential component of spectral density of the velocity field.

As for the magnetic field, clustering of its energy can occur under certain conditions, and we have seen an example of such clustering in Fig. 1.14, page 33, in the case of the simplest potential velocity field. However, general generation of magnetic field occurs even in noncompressible flows of liquid; this generation causes intense variability of the field structure, and different moment functions of both magnetic field by itself and its spatial derivatives increase exponentially in time. As a result, field dissipation related to higher-order derivatives is rapidly increased at a certain time instant, and the effects of dynamic diffusion become governing.

The above equations correspond to the *Eulerian description* of the evolution of the density field and magnetic field.

Equations (11.3), (11.4) are the first-order partial differential equations and can be solved by the method of characteristics. Introducing characteristic curves $\boldsymbol{r}(t)$ satisfying the equations of particle motion

$$\frac{d}{dt}\boldsymbol{r}(t) = \boldsymbol{U}(\boldsymbol{r}, t), \qquad \boldsymbol{r}(0) = \boldsymbol{r}_0, \tag{11.5}$$

we can change over from Eq. (11.3) to the ordinary differential equation

$$\frac{d}{dt}\rho(t) = -\frac{\partial \boldsymbol{U}(\boldsymbol{r}, t)}{\partial \boldsymbol{r}}\rho(t), \quad \rho(0) = \rho_0(\boldsymbol{r}_0). \tag{11.6}$$

Solutions to Eqs. (11.5) and (11.6) have an obvious geometric interpretation. They describe the concentration behavior around a fixed tracer particle moving along trajectory $\boldsymbol{r} = \boldsymbol{r}(t)$. As may be seen from Eq. (11.6), the concentration in divergent flows varies: it increases in regions where medium is dense and decreases in regions where medium is rarefied.

Solutions to system (11.5), (11.6) depend on characteristic parameter $\boldsymbol{r}_0$ (the initial coordinate of the particle)

$$\boldsymbol{r}(t) = \boldsymbol{r}(t|\boldsymbol{r}_0), \qquad \rho(t) = \rho(t|\boldsymbol{r}_0), \tag{11.7}$$

which we will separate by the bar. Components of vector $\boldsymbol{r}_0$ are called the *Lagrangian coordinates* of a particle; they unambiguously specify the position of arbitrary particle. Equations (11.5), (11.6) correspond in this case to the *Lagrangian description* of concentration evolution. The first of equalities (11.7) specify the relationship between the Eulerian and Lagrangian descriptions. Solving it in $\boldsymbol{r}_0$, we obtain the relationship that expresses the Lagrangian coordinates in terms of the Eulerian ones

$$\boldsymbol{r}_0 = \boldsymbol{r}_0(\boldsymbol{r}, t). \tag{11.8}$$

Then, using Eq. (11.8), to eliminate $\boldsymbol{r}_0$ in the last equality in (11.7), we turn back to the concentration in the *Eulerian description*

$$\rho(\boldsymbol{r}, t) = \rho(t|\boldsymbol{r}_0(\boldsymbol{r}, t)) = \int d\boldsymbol{r}_0 \rho(t|\boldsymbol{r}_0) j(t|\boldsymbol{r}_0) \delta\left(\boldsymbol{r}(t|\boldsymbol{r}_0) - \boldsymbol{r}\right), \tag{11.9}$$

where we introduced a new function called *divergence*

$$j(t|\boldsymbol{r}_0) = \det \|j_{ik}(t|\boldsymbol{r}_0)\| = \det \left\| \frac{\partial r_i(t|\boldsymbol{r}_0)}{\partial r_{0k}} \right\|,$$

which is the quantitative measure of the degree of compression (extension) of physically infinitely small liquid particles. One can easily obtain that it satisfies the equation

$$\frac{d}{dt} j(t|\boldsymbol{r}_0) = \frac{\partial \boldsymbol{U}(\boldsymbol{r}, t)}{\partial \boldsymbol{r}} j(t|\boldsymbol{r}_0), \quad j(0|\boldsymbol{r}_0) = 1. \tag{11.10}$$

Comparing Eq. (11.6) with Eq. (11.10), we see that

$$\rho(t|\boldsymbol{r}_0) = \frac{\rho_0(\boldsymbol{r}_0)}{j(t|\boldsymbol{r}_0)}. \tag{11.11}$$

Thus, we can rewrite Eq. (11.9) as the equality

$$\rho(\boldsymbol{r}, t) = \int d\boldsymbol{r}_0 \rho_0(\boldsymbol{r}_0) \delta\left(\boldsymbol{r}(t|\boldsymbol{r}_0) - \boldsymbol{r}\right) \tag{11.12}$$

specifying the relationship between the Lagrangian and Eulerian characteristics. Delta function in the right-hand side of Eq. (11.12) is the *indicator function* for the position of the Lagrangian particle; as a consequence, after averaging Eq. (11.12) over an ensemble of realizations of random velocity field, we arrive at the well-known relationship between the average concentration in the Eulerian description and the one-time probability density

$$P(\boldsymbol{r}, t|\boldsymbol{r}_0) = \left\langle \delta\left(\boldsymbol{r}(t|\boldsymbol{r}_0) - \boldsymbol{r}\right) \right\rangle$$

of the Lagrangian particle

$$\langle \rho(\boldsymbol{r}, t) \rangle = \int d\boldsymbol{r}_0 \rho_0(\boldsymbol{r}_0) P(\boldsymbol{r}, t|\boldsymbol{r}_0).$$

The relationship between the spatial correlation function of the concentration field in the Eulerian description

$$\langle R(\boldsymbol{r}_1, \boldsymbol{r}_2, t)\rangle = \langle \rho(\boldsymbol{r}_1, t)\rho(\boldsymbol{r}_2, t)\rangle,$$

where $R(\boldsymbol{r}_1, \boldsymbol{r}_2, t) = \rho(\boldsymbol{r}_1, t)\rho(\boldsymbol{r}_2, t)$, and the joint probability density function of positions of two particles

$$P(\boldsymbol{r}_1, \boldsymbol{r}_2, t|\boldsymbol{r}_{01}, \boldsymbol{r}_{02}) = \langle \delta\left(\boldsymbol{r}_1(t|\boldsymbol{r}_{01}) - \boldsymbol{r}_1\right)\delta\left(\boldsymbol{r}_2(t|\boldsymbol{r}_{02}) - \boldsymbol{r}_2\right)\rangle$$

can be obtained similarly

$$\langle R(\boldsymbol{r}_1, \boldsymbol{r}_2, t)\rangle = \int d\boldsymbol{r}_{01} \int d\boldsymbol{r}_{02}\rho_0(\boldsymbol{r}_{01})\rho_0(\boldsymbol{r}_{02})P(\boldsymbol{r}_1, \boldsymbol{r}_2, t|\boldsymbol{r}_{01}, \boldsymbol{r}_{02}).$$

For the nondivergent velocity field ($\operatorname{div}\boldsymbol{U}(\boldsymbol{r}, t) = 0$), both particle divergence and particle concentration are invariant, i.e.,

$$j(t|\boldsymbol{r}_0) = 1, \qquad \rho(t|\boldsymbol{r}_0) = \rho_0(\boldsymbol{r}_0).$$

We assume in the general case that the random component of the velocity field is the divergent ($\operatorname{div}\boldsymbol{u}(\boldsymbol{r}, t) \neq 0$) statistically homogeneous Gaussian random field, which is spherically symmetric (but generally not possessing reflection symmetry) in space and stationary in time with zero-valued average ($\langle \boldsymbol{u}(\boldsymbol{r}, t)\rangle = 0$) and correlation and spectral tensors given by the formulas

$$B_{ij}(\boldsymbol{r} - \boldsymbol{r}_1, \tau) = \left\langle u_i(\boldsymbol{r}, t)u_j(\boldsymbol{r}_1, t_1)\right\rangle = \int d\boldsymbol{k}\, E_{ij}(\boldsymbol{k}, \tau)e^{i\boldsymbol{k}(\boldsymbol{r}-\boldsymbol{r}_1)},$$

$$E_{ij}(\boldsymbol{k}, \tau) = \frac{1}{(2\pi)^d}\int d\boldsymbol{r}\, B_{ij}(\boldsymbol{r}, \tau)e^{-i\boldsymbol{k}\boldsymbol{r}},$$

where d is the dimension of space. In view of the suggested symmetry condition, correlation tensor $B_{ij}(\boldsymbol{r}, \tau)$ has vector structure (4.49), page 105, ($\boldsymbol{r} - \boldsymbol{r}_1 \to \boldsymbol{r}$)

$$B_{ij}(\boldsymbol{r}, \tau) = B_{ij}^{\mathrm{iso}}(\boldsymbol{r}, \tau) + C(r, \tau)\varepsilon_{ijk}r_k,$$

where

$$B_{ij}^{\mathrm{iso}}(\boldsymbol{r}, \tau) = A(r, \tau)r_ir_j + B(r, \tau)\delta_{ij}$$

is the isotropic portion of the correlation tensor and ε_{ijk} is the pseudotensor antisymmetric with respect to an arbitrary pair of indices. The isotropic portion of the correlation tensor corresponds to the spatial spectral tensor of the form

$$E_{ij}(\boldsymbol{k}, \tau) = E_{ij}^{\mathrm{s}}(\boldsymbol{k}, \tau) + E_{ij}^{\mathrm{p}}(\boldsymbol{k}, \tau),$$

where the spectral components of the tensor of velocity field have the following structure

$$E_{ij}(\boldsymbol{k},\tau) = E^{\mathrm{s}}(k,\tau)\left(\delta_{ij} - \frac{k_i k_j}{k^2}\right), \quad E^{\mathrm{p}}_{ij}(\boldsymbol{k},\tau) = E^{\mathrm{p}}(k,\tau)\frac{k_i k_j}{k^2}.$$

Here, $E^{\mathrm{s}}(k,t)$ and $E^{\mathrm{p}}(k,t)$ are the solenoidal and potential components of the spectral density of the velocity field, respectively.

Calculating statistical properties of the concentration field and its gradient, we will approximate the velocity field $\boldsymbol{u}(\boldsymbol{r},t)$ by the random delta-correlated (in time) process. In the framework of this approximation, correlation tensor $B_{ij}(\boldsymbol{r},t)$ is approximated by the expression

$$B_{ij}(\boldsymbol{r},t) = 2B_{ij}(\boldsymbol{r})\delta(t), \tag{11.13}$$

where

$$B_{ij}(\boldsymbol{r}) = \frac{1}{2}\int\limits_{-\infty}^{\infty} d\tau B_{ij}(\boldsymbol{r},\tau) = \int\limits_{0}^{\infty} d\tau B_{ij}(\boldsymbol{r},\tau).$$

For the Gaussian field $\boldsymbol{u}(\boldsymbol{r},t)$ correlation with functionals of it can be split by the Furutsu–Novikov formula (8.10), page 193; in the case of delta-correlated field $\boldsymbol{u}(\boldsymbol{r},t)$ this formula becomes simpler and assumes the form

$$\langle u_k(\boldsymbol{r},t)R[t;\boldsymbol{u}(\boldsymbol{r},\tau)]\rangle = \int d\boldsymbol{r}' B_{kl}(\boldsymbol{r}-\boldsymbol{r}')\left\langle \frac{\delta R[t;\boldsymbol{u}(\boldsymbol{r},\tau)]}{\delta u_l(\boldsymbol{r}',t-0)}\right\rangle, \tag{11.14}$$

where $0 \le \tau \le t$.

11.2 Particle Diffusion in Random Velocity Field

11.2.1 One-Point Statistical Characteristics

Thus, in the Lagrangian representation, the behavior of passive tracer is described in terms of ordinary differential Equations (11.5), (11.6), (11.10). We can easily pass on from these equations to the linear Liouville equation in the corresponding phase space. With this goal in view, introduce the indicator function

$$\varphi_{\mathrm{Lag}}(\boldsymbol{r},\rho,j,t|\boldsymbol{r}_0) = \delta(\boldsymbol{r}(t|\boldsymbol{r}_0)-\boldsymbol{r})\delta(\rho(t|\boldsymbol{r}_0)-\rho)\delta(j(t|\boldsymbol{r}_0)-j), \tag{11.15}$$

where we explicitly emphasized the fact that the solution to the initial dynamic equations depends on the Lagrangian coordinates $\boldsymbol{r}_0$. Differentiating Eq. (11.15) with respect to time and using Eqs. (11.5), (11.6), and (11.10), we arrive at the Liouville equation equivalent to the initial problem

$$\left(\frac{\partial}{\partial t}+\frac{\partial}{\partial \boldsymbol{r}}\boldsymbol{U}(\boldsymbol{r},t)\right)\varphi_{\mathrm{Lag}}(\boldsymbol{r},\rho,j,t|\boldsymbol{r}_0)=\frac{\partial \boldsymbol{U}(\boldsymbol{r},t)}{\partial \boldsymbol{r}}\left(\frac{\partial}{\partial \rho}\rho-\frac{\partial}{\partial j}j\right)\varphi_{\mathrm{Lag}}(\boldsymbol{r},\rho,j,t|\boldsymbol{r}_0),$$
$$\varphi_{\mathrm{Lag}}(\boldsymbol{r},\rho,j,t|\boldsymbol{r}_0)=\delta(\boldsymbol{r}_0-\boldsymbol{r})\delta(\rho_0(\boldsymbol{r}_0)-\rho)\delta(j-1). \tag{11.16}$$

The one-time probability density of the solutions to dynamic problems (11.5), (11.6), and (11.10) coincides with the indicator function averaged over an ensemble of realizations

$$P(\boldsymbol{r},\rho,j,t|\boldsymbol{r}_0)=\left\langle\varphi_{\mathrm{Lag}}(t;\boldsymbol{r},\rho,j|\boldsymbol{r}_0)\right\rangle.$$

Averaging Eq. (11.16) over an ensemble of realizations of random field $\boldsymbol{u}(\boldsymbol{r},t)$, using the Furutsu–Novikov formula (11.14), and taking into account the equality

$$\frac{\delta}{\delta u_\beta(\boldsymbol{r}',t-0)}\varphi_{\mathrm{Lag}}(\boldsymbol{r},\rho,j,t|\boldsymbol{r}_0)$$
$$=\left\{-\frac{\partial}{\partial r_\beta}\delta(\boldsymbol{r}-\boldsymbol{r}')+\frac{\partial\delta(\boldsymbol{r}-\boldsymbol{r}')}{\partial r_\beta}\left(\frac{\partial}{\partial \rho}\rho-\frac{\partial}{\partial j}j\right)\right\}\varphi_{\mathrm{Lag}}(t;\boldsymbol{r},\rho,j|\boldsymbol{r}_0),$$

and Eqs. (4.57), page 107, we arrive at the Fokker–Planck equation for the one-time Lagrangian probability density $P(\boldsymbol{r},\rho,j,t|\boldsymbol{r}_0)$ of particle coordinate $\boldsymbol{r}(t|\boldsymbol{r}_0)$, of particle density $\rho(t|\boldsymbol{r}_0)$ and of the particle divergence $j(t|\boldsymbol{r}_0)$:

$$\left(\frac{\partial}{\partial t}-D_0\Delta\right)P(\boldsymbol{r},\rho,j,t|\boldsymbol{r}_0)=D^{\mathrm{p}}\left(\frac{\partial}{\partial \rho}\rho^2\frac{\partial}{\partial \rho}-2\frac{\partial^2}{\partial\rho\partial j}\rho j+\frac{\partial^2}{\partial j^2}j^2\right)P(\boldsymbol{r},\rho,j,t|\boldsymbol{r}_0),$$
$$P(\boldsymbol{r},\rho,j,0|\boldsymbol{r}_0)=\delta(\boldsymbol{r}-\boldsymbol{r}_0)\delta(\rho-\rho_0(\boldsymbol{r}_0))\delta(j-1). \tag{11.17}$$

The solution to Eq. (11.17) has the form

$$P(\boldsymbol{r},\rho,j,t|\boldsymbol{r}_0)=P(\boldsymbol{r},t|\boldsymbol{r}_0)P(j,t|\boldsymbol{r}_0)P(\rho,t|\boldsymbol{r}_0). \tag{11.18}$$

Function $P(\boldsymbol{r},t|\boldsymbol{r}')$ is the probability distribution of particle coordinates. This distribution satisfies the equation following from Eq. (11.17),

$$\frac{\partial}{\partial t}P(\boldsymbol{r},t|\boldsymbol{r}_0)=D_0\frac{\partial^2}{\partial \boldsymbol{r}^2}P(\boldsymbol{r},t|\boldsymbol{r}_0),\qquad P(\boldsymbol{r},0|\boldsymbol{r}_0)=\delta(\boldsymbol{r}-\boldsymbol{r}_0),$$

and, consequently, it is the Gaussian distribution

$$P(\boldsymbol{r},t|\boldsymbol{r}_0)=\exp\left\{D_0t\frac{\partial^2}{\partial \boldsymbol{r}^2}\right\}\delta(\boldsymbol{r}-\boldsymbol{r}_0)=\frac{1}{(4\pi D_0dt)^{d/2}}\exp\left\{-\frac{(\boldsymbol{r}-\boldsymbol{r}_0)^2}{4D_0t}\right\}, \tag{11.19}$$

where d is the dimension of space.

Function $P(j,t|\mathbf{r}_0)$ is the probability distribution of the divergence field in particle's neighborhood and satisfies the Fokker–Planck equation following from Eq. (11.17),

$$\frac{\partial}{\partial t}P(j,t|\mathbf{r}_0) = D^{\mathrm{p}}\frac{\partial^2}{\partial j^2}j^2P(j,t|\mathbf{r}_0), \qquad P(j,0|\mathbf{r}_0) = \delta(j-1), \tag{11.20}$$

which we rewrite in the form

$$\frac{\partial}{\partial t}jP(j,t|\mathbf{r}_0) = D^{\mathrm{p}}j\frac{\partial}{\partial j}\left(j\frac{\partial}{\partial j}+1\right)jP(j,t|\mathbf{r}_0), \quad P(j,0|\mathbf{r}_0) = \delta(j-1). \tag{11.21}$$

It is convenient to change the variables in Eq. (11.21) according to the formulas

$$j = e^{\eta}, \quad \eta = \ln j, \tag{11.22}$$

so that

$$\frac{\partial}{\partial \eta} = j\frac{\partial}{\partial j}.$$

After the variable change (11.22), Eq. (11.21) in function

$$jP(j,t|\mathbf{r}_0)|_{j=e^{\eta}} = F(\eta,t)$$

is reduced to the equation

$$\frac{\partial}{\partial t}F(\eta,t) = D^{\mathrm{p}}\frac{\partial}{\partial \eta}\left(\frac{\partial}{\partial \eta}+1\right)F(\eta,t) \tag{11.23}$$

with the initial condition

$$F(\eta,0) = jP(j,0|\mathbf{r}_0)|_{j=e^{\eta}} = e^{\eta}\delta(e^{\eta}-1) = e^{\eta}\frac{\delta(\eta)}{e^{\eta}} = \delta(\eta). \tag{11.24}$$

Consequently, the solution to problem (11.23), (11.24) is the Gaussian probability distribution

$$F(\eta,t) = \frac{1}{2\sqrt{\pi\tau}}\exp\left\{-\frac{(\eta+\tau)^2}{4\tau}\right\}. \tag{11.25}$$

In Eq. (11.25) and below, we use the dimensionless time $\tau = D^{\mathrm{p}}t$. Considering now function $F(\eta,t)|_{\eta=\ln j} = jP(j,t|\mathbf{r}_0)$, we obtain the solution to Eq. (11.20) in the form

$$\begin{aligned} P(j,t|\mathbf{r}_0) &= \exp\left\{D^{\mathrm{p}}t\frac{\partial^2}{\partial j^2}j^2\right\}\delta(j-1) = \frac{1}{2j\sqrt{\pi\tau}}\exp\left\{-\frac{(\ln j+\tau)^2}{4\tau}\right\} \\ &= \frac{1}{2j\sqrt{\pi\tau}}\exp\left\{-\frac{\ln^2\left(je^{\tau}\right)}{4\tau}\right\}. \end{aligned} \tag{11.26}$$

It should be emphasized that the obtained solution (11.18) means that coordinates $\boldsymbol{r}(t|\boldsymbol{r}_0)$ and divergence $j(t|\boldsymbol{r}_0)$ are statistically independent in the neighborhood of a particle characterized by the Lagrangian coordinate $\boldsymbol{r}_0$. Moreover, the logarithmically normal distribution (11.26) means that quantity $\eta(t|\boldsymbol{r}_0) = \ln j(t|\boldsymbol{r}_0)$ is distributed according to the Gaussian law with the parameters

$$\langle \eta(t|\boldsymbol{r}_0)\rangle = -\tau, \quad \sigma_\eta^2(t) = 2\tau. \tag{11.27}$$

In particular, the following expressions for the moments of the random divergency field follow from distribution (11.26) (and immediately from Eq. (11.20), though)

$$\left\langle j^n(t|\boldsymbol{r}_0)\right\rangle = e^{n(n-1)\tau}, \qquad n = \pm 1, \pm 2, \ldots. \tag{11.28}$$

We note that average divergence is constant, $\langle j(t|\boldsymbol{r}_0)\rangle = 1$, and its higher moments increase with time exponentially.

Note additionally that, according to Eq. (11.11), page 274, and Eq. (11.28), we have the following expression for the Lagrangian moments of density:

$$\left\langle \rho^n(t|\boldsymbol{r}_0)\right\rangle = \rho_0^n(\boldsymbol{r}_0) e^{n(n+1)\tau},$$

according to which both average density and higher moments of density increase exponentially in the Lagrangian representation. Consequently the random process $\rho(t|\boldsymbol{r}_0)$ is the logarithmically normal process and its probability density has the form

$$P(\rho, t|\boldsymbol{r}_0) = \frac{1}{2\rho\sqrt{\pi\tau}} \exp\left\{-\frac{\ln^2\left(\rho e^{-\tau}/\rho_0(\boldsymbol{r}_0)\right)}{4\tau}\right\}. \tag{11.29}$$

This probability density can be obtained also as the solution to the Fokker–Planck equation following from Eq. (11.17)

$$\frac{\partial}{\partial\tau} P(\rho, \tau|\boldsymbol{r}_0) = \frac{\partial}{\partial\rho}\rho^2\frac{\partial}{\partial\rho} P(\rho, \tau|\boldsymbol{r}_0), \quad P(\rho, 0|\boldsymbol{r}_0) = \delta(\rho - \rho_0(\boldsymbol{r}_0)). \tag{11.30}$$

Rewrite Eq. (11.30) in the form

$$\begin{gathered} \frac{\partial}{\partial\tau}\rho P(\rho, \tau|\boldsymbol{r}_0) = \rho\frac{\partial}{\partial\rho}\left(\rho\frac{\partial}{\partial\rho} - 1\right)\rho P(\rho, \tau|\boldsymbol{r}_0), \\ P(\rho, 0|\boldsymbol{r}_0) = \delta(\rho - \rho_0(\boldsymbol{r}_0)). \end{gathered} \tag{11.31}$$

Then, the variable change (11.22) reduces Eq. (11.31) in function

$$\rho P(\rho, \tau|\boldsymbol{r}_0)|_{\rho = e^\eta} = F(\eta, \tau|\boldsymbol{r}_0)$$

to the equation

$$\frac{\partial}{\partial \tau} F(\eta, \tau | \boldsymbol{r}_0) = \frac{\partial}{\partial \eta} \left(\frac{\partial}{\partial \eta} - 1 \right) F(\eta, \tau | \boldsymbol{r}_0) \tag{11.32}$$

with the initial condition

$$F(\eta, 0 | \boldsymbol{r}_0) = \rho P(\rho, 0 | \boldsymbol{r}_0)|_{\rho = e^{\eta}} = e^{\eta} \delta(e^{\eta} - \rho_0(\boldsymbol{r}_0)) = \delta(\eta - \ln \rho_0(\boldsymbol{r}_0)). \tag{11.33}$$

Consequently, the solution of problem (11.32), (11.33) is the Gaussian probability distribution

$$F(\eta, \tau | \boldsymbol{r}_0) = \frac{1}{2\sqrt{\pi \tau}} \exp \left\{ -\frac{(\eta - \ln \rho_0(\boldsymbol{r}_0) - \tau)^2}{4\tau} \right\}. \tag{11.34}$$

Considering now function $F(\eta, \tau | \boldsymbol{r}_0)|_{\eta = \ln \rho} = \rho P(\rho, \tau | \boldsymbol{r}_0)$, we obtain the solution to Eq. (11.30) in form (11.29). In this case, the logarithmically normal distribution (11.34) means that quantity $\eta(t|\boldsymbol{r}_0) = \ln \rho(t|\boldsymbol{r}_0)$ is distributed according to the Gaussian law with the parameters

$$\langle \eta(t|\boldsymbol{r}_0) \rangle = \ln \rho_0(\boldsymbol{r}_0) + \tau, \quad \sigma_\eta^2(t) = 2\tau.$$

The above paradoxical behavior of statistical characteristics of the divergence and concentration (simultaneous growth of all moment functions in time) is a consequence of the lognormal probability distribution. Indeed, the typical realization curve of random divergence is the exponentially decaying curve

$$j^*(\tau) = e^{-\tau},$$

in accordance with Eqs. (4.60), page 108.

Moreover, realizations of the lognormal process satisfy certain majorant estimates. For example, with probability $p = 1/2$, we have

$$j(t|\boldsymbol{r}_0) < 4e^{-\tau/2}$$

throughout arbitrary temporal interval $t \in (t_1, t_2)$.

Similarly, the typical realization curve of concentration and its minorant estimate have the following form

$$\rho^*(t) = \rho_0 e^{\tau}, \quad \rho(t|\boldsymbol{r}_0) > \frac{\rho_0}{4} e^{\tau/2}.$$

We emphasize that the above Lagrangian statistical properties of a particle in flows containing the potential random component are qualitatively different from the statistical properties of a particle in nondivergent flows where $j(t|\boldsymbol{r}_0) \equiv 1$ and particle concentration remains invariant in the vicinity of a fixed particle $\rho(t|\boldsymbol{r}_0) = \rho_0(\boldsymbol{r}_0) =$ const. The above statistical estimates mean that the statistics of random processes $j(t|\boldsymbol{r}_0)$ and $\rho(t|\boldsymbol{r}_0)$ is formed by the realization spikes relative typical realization curves.

At the same time, probability distributions of particle coordinates in essence coincide for both divergent and nondivergent velocity fields.

11.2.2 Two-Point Statistical Characteristics

Consider now the joint dynamics of two particles in the absence of mean flow. In this case, the indicator function

$$\varphi(\boldsymbol{r}_1, \boldsymbol{r}_2, t) = \delta\left(\boldsymbol{r}_1(t) - \boldsymbol{r}_1\right)\delta\left(\boldsymbol{r}_2(t) - \boldsymbol{r}_2\right)$$

satisfies the Liouville equation

$$\frac{\partial}{\partial t}\varphi(\boldsymbol{r}_1, \boldsymbol{r}_2, t) = -\left[\frac{\partial}{\partial \boldsymbol{r}_1}\boldsymbol{u}_1(\boldsymbol{r}, t) + \frac{\partial}{\partial \boldsymbol{r}_2}\boldsymbol{u}_2(\boldsymbol{r}, t)\right]\varphi(\boldsymbol{r}_1, \boldsymbol{r}_2, t).$$

If we average the indicator function over an ensemble of realizations of field $\boldsymbol{u}(\boldsymbol{r}, t)$, use the Furutsu–Novikov formula (11.14), page 276, and the equality

$$\frac{\delta}{\delta u_j(\boldsymbol{r}', t-0)}\varphi(t; \boldsymbol{r}_1, \boldsymbol{r}_2) = -\left[\frac{\partial}{\partial r_{1j}}\delta(\boldsymbol{r}_1 - \boldsymbol{r}') + \frac{\partial}{\partial r_{2j}}\delta(\boldsymbol{r}_2 - \boldsymbol{r}')\right]\varphi(t; \boldsymbol{r}_1, \boldsymbol{r}_2),$$

then we obtain that the joint probability density of positions of two particles for isotropic velocity field

$$P(\boldsymbol{r}_1, \boldsymbol{r}_2, t) = \langle\varphi(t; \boldsymbol{r}_1, \boldsymbol{r}_2)\rangle$$

satisfies the Fokker–Planck equation

$$\begin{aligned}\frac{\partial}{\partial t}P(\boldsymbol{r}_1, \boldsymbol{r}_2, t) = &\left[\frac{\partial^2}{\partial r_{1i}\partial r_{1j}} + \frac{\partial^2}{\partial r_{2i}\partial r_{2j}}\right]B_{ij}(0)P(t; \boldsymbol{r}_1, \boldsymbol{r}_2)\\ &+ 2\frac{\partial^2}{\partial r_{1i}\partial r_{2j}}B_{ij}(\boldsymbol{r}_1 - \boldsymbol{r}_2)P(\boldsymbol{r}_1, \boldsymbol{r}_2, t).\end{aligned} \tag{11.35}$$

Multiplying now Eq. (11.35) by function $\delta(\boldsymbol{r}_1 - \boldsymbol{r}_2 - \boldsymbol{l})$ and integrating over $\boldsymbol{r}_1$ and $\boldsymbol{r}_2$, we obtain that the probability density of relative diffusion of two particles

$$P(\boldsymbol{l}, t) = \langle\delta(\boldsymbol{r}_1(t) - \boldsymbol{r}_2(t) - \boldsymbol{l})\rangle$$

satisfies the Fokker–Planck equation

$$\frac{\partial}{\partial t}P(\boldsymbol{l}, t) = \frac{\partial^2}{\partial l_\alpha \partial l_\beta}D_{\alpha\beta}(\boldsymbol{l})P(\boldsymbol{l}, t), \quad P(\boldsymbol{l}, 0) = \delta(\boldsymbol{l} - \boldsymbol{l}_0), \tag{11.36}$$

where

$$D_{\alpha\beta}(\boldsymbol{l}) = 2\left[B_{\alpha\beta}(0) - B_{\alpha\beta}(\boldsymbol{l})\right]$$

is the structure matrix of vector field $\boldsymbol{u}(\boldsymbol{r}, t)$, and $\boldsymbol{l}_0$ is the initial distance between the particles.

In the general case, Eq. (11.36) hardly can be solved analytically. However, if the initial distance between particles l_0 is sufficiently small, namely, if $l_0 \ll l_{\text{cor}}$, where l_{cor} is the spatial correlation radius of the velocity field $\boldsymbol{u}(\boldsymbol{r},t)$, we can expand functions $D_{\alpha\beta}(\boldsymbol{l})$ in the Taylor series to obtain in the first approximation

$$D_{\alpha\beta}(\boldsymbol{l}) = -\left.\frac{\partial^2 B_{\alpha\beta}(\boldsymbol{l})}{\partial l_i \partial l_j}\right|_{l=0} l_i l_j.$$

The use of representation (4.57), page 107, simplifies the diffusion tensor reducing it to the form

$$D_{\alpha\beta}(\boldsymbol{l}) = \frac{1}{d(d+2)}\left[\left(D^{\text{s}}(d+1)+D^{\text{p}}\right)\delta_{\alpha\beta}l^2 - 2\left(D^{\text{s}}-D^{\text{p}}\right)l_\alpha l_\beta\right], \tag{11.37}$$

where d is the dimension of space.

Substituting now Eq. (11.37) in Eq. (11.36), multiplying both sides of the resulting equation by l^n, and integrating over $\boldsymbol{l}$, we obtain the closed equation

$$\frac{d}{dt}\ln\left\langle l^n(t)\right\rangle = \frac{1}{d(d+2)}\left[\left(D^{\text{s}}(d+1)+D^{\text{p}}\right)n\,(d+n-2) - 2\left(D^{\text{s}}-D^{\text{p}}\right)n(n-1)\right],$$

whose solution shows the exponential growth of all moment functions ($n = 1, 2, \ldots$) in time. In this case, the probability distribution of random process $l(t)/l_0$ will be logarithmic-normal.

Note that, multiplying Eq. (11.36) by $\delta(l(t)-l)$ and integrating the result over $\boldsymbol{l}$, we can easily obtain that the probability density of the modulus of vector $\boldsymbol{l}(t)$

$$P(l,t) = \langle\delta(|\boldsymbol{l}(t)|-l)\rangle = \int d\boldsymbol{l}\,\delta(|\boldsymbol{l}(t)|-l)P(\boldsymbol{l},t),$$

satisfies the equation

$$\frac{\partial}{\partial t}P(l,t) = -\frac{\partial}{\partial l}\frac{D_{ii}(\boldsymbol{l})}{l}P(l,t) + \frac{\partial}{\partial l}\frac{N(l)}{l}P(t,l) + \frac{\partial^2}{\partial l^2}N(l)P(l,t),$$

where $N(l) = l_j l_i D_{ij}(\boldsymbol{l})/l^2$. Using this equation, we can easily derive the equation in function $\langle \ln l(t)\rangle$,

$$\frac{d}{dt}\langle\ln l(t)\rangle = \left\langle\frac{D_{ii}(l)}{l^2} - 2\frac{N(l)}{l^2}\right\rangle,$$

whose solution for the tensor $D_{ij}(\boldsymbol{l})$ of form (11.37) is as follows

$$\left\langle\ln\left(\frac{l(t)}{l_0}\right)\right\rangle = \frac{1}{d(d+2)}\{D^{\text{s}}(d-1)d - D^{\text{p}}(4-d)\}t.$$

As a consequence, in accordance with Eqs. (4.60), page 108, the typical realization curve of the distance between two particles will be the exponential function of time

$$l^*(t) = l_0 \exp\left\{\frac{1}{d(d+2)}\left\{D^s d(d-1) - D^p(4-d)\right\} t\right\}, \tag{11.38}$$

and it is the *Lyapunov exponent* of the lognormal random process $l(t)$.

It appears that this expression in the two-dimensional case ($d = 2$)

$$l^*(t) = l_0 \exp\left\{\frac{1}{4}\left(D^s - D^p\right) t\right\}$$

significantly depends of the sign of the difference ($D^s - D^p$). In particular, for the nondivergent velocity field ($D^p = 0$), we have the exponentially increasing typical realization curve, which means that particle scatter is exponentially fast for small distances between them. This result is valid for times

$$\frac{1}{4} D^s t \ll \ln\left(\frac{l_{cor}}{l_0}\right),$$

for which expansion (11.37) holds. In another limiting case of the potential velocity field ($D^s = 0$), the typical realization curve is the exponentially decreasing curve, which means that particles tend to join. In view of the fact that liquid particles themselves are compressed during this process, we arrive at the conclusion that particles must form *clusters*, i.e., compact particle concentration zones located merely in rarefied regions, which agrees with the evolution of the realization (see Fig. 1.1*b*, page 5) obtained by simulating the behavior of the initially homogeneous particle distribution in random potential velocity field (though, for drastically other statistical model of the velocity field). This means that the phenomenon of clustering by itself is independent of the model of a velocity field, although statistical parameters characterizing this phenomenon surely depend on this model.

Thus, particle clustering requires that inequality

$$D^s < D^p \tag{11.39}$$

be satisfied.

In the three-dimensional case ($d = 3$), Eq. (11.38) grades into

$$l^*(t) = l_0 \exp\left\{\frac{1}{15}\left(6D^s - D^p\right) t\right\},$$

and typical realization curve will exponentially decay in time under the condition

$$D^p > 6D^s,$$

which is stronger than that in the two-dimensional case.

In the one-dimensional case

$$l^*(t) = l_0 e^{-D^{\mathrm{p}} t},$$

and typical realization curve always decays in time.

11.3 Probabilistic Description of Density Field in Random Velocity Field

To describe the local behavior of tracer realizations in random velocity field, we need the probability distribution of tracer concentration, which can be obtained only in the absence of dynamic diffusion.

To describe the concentration field in the Eulerian description, we introduce the indicator function

$$\varphi(\boldsymbol{r}, t; \rho) = \delta(\rho(\boldsymbol{r}, t) - \rho), \tag{11.40}$$

which is similar to function (11.15) and is localized on surface $\rho(\boldsymbol{r}, t) = \rho = \text{const}$ in the three-dimensional case or on a contour in the two-dimensional case. The Liouville equation for this function coincides in form with Eq. (3.18), page 74,

$$\left(\frac{\partial}{\partial t} + \boldsymbol{U}(\boldsymbol{r}, t)\frac{\partial}{\partial \boldsymbol{r}}\right)\varphi(\boldsymbol{r}, t; \rho) = \frac{\partial \boldsymbol{U}(\boldsymbol{r}, t)}{\partial \boldsymbol{r}}\frac{\partial}{\partial \rho}\left[\rho\varphi(\boldsymbol{r}, t; \rho)\right], \tag{11.41}$$

so that the on-point probability density of the solution to dynamic equation (11.13) coincides in this case with the ensemble averaged indicator function of random velocity field which we assume, as earlier, to be the homogeneous and isotropic Gaussian random field delta-correlated in time,

$$P(\boldsymbol{r}, t; \rho) = \langle \delta(\rho(\boldsymbol{r}, t) - \rho) \rangle.$$

Averaging Eq. (11.41) with absent average flow over an ensemble of realizations of field $\boldsymbol{u}(\boldsymbol{r}, t)$ using the Furutsu–Novikov formula (11.14), page 276, and the equality for variational derivative

$$\frac{\delta}{\delta u_j(\boldsymbol{r}', t-0)}\varphi(\boldsymbol{r}, t; \rho) = -\left[\delta(\boldsymbol{r} - \boldsymbol{r}')\frac{\partial}{\partial r_j} + \frac{\partial \delta(\boldsymbol{r} - \boldsymbol{r}')}{\partial r_j}\right]\varphi(\boldsymbol{r}, t; \rho)$$

following from Eq. (11.41) with the use of Eqs. (4.56) and (4.57), page 107, we obtain the equation for probability density of the concentration field

$$\begin{gathered}\left(\frac{\partial}{\partial t} - D_0\Delta\right)P(\boldsymbol{r}, t; \rho|\rho_0(\boldsymbol{r})) = D_\rho \frac{\partial^2}{\partial \rho^2}\rho^2 P(\boldsymbol{r}, t; \rho|\rho_0(\boldsymbol{r})),\\ P(\boldsymbol{r}, 0; \rho|\rho_0(\boldsymbol{r})) = \delta(\rho - \rho_0(\boldsymbol{r})),\end{gathered} \tag{11.42}$$

where the diffusion coefficients D_0 (in $\boldsymbol{r}$-space) and $D_\rho = D^{\rho}$ (in ρ-space) are given by Eqs. (4.54) and (4.59), pages 107 and 108; the vertical bar separates the dependence of the probability density of the Eulerian field $\rho(\boldsymbol{r}, t)$ on the initial distribution $\rho_0(\boldsymbol{r})$. This equation corresponds to Eq. (8.55), page 207, with parameter $\alpha = D_\rho$.

Note that Eq. (11.42) can also be derived by using the following relationship between the one-point probability density of concentration field in the Eulerian representation and the one-time probability density in the Lagrangian description

$$P(\boldsymbol{r}, t; \rho) = \int d\boldsymbol{r}_0 \int\limits_0^\infty j dj P(\boldsymbol{r}, \rho, j, t|\boldsymbol{r}_0). \tag{11.43}$$

The solution to Eq. (11.42) has the form

$$P(\boldsymbol{r}, t; \rho|\rho_0(\boldsymbol{r})) = \exp\left\{D_0 t \frac{\partial^2}{\partial \boldsymbol{r}^2}\right\} \widetilde{P}(\boldsymbol{r}, t; \rho|\rho_0(\boldsymbol{r})),$$

where function $\widetilde{P}(\boldsymbol{r}, t; \rho|\rho_0(\boldsymbol{r}))$ satisfies the equation

$$\begin{aligned} \frac{\partial}{\partial t} \widetilde{P}(\boldsymbol{r}, t; \rho|\rho_0(\boldsymbol{r})) &= D_\rho \frac{\partial^2}{\partial \rho^2} \rho^2 \widetilde{P}(\boldsymbol{r}, t; \rho|\rho_0(\boldsymbol{r})), \\ \widetilde{P}(\boldsymbol{r}, 0; \rho|\rho_0(\boldsymbol{r})) &= \delta(\rho - \rho_0(\boldsymbol{r})) \end{aligned} \tag{11.44}$$

that depends on $\boldsymbol{r}$ parametrically and coincides with Eq. (11.20), page 278, for the Lagrangian probability density of particle divergence (the only difference consists in the initial condition). Consequently, the solution to this equation is as follows

$$\widetilde{P}(\boldsymbol{r}, t; \rho|\rho_0(\boldsymbol{r})) = \frac{1}{2\rho\sqrt{\pi D_\rho t}} \exp\left\{-\frac{\ln^2\left(\rho e^{D_\rho t}/\rho_0(\boldsymbol{r})\right)}{4 D_\rho t}\right\}, \tag{11.45}$$

and probability density of the Eulerian density field assumes the form

$$P(\boldsymbol{r}, t; \rho|\rho_0(\boldsymbol{r})) = \frac{1}{2\rho\sqrt{\pi D_\rho t}} \exp\left\{D_0 t \frac{\partial^2}{\partial \boldsymbol{r}^2}\right\} \exp\left\{-\frac{\ln^2\left(\rho e^{D_\rho t}/\rho_0(\boldsymbol{r})\right)}{4 D_\rho t}\right\}. \tag{11.46}$$

In the case of uniform initial tracer concentration $\rho_0(\boldsymbol{r}) = \rho_0 = \text{const}$ the probability distribution is independent of $\boldsymbol{r}$. Consequently, the Eulerian concentration field is in this case the lognormal field; the corresponding probability density and integral distribution function are as follows

$$\begin{aligned} P(t; \rho|\rho_0) &= \frac{1}{2\rho\sqrt{\pi\tau}} \exp\left\{-\frac{\ln^2\left(\rho e^{\tau}/\rho_0\right)}{4\tau}\right\}, \\ F(t; \rho|\rho_0) &= \Pr\left(\frac{\ln\left(\rho e^{\tau}/\rho_0\right)}{2\sqrt{\tau}}\right), \end{aligned} \tag{11.47}$$

where Pr(z) is the probability integral (4.20), page 94, and we use the dimensionless time $\tau = D^{\text{p}} t$.

In the context of the one-point characteristics of density field $\rho(\boldsymbol{r}, t)$, the problem is statistically equivalent in this case to the analysis of a random process; hence, all moment functions beginning from the second one appear to be exponentially increasing functions of time

$$\langle \rho(\boldsymbol{r}, t) \rangle = \rho_0, \quad \left\langle \rho^n(\boldsymbol{r}, t) \right\rangle = \rho_0^n e^{n(n-1)\tau}, \tag{11.48}$$

and in accordance with Eqs. (4.60), page 108, the typical realization curve of the concentration field exponentially decays with time at any fixed spatial point

$$\rho^*(t) = \rho_0 e^{-\tau},$$

which is evidence of cluster behavior of medium density fluctuations in arbitrary divergent flows. The Eulerian concentration statistics at any fixed point is formed due to concentration fluctuations about this curve.

Even the above discussion of the one-point probability density of tracer concentration in the Eulerian representation revealed several regularities concerning the temporal behavior of concentration field realizations at fixed spatial points. Now we show that this distribution additionally allows us to reveal certain features in the space–time structure of concentration field realizations.

For simplicity, we content ourselves with the two-dimensional case. As was mentioned earlier, the analysis of level lines defined by the equality $\rho(\boldsymbol{r}, t) = \rho = \text{const}$ can give important data on the spatial behavior of realizations, in particular, on different functionals of the concentration field, such as total area $S(t, \rho)$ of regions where $\rho(\boldsymbol{r}, t) > \rho$ and total tracer mass $M(t, \rho)$ within these regions. Average values of these functionals can be expressed in terms of the one-point probability density:

$$\langle S(t, \rho) \rangle = \int\limits_{\rho}^{\infty} d\widetilde{\rho} \int d\boldsymbol{r}\, P(\boldsymbol{r}, t; \widetilde{\rho}), \quad \langle M(t, \rho) \rangle = \int\limits_{\rho}^{\infty} \widetilde{\rho} d\widetilde{\rho} \int d\boldsymbol{r}\, P(\boldsymbol{r}, t; \widetilde{\rho}). \tag{11.49}$$

Substituting the solution to Eq. (11.46) in these expressions and performing some rearrangement, we easily obtain explicit expressions for these quantities

$$\begin{aligned} \langle S(t, \rho) \rangle &= \int d\boldsymbol{r}\, \Pr\left(\frac{1}{\sqrt{2\tau}} \ln\left(\frac{\rho_0(\boldsymbol{r}) e^{-\tau}}{\rho} \right) \right), \\ \langle M(t, \rho) \rangle &= \int d\boldsymbol{r}\, \rho_0(\boldsymbol{r}) \Pr\left(\frac{1}{\sqrt{2\tau}} \ln\left(\frac{\rho_0(\boldsymbol{r}) e^{\tau}}{\rho} \right) \right), \end{aligned} \tag{11.50}$$

where probability integral Pr(z) is defined as Eq. (4.20), page 94. Taking into account the asymptotic representation of function Pr(z) (4.23), page 95, we obtain that, for

$\tau \gg 1$, the average area of regions where concentration exceeds level ρ decreases in time according to the law

$$\langle S(t,\rho)\rangle \approx \frac{1}{\sqrt{\pi\tau\rho}} e^{-\tau/4} \int d\boldsymbol{r}\sqrt{\rho_0(\boldsymbol{r})}, \tag{11.51}$$

while the average tracer mass within these regions

$$\langle M(t,\rho)\rangle \approx M_0 - \sqrt{\frac{\rho}{\pi\tau}} e^{-\tau/4} \int d\boldsymbol{r}\sqrt{\rho_0(\boldsymbol{r})} \tag{11.52}$$

monotonously tends to the total mass

$$M_0 = \int d\boldsymbol{r}\rho_0(\boldsymbol{r}).$$

This is an additional evidence in favor of the above conclusion that tracer particles tend to join in clusters, i.e., in compact regions of enhanced concentration surrounded with rarefied regions.

We illustrate the dynamics of cluster formation with the example of the initially uniform distribution of buoyant tracer over the plane, in which case $\rho_0(\boldsymbol{r}) = \rho_0 =$ const. In this case, the average specific area (per unit surface area) of regions within which $\rho(\boldsymbol{r},t) > \rho$ is

$$s(t,\rho|\rho_0) = \int_{\rho}^{\infty} d\widetilde{\rho}\, P(t;\widetilde{\rho}|\rho_0) = \Pr\left(\frac{\ln\left(\rho_0\, e^{-\tau}/\rho\right)}{\sqrt{2\tau}}\right), \tag{11.53}$$

where $P(t;\rho)$ is the solution to Eq. (11.44) independent of $\boldsymbol{r}$ (i.e., function (11.47), and average specific tracer mass (per unit surface area) within these regions is given by the expression

$$m(t,\rho|\rho_0)/\rho_0 = \frac{1}{\rho_0}\int_{\rho}^{\infty} \widetilde{\rho}\, d\widetilde{\rho}\, P(t;\widetilde{\rho}|\rho_0) = \Pr\left(\frac{\ln\left(\rho_0\, e^{\tau}/\rho\right)}{\sqrt{2\tau}}\right). \tag{11.54}$$

From Eqs. (11.53) and (11.54) it follows that, for sufficiently large times, average specific area of such regions decreases according to the law

$$s(t,\rho) \approx \frac{1}{\sqrt{\pi\tau}} e^{-\tau/4} \tag{11.55}$$

irrespective of ratio ρ/ρ_0; at the same time, these regions concentrate almost all tracer mass

$$m(t,\rho) \approx 1 - \frac{1}{\sqrt{\pi\tau}} e^{-\tau/4}. \tag{11.56}$$

Nevertheless, the time-dependent behavior of the formation of cluster structure essentially depends on ratio ρ/ρ_0. If $\rho/\rho_0 < 1$, then $s(0,\rho) = 1$ and $m(0,\rho) = 1$ at the initial instant. Then, in view of the fact that particles of buoyant tracer initially tend to scatter, there appear small areas within which $\rho(\boldsymbol{r},t) < \rho$ and which concentrate only insignificant portion of the total mass. These regions rapidly grow with time and their mass flows into cluster region relatively quickly approaching asymptotic expressions (11.55) and (11.56) (Fig. 11.1).

Note that $s(t^*,\rho) = 1/2$ at instant $\tau^* = \ln(\rho/\rho_0)$.

In the opposite, more interesting case $\rho/\rho_0 > 1$, we have $s(0,\rho) = 0$ and $m(0,\rho) = 0$ at the initial instant. In view of initial scatter of particles, there appear small cluster regions within which $\rho(\boldsymbol{r},t) > \rho$; these regions remain at first almost invariable in time and intensively absorb a significant portion of total mass. With time, the area of these regions begins to decrease and the mass within them begins to increase according to asymptotic expressions (11.55) and (11.56) (Fig. 11.2).

As we mentioned earlier, a more detailed description of the tracer concentration field in random velocity field requires that spatial gradient $\boldsymbol{p}(\boldsymbol{r},t) = \nabla\rho(\boldsymbol{r},t)$ (and, generally, higher-order derivatives) was additionally included in the consideration.

The concentration gradient satisfies dynamic equation (1.50), page 28; consequently, the extended indicator function

$$\varphi(\boldsymbol{r},t;\rho,\boldsymbol{p}) = \delta\left(\rho(\boldsymbol{r},t) - \rho\right)\delta\left(\boldsymbol{p}(\boldsymbol{r},t) - \boldsymbol{p}\right)$$

for nondivergent velocity field satisfies Eq. (3.24). Averaging now Eq. (3.24) over an ensemble of velocity field realizations in the approximation of delta-correlated velocity field, we obtain the equation for the one-point joint probability density of the concentration and its gradient $P(\boldsymbol{r},t;\rho,\boldsymbol{p}) = \langle\varphi(\boldsymbol{r},t;\rho,\boldsymbol{p})\rangle$ (this probability density is

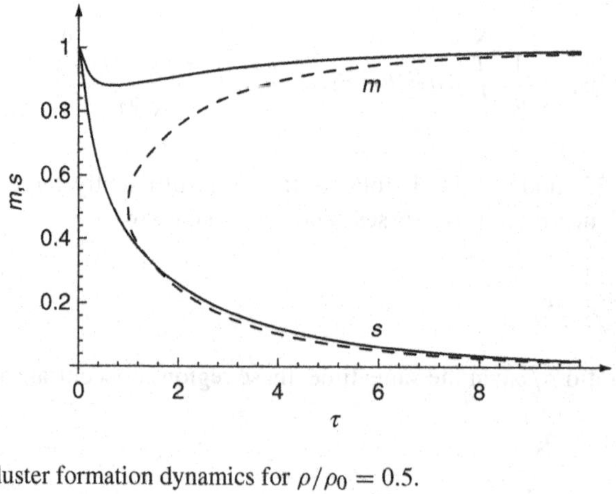

Figure 11.1 Cluster formation dynamics for $\rho/\rho_0 = 0.5$.

a function of the space-time point ($\boldsymbol{r}$, t))

$$\frac{\partial}{\partial t}P(\boldsymbol{r},t;\rho,\boldsymbol{p}) = D_0\Delta P(\boldsymbol{r},t;\rho,\boldsymbol{p}) + \frac{1}{d(d+2)}D^{\rm s}\left((d+1)\frac{\partial^2}{\partial \boldsymbol{p}^2}\boldsymbol{p}^2 - 2\frac{\partial^2}{\partial p_k \partial p_l}p_k p_l\right)P(\boldsymbol{r},t;\rho,\boldsymbol{p}). \tag{11.57}$$

$$P(\boldsymbol{r},0;\rho,\boldsymbol{p}) = \delta\left(\rho_0(\boldsymbol{r})-\rho\right)\delta(\boldsymbol{p}_0(\boldsymbol{r})-\boldsymbol{p}),$$

where the diffusion coefficients D_0 (in $\boldsymbol{r}$-space) and $D^{\rm s}$ (in $\boldsymbol{p}$-space) are given by Eqs. (4.54) and (4.59), pages 107 and 108.

In view of the fact that the random velocity field is nondivergent, we can represent the solution to Eq. (11.57) in the form

$$P(\boldsymbol{r},t;\rho,\boldsymbol{p}) = \int d\boldsymbol{r}_0 P(\boldsymbol{r},t|\boldsymbol{r}_0)P(\boldsymbol{p},t|\boldsymbol{r}_0), \tag{11.58}$$

where $P(\boldsymbol{r},t|\boldsymbol{r}_0)$ and $P(\boldsymbol{p},t|\boldsymbol{r}_0)$ are the corresponding Lagrangian probability densities of particle position and its gradient. The first density is given by Eq. (11.19), and the

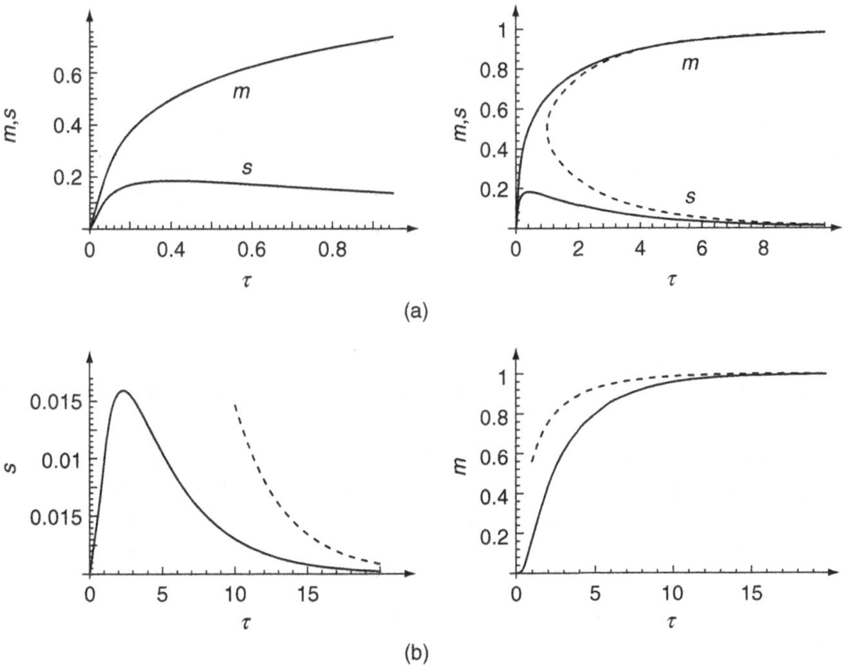

Figure 11.2 Cluster formation dynamics for (a) $\rho/\rho_0 = 1.5$ and (b) $\rho/\rho_0 = 10$.

second one satisfies the equation

$$\frac{\partial}{\partial t}P(\boldsymbol{p},t|\boldsymbol{r}_0) = \frac{1}{d(d+2)}D^{\mathrm{s}}\left((d+1)\frac{\partial^2}{\partial \boldsymbol{p}^2}\boldsymbol{p}^2 - 2\frac{\partial^2}{\partial p_k \partial p_l}p_k p_l\right)P(\boldsymbol{p},t|\boldsymbol{r}_0). \tag{11.59}$$

From Eq. (11.59) follows that average tracer concentration gradient is invariant

$$\langle \boldsymbol{p}(\boldsymbol{r},t)\rangle = \boldsymbol{p}_0(\boldsymbol{r}_0).$$

As regards the moment functions of the concentration gradient modulus, they satisfy the equations

$$\frac{\partial}{\partial t}\left\langle p^n(t|\boldsymbol{r}_0)\right\rangle = \frac{n(d+n)(d-1)}{d(d+2)}D^{\mathrm{s}}\left\langle p^n(t|\boldsymbol{r}_0)\right\rangle, \quad \left\langle p^n(0|\boldsymbol{r}_0)\right\rangle = p_0^n(\boldsymbol{r}_0) \tag{11.60}$$

that follow from (11.59).

Consequently, the concentration gradient modulus in the Lagrangian representation is the logarithmic-normal quantity whose typical realization curve and all moment functions increase exponentially in time. In particular, the first and second moments in the two-dimensional case are given by the equalities

$$\langle |\boldsymbol{p}(t|\boldsymbol{r}_0)|\rangle = |\boldsymbol{p}_0(\boldsymbol{r}_0)|e^{3D^{\mathrm{s}}t/8}, \quad \left\langle \boldsymbol{p}^2(t|\boldsymbol{r}_0)\right\rangle = \boldsymbol{p}_0^2(\boldsymbol{r}_0)e^{D^{\mathrm{s}}t}. \tag{11.61}$$

In addition, from Eq. (11.57) with allowance for Eq. (4.38) it follows that average total length of contour $\rho(\boldsymbol{r},t) = \rho = \text{const}$ (remember that we deal with the two-dimensional case) also exponentially increases in time according to the law

$$\langle l(t,\rho)\rangle = l_0 e^{D^{\mathrm{s}}t},$$

where l_0 is the initial length of the contour. Remember that, in the case of the nondivergent velocity field, the number of contours remains unchanged; the contours cannot appear and disappear in the medium, they only evolve in time depending on their spatial distribution at the initial instant.

Thus, the initially smooth tracer distribution becomes with time increasingly inhomogeneous in space; its spatial gradients sharpen and level lines acquire the fractal behavior. We observed such a pattern in Fig. 1.1*a*, page 5 that shows simulated results (for quite other model of velocity field fluctuations). This means that the above general behavioral characteristics only slightly depend on the fluctuation model.

Above, we studied statistical characteristics of the solution to Eq. (11.3) in the Lagrangian and Eulerian representations and showed that, if the velocity field has a nonzero potential component, clustering of the Eulerian concentration field occurs with a probability of unity and, under certain conditions, clustering of particles in the Lagrangian description can take place.

Along with dynamic equation (11.13), there is certain interest to the equation

$$\left(\frac{\partial}{\partial t} + \boldsymbol{U}(\boldsymbol{r}, t)\frac{\partial}{\partial \boldsymbol{r}}\right)\rho(\boldsymbol{r}, t) = 0, \quad \rho(\boldsymbol{r}, 0) = \rho_0(\boldsymbol{r})$$

that describes transfer of nonconservative tracer. In this case, particle dynamics in the Lagrangian representation is described by the equation coinciding with Eq. (11.5); consequently, particles are clustered. However, in the Eulerian representation, no clustering occurs. In this case, as in the case of nondivergent velocity field, average number of contours, average area of regions within which $\rho(\boldsymbol{r}, t) > \rho$, and average tracer mass $\int dS\rho(\boldsymbol{r}, t)$ within these regions remain invariant.

Above, we considered the simplest statistical problems on tracer diffusion in random velocity field in the absence of regular flows and dynamic diffusion. Moreover, our statistical description used the approximation of random delta-correlated (in time) field. All unaccounted factors begin from a certain time, so that the above results hold only during the initial stage of diffusion. Furthermore, these factors can give rise to new physical effects.

11.4 Probabilistic Description of Magnetic Field and Magnetic Energy in Random Velocity Field

Here, we consider probabilistic description of the magnetic field starting from dynamic equation (11.4). As in the case of the density field, we will assume that random component of velocity field $\boldsymbol{u}(\boldsymbol{r}, t)$ is the divergent ($\operatorname{div}\boldsymbol{u}(\boldsymbol{r}, t) \neq 0$) Gaussian random field, homogeneous and isotropic in space and stationary and delta-correlated in time.

Introduce the indicator function of magnetic field $\boldsymbol{H}(\boldsymbol{r}, t)$,

$$\varphi(\boldsymbol{r}, t; \boldsymbol{H}) = \delta(\boldsymbol{H}(\boldsymbol{r}, t) - \boldsymbol{H}).$$

It satisfies the Liouville equation (3.26), page 76,

$$\left(\frac{\partial}{\partial t} + \boldsymbol{u}(\boldsymbol{r}, t)\frac{\partial}{\partial \boldsymbol{r}}\right)\varphi(\boldsymbol{r}, t; \boldsymbol{H}) = -\frac{\partial}{\partial H_i}\left[\boldsymbol{H}\frac{\partial u_i(\boldsymbol{r}, t)}{\partial \boldsymbol{r}} - H_i\frac{\partial \boldsymbol{u}(\boldsymbol{r}, t)}{\partial \boldsymbol{r}}\right]\varphi(\boldsymbol{r}, t; \boldsymbol{H}) \tag{11.62}$$

with the initial condition

$$\varphi(\boldsymbol{r}, 0; \boldsymbol{H}) = \delta(\boldsymbol{H}_0(\boldsymbol{r}) - \boldsymbol{H}).$$

The solution to this equation is a functional of velocity field $\boldsymbol{u}(\boldsymbol{r}, t)$, i.e.,

$$\varphi(\boldsymbol{r}, t; \boldsymbol{H}) = \varphi[\boldsymbol{r}, t; \boldsymbol{H}; \boldsymbol{u}(\widetilde{\boldsymbol{r}}, \tau)],$$

where $0 \leq \tau \leq t$. This solution obeys the condition of dynamic causality

$$\frac{\delta\varphi[\boldsymbol{r}, t; \boldsymbol{H}; \boldsymbol{u}(\widetilde{\boldsymbol{r}}, \tau)]}{\delta u_j(\boldsymbol{r}', t')} = 0 \text{ for } t' < 0 \text{ and } t' > t.$$

Moreover, the variational derivative at $t' = t - 0$ can be represented as

$$\frac{\delta\varphi(\boldsymbol{r}, t; \boldsymbol{H})}{\delta u_j(\boldsymbol{r}', t-0)} = \widehat{L}_j(\boldsymbol{r}, \boldsymbol{r}', t; \boldsymbol{H})\varphi(\boldsymbol{r}, t; \boldsymbol{H}), \tag{11.63}$$

where operator

$$\widehat{L}_j(\boldsymbol{r}, \boldsymbol{r}', t; \boldsymbol{H}) = -\delta(\boldsymbol{r}-\boldsymbol{r}')\frac{\partial}{\partial r_j} - \frac{\partial\delta(\boldsymbol{r}-\boldsymbol{r}')}{\partial r_l}\frac{\partial}{\partial H_j}H_l + \frac{\partial\delta(\boldsymbol{r}-\boldsymbol{r}')}{\partial r_j}\frac{\partial}{\partial H_l}H_l. \tag{11.64}$$

The one-point probability density of magnetic field is defined as the equality

$$P(\boldsymbol{r}, t; \boldsymbol{H}) = \langle\varphi(\boldsymbol{r}, t; \boldsymbol{H})\rangle_{\boldsymbol{u}}.$$

Let us average Eq. (11.62) over an ensemble of realizations of field $\{\boldsymbol{u}(\boldsymbol{r}, t)\}$ and use the Furutsu–Novikov formula (11.14) to split the appeared correlations. Taking into account equalities (11.63) and (11.64), and Eqs. (4.54), (4.56) and (4.57), page 107, with parameters (4.59), page 108, we obtain then the desired equation

$$\left(\frac{\partial}{\partial t} - D_0\frac{\partial^2}{\partial \boldsymbol{r}^2}\right)P(\boldsymbol{r}, t; \boldsymbol{H}) = \left\{D_1\frac{\partial}{\partial H_l}\frac{\partial}{\partial H_k}H_lH_k + D_2\frac{\partial}{\partial H_l}\frac{\partial}{\partial H_l}H_k^2\right\}P(\boldsymbol{r}, t; \boldsymbol{H}), \tag{11.65}$$

where

$$D_1 = \frac{\left(d^2-2\right)D^{\mathrm{p}} - 2D^{\mathrm{s}}}{d(d+2)}, \quad D_2 = \frac{(d+1)D^{\mathrm{s}} + D^{\mathrm{p}}}{d(d+2)},$$

are the diffusion coefficients and d is the dimension of space. Note that the one-point probability density is independent of variable $\boldsymbol{r}$ in the case of homogeneous initial conditions, and Eq. (11.65) assumes the form

$$\frac{\partial}{\partial t}P(t; \boldsymbol{H}) = \left\{D_1\frac{\partial}{\partial H_l}\frac{\partial}{\partial H_k}H_lH_k + D_2\frac{\partial}{\partial H_l}\frac{\partial}{\partial H_l}H_k^2\right\}P(t; \boldsymbol{H}). \tag{11.66}$$

Derive now an expression for the one-point correlation of the magnetic field

$$\langle W_{ij}(t)\rangle = \langle H_i(\boldsymbol{r}, t)H_j(\boldsymbol{r}, t)\rangle$$

in the case of homogeneous initial conditions. Multiplying Eq. (11.66) by H_i and H_j and integrating over $\boldsymbol{H}$, we obtain the equation

$$\frac{\partial}{\partial t}\langle W_{ij}(t)\rangle = 2D_1\langle W_{ij}(t)\rangle + 2D_2\delta_{ij}\langle E(t)\rangle,$$

from which follows the equation for average energy

$$\frac{\partial}{\partial t}\langle E(t)\rangle = 2\frac{d-1}{d}\left(D^{s}+D^{p}\right)\langle E(t)\rangle,$$

whose solution is

$$\langle E(t)\rangle = E_0 \exp\left\{2\frac{d-1}{d}\left(D^{s}+D^{p}\right)t\right\}. \tag{11.67}$$

Now, the solution to the equation in the magnetic field correlation is easily found and has the form

$$\frac{\langle W_{ij}(t)\rangle}{\langle E(t)\rangle} = \frac{1}{d}d_{ij} + \left(\frac{W_{ij}(0)}{E_0} - \frac{1}{d}d_{ij}\right)\exp\left\{-2\frac{(d+1)D^{s}+D^{p}}{d+2}t\right\}. \tag{11.68}$$

Thus average energy exponentially increases with time, and this process is accompanied by isotropization of magnetic field, which also goes according to the exponential law.

Introduce now the indicator function of energy of the magnetic field $E(\boldsymbol{r},t) = \boldsymbol{H}^2(\boldsymbol{r},t)$,

$$\varphi(\boldsymbol{r},t;E) = \delta(E(\boldsymbol{r},t)-E),$$

in terms of which probability density of energy $P(\boldsymbol{r},t;E)$ is defined as the equality

$$P(\boldsymbol{r},t;E) = \langle\delta(E(\boldsymbol{r},t)-E)\rangle_{\boldsymbol{u}} = \left\langle\delta(\boldsymbol{H}^2(\boldsymbol{r},t)-E)\right\rangle_{\boldsymbol{H}}.$$

To derive an equation for this function, one should multiply Eq. (11.65) by function $\delta(\boldsymbol{H}^2 - E)$ and integrate the result over $\boldsymbol{H}$. The result will be the equation

$$\left(\frac{\partial}{\partial t} - D_0\frac{\partial^2}{\partial \boldsymbol{r}^2}\right)P(\boldsymbol{r},t;E) = \left\{\alpha\frac{\partial}{\partial E}E + D\frac{\partial}{\partial E}E\frac{\partial}{\partial E}E\right\}P(\boldsymbol{r},t;E),$$
$$P(\boldsymbol{r},0;E) = \delta\left(E - E_0(\boldsymbol{r})\right) \tag{11.69}$$

that coincides with Eq. (8.55), page 207, with the parameters

$$\alpha = 2\frac{d-1}{d+2}\left(D^{p}-D^{s}\right), \quad D = 4(d-1)\frac{(d+1)\,D^{p}+D^{s}}{d(d+2)}.$$

Parameter α can be both positive and negative. In the context of the one-point characteristics, the change of the sign of α means transition from E to $1/E$.

The solution to Eq. (11.69) has the form

$$P(\boldsymbol{r},t;E) = \exp\left\{D_0 t\frac{\partial^2}{\partial \boldsymbol{r}^2}\right\}\widetilde{P}(\boldsymbol{r},t;E),$$

where function $\widetilde{P}(\boldsymbol{r}, t; E)$ satisfies the equation

$$\frac{\partial}{\partial t}\widetilde{P}(\boldsymbol{r}, t; E) = \left\{\alpha\frac{\partial}{\partial E}E + D\frac{\partial}{\partial E}E\frac{\partial}{\partial E}E\right\}\widetilde{P}(\boldsymbol{r}, t; E),$$

$$\widetilde{P}(\boldsymbol{r}, 0; E) = \delta\left(E - E_0(\boldsymbol{r})\right).$$

Then, dependence of function $\widetilde{P}(\boldsymbol{r}, t; E)$ on parameter $\boldsymbol{r}$ appears only through the initial value $E_0(\boldsymbol{r})$,

$$\widetilde{P}(\boldsymbol{r}, t; E) \equiv \widetilde{P}(t; E|E_0(\boldsymbol{r}))$$

and, consequently, function $\widetilde{P}(t; E|E_0(\boldsymbol{r}))$ is the lognormal probability density of random process $E(t, |E_0(\boldsymbol{r}))$ parametrically dependent on $\boldsymbol{r}$,

$$\widetilde{P}(t; E|E_0(\boldsymbol{r})) = \frac{1}{2E\sqrt{\pi Dt}}\exp\left\{-\frac{\ln^2\left[Ee^{\alpha t}/E_0(\boldsymbol{r})\right]}{4Dt}\right\}. \tag{11.70}$$

Thus, the solution to Eq. (11.69) has the form

$$P(\boldsymbol{r}, t; E) = \frac{1}{2E\sqrt{\pi Dt}}\exp\left\{D_0 t\frac{\partial^2}{\partial \boldsymbol{r}^2}\right\}\exp\left\{-\frac{\ln^2\left[Ee^{\alpha t}/E_0(\boldsymbol{r})\right]}{4Dt}\right\}. \tag{11.71}$$

Consequences of Eq. (11.69) or Eq. (11.71) are the expressions for spatial integrals of moment functions

$$\int d\boldsymbol{r}\left\langle E^n(\boldsymbol{r}, t)\right\rangle = e^{n(nD-\alpha)t}\int d\boldsymbol{r}E_0^n(\boldsymbol{r}),$$

which are independent of coefficient of diffusion D_0 in $\boldsymbol{r}$-space, and, in particular, the expression for average total energy in the whole space,

$$\int d\boldsymbol{r}\left\langle E(\boldsymbol{r}, t)\right\rangle = e^{\gamma t}\int d\boldsymbol{r}E_0(\boldsymbol{r}), \tag{11.72}$$

where parameter

$$\gamma = D - \alpha = \frac{2(d-1)}{d}\left(D^{\mathrm{p}} + D^{\mathrm{s}}\right). \tag{11.73}$$

In the case of spatially homogeneous initial distribution of energy $E_0(\boldsymbol{r}) = E_0$, probability density (11.71) is independent of $\boldsymbol{r}$ and is described by the formula

$$P(t; E) = \frac{1}{2E\sqrt{\pi Dt}}\exp\left\{-\frac{\ln^2\left[Ee^{\alpha t}/E_0\right]}{4Dt}\right\}. \tag{11.74}$$

Thus, in this case, the one-point statistical characteristics of magnetic energy $E(\boldsymbol{r}, t)$ are statistically equivalent to statistical characteristics of random process

$$E(t;\alpha) = \exp\left\{-\alpha t + \int_0^t d\tau \xi(\tau)\right\},$$

where $\xi(t)$ is the Gaussian process of white noise with the parameters

$$\langle \xi(t)\rangle = 0, \ \langle \xi(t)\xi(t')\rangle = 2D\delta(t-t').$$

It satisfies the stochastic equation

$$\frac{d}{dt}E(t;\alpha) = \{-\alpha + \xi(t)\}\, E(t;\alpha), \quad E(0;\alpha) = E_0,$$

and its one-point probability density $P(t; E, \alpha)$ is described by the Fokker–Planck equation

$$\frac{\partial}{\partial t}P(t;E) = \left(\alpha\frac{\partial}{\partial E}E + D\frac{\partial}{\partial E}E\frac{\partial}{\partial E}E\right)P(t;E), \quad P(0;E) = \delta(E-E_0), \quad (11.75)$$

whose solution is given by Eq. (11.74).

A characteristic feature of distribution (11.74) consists in the appearance of a long flat *tail* for $Dt \gg 1$, which is indicative of an increased role of great peaks of process $E(t;\alpha)$ in the formation of the one-point statistics. For this distribution, all moments of magnetic energy are functions exponentially increasing with time

$$\langle E^n(t)\rangle = E_0^n \exp\left\{-2n\frac{d-1}{d+2}\left(D^{\mathrm{p}} - D^{\mathrm{s}}\right)t + 4n^2(d-1)\frac{(d+1)\,D^{\mathrm{p}} + D^{\mathrm{s}}}{d(d+2)}t\right\}$$

and, in particular case of $n = 1$, the specific average energy is given by Eq. (11.67). Moreover, quantity

$$\langle \ln\left(E(t)/E_0\right)\rangle = -\alpha t = -2\frac{d-1}{d+2}\left(D^{\mathrm{p}} - D^{\mathrm{s}}\right)t,$$

so that parameter $\{-\alpha\}$ is the *characteristic Lyapunov exponent* [14]. In addition, the *typical realization curve* of random process $E(t)$, which determines the behavior of magnetic energy at arbitrary spatial point in separate realizations, is the exponential function

$$E^*(t) = E_0 e^{-\alpha t} = E_0 \exp\left\{-2\frac{d-1}{d+2}\left(D^{\mathrm{p}} - D^{\mathrm{s}}\right)t\right\}$$

that can both increase and decrease with time. Indeed, for $\alpha < 0$ $(D^{\mathrm{p}} < D^{\mathrm{s}})$, the typical realization curve exponentially increases with time, which is evidence of general increase of magnetic energy at every spatial point. Otherwise, for $\alpha > 0$ $(D^{\mathrm{p}} > D^{\mathrm{s}})$, the typical realization curve exponentially decreases at every spatial point, which is indicative of cluster structure of energy field; the increase of moments of magnetic energy is determined in this case by occasional, but great peaks of magnetic energy against the typical realization curve, which are characteristic of lognormal processes. Figure 8.4, page 209, schematically shows random realizations of magnetic field energy in random velocity field for different signs of parameter α.

The indicator function of magnetic field energy yields general understanding of the spatial structure of the energy field. In particular, functionals of the energy field such as the total volume (in the three-dimensional case) or area (in the two-dimensional case) of the region in which $E(\boldsymbol{r}, t) > E$,

$$V(t, E) = \int d\boldsymbol{r}\, \theta\left(E(\boldsymbol{r}, t) - E\right) = \int d\boldsymbol{r} \int_{E}^{\infty} d\widetilde{E}\, \delta\bigl(E(\boldsymbol{r}, t) - \widetilde{E}\bigr),$$

and the total energy contained in this region,

$$\mathcal{E}(t, E) = \int d\boldsymbol{r}\, E(\boldsymbol{r}, t)\theta\left(E(\boldsymbol{r}, t) - E\right) = \int d\boldsymbol{r} \int_{E}^{\infty} \widetilde{E} d\widetilde{E}\, \delta\bigl(E(\boldsymbol{r}, t) - \widetilde{E}\bigr),$$

whose averages are determined in terms of the one-point probability density (11.71), are given in the general case by the equalities

$$\langle V(t, E)\rangle = \int_{E}^{\infty} d\widetilde{E} \int d\boldsymbol{r}\, P(\boldsymbol{r}, t; \widetilde{E}),$$

$$\langle \mathcal{E}(t, E)\rangle = \int_{E}^{\infty} \widetilde{E} d\widetilde{E} \int d\boldsymbol{r}\, P(\boldsymbol{r}, t; \widetilde{E}).$$

The average values of these functionals are independent of the coefficient of diffusion in $\boldsymbol{r}$-space (coefficient D_0), and, using probability density (11.71), we obtain the expressions (8.58), page 210

$$\langle V(t, E)\rangle = \int d\boldsymbol{r}\, \Pr\left(\frac{1}{\sqrt{2Dt}} \ln\left(\frac{E_0(\boldsymbol{r})}{E} e^{-\alpha t}\right)\right),$$

$$\langle \mathcal{E}(t, E)\rangle = e^{\gamma t} \int d\boldsymbol{r}\, E_0(\boldsymbol{r}) \Pr\left(\frac{1}{\sqrt{2Dt}} \ln\left(\frac{E_0(\boldsymbol{r})}{E} e^{(2D-\alpha)t}\right)\right),$$

where function $\Pr(z)$ is defined as (4.20), page 94.

Asymptotic expressions of function $\Pr(z)$ for $z \to \infty$ and $z \to -\infty$ (4.23), page 95, offer a possibility of studying temporal evolution of these functionals. Namely, for $\alpha > 0$, average volume asymptotically decays with time ($t \to \infty$) according to the law

$$\langle V(t, E)\rangle \approx \frac{1}{\alpha}\sqrt{\frac{D}{\pi E^{\alpha/D} t}} e^{-\alpha^2 t/4D} \int d\boldsymbol{r} \sqrt{E_0^{\alpha/D}(\boldsymbol{r})}.$$

On the contrary, for $\alpha < 0$, average volume occupies the whole space for $t \to \infty$.

In both cases, asymptotic behavior of total energy for $t \to \infty$ has the form (because $\alpha < 2D$)

$$\langle \mathcal{E}(t, E)\rangle \approx e^{\gamma t} \int d\boldsymbol{r} E_0(\boldsymbol{r}) \left[1 - \frac{1}{(2D-\alpha)}\sqrt{\frac{D}{\pi t}\left(\frac{E}{E_0(\boldsymbol{r})}\right)^{(2D-\alpha)/D}} e^{-(2D-\alpha)^2 t/4}\right],$$

where parameter γ is given by Eq. (11.73), which means that 100% of the total average energy is contained in clusters for $\alpha > 0$.

In the case of homogeneous initial conditions, the corresponding expressions without integration over $\boldsymbol{r}$ present specific values of volume occupied by large peaks and their total energy per unit volume:

$$\begin{aligned} \langle \mathfrak{V}(t, E)\rangle &= \Pr\left(\frac{1}{\sqrt{2Dt}} \ln\left(\frac{E_0(\boldsymbol{r})}{E} e^{-\alpha t}\right)\right), \\ \langle \mathfrak{E}(t, E)\rangle &= E_0 e^{\gamma t} \Pr\left(\frac{1}{\sqrt{2Dt}} \ln\left(\frac{E_0(\boldsymbol{r})}{E} e^{(2D-\alpha)t}\right)\right), \end{aligned} \tag{11.76}$$

where parameter $\gamma = D - \alpha$.

If we take section at level $E > E_0$, then the initial values of these quantities will be equal to zero at the initial instant, $\langle \mathfrak{V}(0, E)\rangle = 0$ and $\langle \mathfrak{E}(0, E)\rangle = 0$. Then, spatial disturbances of energy field appear with time and, for $t \to \infty$, they are given by the following asymptotic expressions:

$$\langle \mathfrak{V}(t, E)\rangle \approx \begin{cases} \dfrac{1}{\alpha}\sqrt{\dfrac{D}{\pi t}\left(\dfrac{E_0}{E}\right)^{\alpha/D}} e^{-\alpha^2 t/4D} \quad (\alpha > 0), \\ 1 - \dfrac{1}{|\alpha|}\sqrt{\dfrac{D}{\pi t}\left(\dfrac{E}{E_0}\right)^{|\alpha|/D}} e^{-|\alpha|^2 t/4D} \quad (\alpha < 0) \end{cases}$$

and, because $(2D - \alpha) > 0$,

$$\langle \mathfrak{E}(t, E))\rangle \approx E_0 e^{\gamma t} \left[1 - \frac{1}{(2D-\alpha)}\sqrt{\frac{D}{\pi t}\left(\frac{E}{E_0}\right)^{(2D-\alpha)/D}} e^{-(2D-\alpha)^2 t/4D}\right]$$

Thus, for $\alpha > 0$ ($D^{\rm p} > D^{\rm s}$), the specific total volume tends to zero and specific total energy contained in this volume coincides with average energy in the whole space, which is evidence of clustering the magnetic field.

For $\alpha < 0$ ($D^{\rm p} < D^{\rm s}$), no clustering occurs and specific volume occupies the whole space in which specific average energy increases with time.

11.5 Integral One-Point Statistical Characteristics of Passive Vector Fields

Above, we derived the equations for the one-point probability densities of the density field and magnetic field under the assumption that effects of dynamic diffusion are absent. The one-point probability densities allow calculating arbitrary one-point characteristics of these fields. Combined with the ideas of statistical topography, they are sufficient to obtain the conditions of possible formation of cluster structures. However, the analysis of derivatives of these fields requires the knowledge of at least the two-point probability densities. In principle, the equations for such probability densities can be obtained in a standard manner, by using the general procedure for the linear partial differential equations of the first order. However, this derivation requires very cumbersome calculations, and examination of consequences of such description is a very difficult task. Moreover, effects of dynamic diffusion cannot be included in such probabilistic description.

We note however that, in the case of the delta-correlated random velocity field with absent average flow, one can easily pass from linear Equations (11.1) and (11.2) to closed equations for both average values of these fields by themselves and their higher multi-point correlation functions.

For example, averaging Eq. (11.1), page 271, with the use of the Furutsu–Novikov formula (11.14), page 276, and the expression for variational derivative

$$\frac{\delta\rho(\boldsymbol{r},t)}{\delta u_\alpha(\boldsymbol{r}',t-0)} = -\frac{\partial}{\partial r_\alpha}\delta(\boldsymbol{r}-\boldsymbol{r}')\rho(\boldsymbol{r},t)$$

following from Eq. (11.1), page 271, we obtain the equation for tracer average density

$$\frac{\partial}{\partial t}\langle\rho(\boldsymbol{r},t)\rangle = \left(D_0+\mu_\rho\right)\Delta\langle\rho(\boldsymbol{r},t)\rangle, \tag{11.77}$$

where coefficient D_0 is given by Eq. (4.54), page 107. Under condition $D_0 \gg \mu$ ($\mu \ll \sigma_u^2 l_{\rm cor}^2$, where σ_u^2 is the variance and $l_{\rm cor}$ is the spatial correlation radius of random velocity field), Eq. (11.77) coincides with the equation for the probability distribution of particle coordinates (11.19), page 277, and, consequently, diffusion coefficient D_0 characterizes only the scales of the region of tracer concentration in large and give no data about the local structure of density realizations, as it was the case for the diffusion in the nondivergent random velocity field. In the case of the homogeneous initial condition $\rho(\boldsymbol{r}, t) = \rho_0$, we obtain that $\langle \rho(\boldsymbol{r}, t) \rangle = \rho_0$ too, because random field $\rho(\boldsymbol{r}, t)$ is the nonstationary homogeneous random field in this case, and average value $\langle \rho(\boldsymbol{r}, t) \rangle$ is independent of $\boldsymbol{r}$.

For this reason, it remains only one way to analyze the formulated problems, which consists in studying the two-point correlation functions of the density field $R(\boldsymbol{r}, \boldsymbol{r}_1, t) = \langle \rho(\boldsymbol{r}, t)\rho(\boldsymbol{r}_1, t) \rangle$ and magnetic field $W_{ij}(\boldsymbol{r}, \boldsymbol{r}_1, t) = \langle H_i(\boldsymbol{r}, t)H_j(\boldsymbol{r}_1, t) \rangle$. In the case of the homogeneous initial conditions, these correlation functions depend on difference $(\boldsymbol{r} - \boldsymbol{r}_1)$, which significantly simplifies the consideration. As we mentioned earlier, all statistical averages in this case are the integral quantities, namely, the specific (per unit volume) quantities.

Various correlations of spatial derivatives of the fields under consideration can be obtained by successive differentiation of the above correlation functions with respect to spatial variables. The original stochastic equations must contain here dissipative terms.

11.5.1 Spatial Correlation Function of Density Field

We start from Eq. (11.1) and first of all draw a stochastic dynamic equation in function $R(\boldsymbol{r}, \boldsymbol{r}_1, t) = \rho(\boldsymbol{r}, t)\rho(\boldsymbol{r}_1, t)$,

$$\frac{\partial}{\partial t} R(\boldsymbol{r}, \boldsymbol{r}_1, t) = -\left(\frac{\partial}{\partial r_i} u_i(\boldsymbol{r}, t) + \frac{\partial}{\partial r_{1k}} u_k(\boldsymbol{r}_1, t)\right) R(\boldsymbol{r}, \boldsymbol{r}_1, t) + \mu_\rho \left(\frac{\partial^2}{\partial \boldsymbol{r}^2} + \frac{\partial^2}{\partial \boldsymbol{r}_1^2}\right) R(\boldsymbol{r}, \boldsymbol{r}_1, t). \tag{11.78}$$

Average Eq. (11.78) over an ensemble of realizations of random velocity field. Using then Furutsu–Novikov formula (11.14), page 276, we obtain that spatial correlation function

$$\langle R(\boldsymbol{r} - \boldsymbol{r}_1, t) \rangle = \langle R(\boldsymbol{r}, \boldsymbol{r}_1, t) \rangle = \langle \rho(\boldsymbol{r}, t)\rho(\boldsymbol{r}_1, t) \rangle$$

satisfies the following equality

$$\frac{\partial}{\partial t} \langle R(\boldsymbol{r} - \boldsymbol{r}_1, t) \rangle = -\int d\boldsymbol{R} \left(\frac{\partial}{\partial r_i} B_{ij}(\boldsymbol{r} - \boldsymbol{R}) + \frac{\partial}{\partial r_{1k}} B_{kj}(\boldsymbol{r}_1 - \boldsymbol{R})\right) \times \left\langle \frac{\delta R(\boldsymbol{r}, \boldsymbol{r}_1, t)}{\delta u_j(\boldsymbol{R}, t - 0)} \right\rangle + 2\mu_\rho \langle R_{kk}(\boldsymbol{r} - \boldsymbol{r}_1, t) \rangle, \tag{11.79}$$

where $\langle R_{kk}(\boldsymbol{r} - \boldsymbol{r}_1, t) \rangle = \frac{\partial^2}{\partial \boldsymbol{r}^2} \langle R(\boldsymbol{r} - \boldsymbol{r}_1, t) \rangle$, i.e., we use index notation for spatial derivatives of function $\langle R(\boldsymbol{r}, t) \rangle$. Note that $\langle R_{kk}(\boldsymbol{0}, t) \rangle < 0$.

To calculate the variational derivative, omit in Eq. (11.78) the dynamic diffusion term, which has no explicit dependence on the velocity field, and write it in the form

$$\frac{\partial}{\partial t}R(\boldsymbol{r},\boldsymbol{r}_1,t) = -\left(u_i(\boldsymbol{r},t)\frac{\partial}{\partial r_i} + u_k(\boldsymbol{r}_1,t)\frac{\partial}{\partial r_{1k}}\right)R(\boldsymbol{r},\boldsymbol{r}_1,t) - \left(\frac{\partial u_i(\boldsymbol{r},t)}{\partial r_i} + \frac{\partial u_k(\boldsymbol{r}_1,t)}{\partial r_{1k}}\right)R(\boldsymbol{r},\boldsymbol{r}_1,t),$$

from which follows the equality

$$\left\langle\frac{\delta R(\boldsymbol{r},\boldsymbol{r}_1,t)}{\delta u_j(\boldsymbol{R},t-0)}\right\rangle = -\left(\delta(\boldsymbol{r}-\boldsymbol{R})\frac{\partial}{\partial r_j} + \delta(\boldsymbol{r}_1-\boldsymbol{R})\frac{\partial}{\partial r_{1j}}\right)\langle R(\boldsymbol{r}-\boldsymbol{r}_1,t)\rangle - \left(\frac{\partial\delta(\boldsymbol{r}-\boldsymbol{R})}{\partial r_j} + \frac{\partial\delta(\boldsymbol{r}_1-\boldsymbol{R})}{\partial r_{1j}}\right)\langle R(\boldsymbol{r}-\boldsymbol{r}_1,t)\rangle. \tag{11.80}$$

Substituting now Eq. (11.80) in Eq. (11.79) and integrating the result over $\boldsymbol{R}$, we obtain the partial differential equation in the correlation function of the density field. This equation can be rewritten in the form ($\boldsymbol{r}-\boldsymbol{r}_1 \to \boldsymbol{r}$)

$$\begin{aligned}\frac{\partial}{\partial t}\langle R(\boldsymbol{r},t)\rangle = &-\frac{\partial^2[B_{ij}(\boldsymbol{r})+B_{ji}(\boldsymbol{r})]}{\partial r_j\partial r_i}\langle R(\boldsymbol{r},t)\rangle \\ &-\frac{\partial[B_{ij}(\boldsymbol{r})+B_{ji}(\boldsymbol{r})]}{\partial r_i}\langle R_j(\boldsymbol{r},t)\rangle - \frac{\partial[B_{ij}(\boldsymbol{r})+B_{ji}(\boldsymbol{r})]}{\partial r_j}\langle R_i(\boldsymbol{r},t)\rangle \\ &+\left[2B_{ij}(\boldsymbol{0})-B_{ij}(\boldsymbol{r})-B_{ji}(\boldsymbol{r})\right]\langle R_{ji}(\boldsymbol{r},t)\rangle + 2\mu_\rho\langle R_{kk}(\boldsymbol{r},t)\rangle.\end{aligned} \tag{11.81}$$

In the special case of isotropic random velocity field, random density field $\rho(\boldsymbol{r},t)$ will be the homogeneous isotropic random field. In this case the equation for the correlation function becomes simpler and assumes the form

$$\frac{\partial}{\partial t}\langle R(\boldsymbol{r},t)\rangle = 2\mu_\rho\Delta\langle R(\boldsymbol{r},t)\rangle + \frac{\partial^2}{\partial r_i\partial r_j}D_{ij}(\boldsymbol{r})\langle R(\boldsymbol{r},t)\rangle,$$

where

$$D_{ij}(\boldsymbol{r}) = 2\left[B_{ij}(0)-B_{ij}(\boldsymbol{r})\right]$$

is the structure matrix of vector field $\boldsymbol{u}(\boldsymbol{r},t)$.

In the absence of dynamic diffusion, this equation coincides with the equation for the probability density of relative diffusion of two particles.

Correlation function $\langle R(\boldsymbol{r},t)\rangle$ will now depend on the modulus of vector $\boldsymbol{r}$, i.e., $\langle R(\boldsymbol{r},t)\rangle = \langle R(r,t)\rangle$ and, as a function of variables $\{r,t\}$, will satisfy the equation

$$\frac{\partial}{\partial t}\langle R(r,t)\rangle = \frac{1}{r^{d-1}}\frac{\partial}{\partial r}r^{d-1}\left[\frac{\partial D_{ii}(r)}{\partial r} + \left(2\mu + \frac{r_ir_j}{r^2}D_{ij}(\boldsymbol{r})\right)\frac{\partial}{\partial r}\right]\langle R(r,t)\rangle,$$

where d is the dimension of space, as earlier. This equation has the steady-state solution $\langle R(r,t)\rangle = \langle R(r)\rangle$ for $t \to \infty$ [65,66]

$$\langle R(r)\rangle = \rho_0^2 \exp\left\{ \int\limits_r^\infty dr' \frac{\partial D_{ii}(r')/\partial r'}{2\mu + r_i' r_j' D_{ij}(\boldsymbol{r}')/r'^2} \right\},$$

which corresponds to the boundary condition

$$R(\infty) = \rho_0^2.$$

Note that quantity $\langle R(0)\rangle = \left\langle \rho^2(\boldsymbol{r},t)\right\rangle_{t\to\infty}$ at $r = 0$ and, hence,

$$\left\langle \rho^2(\boldsymbol{r},t)\right\rangle_{t\to\infty} = \rho_0^2 \exp\left\{ \int\limits_0^\infty dr' \frac{\partial D_{ii}(r')/\partial r'}{2\mu + r_i' r_j' D_{ij}(\boldsymbol{r}')/r'^2} \right\} > \rho_0^2. \tag{11.82}$$

Now, we turn back to the general case of random velocity field not having specular symmetry (i.e., possessing helicity). Setting $\boldsymbol{r} = \boldsymbol{0}$ in Eq. (11.81) and taking into account Eq. (4.57), page 107, we arrive at the nonclosed equation

$$\left(\frac{\partial}{\partial t} - 2D^{\mathrm{p}}\right)\langle R(\boldsymbol{0},t)\rangle = 2\mu_\rho \langle R_{kk}(\boldsymbol{0},t)\rangle, \quad \langle R(\boldsymbol{0},0)\rangle = \rho_0^2. \tag{11.83}$$

To approximately solve Eq. (11.83), we construct an approximate procedure based on the expansion of its right-hand side in series in small parameter μ_ρ. With this goal in view, mark quantities $\langle R(\boldsymbol{0},t)\rangle_\mu$ and $\langle R_{kk}(\boldsymbol{0},t)\rangle_\mu$ by subscript μ and rewrite Eq. (11.83) in the form of nonclosed integral equation

$$\langle R(\boldsymbol{0},t)\rangle_\mu = \rho_0^2 \exp\left\{ 2D^{\mathrm{p}} t + \int\limits_0^t d\tau \frac{2\mu_\rho}{\langle R(\boldsymbol{0},\tau)\rangle_\mu} \langle R_{kk}(\boldsymbol{0},\tau)\rangle_\mu \right\}. \tag{11.84}$$

At the initial step when dissipation can be neglected, the solution has the form

$$\langle R(\boldsymbol{0},t)\rangle_0 = \left\langle \rho^2(\boldsymbol{r},t)\right\rangle_0 = \rho_0^2 e^{2D^{\mathrm{p}} t}, \tag{11.85}$$

which, naturally, coincides with Eq. (11.48) for $n = 2$ and depends only on the potential component of the spectral function of velocity field.

Then, we represent the right-hand side in the form of a series in parameter μ. In the first approximation the solution of the problem has the following structure

$$\langle R(\boldsymbol{0},t)\rangle_1 = \rho_0^2 \exp\left\{ 2D^{\mathrm{p}} t + \int\limits_0^t d\tau \frac{2\mu_\rho}{\langle R(\boldsymbol{0},\tau)\rangle_0} \langle R_{kk}(\boldsymbol{0},\tau)\rangle_0 \right\}. \tag{11.86}$$

Running a few steps forward, we note that, in view of the exponential increase with time of all moment functions in the problem under consideration, solution (11.86) exponentially increases for small times, reaches a maximum at time instant $t_{\max}$, and then rapidly decreases with time in accordance with physical meaning of the problem under consideration.

Note now that a corollary of the structure of Eq. (11.81) is the fact that helicity of the velocity field has no effect on statistics of the density field gradient. Indeed, all coefficients of this equation contain only symmetric matrix $\left[B_{ij}(\mathbf{r}) + B_{ji}(\mathbf{r})\right]$, and all odd-order derivatives of such a matrix with respect to $\mathbf{r}$ vanish at $\mathbf{r} = \mathbf{0}$ (see Eq. (4.58), page 107). Odd-order derivatives of function $\langle R(\mathbf{r}, t)\rangle$ vanish at $\mathbf{r} = \mathbf{0}$ in exactly the same way. In view of these facts, we can say that helicity of the gradient of tracer density is equal to zero, because it is determined by quantity $\left\langle \dfrac{\partial \rho(\mathbf{r}, t)}{\partial r_i} \dfrac{\partial^2 \rho(\mathbf{r}, t)}{\partial r_j \partial r_k} \right\rangle = \left\langle R_{ijk}(\mathbf{0}, t)\right\rangle$.

11.5.2 Spatial Correlation Tensor of Density Field Gradient and Dissipation of Density Field Variance

Introduce now the vector of density field gradient $p_k(\mathbf{r}, t) = \dfrac{\partial \rho(\mathbf{r}, t)}{\partial r_k}$; then, the spatial correlation tensor of density field gradient is defined by the equality

$$P_{kl}(\mathbf{r} - \mathbf{r}_1, t) = \left\langle \frac{\partial \rho(\mathbf{r}, t)}{\partial r_k} \frac{\partial \rho(\mathbf{r}_1, t)}{\partial r_{1l}} \right\rangle \equiv -\langle R_{kl}(\mathbf{r} - \mathbf{r}_1, t)\rangle,$$

and variance of the density field gradient is given by the formula

$$\sigma_p^2(t) = \left\langle \left(\frac{\partial \rho(\mathbf{r}, t)}{\partial \mathbf{r}} \right)^2 \right\rangle \equiv -\langle R_{kk}(\mathbf{0}, t)\rangle. \tag{11.87}$$

Note that, for $t \to \infty$, expression for steady-state variance of the density field gradient follows from Eq. (11.83)

$$\sigma_p^2(\infty) = \frac{D^{\mathrm{p}}}{\mu_\rho} \langle R(\mathbf{0}, \infty)\rangle.$$

To determine temporal evolution of the spatial correlation tensor of density field gradient, we differentiate Eq. (11.81) with to r_k. As a result, we obtain the equation

$(\boldsymbol{r} - \boldsymbol{r}_1 \to \boldsymbol{r})$

$$\begin{aligned}
\frac{\partial}{\partial t} \langle R_k(\boldsymbol{r}, t)\rangle_0 = &-\frac{\partial^3[B_{ij}(\boldsymbol{r}) + B_{ji}(\boldsymbol{r})]}{\partial r_i \partial r_j \partial r_k} \langle R(\boldsymbol{r}, t)\rangle_0 \\
&- \frac{\partial^2[B_{ij}(\boldsymbol{r}) + B_{ji}(\boldsymbol{r})]}{\partial r_i \partial r_k} \left\langle R_j(\boldsymbol{r}, t)\right\rangle_0 - \frac{\partial^2[B_{ij}(\boldsymbol{r}) + B_{ji}(\boldsymbol{r})]}{\partial r_j \partial r_k} \langle R_i(\boldsymbol{r}, t)\rangle_0 \\
&- \frac{\partial^2[B_{ij}(\boldsymbol{r}) + B_{ji}(\boldsymbol{r})]}{\partial r_i \partial r_j} \langle R_k(\boldsymbol{r}, t)\rangle_0 - \frac{\partial[B_{ij}(\boldsymbol{r}) + B_{ji}(\boldsymbol{r})]}{\partial r_k} \left\langle R_{ji}(\boldsymbol{r}, t)\right\rangle_0 \\
&- \frac{\partial[B_{ij}(\boldsymbol{r}) + B_{ji}(\boldsymbol{r})]}{\partial r_i} \left\langle R_{jk}(\boldsymbol{r}, t)\right\rangle_0 - \frac{\partial[B_{ij}(\boldsymbol{r}) + B_{ji}(\boldsymbol{r})]}{\partial r_j} \langle R_{ik}(\boldsymbol{r}, t)\rangle_0 \\
&- \left[2B_{ij}(\boldsymbol{0}) - B_{ij}(\boldsymbol{r}) - B_{ji}(\boldsymbol{r})\right] \left\langle R_{jik}(\boldsymbol{r}, t)\right\rangle_0 .
\end{aligned}$$

Vector $\langle R_k(\boldsymbol{r}, t)\rangle_0$ describes the spatial correlation of the density field and its gradient, and all terms of this equation are identically equal to zero at $\boldsymbol{r} = 0$.

Differentiating this equation with respect to r_l and setting $\boldsymbol{r} = 0$, we obtain the equation

$$\begin{aligned}
\frac{\partial}{\partial t} \langle R_{kl}(\boldsymbol{0}, t)\rangle_0 = &-2\frac{\partial^4 B_{ij}(\boldsymbol{0})}{\partial r_i \partial r_j \partial r_k \partial r_l} \langle R(\boldsymbol{0}, t)\rangle_0 \\
&- 2\frac{\partial^2 B_{ij}(\boldsymbol{0})}{\partial r_i \partial r_k} \left\langle R_{jl}(\boldsymbol{0}, t)\right\rangle_0 - 2\frac{\partial^2 B_{ij}(\boldsymbol{0})}{\partial r_i \partial r_l} \left\langle R_{jk}(\boldsymbol{0}, t)\right\rangle_0 \\
&- 2\frac{\partial^2 B_{ij}(\boldsymbol{0})}{\partial r_k \partial r_l} \left\langle R_{ji}(\boldsymbol{0}, t)\right\rangle_0 - 2\frac{\partial^2 B_{ij}(\boldsymbol{0})}{\partial r_i \partial r_j} \langle R_{kl}(\boldsymbol{0}, t)\rangle_0 \\
&- 2\frac{\partial^2 B_{ij}(\boldsymbol{0})}{\partial r_j \partial r_k} \langle R_{il}(\boldsymbol{0}, t)\rangle_0 - 2\frac{\partial^2 B_{ij}(\boldsymbol{0})}{\partial r_j \partial r_l} \langle R_{ik}(\boldsymbol{0}, t)\rangle_0 .
\end{aligned}$$

Using Eq. (4.57), page 107, we can rewrite this equation in the form

$$\begin{aligned}
\frac{\partial}{\partial t} \langle R_{kl}(\boldsymbol{0}, t)\rangle_0 = &-2\frac{\partial^4 B_{ij}(\boldsymbol{0})}{\partial r_i \partial r_j \partial r_k \partial r_l} \langle R(\boldsymbol{0}, t)\rangle_0 \\
&+ 2\frac{D^{\mathrm{s}}}{d(d+2)} \left[(d+1)\delta_{kl} \langle R_{ii}(\boldsymbol{0}, t)\rangle_0 - 2\langle R_{kl}(\boldsymbol{0}, t)\rangle_0\right] \\
&+ 2\frac{(4+d)\, D^{\mathrm{p}}}{d} \langle R_{kl}(\boldsymbol{0}, t)\rangle_0 + 2\frac{D^{\mathrm{p}}}{d(d+2)} \left[\delta_{kl} \langle R_{ii}(\boldsymbol{0}, t)\rangle_0 + 2\langle R_{kl}(\boldsymbol{0}, t)\rangle_0\right].
\end{aligned}$$

Then, because the source of the gradient field

$$\begin{aligned}
2\frac{\partial^4 B_{ij}(\boldsymbol{0})}{\partial r_i \partial r_j \partial r_k \partial r_l} &= -\frac{2}{d}\int\limits_0^\infty d\tau \left\langle \frac{\partial \boldsymbol{u}(\boldsymbol{r}, t)}{\partial \boldsymbol{r}} \Delta \frac{\partial \boldsymbol{u}(\boldsymbol{r}, t-\tau)}{\partial \boldsymbol{r}} \right\rangle \delta_{kl} \\
&= \frac{1}{d} D_\rho^{(4)} \delta_{kl}, \quad (D_\rho^{(4)} > 0)
\end{aligned}$$

is isotropic, tensor $\langle R_{kl}(\mathbf{0}, t)\rangle$ is also isotropic, i.e.,

$$\langle R_{kl}(\mathbf{0}, t)\rangle_0 = \frac{1}{d} \langle R_{ii}(\mathbf{0}, t)\rangle_0 \, \delta_{kl}.$$

Consequently, the variance of the density field gradient (11.87) satisfies the equation

$$\frac{\partial}{\partial t} \langle R_{kk}(\mathbf{0}, t)\rangle_0 = -D_\rho^{(4)} \langle R(\mathbf{0}, t)\rangle_0 + 2\frac{D^{\mathrm{s}}(d-1) + (d+5)\, D^{\mathrm{p}}}{d} \langle R_{kk}(\mathbf{0}, t)\rangle_0, \tag{11.88}$$

where the second moment of density field $\langle R(\mathbf{0}, t)\rangle_0$ is given by Eq. (11.85).

The solution to Eq. (11.88) has the form

$$\sigma_p^2(t) = \frac{D_\rho^{(4)} \langle R(\mathbf{0}, t)\rangle_0}{A} \left[e^{At} - 1\right], \tag{11.89}$$

where parameter $A = 2\dfrac{D^{\mathrm{s}}(d-1) + 5D^{\mathrm{p}}}{d}$.

Note that this solution is generated by the potential component of spectral density of random velocity field; however, in the case of weak compressibility ($D^{\mathrm{p}} \ll D^{\mathrm{s}}$) the increment of the exponential increase in time of the solution coincides with the increment of the variance of density field gradient in noncompressible flow of liquid with the inhomogeneous initial distribution, because parameter $A = 2\dfrac{D^{\mathrm{s}}(d-1)}{d}$ in this case.

Now, we turn back to Eq. (11.86) and use Eq. (11.89) to rewrite it in the final form

$$\left\langle \rho^2(t) \right\rangle_1 = \rho_0^2 \exp\left\{2D^{\mathrm{p}} t - 2\frac{\mu_\rho D_\rho^{(4)}}{A^2} \left[e^{At} - 1 - At\right]\right\}. \tag{11.90}$$

This solution has a maximum at

$$At_{\max} \approx \ln \frac{AD^{\mathrm{p}}}{\mu_\rho D_\rho^{(4)}},$$

and quantity $\langle R(\mathbf{0}, t)\rangle_1$ reaches at this moment the value

$$\langle R(\mathbf{0}, t)\rangle_{1\max} \approx \rho_0^2 \left(\frac{AD^{\mathrm{p}}}{\mu_\rho D_\rho^{(4)}}\right)^{\frac{2D^{\mathrm{p}}}{A}} \exp\left\{-\frac{2D^{\mathrm{p}}}{A}\right\}.$$

The condition

$$t \ll t_{\max}$$

is the applicability condition of neglecting the effects of dynamic diffusion. For $t > t_{\max}$ function $\langle R(\mathbf{0}, t)\rangle_1$ rapidly decreases with time. As we mentioned earlier, quantity $\langle R(\mathbf{0}, t)\rangle$ in the general case tends to steady-state value (11.82) for $t \to \infty$.

Let us demonstrate that the suggested method of describing temporal dynamics of statistical characteristics with inclusion of the effects of dynamic diffusion remains valid in more general cases, such as describing dynamics of the variance of density gradient (11.87). To obtain this dynamics, we must include the term with dynamic diffusion in Eq. (11.88) and replace subscript 0 with subscript μ. As a result, we arrive at the equation

$$\frac{\partial}{\partial t}\left\langle R_{kk}(\mathbf{0},t)\right\rangle_{\mu} = -D_{\rho}^{(4)}\left\langle R(\mathbf{0},t)\right\rangle_{\mu} + B\left\langle R_{kk}(\mathbf{0},t)\right\rangle_{\mu} + 2\mu_{\rho}\left\langle R_{kkll}(\mathbf{0},t)\right\rangle_{\mu},$$

where coefficient $B = 2\dfrac{D^{\mathrm{s}}(d-1)+(d+5)\,D^{\mathrm{p}}}{d}$. Rewrite this equation in the form

$$\frac{\partial}{\partial t}\left\langle R_{kk}(\mathbf{0},t)\right\rangle_{\mu} = -D_{\rho}^{(4)}\left\langle R(\mathbf{0},t)\right\rangle_{\mu} + \left[B + 2\mu_{\rho}\frac{\left\langle R_{kkll}(\mathbf{0},t)\right\rangle_{\mu}}{\left\langle R_{kk}(\mathbf{0},t)\right\rangle_{\mu}}\right]\left\langle R_{kk}(\mathbf{0},t)\right\rangle_{\mu}.$$

The corresponding solution has the form

$$\left\langle R_{kk}(\mathbf{0},t)\right\rangle_{\mu} = -D_{\rho}^{(4)}\int_{0}^{t} dt_2\left\langle R(\mathbf{0},t_2)\right\rangle_{\mu}\exp\left\{\int_{t_2}^{t} dt_1\left[B + 2\mu_{\rho}\frac{\left\langle R_{kkll}(\mathbf{0},t_1)\right\rangle_{\mu}}{\left\langle R_{kk}(\mathbf{0},t_1\right\rangle_{\mu}}\right]\right\},$$

and, consequently, in the first approximation in the coefficient of molecular diffusion, we obtain the expression

$$\left\langle R_{kk}(\mathbf{0},t)\right\rangle_{1} = -D_{\rho}^{(4)}\int_{0}^{t} dt_2\left\langle R(\mathbf{0},t_2)\right\rangle_{1}\exp\left\{\int_{t_2}^{t} dt_1\left[B + 2\mu_{\rho}\frac{\left\langle R_{kkll}(\mathbf{0},t_1)\right\rangle_{0}}{\left\langle R_{kk}(\mathbf{0},t_1\right\rangle_{0}}\right]\right\},$$

where quantity $\langle R(\mathbf{0},t)\rangle_1$ is described by Eq. (11.90). As for quantity $\langle R_{kkll}(\mathbf{0},t)\rangle_0$, it satisfies the equation without effects of dynamic diffusion, which can be derived by the standard procedure.

Extension to the Case of Inhomogeneous Initial Conditions

We considered a number of problems on dynamics of statistical characteristics of the density field and its gradient and solved them in the simplest statement (with the homogeneous initial conditions) with the minimal set of governing parameters concerning only statistical characteristics of the homogeneous velocity field, which is assumed to be the field delta-correlated in time. In this case, all fields under study are also spatially homogeneous random fields, but they are nonstationary in time. This means that statistical averages like $F_{ij}(\boldsymbol{r},\boldsymbol{r}_1,t) = \left\langle f_i(\boldsymbol{r},t)f_j(\boldsymbol{r}_1,t)\right\rangle$ depend only on the difference of spatial coordinates $(\boldsymbol{r}-\boldsymbol{r}_1)$ and, consequently,

$$\frac{\partial}{\partial \boldsymbol{r}_1}F_{ij}(\boldsymbol{r},\boldsymbol{r}_1,t) = -\frac{\partial}{\partial \boldsymbol{r}}F_{ij}(\boldsymbol{r},\boldsymbol{r}_1,t),$$

i.e., at $\boldsymbol{r} = \boldsymbol{r}_1$, quantity $\left\langle f_i(\boldsymbol{r}, t)\frac{\partial f_j(\boldsymbol{r}, t)}{\partial \boldsymbol{r}}\right\rangle$, independent of $\boldsymbol{r}$ and satisfies the identity

$$\left\langle f_i(\boldsymbol{r}, t)\frac{\partial f_j(\boldsymbol{r}, t)}{\partial \boldsymbol{r}}\right\rangle = -\left\langle f_j(\boldsymbol{r}, t)\frac{\partial f_i(\boldsymbol{r}, t)}{\partial \boldsymbol{r}}\right\rangle, \tag{11.91}$$

which we widely used in derivation of all equations. This relationship significantly simplified the analysis of both dynamic system by itself and obtained results, because majority of terms vanished at $\boldsymbol{r} = \boldsymbol{r}_1$, which means the absence of advection of statistical characteristics in the considered problems. Namely this fact allowed us to completely solve the considered problems and without very cumbersome calculations.

In the case of inhomogeneous initial conditions, solutions of all problems lose the property of spatial homogeneity, and the equations become very cumbersome. However, the solutions obtained above yield certain data in this case, too. Indeed, property (11.91) holds also for integral (integration by parts)

$$\int d\boldsymbol{r}\, f_i(\boldsymbol{r}, t)\frac{\partial f_j(\boldsymbol{r}, t)}{\partial \boldsymbol{r}} = -\int d\boldsymbol{r}\, f_j(\boldsymbol{r}, t)\frac{\partial f_i(\boldsymbol{r}, t)}{\partial \boldsymbol{r}}.$$

Therefore, it is clear that the density field for inhomogeneous initial conditions with absent dynamic diffusion will be described, instead of Eq. (11.85), by the solution of the form

$$\int d\boldsymbol{r} \left\langle \rho^2(\boldsymbol{r}, t)\right\rangle_0 = \int d\boldsymbol{r}\, \rho_0^2(\boldsymbol{r}) e^{2D^{\mathrm{p}}t}, \tag{11.92}$$

and Eq. (11.89) will be replaced with the expression

$$\int d\boldsymbol{r} \left\langle (\nabla \rho(\boldsymbol{r}, t))^2\right\rangle_0 = \int d\boldsymbol{r}\, (\nabla \rho_0(\boldsymbol{r}))^2\, e^{Bt} + \frac{D_\rho^{(4)}}{A}\int d\boldsymbol{r} \left\langle \rho^2(\boldsymbol{r}, t)\right\rangle_0 \left[e^{At} - 1\right], \tag{11.93}$$

where $B = 2\dfrac{D^{\mathrm{s}}(d-1) + (d+5)D^{\mathrm{p}}}{d}$.

Thus, we can assert that the obtained relationships and associations between different quantities become the integral ones in the context of inhomogeneous problems and form, in figurative words, a skeleton (reference points) that provide a background for the dynamics of complicated stochastic motions. In addition, for inhomogeneous problems, all terms vanished in the above consideration have the divergent ('flow-like') form.

In the same way, we can easily draw an analog of Eq. (11.86) for the variance of density field with inclusion of variance dissipation in the case of inhomogeneous problems. For example, solution (11.86) is replaced with the expression for the whole temporal interval

$$\int d\boldsymbol{r}\left\langle \rho^2(\boldsymbol{r},t)\right\rangle_1 = \int d\boldsymbol{r}\,\rho_0^2(\boldsymbol{r})\exp\left\{2D^{\mathrm{p}}t - 2\mu_\rho\int\limits_0^t d\tau \frac{\int d\boldsymbol{r}\left\langle (\nabla\rho(\boldsymbol{r},\tau))^2\right\rangle_0}{\int d\boldsymbol{r}\left\langle \rho^2(\boldsymbol{r},\tau)\right\rangle_0}\right\},$$

where functions in the right-hand side are given by Eqs. (11.92) and (11.93).

Diffusion of Density Field with Constant Gradient

As we mentioned earlier, in the presence of velocity field fluctuations, the initially smooth distribution of tracer becomes increasingly inhomogeneous with time, there appear changes within increasingly shorter scales, and concentration spatial gradients sharpen. In actuality, dynamic diffusion smooths these processes, so that the mentioned behavior of tracer concentration holds only for limited temporal intervals.

In the presence of dynamic diffusion, the tracer diffusion is described in terms of the second-order partial differential equation (11.1), page 271. Problems in which mean concentration gradient assumes nonzero values allow a more complete analysis for the nondivergent (noncompressible) velocity field.

This case corresponds to solving Eq. (11.1) with the following initial conditions (here, we again content ourselves with the two-dimensional case)

$$\rho_0(\boldsymbol{r}) = \boldsymbol{G}\boldsymbol{r}, \quad \boldsymbol{p}_0(\boldsymbol{r}) = \boldsymbol{G}.$$

Substituting the concentration field in the form

$$\rho(\boldsymbol{r},t) = \boldsymbol{G}\boldsymbol{r} + \widetilde{\rho}(\boldsymbol{r},t),$$

we obtain the equation for fluctuating portion $\widetilde{\rho}(\boldsymbol{r},t)$ of the concentration field

$$\left(\frac{\partial}{\partial t} + \frac{\partial}{\partial \boldsymbol{r}}\boldsymbol{u}(\boldsymbol{r},t)\right)\widetilde{\rho}(\boldsymbol{r},t) = -\boldsymbol{G}\boldsymbol{u}(\boldsymbol{r},t) + \mu\Delta\widetilde{\rho}(\boldsymbol{r},t), \quad \widetilde{\rho}(\boldsymbol{r},0) = 0. \qquad (11.94)$$

Unlike the problems discussed earlier, the solution to this problem is the random field statistically homogeneous in space, so that all one-point statistical averages are independent of $\boldsymbol{r}$ and steady-state probability densities exist for $t \to \infty$ for both concentration field and its gradient.

In this case, Eq. (11.94) easily yields the equation

$$\frac{d}{dt}\left\langle \widetilde{\rho}^n(\boldsymbol{r},t)\right\rangle = n(n-1)D_0\boldsymbol{G}^2\left\langle \widetilde{\rho}^{n-2}(\boldsymbol{r},t)\right\rangle - \mu n(n-1)\left\langle \widetilde{\rho}^{n-2}(\boldsymbol{r},t)\widetilde{\boldsymbol{p}}^2(\boldsymbol{r},t)\right\rangle, \qquad (11.95)$$

where

$$\widetilde{\boldsymbol{p}}(\boldsymbol{r},t) = \frac{\partial}{\partial \boldsymbol{r}}\widetilde{\rho}(\boldsymbol{r},t) = \boldsymbol{p}(\boldsymbol{r},t) - \boldsymbol{G}.$$

In the steady-state regime (for $t \to \infty$), we obtain from Eq. (11.95) that

$$\left\langle \widetilde{\rho}^{n-2}(\boldsymbol{r},t)\widetilde{\boldsymbol{p}}^2(\boldsymbol{r},t)\right\rangle = \frac{D_0\boldsymbol{G}^2}{\mu}\left\langle \widetilde{\rho}^{n-2}(\boldsymbol{r},t)\right\rangle. \tag{11.96}$$

For $n = 2$ in particular, we obtain the expression for the variance of concentration gradient fluctuations

$$\lim_{t\to\infty}\left\langle \widetilde{\boldsymbol{p}}^2(\boldsymbol{r},t)\right\rangle = \frac{D_0\boldsymbol{G}^2}{\mu}. \tag{11.97}$$

Rewrite now Eq. (11.95) in the form

$$\frac{d}{dt}\left\langle \widetilde{\rho}^n(\boldsymbol{r},t)\right\rangle = n(n-1)D_0\boldsymbol{G}^2\left\langle f(\boldsymbol{r},t)\widetilde{\rho}^{n-2}(\boldsymbol{r},t)\right\rangle, \tag{11.98}$$

where

$$f(\boldsymbol{r},t) = 1 - \frac{\mu}{D_0\boldsymbol{G}^2}\widetilde{\boldsymbol{p}}^2(\boldsymbol{r},t).$$

Consequently, the variance of concentration is given by the expression ($\langle\widetilde{\rho}(\boldsymbol{r},t)\rangle = 0$)

$$\left\langle \widetilde{\rho}^2(\boldsymbol{r},t)\right\rangle = 2D_0\boldsymbol{G}^2\int\limits_0^t d\tau\,\langle f(\boldsymbol{r},\tau)\rangle. \tag{11.99}$$

Note that the correlation function of field $\widetilde{\rho}(\boldsymbol{r},t)$,

$$\Gamma(\boldsymbol{r},t) = \langle\widetilde{\rho}(\boldsymbol{r}_1,t)\widetilde{\rho}(\boldsymbol{r}_2,t)\rangle\,, \quad \boldsymbol{r} = \boldsymbol{r}_1 - \boldsymbol{r}_2,$$

satisfies the equation

$$\frac{\partial}{\partial t}\Gamma(\boldsymbol{r},t) = 2G_iG_jB_{ij}(\boldsymbol{r}) + 2\left[B_{ij}(0) - B_{ij}(\boldsymbol{r}) + \mu\delta_{ij}\right]\frac{\partial^2}{\partial r_i\partial r_j}\Gamma(\boldsymbol{r},t)$$

that follows from Eq. (11.94); consequently, its steady-state limit

$$\Gamma(\boldsymbol{r}) = \lim_{t\to\infty}\Gamma(\boldsymbol{r},t)$$

satisfies the equation

$$G_iG_jB_{ij}(\boldsymbol{r}) = -\left[B_{ij}(0) - B_{ij}(\boldsymbol{r}) + \mu\delta_{ij}\right]\frac{\partial^2}{\partial r_i\partial r_j}\Gamma(\boldsymbol{r}). \tag{11.100}$$

Setting $\boldsymbol{r} = 0$ in this equation and taking into account Eqs. (4.54), page 107, we arrive at equality (11.97). If we differentiate Eq. (11.100) two times with respect to $\boldsymbol{r}$

and then set $\boldsymbol{r} = 0$ and take into account Eqs. (4.56) and (4.57), page 107, we obtain the equality

$$\mu^2 \left\langle (\Delta \widetilde{\rho}(\boldsymbol{r}, t))^2 \right\rangle = \frac{1}{2} D^{\mathrm{s}} (D_0 + \mu) \boldsymbol{G}^2. \tag{11.101}$$

Exact equalities (11.97) and (11.101) can appear practicable for testing different numerical schemes and checking simulated results. However, the calculation of the steady-state limit $\langle \widetilde{\rho}^2(\boldsymbol{r}, t) \rangle$ for $t \to \infty$ requires the knowledge of temporal behavior of the second moment $\langle \widetilde{\boldsymbol{p}}^2(\boldsymbol{r}, t) \rangle$, which can be obtained only if dynamic diffusion is absent. In this case, the probability density of the concentration gradient satisfies Eq. (11.57); in the problem under consideration, this equation assumes the form

$$\frac{\partial}{\partial t} P(\boldsymbol{r}, t; \boldsymbol{p}) = \frac{1}{8} D^{\mathrm{s}} \left(3 \frac{\partial^2}{\partial \boldsymbol{p}^2} \boldsymbol{p}^2 - 2 \frac{\partial^2}{\partial p_k \partial p_l} p_k p_l \right) P(\boldsymbol{r}, t; \boldsymbol{p}), \tag{11.102}$$
$$P(0, \boldsymbol{r}; \boldsymbol{p}) = \delta \left(\boldsymbol{p} - \boldsymbol{G} \right).$$

Consequently, according to Eq. (11.61), we have

$$\left\langle |\widetilde{\boldsymbol{p}}(\boldsymbol{r}, t)|^2 \right\rangle = \boldsymbol{G}^2 \left\{ e^{D^{\mathrm{s}} t} - 1 \right\}. \tag{11.103}$$

The exact formula (11.97) and Eq. (11.103) give a possibility of estimating the time required for quantity $\langle \widetilde{\boldsymbol{p}}^2(\boldsymbol{r}, t) \rangle$ to approach the steady-state regime for $t \to \infty$; namely,

$$D^{\mathrm{s}} T_0 \sim \ln \left(\frac{D_0 + \mu}{\mu} \right).$$

As a consequence, we obtain from Eq. (11.99) the following estimate of the steady-state variance of concentration field fluctuations

$$\lim_{t \to \infty} \left\langle \widetilde{\rho}^2(\boldsymbol{r}, t) \right\rangle \sim 2 \frac{D_0}{D^{\mathrm{s}}} \boldsymbol{G}^2 \ln \left(\frac{D_0 + \mu}{\mu} \right).$$

Taking into account the fact that $D_0 \sim \sigma_u^2 \tau_0$ and $D_0 / D^{\mathrm{s}} \sim l_0^2$ (σ_u^2 is the variance of velocity field fluctuations and τ_0 and l_0 are this field temporal and spatial correlation radii, respectively), we see that time T_0, in view of its logarithmic dependence on dynamic diffusion coefficient μ, can be not very long,

$$T_0 \approx \frac{l_0^2}{\sigma_u^2 \tau_0} \ln \left(\frac{\sigma_u^2 \tau_0}{\mu} \right),$$

and quantity

$$\left\langle \widetilde{\rho}^2 \right\rangle \sim \boldsymbol{G}^2 l_0^2 \ln \left(\frac{\sigma_u^2 \tau_0}{\mu} \right) \quad \text{for} \quad \mu \ll \sigma_u^2 \tau_0.$$

11.5.3 Spatial Correlation Function of Magnetic Field

As was mentioned earlier, in the general case of the divergent random velocity field, we can calculate one-point statistical characteristics of magnetic field and magnetic energy, such as their probability densities. Under the neglect of dynamic diffusion, this technique is, in principle, applicable for calculating different quantities related to the spatial derivatives of magnetic field. However, the corresponding equations appear to be very cumbersome, and their use for derivation of comprehensible consequences is hardly possible.

In the analysis of different moment functions, we can take into account the coefficient of dynamic diffusion. However, in the general case of the divergent velocity field, all equations will be very cumbersome again. For this reason, we restrict ourselves to the nondivergent velocity field ($\operatorname{div} \boldsymbol{u}(\boldsymbol{r}, t) = 0$), i.e., we will analyze statistical characteristics of spatial derivatives of magnetic field in a noncompressible turbulent flow. Inclusion of compressibility only changes the coefficients of the corresponding equation, but not the basic tendency of the behavior of moment functions.

Rewrite the vector equation (11.2) in the coordinate form,

$$\frac{\partial}{\partial t}H_i(\boldsymbol{r}, t) = -\frac{\partial}{\partial r_k}u_k(\boldsymbol{r}, t)H_i(\boldsymbol{r}, t) + \frac{\partial u_i(\boldsymbol{r}, t)}{\partial r_k}H_k(\boldsymbol{r}, t) + \mu_H \frac{\partial^2}{\partial \boldsymbol{r}^2}H_i(\boldsymbol{r}, t),$$

and introduce function $W_{ij}(\boldsymbol{r}, \boldsymbol{r}_1; t) = H_i(\boldsymbol{r}, t)H_k(\boldsymbol{r}_1, t)$. This function satisfies the equation

$$\begin{aligned}\frac{\partial}{\partial t}W_{ij}(\boldsymbol{r}, \boldsymbol{r}_1; t) = &\left\{-\left(\frac{\partial}{\partial r_k}u_k(\boldsymbol{r}, t) + \frac{\partial}{\partial r_{1k}}u_k(\boldsymbol{r}_1, t)\right)\delta_{in}\delta_{jm}\right.\\ &\left.+\frac{\partial u_i(\boldsymbol{r}, t)}{\partial r_n}\delta_{jm} + \frac{\partial u_j(\boldsymbol{r}_1, t)}{\partial r_{1m}}\delta_{in}\right\} W_{nm}(\boldsymbol{r}, \boldsymbol{r}_1; t)\\ &+\mu_H\left[\frac{\partial^2}{\partial \boldsymbol{r}^2} + \frac{\partial^2}{\partial \boldsymbol{r}_1^2}\right] W_{ij}(\boldsymbol{r}, \boldsymbol{r}_1; t).\end{aligned} \tag{11.104}$$

In view of solenoidal property of the velocity field, Eq. (11.104) can be rewritten in the form

$$\begin{aligned}\frac{\partial}{\partial t}W_{nm}(\boldsymbol{r}, \boldsymbol{r}_1; t) = &\left\{-\left(u_k(\boldsymbol{r}, t)\frac{\partial}{\partial r_k} + u_k(\boldsymbol{r}_1, t)\frac{\partial}{\partial r_{1k}}\right)\delta_{ns}\delta_{mt}\right.\\ &\left.+\frac{\partial u_n(\boldsymbol{r}, t)}{\partial r_s}\delta_{mt} + \frac{\partial u_m(\boldsymbol{r}_1, t)}{\partial r_{1t}}\delta_{ns}\right\} W_{st}(\boldsymbol{r}, \boldsymbol{r}_1; t)\\ &+\mu_H\left[\frac{\partial^2}{\partial \boldsymbol{r}^2} + \frac{\partial^2}{\partial \boldsymbol{r}_1^2}\right] W_{ij}(\boldsymbol{r}, \boldsymbol{r}_1; t).\end{aligned} \tag{11.105}$$

Equation (11.104) is convenient for immediate averaging, and Eq. (11.105) is convenient for calculating the variational derivative.

As earlier, we assume that field $\boldsymbol{u}(\boldsymbol{r}, t)$ is the homogeneous (but generally not isotropic) Gaussian field stationary and delta-correlated in time with correlation function (11.13), page 276 and use the Furutsu–Novikov formula (11.14), page 276 to split correlations. Then, averaging Eq. (11.104) over an ensemble of realizations of random field $\boldsymbol{u}(\boldsymbol{r}, t)$ and using the variational derivative in the form

$$\left\langle \frac{\delta W_{nm}(\boldsymbol{r}, \boldsymbol{r}_1; t)}{\delta u_q(\boldsymbol{R}, t-0)} \right\rangle = \left\{ -\left(\delta(\boldsymbol{r}-\boldsymbol{R}) \frac{\partial}{\partial r_q} + \delta(\boldsymbol{r}_1 - \boldsymbol{R}) \frac{\partial}{\partial r_{1q}} \right) \delta_{ns}\delta_{mt} \right.$$
$$\left. + \frac{\partial \delta(\boldsymbol{r}-\boldsymbol{R})}{\partial r_s} \delta_{nq}\delta_{mt} + \frac{\partial \delta(\boldsymbol{r}_1-\boldsymbol{R})}{\partial r_{1t}} \delta_{mq}\delta_{ns} \right\} \langle W_{st}(\boldsymbol{r}, \boldsymbol{r}_1; t) \rangle$$

following from Eq. (11.105), we obtain the partial differential equation ($\boldsymbol{r} - \boldsymbol{r}_1 \to \boldsymbol{r}$)

$$\begin{aligned} \frac{\partial}{\partial t} \left\langle W_{ij}(\boldsymbol{r}; t) \right\rangle = &-2 \frac{\partial^2 B_{ij}(\boldsymbol{r})}{\partial r_k \partial r_m} \langle W_{km}(\boldsymbol{r}; t) \rangle - 2 \frac{\partial [B_{ik}(\boldsymbol{0}) - B_{ik}(\boldsymbol{r})]}{\partial r_s} \frac{\partial}{\partial r_k} \left\langle W_{sj}(\boldsymbol{r}; t) \right\rangle \\ &- 2 \frac{\partial [B_{kj}(\boldsymbol{0}) - B_{kj}(\boldsymbol{r})]}{\partial r_s} \frac{\partial}{\partial r_k} \langle W_{is}(\boldsymbol{r}; t) \rangle \\ &+ \left[2B_{kq}(\boldsymbol{0}) - B_{kq}(\boldsymbol{r}) - B_{qk}(\boldsymbol{r}) \right] \frac{\partial^2}{\partial r_q \partial r_k} \left\langle W_{ij}(\boldsymbol{r}; t) \right\rangle + 2\mu_H \left\langle W_{ij;ss}(\boldsymbol{r}; t) \right\rangle, \end{aligned} \tag{11.106}$$

where $\left\langle W_{ij;ss}(\boldsymbol{r}; t) \right\rangle$ is the dissipative tensor,

$$\left\langle W_{ij;ss}(\boldsymbol{r}; t) \right\rangle = \frac{\partial^2}{\partial \boldsymbol{r}^2} \left\langle W_{ij}(\boldsymbol{r}; t) \right\rangle.$$

Setting $\boldsymbol{r} = 0$ in Eq. (11.106), we obtain the unclosed equation in the one-point correlation,

$$\frac{\partial}{\partial t} \left\langle W_{ij}(t) \right\rangle = -2 \frac{\partial^2 B_{ij}(\boldsymbol{0})}{\partial r_k \partial r_m} \langle W_{km}(t) \rangle + 2\mu_H \left\langle W_{ij;ss}(\boldsymbol{0}; t) \right\rangle.$$

Then, taking into account the solenoidal portion of Eq. (4.57), page 107, ($D^{\text{p}} = 0$), we arrive at the unclosed equation in the correlation of the form

$$\frac{\partial}{\partial t} \left\langle W_{ij}(t) \right\rangle = \frac{2(d+1)D^{\text{s}}}{d(d+2)} \delta_{ij} \langle E(t) \rangle - \frac{4D^{\text{s}}}{d(d+2)} \left\langle W_{ij}(t) \right\rangle + 2\mu_H \left\langle W_{ij;ss}(\boldsymbol{0}; t) \right\rangle, \tag{11.107}$$

where $\langle E(t) \rangle = \left\langle \boldsymbol{H}^2(\boldsymbol{r}, t) \right\rangle$ is the average energy of magnetic field. Setting $i = j$ in Eq. (11.107), we obtain the equation in average energy

$$\frac{\partial}{\partial t} \langle E(t) \rangle = \frac{2(d-1)D^{\text{s}}}{d} \langle E(t) \rangle - 2\mu_H D(t),$$

where quantity $D(t)$ describing dissipation of average energy of the nonstationary random magnetic field is defined by the equality,

$$D(t) = \left\langle [\mathrm{rot}\,\boldsymbol{H}(\boldsymbol{r},t)]^2 \right\rangle = -\frac{\partial^2}{\partial \boldsymbol{r}^2} \langle \boldsymbol{H}(\boldsymbol{r},t)\boldsymbol{H}(\boldsymbol{r}_1,t)\rangle_{\boldsymbol{r}_1=\boldsymbol{r}}$$

$$= -\frac{\partial^2}{\partial \boldsymbol{r}^2} \langle W_{ii}(\boldsymbol{r}-\boldsymbol{r}_1,t)\rangle_{\boldsymbol{r}_1=\boldsymbol{r}} = -\left\langle W_{ii;jj}(\boldsymbol{0},t)\right\rangle, \tag{11.108}$$

where $W_{ik}(\boldsymbol{r},\boldsymbol{r}_1,t) = H_i(\boldsymbol{r},t)H_k(\boldsymbol{r}_1,t)$ and the derivatives of this function are denoted by additional indices after semicolon ';'. As in the case of the analysis of statistical characteristics of the density field, we mark all quantities by subscript μ and rewrite this equation in the form

$$\frac{\partial}{\partial t}\ln\frac{\langle E(t)\rangle_\mu}{E_0} = \frac{2(d-1)D^{\mathrm{s}}}{d} - \frac{2\mu_H}{\langle E(t)\rangle_\mu} D_\mu(t).$$

Then, we expand the right-hand side of this equation in series in parameter μ. In the first approximation, we have the equation

$$\frac{\partial}{\partial t}\ln\frac{\langle E(t)\rangle_1}{E_0} = \frac{2(d-1)D^{\mathrm{s}}}{d} - \frac{2\mu_H}{\langle E(t)\rangle_0} D_0(t),$$

where quantities with subscript 0 correspond to the solution of the problem under the neglect of the effect of dynamic diffusion. Thus, in the first approximation, the solution to the problem has the structure

$$\langle E(t)\rangle_1 = E_0 \exp\left\{ \frac{2(d-1)D^{\mathrm{s}}t}{d} - \int\limits_0^t d\tau \frac{2\mu_H}{\langle E(\tau)\rangle_0} D_0(\tau) \right\}. \tag{11.109}$$

At the initial evolutionary stage when dissipation can be neglected, the solution to the problem exponentially increases,

$$\langle E(t)\rangle_0 = E_0 \exp\left\{ \frac{2(d-1)}{d} D^{\mathrm{s}} t \right\}, \tag{11.110}$$

and the equation in correlation (11.107) in the absence of dynamic diffusion grades into the equation with a given source, and the solution of the latter equation has the form

$$\frac{\left\langle W_{ij}(t)\right\rangle_0}{\langle E(t)\rangle_0} = \frac{1}{d}\delta_{ij} + \left(\frac{W_{ij}(0)}{E_0} - \frac{1}{d}\delta_{ij} \right) \exp\left\langle -2\left(\frac{d+1}{d+2}\right) D^{\mathrm{s}} t \right\rangle, \tag{11.111}$$

i.e., magnetic field in noncompressible turbulent flow of a liquid becomes rapidly isotropic (fast isotropization of magnetic field). Note that compressibility of the random flow increases the rate of average energy increase and makes isotropization of magnetic field faster (cf. with Eq. (11.68), page 293).

11.5.4 On the Magnetic Field Helicity

In what follows, we neglect dynamic diffusion and omit subscript 0. Here, we derive an equation for quantity

$$\left\langle W_{kp;j}(\boldsymbol{r};t)\right\rangle = \frac{\partial}{\partial r_j}\left\langle W_{kp}(\boldsymbol{r};t)\right\rangle = \left\langle \frac{\partial H_k(\boldsymbol{r},t)}{\partial r_j} H_p(\boldsymbol{r}_1,t)\right\rangle$$

with the goal of calculating magnetic field helicity

$$H(t) = \varepsilon_{ijk}\left\langle W_{ki;j}(\boldsymbol{0};t)\right\rangle = \langle \boldsymbol{H}\cdot \operatorname{rot}\boldsymbol{H}\rangle.$$

Rewriting Eq. (11.106) in the form

$$\begin{aligned}
\frac{\partial}{\partial t}\left\langle W_{kp}(\boldsymbol{r};t)\right\rangle = &-2\frac{\partial^2 B_{kp}(\boldsymbol{r})}{\partial r_n \partial r_m}\langle W_{nm}(\boldsymbol{r};t)\rangle - 2\frac{\partial[B_{kn}(\boldsymbol{0}) - B_{kn}(\boldsymbol{r})]}{\partial r_s}\frac{\partial}{\partial r_n}\left\langle W_{sp}(\boldsymbol{r};t)\right\rangle \\
&- 2\frac{\partial[B_{np}(\boldsymbol{0}) - B_{np}(\boldsymbol{r})]}{\partial r_s}\frac{\partial}{\partial r_n}\langle W_{ks}(\boldsymbol{r};t)\rangle \\
&+ \left[2B_{nq}(\boldsymbol{0}) - B_{nq}(\boldsymbol{r}) - B_{qn}(\boldsymbol{r})\right]\frac{\partial^2}{\partial r_q \partial r_n}\left\langle W_{kp}(\boldsymbol{r};t)\right\rangle
\end{aligned}$$

and differentiating it with respect to r_j, we obtain the equation

$$\begin{aligned}
\frac{\partial}{\partial t}\left\langle W_{kp;j}(\boldsymbol{r};t)\right\rangle = &-\frac{\partial[B_{nq}(\boldsymbol{r}) + B_{qn}(\boldsymbol{r})]}{\partial r_j}\frac{\partial^2}{\partial r_q \partial r_n}\left\langle W_{kp}(\boldsymbol{r};t)\right\rangle \\
&- 2\frac{\partial^3 B_{kp}(\boldsymbol{r})}{\partial r_n \partial r_m \partial r_j}\langle W_{nm}(\boldsymbol{r};t)\rangle - 2\frac{\partial^2 B_{kp}(\boldsymbol{r})}{\partial r_n \partial r_m}\left\langle W_{nm;j}(\boldsymbol{r};t)\right\rangle \\
&+ 2\frac{\partial^2 B_{kn}(\boldsymbol{r})}{\partial r_m \partial r_j}\left\langle W_{mp;n}(\boldsymbol{r};t)\right\rangle - 2\frac{\partial[B_{kn}(\boldsymbol{0}) - B_{kn}(\boldsymbol{r})]}{\partial r_m}\frac{\partial}{\partial r_n}\left\langle W_{mp;j}(\boldsymbol{r};t)\right\rangle \\
&+ 2\frac{\partial^2 B_{np}(\boldsymbol{r})}{\partial r_m \partial r_j}\left\langle W_{km;n}(\boldsymbol{r};t)\right\rangle - 2\frac{\partial[B_{np}(\boldsymbol{0}) - B_{np}(\boldsymbol{r})]}{\partial r_m}\frac{\partial}{\partial r_n}\left\langle W_{km;j}(\boldsymbol{r};t)\right\rangle \\
&+ \left[2B_{nq}(\boldsymbol{0}) - B_{nq}(\boldsymbol{r}) - B_{qn}(\boldsymbol{r})\right]\frac{\partial^2}{\partial r_q \partial r_n}\left\langle W_{kp;j}(\boldsymbol{r};t)\right\rangle.
\end{aligned} \tag{11.112}$$

Then, setting $\boldsymbol{r} = 0$, we pass to the equation in the one-point correlation

$$\begin{aligned}
\frac{\partial}{\partial t}\left\langle W_{kp;j}(\boldsymbol{0};t)\right\rangle = &-2\frac{\partial^3 B_{kp}(\boldsymbol{0})}{\partial r_n \partial r_m \partial r_j}\langle W_{nm}(\boldsymbol{0};t)\rangle - 2\frac{\partial^2 B_{kp}(\boldsymbol{0})}{\partial r_n \partial r_m}\left\langle W_{nm;j}(\boldsymbol{0};t)\right\rangle \\
&+ 2\frac{\partial^2 B_{kn}(\boldsymbol{0})}{\partial r_m \partial r_j}\left\langle W_{mp;n}(\boldsymbol{0};t)\right\rangle + 2\frac{\partial^2 B_{np}(\boldsymbol{0})}{\partial r_m \partial r_j}\left\langle W_{km;n}(\boldsymbol{0};t)\right\rangle.
\end{aligned}$$

Using Eq. (4.57), page 107, for the solenoidal portion of the correlation function, we obtain the equation

$$\frac{\partial}{\partial t}\left\langle W_{kp;j}(\mathbf{0};t)\right\rangle = -2\frac{\partial^3 B_{kp}(\mathbf{0})}{\partial r_n \partial r_m \partial r_j}\left\langle W_{nm}(\mathbf{0};t)\right\rangle + 4\frac{D^{\mathrm{s}}}{d(d+2)}\left\langle W_{kp;j}(\mathbf{0};t)\right\rangle$$
$$- 2\frac{(d+1)D^{\mathrm{s}}}{d(d+2)}\left[\left\langle W_{jp;k}(\mathbf{0};t)\right\rangle + \left\langle W_{kj;p}(\mathbf{0};t)\right\rangle\right].$$

Now, we must draw an equation for function $\left\langle W_{jp;k}(\mathbf{0};t)\right\rangle + \left\langle W_{kj;p}(\mathbf{0};t)\right\rangle$. It has the form

$$\frac{\partial}{\partial t}\left[\left\langle W_{jp;k}(\mathbf{0};t)\right\rangle + \left\langle W_{kj;p}(\mathbf{0};t)\right\rangle\right] = -2\left[\frac{\partial^3 B_{jp}(\mathbf{0})}{\partial r_n \partial r_m \partial r_k} + \frac{\partial^3 B_{kj}(\mathbf{0})}{\partial r_n \partial r_m \partial r_p}\right]\left\langle W_{nm}(\mathbf{0};t)\right\rangle$$
$$+ \frac{2(d+3)D^{\mathrm{s}}}{d(d+2)}\left[\left\langle W_{jp;k}(\mathbf{0};t)\right\rangle + \left\langle W_{kj;p}(\mathbf{0};t)\right\rangle\right] - 4\frac{(d+1)D^{\mathrm{s}}}{d(d+2)}\left\langle W_{kp;j}(\mathbf{0};t)\right\rangle.$$

Thus, we arrive at the system of equations in functions $\left\langle W_{kp;j}(\mathbf{0};t)\right\rangle$ and $\left[\left\langle W_{jp;k}(\mathbf{0};t)\right\rangle + \left\langle W_{kj;p}(\mathbf{0};t)\right\rangle\right]$ with zero-valued initial conditions, or to the second-order equation

$$\left(\frac{\partial^2}{\partial t^2} - \frac{2(d+5)D^{\mathrm{s}}}{d(d+2)}\frac{\partial}{\partial t} - 8\frac{(d-1)}{d^2(d+2)}\left[D^{\mathrm{s}}\right]^2\right)\left\langle W_{kp;j}(\mathbf{0};t)\right\rangle$$
$$= -2\frac{\partial^3 B_{kp}(\mathbf{0})}{\partial r_n \partial r_m \partial r_j}\left(\frac{\partial}{\partial t} - \frac{2(d+3)D^{\mathrm{s}}}{d(d+2)}\right)\left\langle W_{nm}(\mathbf{0};t)\right\rangle$$
$$+ \frac{4(d+1)D^{\mathrm{s}}}{d(d+2)}\left(\frac{\partial^3 B_{jp}(\mathbf{0})}{\partial r_n \partial r_m \partial r_k} + \frac{\partial^3 B_{kj}(\mathbf{0})}{\partial r_n \partial r_m \partial r_p}\right)\left\langle W_{nm}(\mathbf{0};t)\right\rangle \tag{11.113}$$

with the source resulting from to the absence of reflection symmetry in the right-hand side.

Expand function $C(r)$ in the correlation function of velocity field (4.59), page 108, $C(r) = C(0) - \alpha r^2 + \cdots$, and use Eq. (4.58), page 107. We will solve Eq. (11.113) proceeding from the increasing exponents of the one-point correlation function of magnetic field (11.110) and (11.111)

$$\left\langle W_{nm}(\mathbf{0};t)\right\rangle = \frac{1}{d}\delta_{nm}\left\langle E(t)\right\rangle_0.$$

In this case, Eq. (11.113) becomes simpler and assumes the form of the equation

$$\left(\frac{\partial^2}{\partial t^2} - \frac{2(d+5)D^{\mathrm{s}}}{d(d+2)}\frac{\partial}{\partial t} - 8\frac{(d-1)}{d^2(d+2)}\left[D^{\mathrm{s}}\right]^2\right)\left\langle W_{kp;j}(\mathbf{0};t)\right\rangle$$
$$= 8\alpha D^{\mathrm{s}}\frac{(d+3)(d-1)}{d}\varepsilon_{kpj}\left\langle E(t)\right\rangle_0,$$

which can be rewritten in the form

$$\left(\frac{\partial}{\partial t}-\frac{4D^{s}}{d}\right)\left(\frac{\partial}{\partial t}+\frac{2(d-1)D^{s}}{d(d+2)}\right)\left\langle W_{kp;j}(\mathbf{0};t)\right\rangle = 8\alpha D^{s}\frac{(d+3)\,(d-1)}{d}\varepsilon_{kpj}\left\langle E(t)\right\rangle_{0}. \tag{11.114}$$

Equation (11.114) has two characteristic exponents, positive one, corresponding to the increasing exponent, and negative, corresponding to the damped solution. We will seek the increasing solution to Eq. (11.114) in the form

$$\left\langle W_{kp;j}(\mathbf{0};t)\right\rangle = U(t)\exp\left\{\frac{4D^{s}}{d}t\right\}.$$

Here, function $U(t)$ satisfies the simpler 'abridged' equation

$$\frac{\partial U(t)}{\partial t} = 4\alpha\varepsilon_{kpj}\frac{(d+2)(d-1)(d+3)}{3\,(d+1)}E_{0}\exp\left\{\frac{2(d-3)D^{s}}{d}t\right\}.$$

We mentioned that helicity is equal to zero in the two-dimensional case. In the three-dimensional case, we have

$$U(t) = 20\alpha\varepsilon_{kpj}E_{0}t$$

and, consequently, the main exponentially increasing solution assumes the form

$$\left\langle W_{kp;j}(\mathbf{0};t)\right\rangle = U(t)\exp\left\{\frac{4D^{s}}{3}t\right\} = 20\alpha\varepsilon_{kpj}\left\langle E(t)\right\rangle_{0}t. \tag{11.115}$$

Taking into account Eq. (1.54) (in the three-dimensional case $\varepsilon_{ijk}\varepsilon_{kij} = 6$), we obtain the final expression for helicity of magnetic field

$$H_{0}(t) = 120\alpha\left\langle E(t)\right\rangle_{0}t. \tag{11.116}$$

11.5.5 On the Magnetic Field Dissipation

Dissipation of magnetic field is defined by Eq. (11.108), and calculation of its value requires knowledge of quantity

$$\left\langle W_{kp;js}(\mathbf{0};t)\right\rangle = \frac{\partial}{\partial r_{s}}\left\langle W_{kp;j}(\mathbf{0};t)\right\rangle = \left\langle\frac{\partial H_{k}(\boldsymbol{r},t)}{\partial r_{s}\partial r_{j}}H_{p}(\boldsymbol{r}_{1},t)\right\rangle_{\boldsymbol{r}=\boldsymbol{r}_{1}}.$$

To derive an equation for this quantity, we differentiate Eq. (11.112) with respect to r_s and set $\boldsymbol{r} = 0$. The result is the equation for the one-point correlation

$$
\begin{aligned}
\frac{\partial}{\partial t}\left\langle W_{kp;js}(\mathbf{0};t)\right\rangle = &-2\frac{\partial^4 B_{kp}(\mathbf{0})}{\partial r_n \partial r_m \partial r_j \partial r_s}\left\langle W_{nm}(\mathbf{0};t)\right\rangle \\
&-2\frac{\partial^3 B_{kp}(\mathbf{0})}{\partial r_n \partial r_m \partial r_j}\left\langle W_{nm;s}(\mathbf{0};t)\right\rangle - 2\frac{\partial^3 B_{kp}(\mathbf{0})}{\partial r_n \partial r_m \partial r_s}\left\langle W_{nm;j}(\mathbf{0};t)\right\rangle \\
&+2\frac{\partial^3 B_{kn}(\mathbf{0})}{\partial r_m \partial r_j \partial r_s}\left\langle W_{mp;n}(\mathbf{0};t)\right\rangle + 2\frac{\partial^3 B_{np}(\mathbf{0})}{\partial r_m \partial r_j \partial r_s}\left\langle W_{km;n}(\mathbf{0};t)\right\rangle \\
&-2\frac{\partial^2 B_{kp}(\mathbf{0})}{\partial r_n \partial r_m}\left\langle W_{nm;js}(\mathbf{0};t)\right\rangle + 2\frac{\partial^2 B_{kn}(\mathbf{0})}{\partial r_m \partial r_j}\left\langle W_{mp;ns}(\mathbf{0};t)\right\rangle
\end{aligned}
$$

$$
\begin{aligned}
&+2\frac{\partial^2 B_{kn}(\mathbf{0})}{\partial r_m \partial r_s}\left\langle W_{mp;jn}(\mathbf{0};t)\right\rangle + 2\frac{\partial^2 B_{np}(\mathbf{0})}{\partial r_m \partial r_j}\left\langle W_{km;ns}(\mathbf{0};t)\right\rangle \\
&+2\frac{\partial^2 B_{np}(\mathbf{0})}{\partial r_m \partial r_s}\left\langle W_{km;jn}(\mathbf{0};t)\right\rangle - 2\frac{\partial^2 B_{qn}(\mathbf{0})}{\partial r_j \partial r_s}\left\langle W_{kp;qn}(\mathbf{0};t)\right\rangle.
\end{aligned}
$$

Convolve all functions with respect to indices $k = p$ and $j = s$ and take into account Eq. (4.57), page 107, for the noncompressible flow of liquid. As a result, we arrive at the equation of the form

$$
\begin{aligned}
\frac{\partial}{\partial t}\left\langle W_{kk;ss}(\mathbf{0};t)\right\rangle = &-2\frac{\partial^4 B_{kk}(\mathbf{0})}{\partial r_n \partial r_m \partial r_s \partial r_s}\left\langle W_{nm}(\mathbf{0};t)\right\rangle \\
&-4\frac{\partial^3 B_{kk}(\mathbf{0})}{\partial r_n \partial r_m \partial r_s}\left\langle W_{nm;s}(\mathbf{0};t)\right\rangle + 2\frac{\partial^3 B_{kn}(\mathbf{0})}{\partial r_m \partial r_s \partial r_s}\left\langle W_{mk;n}(\mathbf{0};t)\right\rangle \\
&+2\frac{\partial^3 B_{nk}(\mathbf{0})}{\partial r_m \partial r_s \partial r_s}\left\langle W_{km;n}(\mathbf{0};t)\right\rangle + \frac{4\,(d+1)\,D^{\mathrm{s}}}{(d+2)}\left\langle W_{mm;ss}(\mathbf{0};t)\right\rangle.
\end{aligned}
$$

Now, we content ourselves with excitation sources increasing exponentially in time,

$$
-2\frac{\partial^4 B_{kk}(\mathbf{0})}{\partial r_n \partial r_m \partial r_s \partial r_s}\left\langle W_{nm}(\mathbf{0};t)\right\rangle_0 = -2\frac{\partial^4 B_{kk}(\mathbf{0})}{\partial r_n \partial r_n \partial r_s \partial r_s}\left\langle E(t)\right\rangle_0
$$

$$
-2\int\limits_0^\infty d\tau\,\left\langle [\Delta \boldsymbol{u}(\boldsymbol{r},t+\tau)]\,[\Delta \boldsymbol{u}(\boldsymbol{r},t)]\right\rangle \left\langle E(t)\right\rangle_0 = -2D_H^{(4)}\left\langle E(t)\right\rangle_0,
$$

where $D_H^{(4)} = \int d\boldsymbol{k}\,\boldsymbol{k}^4 E^{\mathrm{s}}(k)$.

Consider now the terms related to helicity (they are nonzero only in the three-dimensional case). Using Eqs. (11.115) and (1.54), we obtain the equation of the form

$$
\left(\frac{\partial}{\partial t} - \frac{4\,(d+1)\,D^{\mathrm{s}}}{(d+2)}\right)\left\langle W_{kk;ss}(\mathbf{0};t)\right\rangle = -2D_H^{(4)}\left\langle E(t)\right\rangle_0 - 4800(d-2)\alpha^2\left\langle E(t)\right\rangle_0 t,
$$

where we introduced factor $(d-2)$ in the last term to emphasize that it vanishes in the two-dimensional case. In accordance with Eq. (11.108), we can rewrite this equation in terms of dissipation $D_0(t)$,

$$\left(\frac{\partial}{\partial t} - A\right) D_0(t) = \left[2D_H^{(4)} + 4800(d-2)\alpha^2 \frac{\partial}{\partial g}\right] E_0 e^{gt}, \tag{11.117}$$

where

$$A = \frac{4\,(d+1)\,D^{\mathrm{s}}}{(d+2)}, \quad g = \frac{2(d-1)}{d} D^{\mathrm{s}}.$$

Integrating Eq. (11.117), we obtain the expression for dissipation

$$D_0(t) = 2D_H^{(4)}\,\langle E(t)\rangle_0\,\frac{1}{(A-g)}\left[e^{(A-g)t} - 1\right]$$
$$+\,4800(d-2)\alpha^2\,\langle E(t)\rangle_0\,\frac{1}{(A-g)^2}\left[e^{(A-g)t} - 1 - (A-g)\,t\right],$$

where positive parameter

$$A - g = \frac{2\left(d^2 + d + 2\right) D^{\mathrm{s}}}{d(d+2)}.$$

Retaining here only increasing exponents, we obtain the final expression for temporal evolution of dissipation,

$$D_0(t) \approx \left[2D_H^{(4)} + \frac{4800(d-2)\alpha^2}{(A-g)}\right] \frac{\langle E(t)\rangle_0}{(A-g)} e^{(A-g)t}. \tag{11.118}$$

In the three-dimensional case $d = 3$ and, hence,

$$D_0^{(3)}(t) \approx \left[2D_H^{(4)} + 2571\frac{\alpha^2}{D^{\mathrm{s}}}\right] \frac{15\,\langle E(t)\rangle_0}{28D^{\mathrm{s}}} \exp\left\{\frac{28}{15}D^{\mathrm{s}}t\right\}, \tag{11.119}$$

The structure of the three-dimensional solution shows that dissipation of magnetic field for great times is determined by helicity of the velocity field and average magnetic field energy (11.110) in absence of helicity.

In the two-dimensional case $d = 2$, helicity is absent and, consequently,

$$D_0^{(2)}(t) \approx \frac{D_H^{(4)}}{D^{\mathrm{s}}}\,\langle E(t)\rangle_0\, e^{D^{\mathrm{s}}t}. \tag{11.120}$$

In this case, energy dissipation is determined only by average energy.

As may be seen from Eqs. (11.119) and (11.120), dissipation increases in time much faster than average energy does.

Now, we turn back to the original problem on dynamics of average energy subject to dissipation. Substituting Eq (11.118) in Eq. (11.109) and integrating over time, we obtain the solution to the problem in the form

$$\langle E(t)\rangle_1 = E_0 \exp\left\{\frac{2(d-1)D^s}{d}t - 4\mu_H\left[D_H^{(4)} + \frac{2400(d-2)\alpha^2}{(A-g)}\right]\right.$$
$$\left.\times \frac{1}{(A-g)^2}\left[e^{(A-g)t} - 1\right]\right\}.$$

As may be seen from this formula, average energy decays very rapidly for $t \to \infty$. Maximum value of average energy is achieved at instant

$$D^s t_{\max} \approx \frac{1}{(A-g)} \ln \frac{(d-1)(A-g)^2 D^s}{2d\mu_H\left[D_H^{(4)}(A-g) + 2400(d-2)\alpha^2\right]}. \tag{11.121}$$

In the three-dimensional case, we have

$$D^s t_{\max} = -\frac{15}{28} \ln \mu_H \left[\frac{1,6D_H^{(4)}}{[D^s]^2} + \frac{2066a^2}{[D^s]^3}\right] \sim$$

$$\sim \begin{cases} \frac{1}{2} \ln \frac{\sigma^2 \tau_0}{\mu_H} - 3,8 & \text{in the presence of helicity,} \\ \frac{1}{2} \ln \frac{\sigma^2 \tau_0}{2\mu_H} & \text{in the absence of helicity,} \end{cases}$$

i.e., average energy reaches the maximum in the presence of helicity much faster than in the absence of helicity.

In the two-dimensional case of plane-parallel flow of a liquid, we have

$$D^s t_{\max} = \frac{1}{2} \ln \frac{[D^s]^2}{2\mu_H D_H^{(4)}} \sim \frac{1}{2} \ln \frac{\sigma^2 \tau_0}{2\mu_H}.$$

Applicability condition of the neglect of the effect of dynamic diffusion is, evidently, the condition

$$t \ll t_{\max}.$$

Above, we considered problems with the homogeneous initial conditions. In the case of the inhomogeneous initial conditions, solutions to all these problems cease to possess the property of spatial homogeneity, and equations assume very cumbersome forms. However, similar to the case of the density field, the obtained solutions give certain data in this case, too. For example, for the average energy of magnetic field in the case of inhomogeneous initial conditions under the neglect of dynamic diffusion, we obtain the expression (11.72), page 294,

$$\int d\boldsymbol{r}\, \langle E(\boldsymbol{r},t)\rangle_0 = \int d\boldsymbol{r}\, E_0(\boldsymbol{r}) \exp\left\{2\frac{d-1}{d}\left(D^s + D^p\right)t\right\}.$$

Problems

Problem 11.1

Derive the Fokker–Planck equation for particle position in random nondivergent velocity field with allowance for the two-dimensional plane-parallel average flow $\boldsymbol{u}_0(\boldsymbol{r}, t) = v(y)\boldsymbol{l}$, *where* $\boldsymbol{r} = (x, y)$, $\boldsymbol{l} = (1, 0)$.

Instruction *Particle motion is described by the stochastic system of equations*

$$\frac{d}{dt}x(t) = v(y) + u_1(\boldsymbol{r}, t), \quad \frac{d}{dt}y(t) = u_2(\boldsymbol{r}, t).$$

Solution

$$\left(\frac{\partial}{\partial t} + v(y)\frac{\partial}{\partial x}\right) P(t; \boldsymbol{r}) = D_0 \Delta P(t; \boldsymbol{r}). \tag{11.122}$$

Problem 11.2

Show that Eq. (11.122) is statistically equivalent to a particle whose dynamics is described by the equations

$$\frac{d}{dt}x(t) = v(y) + u_1(t), \quad \frac{d}{dt}y(t) = u_2(t), \tag{11.123}$$

where $u_i(t)$, $i = 1, 2$ *are the statistically independent Gaussian delta-correlated processes with the characteristics*

$$\langle \boldsymbol{u}(t) \rangle = 0, \quad \left\langle u_i(t) u_j(t') \right\rangle = 2\delta_{ij} D_0 \delta(t - t').$$

Problem 11.3

Integrate Eqs. (11.123) and evaluate statistical characteristics of particle position in the case of linear shear $v(y) = \alpha y$.

Solution

$$y(t) = y_0 + w_2(t), \quad x(t) = x_0 + w_1(t) + \int_0^t d\tau\, v\left(y + w_2(\tau)\right), \tag{11.124}$$

where $w_i(t) = \int_0^t d\tau u_i(\tau)$ are independent Wiener processes with the characteristics

$$\langle \boldsymbol{w}(t)\rangle = 0, \quad \langle w_i(t) w_j(t')\rangle = 2D_0\delta_{ij}\min\{t, t'\}.$$

In all cases, coordinate $y(t)$ has the Gaussian probability density with the parameters

$$\langle y(t)\rangle = y_0, \quad \left\langle y^2(t)\right\rangle = y_0^2 + 2D_0 t.$$

For the linear shear, Eqs. (11.124) correspond to the joint Gaussian probability density with the parameters

$$\begin{aligned}
&\langle y(t)\rangle = y_0, \quad \langle x(t)\rangle = x_0 + \alpha y_0 t,\\
&\sigma_{xx}^2(t) = \left\langle (x(t) - \langle x(t)\rangle)^2\right\rangle = 2D_0 t\left(1 + \alpha t + \frac{1}{3}\alpha^2 t^2\right),\\
&\sigma_{yy}^2(t) = \left\langle (y(t) - \langle y(t)\rangle)^2\right\rangle = 2D_0 t,\\
&\sigma_{xy}^2(t) = \langle (x(t) - \langle x(t)\rangle)(y(t) - \langle y(t)\rangle)\rangle = 2D_0 t(1 + \alpha t).
\end{aligned}$$

Problem 11.4

Derive the equation for the joint Eulerian probability density of the density and density gradient fields in random Gaussian divergent, homogeneous, and isotropic in space and delta-correlated in time velocity field.

Solution Averaging the Liouville equation (3.24), page 75, over an ensemble of realizations of the velocity field, we obtain the equation of the form

$$\begin{aligned}
\frac{\partial}{\partial t}P(\boldsymbol{r}, t; \rho, \boldsymbol{p}) = \Bigg[& D_0\Delta + \frac{2}{d}D^{\mathrm{p}}\frac{\partial^2}{\partial \boldsymbol{r}\partial \boldsymbol{p}}\rho + D^{\mathrm{p}}\frac{\partial^2}{\partial \rho^2}\rho^2 + \frac{1}{d(d+2)}D^{\mathrm{s}}\widehat{L}^{\mathrm{s}}(\boldsymbol{p})\\
& + \frac{1}{d(d+2)}D^{\mathrm{p}}\widehat{L}^{\mathrm{p}}(\boldsymbol{p}) + \frac{2(d+1)}{d}D^{\mathrm{p}}\frac{\partial}{\partial \boldsymbol{p}}\boldsymbol{p}\frac{\partial}{\partial \rho}\rho + D_4^{\mathrm{p}}\frac{\partial^2}{\partial \boldsymbol{p}^2}\rho^2\Bigg] P(\boldsymbol{r}, t; \rho, \boldsymbol{p}),
\end{aligned} \tag{11.125}$$

where operators $\widehat{L}^{\mathrm{s}}(\boldsymbol{p})$ and $\widehat{L}^{\mathrm{p}}(\boldsymbol{p})$ are defined by the formulas

$$\begin{aligned}
\widehat{L}^{\mathrm{s}}(\boldsymbol{p}) &= (d+1)\frac{\partial^2}{\partial \boldsymbol{p}^2}\boldsymbol{p}^2 - 2\frac{\partial}{\partial \boldsymbol{p}}\boldsymbol{p} - 2\left(\frac{\partial}{\partial \boldsymbol{p}}\boldsymbol{p}\right)^2 = (d+1)\frac{\partial^2}{\partial \boldsymbol{p}^2}\boldsymbol{p}^2 - 2\frac{\partial^2}{\partial p_k \partial p_l}p_k p_l,\\
\widehat{L}^{\mathrm{p}}(\boldsymbol{p}) &= \frac{\partial^2}{\partial \boldsymbol{p}^2}\boldsymbol{p}^2 + (d^2 + 4d + 6)\left(\frac{\partial}{\partial \boldsymbol{p}}\boldsymbol{p}\right)^2 + (d^2 + 2d + 2)\frac{\partial}{\partial \boldsymbol{p}}\boldsymbol{p}
\end{aligned}$$

and diffusion coefficients D_0, D^{p}, and D^{s} are given by Eqs. (4.54) and (4.59), pages 107 and 108.

For the nondivergent velocity field ($D^{\mathrm{p}} = 0$) Eq. (11.125) becomes simpler and assumes the form

$$\left(\frac{\partial}{\partial t} - D_0\Delta\right) P(\boldsymbol{r}, t; \rho, \boldsymbol{p}) = \frac{1}{d(d+2)} D^{\mathrm{s}} \widehat{L^{\mathrm{s}}}(\boldsymbol{p}) P(\boldsymbol{r}, t; \rho, \boldsymbol{p}). \qquad (11.126)$$

Problem 11.5

Derive the Fokker–Planck equation for the gradient of particle concentration in nondivergent random velocity field in the case of the two-dimensional average linear shear $\boldsymbol{u}_0(\boldsymbol{r}, t) = \alpha y \boldsymbol{l}$, where $\boldsymbol{r} = (x, y)$, $\boldsymbol{l} = (1, 0)$.

Instruction *In the Lagrangian description, the particle concentration gradient in nondivergent velocity field satisfies the equations*

$$\frac{d}{dt} p_x(t|\boldsymbol{r}_0) = -p_k(t|\boldsymbol{r}_0) \frac{\partial u_k(\boldsymbol{r}, t)}{\partial x},$$
$$\frac{d}{dt} p_y(t|\boldsymbol{r}_0) = -\alpha p_x(t|\boldsymbol{r}_0) - p_k(t|\boldsymbol{r}_0) \frac{\partial u_k(\boldsymbol{r}, t)}{\partial y},$$

$$\boldsymbol{p}(0|\boldsymbol{r}_0) = \boldsymbol{p}_0(\boldsymbol{r}_0) = \nabla \rho_0(\boldsymbol{r}_0).$$

Solution

$$\frac{\partial}{\partial t} P(\boldsymbol{p}, t|\boldsymbol{r}_0) = \left\{\alpha p_x \frac{\partial}{\partial p_y} + \frac{1}{8} D^{\mathrm{s}} \left(3 \frac{\partial^2}{\partial \boldsymbol{p}^2} \boldsymbol{p}^2 - 2 \frac{\partial^2}{\partial p_k \partial p_l} p_k p_l\right)\right\} P(\boldsymbol{p}, t|\boldsymbol{r}_0),$$

$$P(\boldsymbol{p}, 0|\boldsymbol{r}_0) = \delta\left(\boldsymbol{p} - \boldsymbol{p}_0(\boldsymbol{r}_0)\right). \qquad (11.127)$$

Problem 11.6

Starting from Eq. (11.127), study the behavior of statistical characteristics of particle concentration gradient in nondivergent velocity field in the case of the two-dimensional average linear shear $\boldsymbol{u}_0(\boldsymbol{r}, t) = \alpha y \boldsymbol{l}$, where $\boldsymbol{r} = (x, y)$, $\boldsymbol{l} = (1, 0)$.

Solution Average gradient of tracer concentration field coincides with the problem solution for absent fluctuations of the velocity field

$$\langle p_x(t)\rangle = p_x(0), \quad \langle p_y(t)\rangle = p_y(0) - \alpha p_x(0) t.$$

The second moments of the gradient satisfy a closed system of equations with the cubic characteristic equation in parameter λ

$$\left(\lambda+\frac{1}{2}D^{s}\right)^{2}\left(\lambda-D^{s}\right)=\frac{3}{2}\alpha^{2}D^{s}$$

whose roots essentially depend on parameter α/D^{s}. For $\alpha/D^{s}\ll 1$, the approximate expressions for the roots are as follows

$$\lambda_1=D^{s}+\frac{2\alpha^2}{3D^{s}},\quad \lambda_2=-\frac{1}{2}D^{s}+i|\alpha|,\quad \lambda_3=-\frac{1}{2}D^{s}-i|\alpha|.$$

Consequently, random factor completely determines the problem solution for times $D^{s}t\gg 2$. This means that the effects of velocity field fluctuations predominate the effects of the weak gradient of linear shear. In the opposite limit $\alpha/D^{s}\gg 1$, the approximate expressions for the roots have the form

$$\lambda_1=\left(\frac{3}{2}\alpha^2D^{s}\right)^{1/3},\quad \lambda_2=\left(\frac{3}{2}\alpha^2D^{s}\right)^{1/3}e^{i(2/3)\pi},$$

$$\lambda_2=\left(\frac{3}{2}\alpha^2D^{s}\right)^{1/3}e^{-i(2/3)\pi}.$$

Because real parts of λ_2 and λ_3 are negative, the asymptotic solution of the system for second moments for $\left(\frac{3}{2}\alpha^2D^{s}\right)^{1/3}t\gg 1$ has the form

$$\left\langle p^2(t)\right\rangle\sim\exp\left\{\left(\frac{3}{2}\alpha^2D^{s}\right)^{1/3}t\right\},$$

from which follows that even small fluctuations of velocity field appear the governing factor in the case of shears characterized by strong gradients.

Problem 11.7

Derive the equation for the joint Eulerian probability density of the concentration and concentration gradient fields in random nondivergent velocity field in the case of the two-dimensional average linear shear $\boldsymbol{u}_0(\boldsymbol{r}, t) = \alpha y \boldsymbol{l}$, *where* $\boldsymbol{r} = (x, y)$, $\boldsymbol{l} = (1, 0)$.

Solution

$$\frac{\partial}{\partial t} P(\boldsymbol{r}, t; \rho, \boldsymbol{p}) = \left[-\alpha y \frac{\partial}{\partial x} + D_0 \Delta \right] P(\boldsymbol{r}, t; \rho, \boldsymbol{p})$$
$$+ \left\{ \alpha p_x \frac{\partial}{\partial p_y} + \frac{1}{8} D^{\mathrm{s}} \left(3 \frac{\partial^2}{\partial \boldsymbol{p}^2} \boldsymbol{p}^2 - 2 \frac{\partial^2}{\partial p_k \partial p_l} p_k p_l \right) \right\} P(\boldsymbol{r}, t; \rho, \boldsymbol{p}).$$

The solution to this equation can be represented in the form of integral (11.58), page 289, where Lagrangian probability densities $P(\boldsymbol{r}, t|\boldsymbol{r}_0)$ and $P(\boldsymbol{p}, t|\boldsymbol{r}_0)$ satisfy Eqs. (11.123) and (11.127).

Lecture 12

Wave Localization in Randomly Layered Media

12.1 General Remarks

The problem of plane wave propagation in layered media is formulated in terms of the one-dimensional boundary-value problem. It attracts the attention of many researchers because it is much simpler in comparison with the corresponding two- and three-dimensional problems and provides a deeper insight into wave propagation in random media. In view of the fact that the one-dimensional problem allows an exact asymptotic solution, we can use it to trace the effect of different models, medium parameters, and boundary conditions on statistical characteristics of the wavefield.

The problem in the one-dimensional statement was given in Lecture 1.

12.1.1 *Wave Incidence on an Inhomogeneous Layer*

Assume the layer of inhomogeneous medium occupies the portion of space $L_0 < x < L$. The unit-amplitude plane wave is incident on this layer from region $x > L$. The wavefield in the inhomogeneous layer satisfies the Helmholtz equation

$$\frac{d^2}{dx^2}u(x) + k^2(x)u(x) = 0, \tag{12.1}$$

where $k^2(x) = k^2[1 + \varepsilon(x)]$ and function $\varepsilon(x)$ describes the inhomogeneities of the medium. In the simplest case of unmatched boundary, we assume that $k(x) = k$, i.e., $\varepsilon(x) = 0$ outside the layer and $\varepsilon(x) = \varepsilon_1(x) + i\gamma$ inside the layer, where $\varepsilon_1(x)$ is the real part responsible for wave scattering in the medium and $\gamma \ll 1$ is the imaginary part responsible for wave absorption in the medium.

The boundary conditions for Eq. (12.1) are formulated as the continuity of function $u(x)$ and derivative $\frac{d}{dx}u(x)$ at layer boundaries; these conditions can we written in the form

$$\left(1 + \frac{i}{k}\frac{d}{dx}\right)u(x)\bigg|_{x=L} = 2, \qquad \left(1 - \frac{i}{k}\frac{d}{dx}\right)u(x)\bigg|_{x=L_0} = 0. \tag{12.2}$$

Lectures on Dynamics of Stochastic Systems. DOI: 10.1016/B978-0-12-384966-3.00012-X

For $x < L$, from Eq. (12.1) follows the equality

$$k\gamma I(x) = \frac{d}{dx}S(x), \tag{12.3}$$

where $S(x)$ is the energy flux density,

$$S(x) = \frac{i}{2k}\left[u(x)\frac{d}{dx}u^*(x) - u^*(x)\frac{d}{dx}u(x)\right],$$

and $I(x)$ is the wavefield intensity, $I(x) = |u(x)|^2$. In addition,

$$S(L) = 1 - |R_L|^2, \quad S(L_0) = |T_L|^2,$$

where R_L is the complex reflection coefficient from the medium layer and T_L is the complex transmission coefficient of the wave. Integrating Eq. (12.3) over the inhomogeneous layer, we obtain the equality

$$|R_L|^2 + |T_L|^2 + k\gamma\int_{L_0}^{L} dxI(x) = 1. \tag{12.4}$$

If the medium causes no wave attenuation ($\gamma = 0$), then conservation of the energy flux density is expressed by the equality

$$|R_L|^2 + |T_L|^2 = 1.$$

The imbedding method provides a possibility of reformulating boundary-value problem (12.1), (12.2) in terms of the dynamic initial-value problem with respect to parameter L (geometric position of the right-hand boundary of the layer) by considering the solution to the problem as a function of this parameter. For example, reflection coefficient R_L satisfies the Riccati equation

$$\frac{d}{dL}R_L = 2ikR_L + \frac{ik}{2}\varepsilon(L)\left(1 + R_L\right)^2, \quad R_{L_0} = 0, \tag{12.5}$$

and the wavefield in medium layer $u(x) \equiv u(x; L)$ satisfies the linear equation

$$\frac{\partial}{\partial L}u(x; L) = iku(x; L) + \frac{ik}{2}\varepsilon(L)\left(1 + R_L\right)u(x; L), \quad u(x; x) = 1 + R_x. \tag{12.6}$$

From Eqs. (12.5) and (12.6) follow the equations for the squared modulus of the reflection coefficient $W_L = |R_L|^2$ and the wavefield intensity $I(x; L) = |u(x; L)|^2$

$$\begin{aligned}
\frac{d}{dL}W_L &= -\frac{k\gamma}{2}\left[4W_L + \left(R_L + R_L^*\right)(1 + W_L)\right] \\
&\quad - \frac{ik}{2}\varepsilon_1(L)\left(R_L - R_L^*\right)(1 - W_L), \quad W_{L_0} = 0, \\
\frac{\partial}{\partial L}I(x; L) &= -\frac{k\gamma}{2}\left(2 + R_L + R_L^*\right)I(x; L) \\
&\quad + \frac{ik}{2}\varepsilon_1(L)\left(R_L - R_L^*\right)I(x; L), \quad I(x; x) = |1 + R_x|^2,
\end{aligned} \tag{12.7}$$

or, after rearrangement,

$$\begin{aligned}
-\frac{d}{dL}\ln(1 - W_L) &= -\frac{k\gamma}{2}\frac{4W_L + \left(R_L + R_L^*\right)(1 + W_L)}{1 - W_L} \\
&\quad - \frac{ik}{2}\varepsilon_1(L)\left(R_L - R_L^*\right),
\end{aligned} \tag{12.8}$$

$$\frac{\partial}{\partial L}\ln I(x; L) = -\frac{k\gamma}{2}\left(2 + R_L + R_L^*\right) + \frac{ik}{2}\varepsilon_1(L)\left(R_L - R_L^*\right).$$

In the case of non-absorptive medium, Eq. (12.8) can be integrated in the analytic form; the resulting wavefield intensity inside the inhomogeneous layer is explicitly expressed in terms of the layer reflection coefficient.

$$I(x; L) = \frac{|1 + R_x|^2 (1 - W_L)}{1 - W_x}. \tag{12.9}$$

12.1.2 Source Inside an Inhomogeneous Layer

Similarly, the field of the point source located in the layer of random medium is described in terms of the boundary-value problem for Green's function of the Helmholtz equation

$$\begin{aligned}
&\frac{d^2}{dx^2}G(x; x_0) + k^2[1 + \varepsilon(x)]G(x; x_0) = 2ik\delta(x - x_0), \\
&\left(\frac{d}{dx} + ik\right)G(x; x_0)\Big|_{x=L_0} = 0, \quad \left(\frac{d}{dx} - ik\right)G(x; x_0)\Big|_{x=L} = 0.
\end{aligned} \tag{12.10}$$

Note that the source at the layer boundary $x_0 = L$ corresponds to boundary-value problem (12.1), (12.2) on wave incidence on the layer, i.e.,

$$G(x; L) = u(x; L).$$

The solution of boundary-value problem (12.10) has the structure

$$G(x;x_0) = G(x_0;x_0)\begin{cases}\exp\left[ik\int\limits_{x}^{x_0}\psi_1(\xi)\,d\xi\right], & x_0 \ge x,\\ \exp\left[ik\int\limits_{x_0}^{x}\psi_2(\xi)\,d\xi\right], & x_0 \le x,\end{cases} \tag{12.11}$$

where the field at the source location, by virtue of the derivative gap condition

$$\left.\frac{dG(x;x_0)}{dx}\right|_{x=x_0+0} - \left.\frac{dG(x;x_0)}{dx}\right|_{x=x_0-0} = 2ik,$$

is determined by the formula

$$G(x_0;x_0) = \frac{2}{\psi_1(x_0)+\psi_2(x_0)}$$

and functions $\psi_i(x)$ satisfy the Riccati equations

$$\begin{aligned}\frac{d}{dx}\psi_1 &= ik\left[\psi_1^2 - 1 - \varepsilon(x)\right], \quad \psi_1(L_0) = 1,\\ \frac{d}{dx}\psi_2 &= -ik\left[\psi_2^2 - 1 - \varepsilon(x)\right], \quad \psi_2(L) = 1.\end{aligned} \tag{12.12}$$

Introduce new functions $R_i(x)$ related to functions $\psi_i(x)$ by the formula

$$\psi_i(x) = \frac{1-R_i(x)}{1+R_i(x)}, \qquad i = 1, 2.$$

With these functions, the wavefield in region $x < x_0$ can be written in the form

$$G(x;x_0) = \frac{[1+R_1(x_0)][1+R_2(x_0)]}{1-R_1(x_0)R_2(x_0)}\exp\left[ik\int\limits_{x}^{x_0}d\xi\frac{1-R_1(\xi)}{1+R_1(\xi)}\right], \tag{12.13}$$

where parameter $R_1(L) = R_L$ is the reflection coefficient of the plane wave incident on the layer from region $x > L$. In a similar way, quantity $R_2(x_0)$ is the reflection coefficient of the wave incident on the medium layer (x_0, L) from the homogeneous half-space $x < x_0$ (i.e., from region with $\varepsilon = 0$).

Problems with perfectly reflecting boundaries at which either $G(x;x_0)$ or $\frac{d}{dx}G(x;x_0)$ vanishes are of great interest for applications. Indeed, in the latter case, we have $R_2(x_0) = 1$ for the source located at this boundary; consequently,

$$G_{\text{ref}}(x;x_0) = \frac{2}{1-R_1(x_0)}\exp\left[ik\int\limits_{x}^{x_0}d\xi\frac{1-R_1(\xi)}{1+R_1(\xi)}\right], \quad x \le x_0. \tag{12.14}$$

In addition, the expression for wavefield intensity

$$I(x;x_0) = |G(x;x_0)|^2$$

follows from Eq. (12.10) for $x < x_0$

$$k\gamma I(x;x_0) = \frac{d}{dx}S(x;x_0), \tag{12.15}$$

where energy flux density $S(x;x_0)$ is given by the expression

$$S(x;x_0) = \frac{i}{2k}\left[G(x;x_0)\frac{d}{dx}G^*(x;x_0) - G^*(x;x_0)\frac{d}{dx}G(x;x_0)\right].$$

Using Eq. (12.13), we can represent $S(x;x_0)$ in the form ($x \le x_0$)

$$S(x;x_0) = S(x_0;x_0)\exp\left[-k\gamma\int_x^{x_0} d\xi \frac{|1+R_1(\xi)|^2}{1-|R_1(\xi)|^2}\right],$$

where the energy flux density at the point of source location is

$$S(x_0;x_0) = \frac{\left[1-|R_1(x_0)|^2\right]|1+R_2(x_0)|^2}{|1-R_1(x_0)R_2(x_0)|^2}. \tag{12.16}$$

Below, our concern will be with statistical problems on waves incident on random half-space ($L_0 \to -\infty$) and source-generated waves in infinite space ($L_0 \to -\infty$, $L \to \infty$) for sufficiently small absorption ($\gamma \to 0$). One can see from Eq. (12.15) that these limit processes are not commutable in the general case. Indeed, if $\gamma = 0$, then energy flux density $S(x;x_0)$ is conserved in the whole half-space $x < x_0$. However, integrating Eq. (12.15) over half-space $x < x_0$ in the case of small but finite absorption, we obtain the restriction on the energy confined in this half-space

$$k\gamma\int_{-\infty}^{x_0} dx I(x;x_0) = S(x_0;x_0) = \frac{\left[1-|R_1(x_0)|^2\right]|1+R_2(x_0)|^2}{|1-R_1(x_0)R_2(x_0)|^2}. \tag{12.17}$$

Three simple statistical problems are of interest:

- Wave incidence on medium layer (of finite and infinite thickness).
- Wave source in the medium layer or infinite medium.
- Effect of boundaries on statistical characteristics of the wavefield.

All these problems can be exhaustively solved in the analytic form. One can easily simulate these problems numerically and compare the simulated and analytic results.

We will assume that $\varepsilon_1(x)$ is the Gaussian delta-correlated random process with the parameters

$$\langle\varepsilon_1(L)\rangle = 0, \quad \left\langle\varepsilon_1(L)\varepsilon_1(L')\right\rangle = B_\varepsilon(L-L') = 2\sigma_\varepsilon^2 l_0\delta(L-L'), \tag{12.18}$$

where $\sigma_\varepsilon^2 \ll 1$ is the variance and l_0 is the correlation radius of random function $\varepsilon_1(L)$. This approximation means that asymptotic limit process $l_0 \to 0$ in the exact problem solution with a finite correlation radius l_0 must give the result coinciding with the solution to the statistical problem with parameters (12.18).

In view of smallness of parameter σ_ε^2, all statistical effects can be divided into two types, local and accumulated due to multiple wave reflections in the medium. Our concern will be with the latter.

The statement of boundary-value wave problems in terms of the imbedding method clearly shows that two types of wavefield characteristics are of immediate interest. The first type of characteristics deals with quantities, such as values of the wavefield at layer boundaries (reflection and transmission coefficients R_L and T_L), field at the point of source location $G(x_0; x_0)$, and energy flux density at the point of source location $S(x_0; x_0)$. The second type of characteristics deals with statistical characteristics of wavefield intensity in the medium layer, which is the subject matter of the statistical theory of radiative transfer.

12.2 Statistics of Scattered Field at Layer Boundaries

12.2.1 Reflection and Transmission Coefficients

Complex reflection coefficient of wave reflection from a medium layer satisfies the closed Riccati equation (12.5).

Represent reflection coefficient in the form $R_L = \sqrt{W_L} e^{i\phi_L}$, where $\sqrt{W_L}$ is the modulus and ϕ_L is the phase of the reflection coefficient. Then, starting from Eq. (12.5), we obtain the system of equations for squared modulus of the reflection coefficient $W_L = \rho_L^2 = |R_L|^2$ and its phase

$$\begin{aligned}
&\frac{d}{dL} W_L = -2k\gamma W_L + k\varepsilon_1(L)\sqrt{W_L}\,(1 - W_L)\sin\phi_L, \quad W_{L_0} = 0,\\
&\frac{d}{dL}\phi_L = 2k + k\varepsilon_1(L)\left\{1 + \frac{1 + W_L}{2\sqrt{W_L}}\cos\phi_L\right\}, \quad \phi_{L_0} = 0.
\end{aligned} \tag{12.19}$$

Fast functions producing only little contribution to accumulated effects are omitted in the dissipative terms of system (12.19) (cf. with Eq. (12.7)).

Introduce the indicator function

$$\varphi(L; W) = \delta(W_L - W)$$

that satisfies the Liouville equation

$$\begin{aligned}
\frac{\partial}{\partial L}\varphi(L; W) = {} & 2k\gamma\frac{\partial}{\partial W}\{W\varphi(L; W)\}\\
& - k\varepsilon_1(L)\frac{\partial}{\partial W}\left\{\sqrt{W}\,(1 - W)\sin\phi_L\varphi(L; W)\right\}.
\end{aligned} \tag{12.20}$$

Average this equation over an ensemble of realizations of function $\varepsilon_1(L)$ and and split the correlations using the Furutsu–Novikov formula (8.10), page 193, that takes in the case under consideration the form

$$\langle \varepsilon_1(L)R[L,\ L_0;\ \varepsilon_1(x)]\rangle = \int\limits_{L_0}^{L} dL' B_\varepsilon(L-L') \left\langle \frac{\delta}{\delta\varepsilon_1(L')} R[L,\ L_0;\ \varepsilon_1(x)] \right\rangle. \qquad (12.21)$$

This formula holds for arbitrary functional $R[L,\ L_0;\ \varepsilon_1(x)]$ of random process $\varepsilon_1(x)$ for $L_0 \le x \le L$. In the case of the Gaussian delta-correlated process $\varepsilon_1(x)$ with parameters (12.18), Eq. (12.21) becomes simpler and assumes the form

$$\langle \varepsilon_1(L)R[L,\ L_0;\ \varepsilon_1(x)]\rangle = \sigma_\varepsilon^2 l_0 \left\langle \frac{\delta}{\delta\varepsilon_1(L-0)} R[L,\ L_0;\ \varepsilon_1(x)] \right\rangle. \qquad (12.22)$$

As a result we obtain the equation for the probability density of reflection coefficient squared modulus $P(L;\ W) = \langle \varphi(L;\ W)\rangle$

$$\frac{\partial}{\partial L} P(L;\ W) = 2k\gamma \frac{\partial}{\partial W} \{WP(L;\ W)\} - k\frac{\partial}{\partial W} \int\limits_{L_0}^{L} dL' B_\varepsilon(L-L') \sqrt{W}(1-W)$$
$$\times \left\langle \cos\phi_L \frac{\delta\phi_L}{\delta\varepsilon_1(L')} \varphi(L;\ W) + \sin\phi_L \frac{\delta\varphi(L;\ W)}{\delta\varepsilon_1(L')} \right\rangle, \qquad (12.23)$$

where $B_\varepsilon(L-L')$ is the correlation function of random process $\varepsilon_1(L)$. Substituting the correlation function (12.18) in this equation and taking into account the equalities

$$\frac{\delta\varphi(L;\ W)}{\delta\varepsilon_1(L-0)} = -k\frac{\partial}{\partial W} \left\{ \sqrt{W}(1-W) \sin\phi_L \varphi(L;\ W) \right\},$$

$$\frac{\delta\phi_L}{\delta\varepsilon_1(L-0)} = k\left\{ 1 + \frac{1+W_L}{2\sqrt{W_L}} \cos\phi_L \right\}$$

following immediately from Eqs. (12.20) and (12.19), we obtain the unclosed equation for probability density $P(L;\ W)$

$$\frac{\partial}{\partial L} P(L;\ W) = 2k\gamma \frac{\partial}{\partial W} \{WP(L;\ W)\}$$
$$- k^2\sigma_\varepsilon^2 l_0 \frac{\partial}{\partial W} (1-W) \left\langle \left[\sqrt{W}\cos\phi_L + \frac{1}{2}(1+W)\cos^2\phi_L \right] \varphi(L;\ W) \right\rangle$$
$$+ k^2\sigma_\varepsilon^2 l_0 \frac{\partial}{\partial W} \left\{ \sqrt{W}(1-W) \frac{\partial}{\partial W} \left[\sqrt{W}(1-W) \left\langle \sin^2\phi_L \varphi(L;\ W) \right\rangle \right] \right\}.$$

In view of the fact that the phase of the reflection coefficient

$$\phi_L = k(L-L_0) + \widetilde{\phi}_L,$$

rapidly varies within distances about the wavelength, we can additionally average this equation over fast oscillations, which will be valid under the natural restriction $k/D \gg 1$. Thus we arrive at the Fokker–Planck equation

$$\frac{\partial}{\partial L}P(L;W) = 2k\gamma\frac{\partial}{\partial W}WP(L;W) - 2D\frac{\partial}{\partial W}W(1-W)P(L;W)$$
$$+D\frac{\partial}{\partial W}W(1-W)^2\frac{\partial}{\partial W}P(L;W), \quad P(L_0,W) = \delta(W-1) \tag{12.24}$$

with the diffusion coefficient

$$D = \frac{k^2\sigma_\varepsilon^2 l_0}{2}.$$

Representation of quantity W_L in the form

$$W_L = \frac{u_L - 1}{u_L + 1}, \quad u_L = \frac{1 + W_L}{1 - W_L}, \quad u_L \geq 1. \tag{12.25}$$

appears to be more convenient in some cases. Quantity u_L satisfies the stochastic system of equations

$$\frac{d}{dL}u_L = -k\gamma\left(u_L^2 - 1\right) + k\varepsilon_1(L)\sqrt{u_L^2 - 1}\sin\phi_L, \quad u_{L_0} = 1,$$

$$\frac{d}{dL}\phi_L = 2k + k\varepsilon_1(L)\left\{1 + \frac{u_L}{\sqrt{u_L^2 - 1}}\cos\phi_L\right\}, \quad \phi_{L_0} = 0,$$

and we obtain that probability density $P(L;u) = \langle\delta(u_L - u)\rangle$ of random quantity u_L satisfies the Fokker–Planck equation

$$\frac{\partial}{\partial L}P(L;u) = k\gamma\frac{\partial}{\partial u}\left(u^2 - 1\right)P(L;u) + D\frac{\partial}{\partial u}\left(u^2 - 1\right)\frac{\partial}{\partial u}P(L;u). \tag{12.26}$$

Note that quantity inverse to the diffusion coefficient defines the natural spatial scale related to medium inhomogeneities and is usually called the *localization length*

$$l_{\text{loc}} = 1/D.$$

In further analysis of wavefield statistics, we will see that this quantity determines the scale of the *dynamic wave localization* in separate wavefield realizations, although the *statistical localization* related to statistical characteristics of the wavefield may not occur in some cases.

Nondissipative Medium

If the medium is non-absorptive (i.e., if $\gamma = 0$), then Eq. (12.26)) assumes the form

$$\frac{\partial}{\partial \eta} P(\eta; u) = \frac{\partial}{\partial u}\left(u^2 - 1\right)\frac{\partial}{\partial u} P(\eta; u), \tag{12.27}$$

where we introduced the dimensionless layer thickness $\eta = D(L - L_0)$.

The solution to this equation can be easily obtained using the integral *Meler–Fock transform* (see Problem 9.1, page 247). This solution has the form

$$P(\eta, u) = \int_0^\infty d\mu\, \mu \tanh(\pi\mu) \exp\left\{-\left(\mu^2 + \frac{1}{4}\right)\eta\right\} P_{-\frac{1}{2}+i\mu}(u), \tag{12.28}$$

where $P_{-1/2+i\mu}(x)$ is the first-order *Legendre function* (*conal function*).

In view of the formula

$$\int_1^\infty \frac{dx}{(1+x)^n} P_{-\frac{1}{2}+i\mu}(x) = \frac{\pi}{\cosh(\mu\pi)} K_n(\mu),$$

where

$$K_{n+1}(\mu) = \frac{1}{2n}\left[\mu^2 + \left(n - \frac{1}{2}\right)^2\right] K_n(\mu), \quad K_1(\mu) = 1,$$

representation (12.28) offers a possibility of calculating statistical characteristics of reflection and transmission coefficients $W_L = |R_L|^2$ and $|T_L|^2 = 1 - |R_L|^2 = 2/(1 + u_L)$; in particular, we obtain the expression for the moments of the transmission coefficient squared modulus

$$\left\langle |T_L|^{2n} \right\rangle = 2^n \pi \int_0^\infty d\mu \frac{\mu \sinh(\mu\pi)}{\cosh^2(\mu\pi)} K_n(\mu) e^{-(\mu^2+1/4)\eta}. \tag{12.29}$$

Figure 12.1 shows coefficients

$$\langle W_L \rangle = \left\langle |R_L|^2 \right\rangle \quad \text{and} \quad \left\langle |T_L|^2 \right\rangle = 1 - \left\langle |R_L|^2 \right\rangle$$

as functions of layer thickness.

For sufficiently thick layers, namely, if $\eta = D(L - L_0) \gg 1$, Eq. (12.29) yields the asymptotic formula for the moments of the reflection coefficient squared modulus

$$\left\langle |T_L|^{2n} \right\rangle \approx \frac{[(2n-3)!!]^2 \pi^2 \sqrt{\pi}}{2^{2n-1}(n-1)!} \frac{1}{\eta\sqrt{\eta}} e^{-\eta/4}.$$

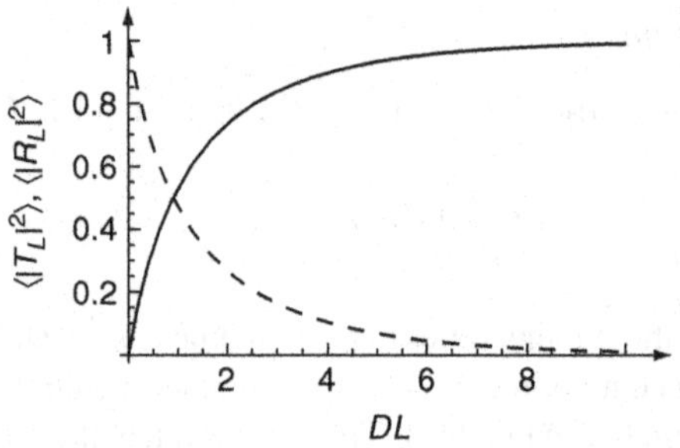

Figure 12.1 Quantities $\langle |R_L|^2 \rangle$ and $\langle |T_L|^2 \rangle$ versus layer thickness.

As may be seen, all moments of the reflection coefficient modulus $|T_L|$ vary with layer thickness according to the universal law (only the numerical factor is changed).

The fact that all moments of quantity $|T_L|$ tend to zero with increasing layer thickness means that $|R_L| \to 1$ with a probability equal to unity, i.e., *the half-space of randomly layered nondissipative medium completely reflects the incident wave.* It is clear that this phenomenon is independent of the statistical model of medium and the condition of applicability of the description based on the additional averaging over fast oscillations related to the reflection coefficient phase.

In the approximation of the delta-correlated random process $\varepsilon_1(L)$, random processes W_L and u_L are obviously the Markovian processes with respect to parameter L. It is obvious that the transition probability density

$$p(u, L|u', L') = \left\langle \delta(u_L - u|u_{L'} = u') \right\rangle$$

also satisfies in this case Eq. (12.27), i.e.,

$$\frac{\partial}{\partial L} p(u, L|u', L') = D \frac{\partial}{\partial u} \left(u^2 - 1 \right) \frac{\partial}{\partial u} p(u, L|u', L')$$

with the initial condition

$$p(u, L''|u', L') = \delta(u - u').$$

The corresponding solution has the form

$$p(u, L|u', L') = \int_0^\infty d\mu\, \mu \tanh(\pi \mu) e^{-D(\mu^2+1/4)(L-L')} P_{-\frac{1}{2}+i\mu}(u) P_{-\frac{1}{2}+i\mu}(u'). \tag{12.30}$$

At $L' = L_0$ and $u' = 1$, expression (12.30) grades into the one-point probability density (12.28).

Dissipative Medium

In the case of absorptive medium, Eqs. (12.24) and (12.26) cannot be solved analytically for the layer of finite thickness. Nevertheless, in the limit of half-space ($L_0 \to -\infty$), quantities W_L and u_L have the steady-state probability density independent of L and satisfying the equations

$$\begin{aligned} 2(\beta - 1 + W)P(W) + (1-W)^2 \frac{d}{dW}P(W) = 0, \quad 0 < W < 1, \\ \beta P(u) + \frac{d}{du}P(u) = 0, \quad u > 1, \end{aligned} \tag{12.31}$$

where $\beta = k\gamma/D$ is the dimensionless absorption coefficient.

Solutions to Eqs. (12.31) have the form

$$P(W) = \frac{2\beta}{(1-W)^2}\exp\left\{-\frac{2\beta W}{1-W}\right\}, \quad P(u) = \beta e^{-\beta(u-1)}, \tag{12.32}$$

and Fig. 12.2 shows function $P(W)$ for different values of parameter β.

The physical meaning of probability density (12.32) is obvious. It describes the statistics of the reflection coefficient of the random layer sufficiently thick for the incident wave could not reach its end because of dynamic absorption in the medium.

Using distributions (12.32), we can calculate all moments of quantity $W_L = |R_L|^2$. For example, the average square of reflection coefficient modulus is given by the formula

$$\langle W \rangle = \int\limits_0^1 dW\, W P(W) = \int\limits_1^\infty du \frac{u-1}{u+1} P(u) = 1 + 2\beta e^{2\beta}\,\mathrm{Ei}(-2\beta),$$

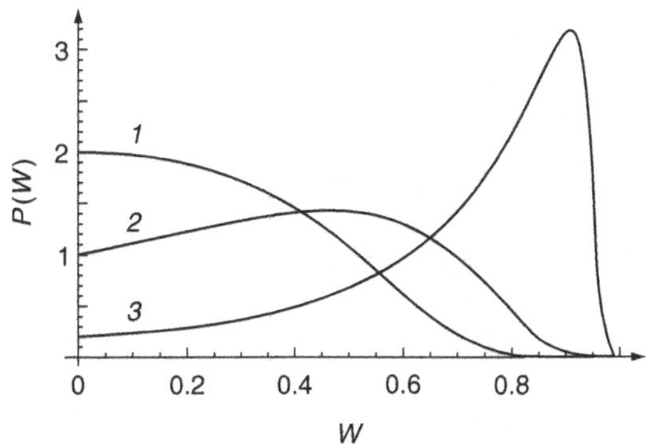

Figure 12.2 Probability density of squared reflection coefficient modulus $P(W)$. Curves *1* to *3* correspond to $\beta = 1, 0.5$, and 0.1, respectively.

where $\mathrm{Ei}(-x) = -\int_x^\infty \frac{dt}{t} e^{-t}$ $(x > 0)$ is the *integral exponent*. Using asymptotic expansions of function $\mathrm{Ei}(-x)$

$$\mathrm{Ei}(-x) = \begin{cases} \ln x & (x \ll 1), \\ -e^{-x}\dfrac{1}{x}\left(1 - \dfrac{1}{x}\right) & (x \gg 1), \end{cases}$$

we obtain the asymptotic expansions of quantity $\langle W \rangle = \langle |R_L|^2 \rangle$

$$\langle W \rangle \approx \begin{cases} 1 - 2\beta \ln(1/\beta), & \beta \ll 1, \\ 1/(2\beta), & \beta \gg 1. \end{cases} \tag{12.33}$$

To determine higher moments of quantity $W_L = |R_L|^2$, we multiply the first equation in (12.31) by W^n and integrate the result over W from 0 to 1. As a result, we obtain recurrence equation

$$n\left\langle W^{n+1}\right\rangle - 2(\beta + n)\left\langle W^n\right\rangle + n\left\langle W^{n-1}\right\rangle = 0 \quad (n = 1, 2, \ldots). \tag{12.34}$$

Using this equation, we can recursively calculate all higher moments. For example, we have for $n = 1$

$$\left\langle W^2\right\rangle = 2(\beta + 1)\langle W \rangle - 1.$$

The steady-state probability distribution can be obtained not only by limiting process $L_0 \to -\infty$, but also $L \to \infty$. Equation (12.24) was solved numerically for two values $\beta = 1.0$ and $\beta = 0.08$ for different initial conditions. Figure 12.3 shows moments $\langle W_L \rangle$, $\langle W_L^2 \rangle$ calculated from the obtained solutions versus dimensionless layer thickness $\eta = D(L - L_0)$. The curves show that the probability distribution

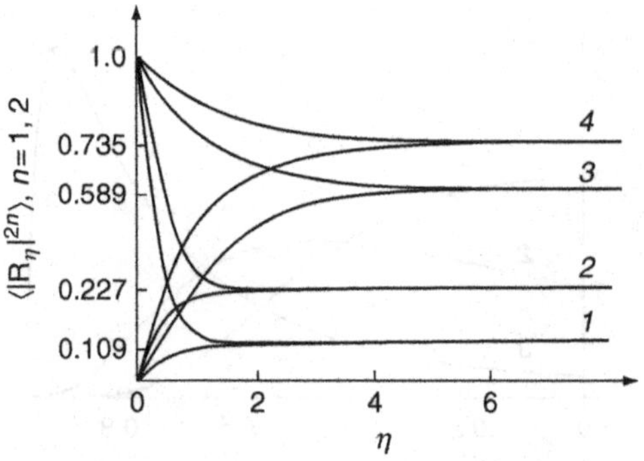

Figure 12.3 Statistical characteristics of quantity $W_L = |R_L|^2$. Curves *1* and *2* show the second and first moments at $\beta = 1$, and curves *3* and *4* show the second and first moments at $\beta = 0.08$.

approaches its steady-state behavior relatively rapidly ($\eta \sim 1.5$) for $\beta \geq 1$ and much slower ($\eta \geq 5$) for strongly stochastic problem with $\beta = 0.08$.

Note that, in the problem under consideration, energy flux density and wavefield intensity at layer boundary $x = L$ can be expressed in terms of the reflection coefficient. Consequently, we have for $\beta \ll 1$

$$\langle S(L, L)\rangle = 1 - \langle W_L\rangle = 2\beta \ln(1/\beta), \quad \langle I(L, L)\rangle = 1 + \langle W_L\rangle = 2. \tag{12.35}$$

Taking into account that $|T_L| = 0$ in the case of random half-space and using Eq. (12.4), we obtain that the wavefield energy contained in this half-space

$$E = D \int\limits_{-\infty}^{L} dx I(x; L),$$

has the probability distribution

$$P(E) = \beta P(W)|_{W=(1-\beta E)} = \frac{2}{E^2} \exp\left\{-\frac{2}{E}(1 - \beta E)\right\} \theta(1 - \beta E), \tag{12.36}$$

so that we have, in particular,

$$\langle E\rangle = 2 \ln(1/\beta) \tag{12.37}$$

for $\beta \ll 1$.

Note that probability distribution (12.36) allows the limit process to $\beta = 0$; as a result, we obtain the limiting probability density

$$P(E) = \frac{2}{E^2} \exp\left\{-\frac{2}{E}\right\} \tag{12.38}$$

that decays according to the power law for large energies E. The corresponding integral distribution function has the form

$$F(E) = \exp\left\{-\frac{2}{E}\right\}.$$

A consequence of Eq. (12.38) is the fact that all moments of the total wave energy appear infinite. Nevertheless, the total energy in separate wavefield realizations can be limited to arbitrary value with a finite probability.

12.2.2 Source Inside the Layer of a Medium

If the source of plane waves is located inside the medium layer, the wavefield and energy flux density at the point of source location are given by Eqs. (12.14) and

(12.16). Quantities $R_1(x_0)$ and $R_2(x_0)$ are statistically independent within the framework of the model of the delta-correlated fluctuations of $\varepsilon_1(x)$, because they satisfy dynamic equations (12.12) for nonoverlapping space portions. In the case of the infinite space ($L_0 \to -\infty, L \to \infty$), probability densities of quantities $R_1(x_0)$ and $R_2(x_0)$ are given by Eq. (12.32); as a result, average intensity of the wavefield and average energy flux density at the point of source location are given by the expressions

$$\langle I(x_0; x_0)\rangle = 1 + \frac{1}{\beta}, \qquad \langle S(x_0; x_0)\rangle = 1. \tag{12.39}$$

The infinite increase of average intensity at the point of source location for $\beta \to 0$ is evidence of wave energy accumulation in a randomly layered medium; at the same time, average energy flux density at the point of source location is independent of medium parameter fluctuations and coincides with energy flux density in free space.

For the source located at perfectly reflecting boundary $x_0 = L$, we obtain from Eqs. (12.14) and (12.16)

$$\langle I_{\text{ref}}(L; L)\rangle = 4\left(1 + \frac{2}{\beta}\right), \qquad \langle S_{\text{ref}}(L; L)\rangle = 4, \tag{12.40}$$

i.e., average energy flux density of the source located at the reflecting boundary is also independent of medium parameter fluctuations and coincides with energy flux density in free space.

Note the singularity of the above formulas (12.39), (12.40) for $\beta \to 0$, which shows that absorption (even arbitrarily small) serves the regularizing factor in the problem on the point source.

Using Eq. (12.17), we can obtain the probability distribution of wavefield energy in the half-space

$$E = D\int_{-\infty}^{x_0} dx I(x; x_0).$$

In particular, for the source located at reflecting boundary, we obtain the expression

$$P_{\text{ref}}(E) = \sqrt{\frac{2}{\pi}}\frac{1}{E\sqrt{E}}\exp\left\{-\frac{2}{E}\left(1 - \frac{\beta E}{4}\right)^2\right\},$$

that allows limiting process $\beta \to 0$, which is similar to the case of the wave incidence on the half-space of random medium.

12.2.3 Statistical Energy Localization

In view of Eq. (12.17), the obtained results related to wave field at fixed spatial points (at layer boundaries and at the point of source location) offer a possibility of making certain general conclusions about the behavior of the wavefield average intensity inside the random medium.

For example, Eq. (12.17) yields the expression for average energy contained in the half-space $(-\infty, x_0)$

$$\langle E\rangle = D\int_{-\infty}^{x_0} dx\,\langle I(x;x_0)\rangle = \frac{1}{\beta}\,\langle S(x_0;x_0)\rangle. \tag{12.41}$$

In the case of the plane wave ($x_0 = L$) incident on the half-space $x \le L$, Eqs. (12.35) and (12.41) result for $\beta \ll 1$ in the expressions

$$\langle E\rangle = 2\ln(1/\beta), \quad \langle I(L;L\rangle = 2. \tag{12.42}$$

Consequently, the space segment of length

$$Dl_\beta \cong \ln(1/\beta),$$

concentrates the most portion of average energy, which means that there occurs the *wavefield statistical localization* caused by wave absorption. Note that, in the absence of medium parameter fluctuations, energy localization occurs on scales about absorption length $Dl_{\text{abs}} \cong 1/\beta$. However, we have $l_{\text{abs}} \gg l_\beta$ for $\beta \ll 1$. If $\beta \to 0$, then $l_\beta \to \infty$, and statistical localization of the wavefield disappears in the limiting case of non-absorptive medium.

In the case of the source in unbounded space, we have

$$\langle E\rangle = \frac{1}{\beta}, \quad \langle I(x_0;x_0)\rangle = 1 + \frac{1}{\beta},$$

and average energy localization is characterized, as distinct from the foregoing case, by spatial scale $D|x - x_0| \cong 1$ for $\beta \to 0$.

In a similar way, we have for the source located at reflecting boundary

$$\langle E\rangle = \frac{4}{\beta}, \quad \langle I_{\text{ref}}(L;L)\rangle = 4\left(1 + \frac{2}{\beta}\right),$$

from which follows that average energy localization is characterized by a half spatial scale $D(L - x)| \cong 1/2$ for $\beta \to 0$.

In the considered problems, wavefield average energy essentially depends on parameter β and tends to infinity for $\beta \to 0$. However, this is the case only for average quantities. In our further analysis of the wavefield in random medium, we will show that the field is localized in separate realization due to the *dynamic localization* even in non-absorptive media.

12.3 Statistical Theory of Radiative Transfer

Now, we dwell on the statistical description of a wavefield in random medium (statistical theory of radiative transfer). We consider two problems of which the first concerns waves incident on the medium layer and the second concerns waves generated by a source located in the medium.

12.3.1 Normal Wave Incidence on the Layer of Random Media

In the general case of absorptive medium, the wavefield is described by the boundary-value problem (12.1), (12.2). We introduce complex opposite waves

$$u(x) = u_1(x) + u_2(x), \quad \frac{d}{dx}u(x) = -ik[u_1(x) - u_2(x)],$$

related to the wavefield through the relationships

$$\begin{aligned} u_1(x) &= \frac{1}{2}\left[1 + \frac{i}{k}\frac{d}{dx}\right]u(x), \quad u_1(L) = 1, \\ u_2(x) &= \frac{1}{2}\left[1 - \frac{i}{k}\frac{d}{dx}\right]u(x), \quad u_2(L_0) = 0, \end{aligned}$$

so that the boundary-value problem (12.1), (12.2) can be rewritten in the form

$$\begin{aligned} \left(\frac{d}{dx} + ik\right)u_1(x) &= -\frac{ik}{2}\varepsilon(x)\left[u_1(x) + u_2(x)\right], \quad u_1(L) = 1, \\ \left(\frac{d}{dx} - ik\right)u_2(x) &= -\frac{ik}{2}\varepsilon(x)\left[u_1(x) + u_2(x)\right], \quad u_2(L_0) = 0. \end{aligned}$$

The wavefield as a function of parameter L satisfies imbedding equation (12.6). It is obvious that the opposite waves will also satisfy Eq. (12.6), but with different initial conditions:

$$\begin{aligned} \frac{\partial}{\partial L}u_1(x;L) &= ik\left(1 + \frac{1}{2}\varepsilon(L)(1 + R_L)\right)u_1(x;L), \quad u_1(x;x) = 1, \\ \frac{\partial}{\partial L}u_2(x;L) &= ik\left(1 + \frac{1}{2}\varepsilon(L)(1 + R_L)\right)u_2(x;L), \quad u_2(x;x) = R_x, \end{aligned}$$

where reflection coefficient R_L satisfies Eq. (12.5).

Introduce now opposite wave intensities

$$W_1(x;L) = |u_1(x;L)|^2 \quad \text{and} \quad W_2(x;L) = |u_2(x;L)|^2.$$

They satisfy the equations

$$\begin{aligned} \frac{\partial}{\partial L}W_1(x;L) &= -k\gamma W_1(x;L) + \frac{ik}{2}\varepsilon(L)\left(R_L - R_L^*\right)W_1(x;L), \\ \frac{\partial}{\partial L}W_2(x;L) &= -k\gamma W_2(x;L) + \frac{ik}{2}\varepsilon(L)\left(R_L - R_L^*\right)W_2(x;L), \\ W_1(x;x) &= 1, \quad W_2(x;x) = |R_x|^2. \end{aligned} \tag{12.43}$$

Quantity $W_L = |R_L|^2$ appeared in the initial condition of Eqs. (12.43) satisfies Eq. (12.7), page 327, or the equation

$$\frac{d}{dL}W_L = -2k\gamma W_L - \frac{ik}{2}\varepsilon_1(L)\left(R_L - R_L^*\right)(1 - W_L), \quad W_{L_0} = 0. \tag{12.44}$$

In Eqs. (12.43) and (12.44), we omitted dissipative terms producing no contribution in accumulated effects.

As earlier, we will assume that $\varepsilon_1(x)$ is the Gaussian delta-correlated process with correlation function (12.18). In view of the fact that Eqs. (12.43), (12.44) are the first-order equations with initial conditions, we can use the standard procedure of deriving the Fokker–Planck equation for the joint probability density of quantities $W_1(x; L)$, $W_2(x; L)$, and W_L

$$P(x; L; W_1, W_2, W) = \langle \delta(W_1(x; L) - W_1)\delta(W_2(x; L) - W_2)\delta(W_L - W)\rangle.$$

As a result, we obtain the Fokker–Planck equation

$$\begin{aligned}
&\frac{\partial}{\partial L}P(x; L; W_1, W_2, W)\\
&\quad = k\gamma\left(\frac{\partial}{\partial W_1}W_1 + \frac{\partial}{\partial W_2}W_2 + 2\frac{\partial}{\partial W}W\right)P(x; L; W_1, W_2, W)\\
&\quad + D\left[\frac{\partial}{\partial W_1}W_1 + \frac{\partial}{\partial W_2}W_2 - \frac{\partial}{\partial W}(1 - W)\right]P(x; L; W_1, W_2, W)\\
&\quad + D\left[\frac{\partial}{\partial W_1}W_1 + \frac{\partial}{\partial W_2}W_2 - \frac{\partial}{\partial W}(1 - W)\right]^2 WP(x; L; W_1, W_2)
\end{aligned} \tag{12.45}$$

with the initial condition

$$P(x; x; W_1, W_2, W) = \delta(W_1 - 1)\delta(W_2 - W)P(x; W),$$

where function $P(L; W)$ is the probability density of reflection coefficient squared modulus W_L, which satisfies Eq. (12.24). As earlier, the diffusion coefficient in Eq. (12.45) is $D = k^2\sigma_\varepsilon^2 l_0/2$. Deriving this equation, we used an additional averaging over fast oscillations ($u(x) \sim e^{\pm ikx}$) that appear in the solution of the problem at $\varepsilon = 0$.

In view of the fact that Eqs. (12.43) are linear in $W_n(x; L)$, we can introduce the generating function of moments of opposite wave intensities

$$Q(x; L; \mu, \lambda, W) = \int\limits_0^1 dW_1 \int\limits_0^1 dW_2\, W_1^{\mu-\lambda} W_2^{\lambda} P(x; L; W_1, W_2, W), \tag{12.46}$$

which satisfies the simpler equation

$$\frac{\partial}{\partial L}Q(x;L;\mu,\lambda,W) = -k\gamma\left(\mu - 2\frac{\partial}{\partial W}W\right)Q(x;L;\mu,\lambda,W)$$
$$-D\left[\mu + \frac{\partial}{\partial W}(1-W)\right]Q(x;L;\mu,\lambda,W)$$
$$+D\left[\mu - \frac{\partial}{\partial W}(1-W)\right]^2 WQ(x;L;\mu,\lambda,W) \qquad (12.47)$$

with the initial condition

$$Q(x;x;\mu,\lambda,W) = W^{\lambda}P(x;W).$$

In terms of function $Q(x;L;\mu,\lambda,W)$, we can determine the moment functions of opposite wave intensities by the formula

$$\left\langle W_1^{\mu-\lambda}(x;L)W_2^{\lambda}(x;L)\right\rangle = \int\limits_0^1 dWQ(x;L;\mu,\lambda,W). \qquad (12.48)$$

Equation (12.47) describes statistics of the wavefield in medium layer $L_0 \le x \le L$. In particular, at $x = L_0$, it describes the transmission coefficient of the wave.

In the limiting case of the half-space ($L_0 \to -\infty$), Eq. (12.47) grades into the equation

$$\frac{\partial}{\partial \xi}Q(\xi;\mu,\lambda,W) = -\beta\left(\mu - 2\frac{\partial}{\partial W}W\right)Q(\xi;\mu,\lambda,W)$$
$$-\left[\mu + \frac{\partial}{\partial W}(1-W)\right]Q(\xi;\mu,\lambda,W)$$
$$+\left[\mu - \frac{\partial}{\partial W}(1-W)\right]^2 WQ(\xi;\mu,\lambda,W) \qquad (12.49)$$

with the initial condition

$$Q(0;\mu,\lambda,W) = W^{\lambda}P(W),$$

where $\xi = D(L-x) > 0$ is the dimensionless distance, and steady-state (independent of L) probability density of the reflection coefficient modulus $P(W)$ is given by Eq. (12.32). In this case, Eq. (12.48) assumes the form

$$\left\langle W_1^{\mu-\lambda}(\xi)W_2^{\lambda}(\xi)\right\rangle = \int\limits_0^1 dWQ(\xi;\mu,\lambda,W). \qquad (12.50)$$

Further discussion will be more convenient if we consider separately the cases of absorptive (dissipative) and non-absorptive (nondissipative) random medium.

Nondissipative Medium (Stochastic Wave Parametric Resonance and Dynamic Wave Localization)

Let us now discuss the equations for moments of opposite wave intensities in the non-absorptive medium, i.e., to Eq. (12.49) at $\beta = 0$ in the limit of the half-space ($L_0 \to -\infty$). In this case, $W_L = 1$ with a probability equal to unity, and the solution to Eq. (12.49) has the form

$$Q(x, L; \mu, \lambda, W) = \delta(W - 1)e^{D\lambda(\lambda-1)(L-x)},$$

so that

$$\left\langle W_1^{\lambda-\mu}(x; L)W_2^{\mu}(x; L)\right\rangle = e^{D\lambda(\lambda-1)(L-x)}.$$

In view of arbitrariness of parameters λ and μ, this means that

$$W_1(x; L) = W_2(x; L) = W(x; L)$$

with a probability equal to unity, and quantity $W(x; L)$ has the lognormal probability density. In addition, the mean value of this quantity is equal to unity, and its higher moments beginning from the second one exponentially increase with the distance in the medium

$$\langle W(x; L)\rangle = 1, \quad \left\langle W^n(x; L)\right\rangle = e^{Dn(n-1)(L-x)}, \quad n = 2, 3, \ldots .$$

Note that wavefield intensity $I(x; L)$ has in this case the form

$$I(x; L) = 2W(x; L)\,(1 + \cos\phi_x), \tag{12.51}$$

where ϕ_x is the phase of the reflection coefficient.

In view of the lognormal probability distribution, the typical realization curve of function $W(x, L)$ is, according to Eq. (4.61), page 109, the function exponentially decaying with distance in the medium

$$W^*(x, L) = e^{-D(L-x)}, \tag{12.52}$$

and this function is related to the *Lyapunov exponent*.

Moreover, realizations of function $W(x, L)$ satisfy with a probability of $1/2$ the inequality

$$W(x, L) < 4e^{-D(L-x)/2}$$

within the whole half-space.

In physics of disordered systems, the exponential decay of typical realization curve (12.52) with increasing $\xi = D(L-x)$ is usually identified with the property of *dynamic localization*, and quantity

$$l_{\text{loc}} = \frac{1}{D}$$

is usually called the *localization length*. Here,

$$l_{\text{loc}}^{-1} = -\frac{\partial}{\partial L}\langle \varkappa(x; L)\rangle,$$

where

$$\varkappa(x; L) = \ln W(x; L).$$

Physically, the lognormal property of the wavefield intensity $W(x; L)$ implies the existence of large spikes relative typical realization curve (12.52) towards both large and small intensities. This result agrees with the example of simulations given in Lecture I (see Fig. 1.10, page 20). However, these spikes of intensity contain only small portion of energy, because random area below curve $W(x; L)$,

$$S(L) = D\int_{-\infty}^{L} dx W(x; L),$$

has, in accordance with the lognormal probability distribution attribute, the steady-state (independent of L) probability density

$$P(S) = \frac{1}{S^2}\exp\left\{-\frac{1}{S}\right\}$$

that coincides with the distribution of total energy of the wave field in the half-space (12.38) if we set $E = 2S$. This means that the term dependent on fast phase oscillations of reflection coefficient in Eq. (12.51) only slightly contributes to total energy.

Thus, the knowledge of the one-point probability density provides an insight into the evolution of separate realizations of wavefield intensity in the whole space and allows us to estimate the parameters of this evolution in terms of statistical characteristics of fluctuating medium.

Dissipative Medium

In the presence of a finite (even arbitrary small) absorption in the medium occupying the half-space, the exponential growth of moment functions must cease and give place to attenuation. If $\beta \gg 1$ (i.e., if the effect of absorption is great in comparison with the effect of diffusion), then

$$P(W) = 2\beta e^{-2\beta W},$$

and, as can be easily seen from Eq. (12.49), opposite wave intensities $W_1(x; L)$ and $W_2(x; L)$ appear statistically independent, i.e., uncorrelated. In this case,

$$\langle W_1(\xi)\rangle = \exp\left\{-\beta\xi\left(1+\frac{1}{\beta}\right)\right\}, \quad \langle W_2(\xi)\rangle = \frac{1}{2\beta}\exp\left\{-\beta\xi\left(1+\frac{1}{\beta}\right)\right\}.$$

Figures 12.4–12.7 show the examples of moment functions of random processes obtained by numerical solution of Eq. (12.49) and calculation of quadrature (12.50) for different values of parameter β. Different figures mark the curves corresponding

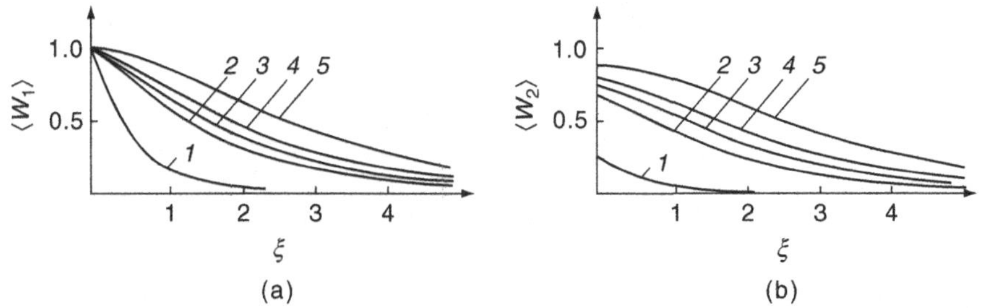

Figure 12.4 Distribution of wavefield average intensity along the medium; (a) the transmitted wave and (b) the reflected wave. Curves *1* to *5* correspond to parameter $\beta = 1, 0.1, 0.06, 0.04$ and 0.02, respectively.

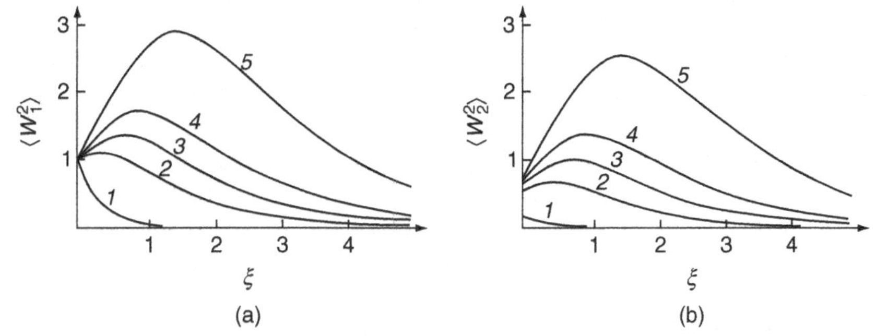

Figure 12.5 Distribution of the second moment of wavefield intensity along the medium; (a) the transmitted wave and (b) the reflected wave. Curves *1* to *5* correspond to parameter $\beta = 1, 0.1, 0.06, 0.04$ and 0.02, respectively.

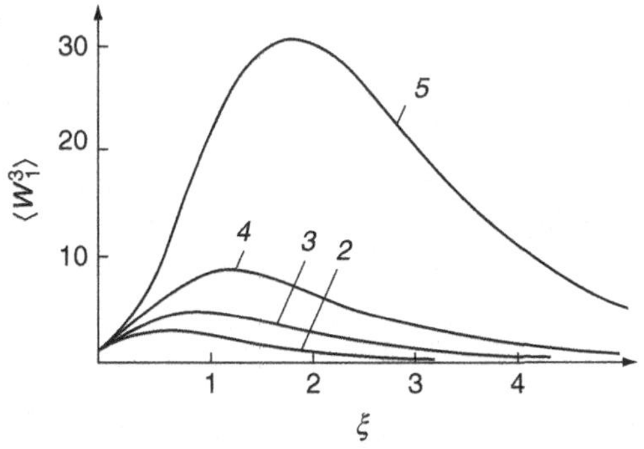

Figure 12.6 Distribution of the third moment of transmitted wave intensity. Curves *1* to *5* correspond to parameter $\beta = 1, 0.1, 0.06, 0.04$ and 0.02, respectively.

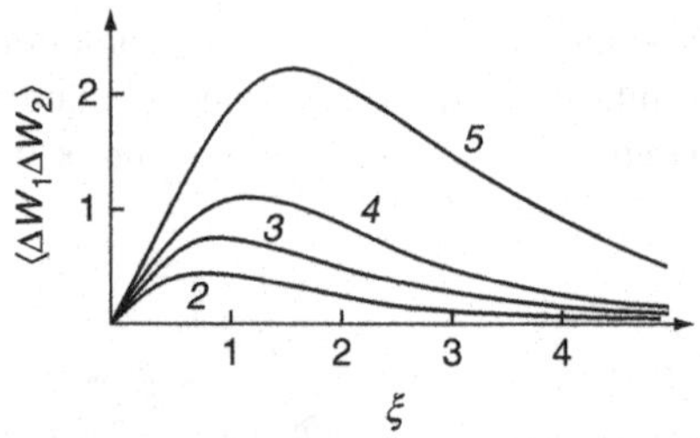

Figure 12.7 Correlation between the intensities of transmitted and reflected waves. Curves *1* to *5* correspond to parameter $\beta = 1, 0.1, 0.06, 0.04$ and 0.02, respectively.

to different values of parameter β. Figure 12.4 shows average intensities of the transmitted and reflected waves. The curves monotonically decrease with increasing ξ.

Figure 12.5 shows the corresponding curves for second moments. We see that $\langle W_1^2(0)\rangle = 1$ and $\langle W_2^2(0)\rangle = \langle |R_L|^4\rangle$ at $\xi = 0$. For $\beta < 1$, the curves as functions of ξ become nonmonotonic; the moments first increase, then pass the maximum, and finally monotonically decay. With decreasing parameter β, the position of the maximum moves to the right and the maximum value increases.

The fact that moments of intensity behave in the layer as exponentially increasing functions is evidence of the *phenomenon of stochastic wave parametric resonance*, which is similar to the ordinary parametric resonance. The only difference consists in the fact that values of intensity moments at layer boundary are asymptotically predetermined; as a result, the wavefield intensity exponentially increases inside the layer and its maximum occurs inside the layer. Figure 12.6 shows the similar curves for the third moment $\langle W_1^3(\xi)\rangle$, and Fig. 12.7 shows curves for mutual correlation of intensities of the transmitted and reflected waves $\langle \Delta W_1(\xi)\Delta W_2(\xi)\rangle$ (here, $\Delta W_n(\xi) = W_n(\xi) - \langle W_n(\xi)\rangle$). For $\beta \geq 1$, this correlation disappears. For $\beta < 1$, the correlation is strong, and wave division into opposite waves appears physically senseless, but mathematically useful technique. For $\beta \geq 1$, such a division is justified in view of the lack of mutual correlation.

As was shown earlier, in the case of the half-space of random medium with $\beta = 0$, all wavefield moments beginning from the second one exponentially increase with the distance the wave travels in the medium. It is clear, that problem solution for small β ($\beta \ll 1$) must show the singular behavior in β in order to vanish the solution for sufficiently long distances.

One can show that, in asymptotic limit $\beta \ll 1$, intensities of the opposite waves are equal with a probability equal to unity, and the solution for small distances from the boundary coincides with the solution corresponding to the stochastic parametric resonance.

For sufficiently great distances ξ, namely

$$\xi \gg 4\left(n - \frac{1}{2}\right)\ln\left(\frac{n}{\beta}\right),$$

quantities $\langle W^n(\xi)\rangle$ are characterized by the universal spatial localization behavior [67]

$$\left\langle W^n(\xi)\right\rangle \cong A_n \frac{1}{\beta^{n-1/2}} \ln\left(\frac{1}{\beta}\right) \frac{1}{\xi\sqrt{\xi}} e^{-\xi/4},$$

which coincides to a numerical factor with the asymptotic behavior of moments of the transmission coefficient of a wave passed through the layer of thickness ξ in the case $\beta = 0$.

Thus, there are three regions where moments of opposite wave intensities show essentially different behavior. In the first region (it corresponds to the stochastic wave parametric resonance), the moments exponentially increase with the distance in medium and wave absorption plays only insignificant role. In the second region, absorption plays the most important role, because absorption ceases the exponential growth of moments. In the third region, the decrease of moment functions of opposite wave intensities is independent of absorption. The boundaries of these regions depend on parameter β and tend to infinity for $\beta \to 0$.

12.3.2 Plane Wave Source Located in Random Medium

Consider now the asymptotic solution of the problem on the plane wave source in infinite space ($L_0 \to -\infty$, $L \to \infty$) under the condition $\beta \to 0$. In this case, it appears convenient to calculate average wavefield intensity in region $x < x_0$ using relationships (12.15) and (12.16)

$$\beta\left\langle I(x;x_0)\right\rangle = \frac{1}{D}\frac{\partial}{\partial x}\left\langle S(x;x_0)\right\rangle = \frac{1}{D}\frac{\partial}{\partial x}\left\langle \psi(x;x_0)\right\rangle,$$

where quantity

$$\psi(x;x_0) = \exp\left\{-\beta D\int_x^{x_0} d\xi \frac{|1+R_\xi|^2}{1-|R_\xi|^2}\right\}$$

satisfies, as a function of parameter x_0, the stochastic equation

$$\frac{\partial}{\partial x_0}\psi(x;x_0) = -\beta D\frac{|1+R_{x_0}|^2}{1-|R_{x_0}|^2}\psi(x;x_0), \quad \psi(x;x) = 1.$$

Introduce function

$$\varphi(x;x_0;u) = \psi(x;x_0)\delta(u_{x_0} - u), \tag{12.53}$$

where function $u_L = (1+W_L)/(1-W_L)$ satisfies the stochastic system of equations (12.25). Differentiating Eq. (12.53) with respect to x_0, we obtain the stochastic equation

$$\frac{\partial}{\partial x_0}\varphi(x;x_0;u) = -\beta D\left\{u+\sqrt{u^2-1}\cos\phi_{x_0}\right\}\varphi(x;x_0;u)$$
$$+\beta D\frac{\partial}{\partial u}\left\{\left(u^2-1\right)\varphi(x;x_0;u)\right\}$$
$$-k\varepsilon_1(x_0)\frac{\partial}{\partial u}\left\{\sqrt{u^2-1}\sin\phi_{x_0}\varphi(x;x_0;u)\right\}. \quad (12.54)$$

Average now Eq. (12.54) over an ensemble of realizations of random process $\varepsilon_1(x_0)$ assuming it, as earlier, the Gaussian process delta-correlated in x_0. Using the Furutsu–Novikov formula (12.22), page 331, the following expression for the variational derivatives

$$\frac{\delta\varphi(x;x_0;u)}{\delta\varepsilon_1(x_0)} = -k\frac{\partial}{\partial u}\left\{\sqrt{u^2-1}\sin\phi_{x_0}\varphi(x;x_0;u)\right\},$$
$$\frac{\delta\phi_{x_0}}{\delta\varepsilon_1(x_0)} = k\left[1+\frac{u_{x_0}}{\sqrt{u_{x_0}^2-1}}\right]\cos\phi_{x_0},$$

and additionally averaging over fast oscillations (over the phase of reflection coefficient), we obtain that function

$$\Phi(\xi;u) = \langle\varphi(x;x_0;u)\rangle = \left\langle\psi(x;x_0)\delta(u_{x_0}-u)\right\rangle,$$

where $\xi = D|x-x_0|$, satisfies the equation

$$\frac{\partial}{\partial\xi}\Phi(\xi;u) = -\beta u\Phi(\xi;u) + \beta\frac{\partial}{\partial u}\left(u^2-1\right)\Phi(\xi;u)$$
$$+\frac{\partial}{\partial u}\left(u^2-1\right)\frac{\partial}{\partial u}\Phi(\xi;u) \quad (12.55)$$

with the initial condition

$$\Phi(0;u) = P(u) = \beta e^{-\beta(u-1)}.$$

The average intensity can now be represented in the form

$$\beta\langle I(x;x_0)\rangle = -\frac{\partial}{\partial\xi}\int\limits_1^\infty du\Phi(\xi;u) = \beta\int\limits_1^\infty duu\Phi(\xi;u).$$

Equation (12.55) allows limiting process $\beta\to 0$. As a result, we obtain a simpler equation

$$\frac{\partial}{\partial\xi}\widetilde{\Phi}(\xi;u) = -u\widetilde{\Phi}(\xi;u) + \frac{\partial}{\partial u}u^2\widetilde{\Phi}(\xi;u) + \frac{\partial}{\partial u}u^2\frac{\partial}{\partial u}\widetilde{\Phi}(\xi;u),$$
$$\widetilde{\Phi}(0;u) = e^{-u}. \quad (12.56)$$

Consequently, localization of average intensity in space is described by the quadrature

$$\Phi_{\text{loc}}(\xi) = \int\limits_1^\infty du u \widetilde{\Phi}(\xi; u),$$

where

$$\Phi_{\text{loc}}(\xi) = \lim_{\beta \to 0} \beta \left\langle I(x; x_0) \right\rangle = \lim_{\beta \to 0} \frac{\left\langle I(x; x_0) \right\rangle}{\left\langle I(x_0; x_0) \right\rangle}.$$

Thus, the average intensity of the wavefield generated by the point source has for $\beta \ll 1$ the following asymptotic behavior

$$\left\langle I(x; x_0) \right\rangle = \frac{1}{\beta} \Phi_{\text{loc}}(\xi).$$

Equation (12.56) can be easily solved with the use of the Kontorovich–Lebedev transform; as a result, we obtain the expression for the *localization curve*

$$\Phi_{\text{loc}}(\xi) = 2\pi \int\limits_0^\infty d\tau\, \tau \left(\tau^2 + \frac{1}{4}\right) \frac{\sinh(\pi\tau)}{\cosh^2(\pi\tau)} e^{-\left(\tau^2 + \frac{1}{4}\right)\xi}. \tag{12.57}$$

For small distances ξ, the localization curve decays according to relatively fast law

$$\Phi_{\text{loc}}(\xi) \approx e^{-2\xi}. \tag{12.58}$$

For great distances ξ (namely, for $\xi \gg \pi^2$), it decays significantly slower, according to the universal law

$$\Phi_{\text{loc}}(\xi) \approx \frac{\pi^2 \sqrt{\pi}}{8} \frac{1}{\xi \sqrt{\xi}} e^{-\xi/4}, \tag{12.59}$$

but for all that

$$\int\limits_0^\infty d\xi\, \Phi_{\text{loc}}(\xi) = 1.$$

Function (12.57) is given in Fig. 12.8, where asymptotic curves (12.58) and (12.59) are also shown for comparison purposes.

A similar situation occurs in the case of the plane wave source located at the reflecting boundary.

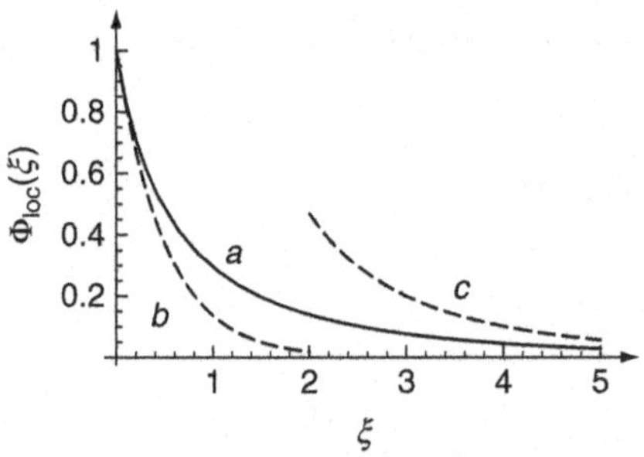

Figure 12.8 Localization curve for a source in infinite space (12.57) (curve a). Curves b and c correspond to asymptotic expressions for small and large distances from the source.

In this case, we obtain the expression

$$\lim_{\beta\to 0}\frac{\langle I_{\text{ref}}(x;L)\rangle}{\langle I_{\text{ref}}(L;L)\rangle}=\frac{1}{2}\Phi_{\text{loc}}(\xi),\quad \xi=D(L-x). \tag{12.60}$$

12.4 Numerical Simulation

The above theory rests on two simplifications – on using the delta-correlated approximation of function $\varepsilon_1(x)$ (or the diffusion approximation) and eliminating slow (within the scale of a wavelength) variations of statistical characteristics by averaging over fast oscillations. Averaging over fast oscillations is validated only for statistical characteristics of the reflection coefficient in the case of random medium occupying a half-space. For statistical characteristics of the wavefield intensity in medium, the corresponding validation appears very difficult if at all possible (this method is merely physical than mathematical). Numerical simulation of the exact problem offers a possibility of both verifying these simplifications and obtaining the results concerning more difficult situations for which no analytic results exists.

In principle, such numerical simulation could be performed by way of multiply solving the problem for different realizations of medium parameters followed by averaging the obtained solutions over an ensemble of realizations. However, such an approach is not very practicable because it requires a vast body of realizations of medium parameters. Moreover, it is unsuitable for real physical problems, such as wave propagation in Earth's atmosphere and ocean, where only a single realization is usually available. A more practicable approach is based on the ergodic property of boundary-value problem solutions with respect to the displacement of the problem along the single realization of function $\varepsilon_1(x)$ defined along the half-axis (L_0,∞) (see Fig. 12.9).

This approach assumes that statistical characteristics are calculated by the formula

$$\langle F(L_0;x,x_0;L)\rangle=\lim_{\delta\to\infty}F_\delta(L_0;x,x_0;L),$$

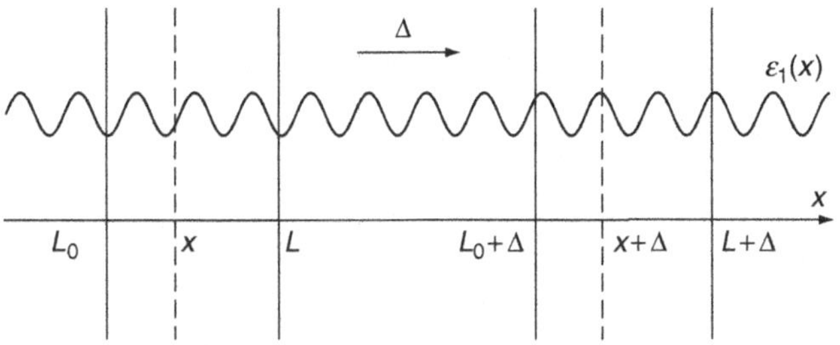

Figure 12.9 Averaging over parameter Δ by the procedure based on ergodicity of imbedding equations for a half-space of random medium.

where

$$F_\delta(L_0; x, x_0; L) = \frac{1}{\delta}\int_0^\delta d\Delta F(L_0 + \Delta; x + \Delta, x_0 + \Delta; L + \Delta).$$

In the limit of a half-space ($L_0 \to -\infty$), statistical characteristics are independent of L_0, and, consequently, the problem have ergodic property with respect to the position of the right-hand layer boundary L (simultaneously, parameter L is the variable of the imbedding method), because this position is identified in this case with the displacement parameter. As a result, having solved the imbedding equation for the sole realization of medium parameters, we simultaneously obtain all desired statistical characteristics of this solution by using the obvious formula

$$\langle F(x, x_0; L)\rangle = \frac{1}{\delta}\int_0^\delta d\xi F\left(\xi, \xi + x_0 - x; \xi + (L - x_0) + (x_0 - x)\right)$$

for sufficiently large interval $(0, \delta)$. This approach offers a possibility of calculating even the wave statistical characteristics that cannot be obtained within the framework of current statistical theory, and this calculation requires no additional simplifications.

In the case of the layer of finite thickness, the problem is not ergodic with respect to parameter L. However, the corresponding solution can be expressed in terms of two independent solutions of the problem on the half-space and, consequently, it can be reduced to the case ergodic with respect to L.

As an example let us discuss the case of the point source located at reflecting boundary $x_0 = L$ with boundary condition $dG(x; x_0; L)/dx|_{x=L} = 0$. Figure 12.10 shows the quantity $\langle I_{\text{ref}}(x; x_0)\rangle$ simulated with $\beta = 0.08$ and $k/D = 25$. In region $\xi = D(L-x) < 0.3$, one can see oscillations of period $T = 0.13$. For larger ξ, simulated results agree well with localization curve (12.60).

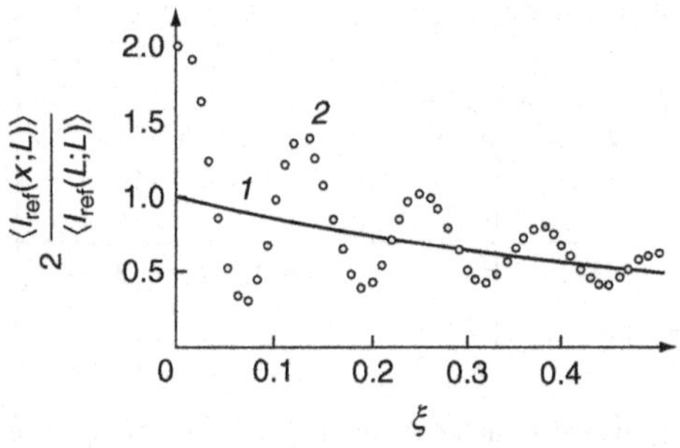

Figure 12.10 Curve *1* shows the localization curve (12.57) and circles *2* show the simulated result.

Problems

Problem 12.1

Derive the probability distribution of the reflection coefficient phase in the problem on a plane wave incident on the half-space of nondissipative medium (see, for example, [67])

Instruction *In this case, reflection coefficient has the form* $R_L = \exp\{i\phi_L\}$, *where phase* ϕ_L *satisfies the imbedding equation following from Eq. (12.19)*

$$\frac{d}{dL}\phi_L = 2k + k\varepsilon_1(L)\left(1 + \cos\phi_L\right), \quad \phi_{L_0} = 0. \tag{12.61}$$

The solution to Eq. (12.61) is defined over the whole axis $\phi_L\,(-\infty, \infty)$. *A more practicable characteristic is the probability distribution in interval* $(-\pi, \pi)$, *which must be independent of L in the case of the half-space. To obtain such a distribution, it is convenient to consider singular function* $z_L = \tan(\phi_L/2)$ *instead of phase* ϕ_L. *The corresponding dynamic equation has the form*

$$\frac{d}{dL}z_L = k\left(1 + z_L^2\right) + k\varepsilon_1(L), \quad z_{L_0} = 0.$$

Solution Assuming that $\varepsilon_1(L)$ is the Gaussian delta-correlated random function with the parameters given in Eq. (12.18)), we obtain that probability density $P(z, L) = \langle\delta(z_L - z)\rangle$ defined over the whole axis $(-\infty, \infty)$ satisfies the Fokker–Planck equation

$$\frac{\partial}{\partial L}P(z, L) = -k\frac{\partial}{\partial z}\left(1 + z_L^2\right)P(z, L) + 2D\frac{\partial^2}{\partial z^2}P(z, L).$$

For the random half-space ($L_0 \to -\infty$), the steady-state (independent of L) solution to the Fokker–Planck equation can be obtained as $P(z) = \lim_{L_0\to-\infty} P(z, L)$

and satisfies the equation

$$-\kappa\frac{d}{dz}\left(1+z^2\right)P(z)+\frac{d^2}{dz^2}P(z)=0, \qquad (12.62)$$

where

$$\kappa=\frac{k}{2D}, \quad D=\frac{k^2\sigma_\varepsilon^2 l_0}{2}.$$

The solution to Eq. (12.62) corresponding to constant probability flux density has the form

$$P(z)=J(\kappa)\int\limits_z^\infty d\xi\,\exp\left\{-\kappa\xi\left[1+\frac{\xi^3}{3}+z(z+\xi)\right]\right\},$$

where $J(\kappa)$ is the steady-state probability flux density,

$$J^{-1}(\kappa)=\sqrt{\frac{\pi}{\kappa}}\int\limits_0^\infty \xi^{-1/2}d\xi\,\exp\left\{-\kappa\left(\xi+\frac{\xi^3}{12}\right)\right\}.$$

The corresponding probability distribution of the wave phase over interval $(-\pi,\pi)$ is as follows

$$P(\phi)=\left.\frac{1+z^2}{2}P(z)\right|_{z=\tan(\phi/2)}.$$

For $\kappa\gg 1$, we have asymptotically

$$P(z)=\frac{1}{\pi(1+z^2)},$$

which corresponds to the uniform distribution of the reflection coefficient phase

$$P(\phi)=\frac{1}{2\pi}, \quad -\pi<\phi<\pi.$$

In the opposite limiting case $\kappa\ll 1$, we obtain

$$P(z)=\kappa^{1/3}\left(\frac{3}{4}\right)^{1/6}\frac{1}{\sqrt{\pi}\Gamma(1/6)}\Gamma\left(\frac{1}{3},\frac{\kappa z^3}{3}\right),$$

where $\Gamma(\mu,z)$ is the *incomplete gamma function*. From this expression follows that

$$P(z)=\kappa^{1/3}\left(\frac{3}{4}\right)^{1/6}\frac{1}{\sqrt{\pi}\Gamma(1/6)}\left(\frac{3}{\kappa z^3}\right)^{2/3}$$

for $\kappa|z|^3\gg 3$ and $|z|\to\infty$.

Problem 12.2

In the diffusion approximation, derive the equation for the reflection coefficient probability density.

Solution The equation for the reflection coefficient probability density has the form of the Fokker–Planck equation (12.22) with the diffusion coefficient

$$D(k, l_0) = \frac{k^2}{4} \int_{-\infty}^{\infty} d\xi B_\varepsilon(\xi) \cos(2k\xi) = \frac{k^2}{4} \Phi_\varepsilon(2k),$$

where $\Phi_\varepsilon(q) = \int_{-\infty}^{\infty} d\xi B_\varepsilon(\xi) e^{iq\xi}$ is the spectral function of random process $\varepsilon_1(x)$. Applicability range of this equation is given by the conditions

$$D(k, l_0) l_0 \ll 1, \quad \alpha = \frac{k}{D(k, l_0)} \gg 1.$$

Problem 12.3

Derive the equation for the reflection coefficient probability density in the case of matched boundary.

Instruction *In the case of matched boundary, reflection coefficient satisfies the Riccati equation*

$$\frac{d}{dL} R_L = 2ikR_L - k\gamma R_L + \frac{\xi(L)}{2}\left(1 - R_L^2\right), \quad R_{L_0} = 0,$$

where $\xi(L) = \varepsilon_1'(L)$. For the Gaussian process with correlation function $B_\varepsilon(x)$, random process $\xi(x)$ is also the Gaussian process with correlation function

$$B_\xi(x - x') = \langle \xi(x)\xi(x') \rangle = -\frac{\partial^2}{\partial x^2} B_\varepsilon(x - x').$$

Solution The equation for the reflection coefficient probability density has the form of the Fokker–Planck equation (12.22) with the diffusion coefficient

$$D(k, l_0) = \frac{k^2}{4} \Phi_\varepsilon(2k).$$

Lecture 13

Caustic Structure of Wavefield in Random Media

Fluctuations of wavefield propagating in a medium with random large-scale (in comparison with wavelength) inhomogeneities rapidly grow with distance because of multiple forward scattering. Perturbation theory in any version fails from a certain distance (the boundary of the region of strong fluctuation). Strong fluctuations of intensity can appear in radiowaves propagating through the ionosphere, solar corona, or interstellar medium, in occultation experiments on transilluminating planet's atmospheres when planets shadow natural or artificial radiation sources, and in a number of other cases.

The current state of the theory of wave propagation in random media can be found in monographs and reviews [2, 68, 69]. Below, we will follow works [68, 69] to describe wave propagation in random media within the framework of the parabolic equation of quasi-optics and delta-correlated approximation of medium parameter fluctuations and discuss the applicability of such an approach.

13.1 Input Stochastic Equations and Their Implications

We will describe the propagation of a monochromatic wave in the medium with large-scale inhomogeneities in terms of the complex scalar parabolic equation (1.89), page 39,

$$\frac{\partial}{\partial x}u(x,\boldsymbol{R}) = \frac{i}{2k}\Delta_{\boldsymbol{R}}u(x,\boldsymbol{R}) + i\frac{k}{2}\varepsilon(x,\boldsymbol{R})u(x,\boldsymbol{R}), \tag{13.1}$$

where function $\varepsilon(x,\boldsymbol{R})$ is the fluctuating portion (deviation from unity) of dielectric permittivity, x-axis is directed along the initial direction of wave propagation, and vector $\boldsymbol{R}$ denotes the coordinates in the transverse plane. The initial condition to Eq. (13.1) is the condition

$$u(0,\boldsymbol{R}) = u_0(\boldsymbol{R}). \tag{13.2}$$

Because Eq. (13.1) is the first-order equation in x and satisfies initial condition (13.2) at $x = 0$, it possesses the causality property with respect to the x-coordinate (it

Lectures on Dynamics of Stochastic Systems. DOI: 10.1016/B978-0-12-384966-3.00013-1

plays here the role of time), i.e., its solution satisfies the relationship

$$\frac{\delta u(x, \boldsymbol{R})}{\delta \varepsilon(x', \boldsymbol{R}')} = 0 \quad \text{for} \quad x' < 0,\ x' > x. \tag{13.3}$$

The variational derivative at $x' \to x - 0$ can be obtained according to the standard procedure,

$$\frac{\delta u(x, \boldsymbol{R})}{\delta \varepsilon(x - 0, \boldsymbol{R}')} = \frac{ik}{2} \delta\left(\boldsymbol{R} - \boldsymbol{R}'\right) u(x, \boldsymbol{R}). \tag{13.4}$$

Consider now the statistical description of the wavefield. We will assume that random field $\varepsilon(x, \boldsymbol{R})$ is the homogeneous and isotropic Gaussian field with the parameters

$$\langle \varepsilon(x, \boldsymbol{R}) \rangle = 0, \quad B_\varepsilon(x - x', \boldsymbol{R} - \boldsymbol{R}') = \left\langle \varepsilon(x, \boldsymbol{R}) \varepsilon(x', \boldsymbol{R}') \right\rangle.$$

As was noted, field $u(x, \boldsymbol{R})$ depends functionally only on preceding values of field $\varepsilon(x, \boldsymbol{R})$. Nevertheless, statistically, field $u(x, \boldsymbol{R})$ can depend on subsequent values $\varepsilon(x_1, \boldsymbol{R})$ for $x_1 > x$ due to nonzero correlation between values $\varepsilon(x', \boldsymbol{R}')$ for $x' < x$ and values $\varepsilon(\xi, \boldsymbol{R})$ for $\xi > x$. It is clear that correlation of field $u(x, \boldsymbol{R})$ with subsequent values $\varepsilon(x', \boldsymbol{R}')$ is appreciable only if $x' - x \sim l_\parallel$, where $l_\parallel$ is the longitudinal correlation radius of field $\varepsilon(x, \boldsymbol{R})$. At the same time, the characteristic correlation radius of field $u(x, \boldsymbol{R})$ in the longitudinal direction is estimated approximately as x (see, e.g., [33,34]). Therefore, the problem under consideration has small parameter $l_\parallel/x$, and we can use it to construct an approximate solution.

In the first approximation, we can set $l_\parallel/x \to 0$. In this case, field values $u(\xi_i, \boldsymbol{R})$ for $\xi_i < x$ will be independent of field values $\varepsilon(\eta_j, \boldsymbol{R})$ for $\eta_j > x$ not only functionally, but also statistically. This is equivalent to approximating the correlation function of field $\varepsilon(x, \boldsymbol{R})$ by the delta function of longitudinal coordinate, i.e., to the replacement of the correlation function $B_\varepsilon(x, \boldsymbol{R})$ whose three-dimensional spectral function is

$$\Phi_\varepsilon(q_1, \boldsymbol{q}) = \frac{1}{(2\pi)^3} \int_{-\infty}^{\infty} dx \int d\boldsymbol{R}\, B_\varepsilon(x, \boldsymbol{R}) e^{-iq_1 x - i\boldsymbol{q}\boldsymbol{R}} \tag{13.5}$$

with the effective function $B_\varepsilon^{\text{eff}}(x, \boldsymbol{R})$,

$$B_\varepsilon(x, \boldsymbol{R}) = B_\varepsilon^{\text{eff}}(x, \boldsymbol{R}) = \delta(x) A(\boldsymbol{R}), \quad A(\boldsymbol{R}) = \int_{-\infty}^{\infty} dx B_\varepsilon(x, \boldsymbol{R}) \tag{13.6}$$

Using this approximation, we derive the equations for moment functions

$$M_{mn}(x; \boldsymbol{R}_1, \ldots, \boldsymbol{R}_m; \boldsymbol{R}'_1, \ldots, \boldsymbol{R}'_n) = \left\langle \prod_{p=1}^{m} \prod_{q=1}^{n} u\left(x; \boldsymbol{R}_p\right) u^*\left(x; \boldsymbol{R}'_q\right) \right\rangle. \tag{13.7}$$

In the case of $m = n$, these functions are usually called the *coherence functions* of order $2n$.

Differentiating function (13.7) with respect to x and using Eq. (13.1) and its complex conjugated version, we obtain the equation

$$\begin{aligned}
&\frac{\partial}{\partial x} M_{mn}(x; \boldsymbol{R}_1, \ldots, \boldsymbol{R}_m; \boldsymbol{R}'_1, \ldots, \boldsymbol{R}'_n) \\
&= \frac{i}{2k}\left(\sum_{p=1}^{m} \Delta_{\boldsymbol{R}_p} - \sum_{q=1}^{n} \Delta_{\boldsymbol{R}'_q}\right) M_{mn}(x; \boldsymbol{R}_1, \ldots, \boldsymbol{R}_m; \boldsymbol{R}'_1, \ldots, \boldsymbol{R}'_n) \\
&\quad + i\frac{k}{2}\left\langle \left(\sum_{p=1}^{m} \varepsilon(x, \boldsymbol{R}_p) - \sum_{q=1}^{n} \varepsilon(x, \boldsymbol{R}'_q)\right)\left[\prod_{p=1}^{m}\prod_{q=1}^{n} u\left(x; \boldsymbol{R}_p\right) u^*\left(x; \boldsymbol{R}_q\right)\right]\right\rangle .
\end{aligned} \tag{13.8}$$

To split the correlator in the right-hand side of Eq. (13.8), we use the Furutsu–Novikov formula together with the delta-correlated approximation of medium parameter fluctuations with effective correlation function (13.6) to obtain the following relationships

$$\begin{aligned}
\left\langle \varepsilon(x, \boldsymbol{R}) u\left(x; \boldsymbol{R}_p\right)\right\rangle &= \frac{1}{2}\int dR'\, A(R - R')\left\langle \frac{\delta u\left(x; \boldsymbol{R}_p\right)}{\delta\varepsilon(x - 0, \boldsymbol{R}')}\right\rangle, \\
\left\langle \varepsilon(x, \boldsymbol{R}) u^*\left(x; \boldsymbol{R}'_q\right)\right\rangle &= \frac{1}{2}\int dR'\, A(R - R')\left\langle \frac{\delta u^*\left(x; \boldsymbol{R}'_q\right)}{\delta\varepsilon(x - 0, \boldsymbol{R}')}\right\rangle .
\end{aligned} \tag{13.9}$$

If we additionally take into account Eq. (13.4) and its complex conjugated version, then we arrive at the closed equation for the wavefield moment function

$$\begin{aligned}
&\frac{\partial}{\partial x} M_{mn}(x; \boldsymbol{R}_1, \ldots, \boldsymbol{R}_m; \boldsymbol{R}'_1, \ldots, \boldsymbol{R}'_n) \\
&= \frac{i}{2k}\left(\sum_{p=1}^{m} \Delta_{\boldsymbol{R}_p} - \sum_{q=1}^{n} \Delta_{\boldsymbol{R}'_q}\right) M_{mn}(x; \boldsymbol{R}_1, \ldots, \boldsymbol{R}_m; \boldsymbol{R}'_1, \ldots, \boldsymbol{R}'_n) \\
&\quad - \frac{k^2}{8} Q(\boldsymbol{R}_1, \ldots, \boldsymbol{R}_m; \boldsymbol{R}'_1, \ldots, \boldsymbol{R}'_n) M_{mn}(x; \boldsymbol{R}_1, \ldots, \boldsymbol{R}_m; \boldsymbol{R}'_1, \ldots, \boldsymbol{R}'_n),
\end{aligned} \tag{13.10}$$

where

$$\begin{aligned}
Q(\boldsymbol{R}_1, \ldots, \boldsymbol{R}_m; \boldsymbol{R}'_1, \ldots, \boldsymbol{R}'_n) &= \sum_{i=1}^{m}\sum_{j=1}^{m} A\left(\boldsymbol{R}_i - \boldsymbol{R}_j\right) - 2\sum_{i=1}^{m}\sum_{j=1}^{n} A\left(\boldsymbol{R}_i - \boldsymbol{R}'_j\right) \\
&\quad + \sum_{i=1}^{n}\sum_{j=1}^{n} A\left(\boldsymbol{R}'_i - \boldsymbol{R}'_j\right).
\end{aligned} \tag{13.11}$$

An equation for the characteristic functional of random field $u(x,\boldsymbol{R})$ also can be obtained; however it will be the linear variational derivative equation.

Draw explicitly equations for average field $\langle u(x,\boldsymbol{R})\rangle$, second-order coherence function

$$\Gamma_2(x,\boldsymbol{R},\boldsymbol{R}') = \left\langle \gamma_2(x,\boldsymbol{R},\boldsymbol{R}')\right\rangle, \quad \gamma_2(x,\boldsymbol{R},\boldsymbol{R}') = u(x,\boldsymbol{R})u^*(x,\boldsymbol{R}'),$$

and fourth-order coherence function

$$\Gamma_4(x,\boldsymbol{R}_1,\boldsymbol{R}_2,\boldsymbol{R}_1'\boldsymbol{R}_2') = \left\langle u(x,\boldsymbol{R}_1)u\,(x,\boldsymbol{R}_2)\,u^*(x,\boldsymbol{R}_1')u^*(x,\boldsymbol{R}_2')\right\rangle,$$

which follow from Eqs. (13.10) and (13.1) for $m = 1$, $n = 0$; $m = n = 1$; and $m = n = 2$. They have the forms

$$\frac{\partial}{\partial x}\langle u(x,\boldsymbol{R})\rangle = \frac{i}{2k}\Delta_{\boldsymbol{R}}\langle u(x,\boldsymbol{R})\rangle - \frac{k^2}{8}A(0)\langle u(x,\boldsymbol{R})\rangle,$$
$$\langle u(0,\boldsymbol{R})\rangle = u_0(\boldsymbol{R}), \tag{13.12}$$

$$\frac{\partial}{\partial x}\Gamma_2(x,\boldsymbol{R},\boldsymbol{R}') = \frac{i}{2k}\left(\Delta_{\boldsymbol{R}} - \Delta_{\boldsymbol{R}'}\right)\Gamma_2(x,\boldsymbol{R},\boldsymbol{R}') - \frac{k^2}{4}D(\boldsymbol{R}-\boldsymbol{R}_1)\Gamma_2(x,\boldsymbol{R},\boldsymbol{R}'),$$
$$\Gamma_2(0,\boldsymbol{R},\boldsymbol{R}') = u_0(\boldsymbol{R})u_0^*(\boldsymbol{R}'), \tag{13.13}$$

$$\begin{aligned}
&\frac{\partial}{\partial x}\Gamma_4(x,\boldsymbol{R}_1,\boldsymbol{R}_2,\boldsymbol{R}_1',\boldsymbol{R}_2') \\
&\quad = \frac{i}{2k}\left(\Delta_{\boldsymbol{R}_1} + \Delta_{\boldsymbol{R}_2} - \Delta_{\boldsymbol{R}_1'} - \Delta_{\boldsymbol{R}_2'}\right)\Gamma_4(x,\boldsymbol{R}_1,\boldsymbol{R}_2,\boldsymbol{R}_1',\boldsymbol{R}_2') \\
&\qquad - \frac{k^2}{8}Q(\boldsymbol{R}_1,\boldsymbol{R}_2,\boldsymbol{R}_1',\boldsymbol{R}_2')\Gamma_4(x,\boldsymbol{R}_1,\boldsymbol{R}_2,\boldsymbol{R}_1',\boldsymbol{R}_2'), \\
&\Gamma_4(0,\boldsymbol{R}_1,\boldsymbol{R}_2,\boldsymbol{R}_1',\boldsymbol{R}_2') = u_0(\boldsymbol{R}_1)u_0\,(\boldsymbol{R}_2)\,u_0^*(\boldsymbol{R}_1')u_0^*(\boldsymbol{R}_2'),
\end{aligned} \tag{13.14}$$

where we introduced new functions

$$\begin{aligned}
D(\boldsymbol{R}) &= A(0) - A(\boldsymbol{R}), \\
Q(\boldsymbol{R}_1,\boldsymbol{R}_2,\boldsymbol{R}_1',\boldsymbol{R}_2') &= D(\boldsymbol{R}_1-\boldsymbol{R}_1') + D(\boldsymbol{R}_2-\boldsymbol{R}_2') + D(\boldsymbol{R}_1-\boldsymbol{R}_2') \\
&\quad + D(\boldsymbol{R}_2-\boldsymbol{R}_1') - D(\boldsymbol{R}_2-\boldsymbol{R}_1) - D(\boldsymbol{R}_2'-\boldsymbol{R}_1')
\end{aligned}$$

related to the structure function of random field $\varepsilon(x,\boldsymbol{R})$.

Introducing new variables

$$\boldsymbol{R} \to \boldsymbol{R} + \frac{1}{2}\boldsymbol{\rho}, \quad \boldsymbol{R}' \to \boldsymbol{R} - \frac{1}{2}\boldsymbol{\rho},$$

Eq. (13.13) can be rewritten in the form

$$\left(\frac{\partial}{\partial x} - \frac{i}{k}\nabla_{\boldsymbol{R}}\nabla_{\boldsymbol{\rho}}\right)\Gamma_2(x, \boldsymbol{R}, \boldsymbol{\rho}) = -\frac{k^2}{4}D(\boldsymbol{\rho})\Gamma_2(x, \boldsymbol{R}, \boldsymbol{\rho}),$$
$$\Gamma_2(0, \boldsymbol{R}, \boldsymbol{\rho}) = \gamma_0(\boldsymbol{R}, \boldsymbol{\rho}) = u_0\left(\boldsymbol{R} + \frac{1}{2}\boldsymbol{\rho}\right)u_0^*\left(\boldsymbol{R}' - \frac{1}{2}\boldsymbol{\rho}\right). \tag{13.15}$$

Equations (13.12) and (13.15) can be easily solved for arbitrary function $D(\boldsymbol{\rho})$ and arbitrary initial conditions. Indeed, the average field is given by the expression

$$\langle u(x, \boldsymbol{R})\rangle = u_0(x, \boldsymbol{R})e^{-\frac{\gamma}{2}x}, \tag{13.16}$$

where $u_0(x, \boldsymbol{R})$ is the solution to the problem with absent fluctuations of medium parameters,

$$u_0(x, \boldsymbol{R}) = \int d\boldsymbol{R}' g(x, \boldsymbol{R} - \boldsymbol{R}')u_0(\boldsymbol{R}').$$

The function $g(x, \boldsymbol{R})$ is free space Green's function for $x > x'$

$$g\left(x, \boldsymbol{R}; x', \boldsymbol{R}'\right) = \exp\left(\frac{i(x - x')}{2k}\Delta_{\boldsymbol{R}}\right)\delta\left(\boldsymbol{R} - \boldsymbol{R}'\right)$$
$$= \frac{k}{2\pi i(x - x')}\exp\left(\frac{ik\left(\boldsymbol{R} - \boldsymbol{R}'\right)^2}{2(x - x')}\right) \tag{13.17}$$

and quantity $\gamma = \frac{k^2}{4}A(0)$ is the extinction coefficient.

Correspondingly, the second-order coherence function is given by the expression

$$\Gamma_2(x, \boldsymbol{R}, \boldsymbol{\rho}) = \int d\boldsymbol{q}\gamma_0\left(\boldsymbol{q}, \boldsymbol{\rho} - \boldsymbol{q}\frac{x}{k}\right)\exp\left\{i\boldsymbol{q}\boldsymbol{R} - \frac{k^2}{4}\int\limits_0^x d\xi D\left(\boldsymbol{\rho} - \boldsymbol{q}\frac{\xi}{k}\right)\right\}, \tag{13.18}$$

where

$$\gamma_0(\boldsymbol{q}, \boldsymbol{\rho}) = \frac{1}{(2\pi)^2}\int d\boldsymbol{R}\gamma_0(\boldsymbol{R}, \boldsymbol{\rho})e^{-i\boldsymbol{q}\boldsymbol{R}}.$$

The further analysis depends on the initial conditions to Eq. (13.1) and the fluctuation nature of field $\varepsilon(x, \boldsymbol{R})$. Three types of initial data are usually used in practice. They are:

- plane incident wave, in which case $u_0(\boldsymbol{R}) = u_0$;
- spherical divergent wave, in which case $u_0(\boldsymbol{R}) = \delta(\boldsymbol{R})$; and

- incident wave beam with the initial field distribution

$$u_0(\boldsymbol{R}) = u_0 \exp\left\{-\frac{\boldsymbol{R}^2}{2a^2} + i\frac{k\boldsymbol{R}^2}{2F}\right\}, \tag{13.19}$$

where a is the effective beam width, F is the distance to the radiation center (in the case of free space, value $F = \infty$ corresponds to the collimated beam and value $F < 0$ corresponds to the beam focused at distance $x = |F|$).

In the case of the plane incident wave, we have

$$u_0(\boldsymbol{R}) = u_0 = \text{const}, \quad \gamma_0(\boldsymbol{R}, \boldsymbol{\rho}) = |u_0|^2, \quad \gamma_0(\boldsymbol{q}, \boldsymbol{\rho}) = |u_0|^2\delta(\boldsymbol{q}),$$

and Eqs. (13.16) and (13.18) become significantly simpler

$$\langle u(x, \boldsymbol{R})\rangle = u_0 e^{-\frac{1}{2}\gamma x}, \quad \Gamma_2(x, \boldsymbol{R}, \boldsymbol{\rho}) = |u_0|^2 e^{-\frac{1}{4}k^2 x D(\boldsymbol{\rho})} \tag{13.20}$$

and appear to be independent of plane wave diffraction in random medium. Moreover, the expression for the coherence function shows the appearance of new statistical scale ρ_{coh} defined by the condition

$$\frac{1}{4}k^2 x D\left(\rho_{\text{coh}}\right) = 1. \tag{13.21}$$

This scale is called the *coherence radius* of field $u(x, \boldsymbol{R})$. Its value depends on the wavelength, distance the wave travels in the medium, and medium statistical parameters.

In the case of the wave beam (13.19), we easily obtain that

$$\gamma_0(\boldsymbol{q}, \boldsymbol{\rho}) = \frac{|u_0|^2 a^2}{4\pi} \exp\left\{-\frac{1}{4}\left[\frac{\boldsymbol{\rho}^2}{a^2} + \left(\frac{k\boldsymbol{\rho}}{F} - \boldsymbol{q}\right)^2 a^2\right]\right\}.$$

Using this expression and considering the turbulent atmosphere as an example of random medium for which the structure function $D(\boldsymbol{R})$ is described by the *Kolmogorov–Obukhov law* (see, e.g., [33, 34])

$$D(\boldsymbol{R}) = NC_\varepsilon^2 R^{5/3} \quad (R_{\min} \ll R \ll R_{\max}),$$

where $N = 1.46$, and C_ε^2 is the structure characteristic of dielectric permittivity fluctuations, we obtain that average intensity in the beam

$$\langle I(x, \boldsymbol{R})\rangle = \Gamma_2(x, \boldsymbol{R}, 0)$$

is given by the expression

$$\langle I(x, \boldsymbol{R})\rangle$$
$$= \frac{2|u_0|^2 k^2 a^4}{x^2 g^2(x)} \int_0^\infty dt J_0\left(\frac{2kaRt}{xg(x)}\right) \exp\left\{-t^2 - \frac{3\pi N}{32} C_\varepsilon^2 k^2 x \left(\frac{2a}{g(x)}\right)^{5/3} t^{5/3}\right\},$$

where

$$g(x) = \sqrt{1 + k^2 a^4 \left(\frac{1}{x} + \frac{1}{F}\right)^2}$$

and $J_0(t)$ is the Bessel function. Many full-scale experiments testified this formula in turbulent atmosphere and showed a good agreement between measured data and theory.

Equation (13.14) for the fourth-order coherence function cannot be solved in the analitic form; the analysis of this function requires either numerical, or approximate techniques. It describes intensity fluctuations and reduces to the intensity variance for equal transverse coordinates.

In the case of the plane incident wave, Eq. (13.14) can be simplified by introducing new transverse coordinates

$$\widetilde{\boldsymbol{R}}_1 = \boldsymbol{R}'_1 - \boldsymbol{R}_1 = \boldsymbol{R}_2 - \boldsymbol{R}'_2, \quad \widetilde{\boldsymbol{R}}_2 = \boldsymbol{R}'_2 - \boldsymbol{R}_1 = \boldsymbol{R}_2 - \boldsymbol{R}'_1.$$

In this case, Eq. (13.14) assumes the form (we omit here the tilde sign)

$$\frac{\partial}{\partial x}\Gamma_4(x, \boldsymbol{R}_1, \boldsymbol{R}_2) = \frac{i}{k}\frac{\partial^2}{\partial \boldsymbol{R}_1 \partial \boldsymbol{R}_2}\Gamma_4(x, \boldsymbol{R}_1, \boldsymbol{R}_2) - \frac{k^2}{4}F(\boldsymbol{R}_1, \boldsymbol{R}_2)\Gamma_4(x, \boldsymbol{R}_1, \boldsymbol{R}_2), \tag{13.22}$$

where

$$F(\boldsymbol{R}_1, \boldsymbol{R}_2) = 2D(\boldsymbol{R}_1) + 2D(\boldsymbol{R}_2) - D(\boldsymbol{R}_1 + \boldsymbol{R}_2) - D(\boldsymbol{R}_1 - \boldsymbol{R}_2).$$

We will give the asymptotic solution to this equation in Sect. 13.3.1, page 371.

13.2 Wavefield Amplitude–Phase Fluctuations. Rytov's Smooth Perturbation Method

Here, we consider the statistical description of wave amplitude–phase fluctuations.

We introduce the amplitude and phase (and the complex phase) of the wavefield by the formula

$$u(x, \boldsymbol{R}) = A(x, \boldsymbol{R})e^{iS(x,\boldsymbol{R})} = e^{\phi(x,\boldsymbol{R})},$$

where

$$\phi(x, \boldsymbol{R}) = \chi(x, \boldsymbol{R}) + iS(x, \boldsymbol{R}),$$

$\chi(x, \boldsymbol{R}) = \ln A(x, \boldsymbol{R})$ being the level of the wave and $S(x, \boldsymbol{R})$ being the wave random phase addition to the phase of the incident wave kx. Starting from parabolic equation (13.1), we can obtain that the complex phase satisfies the nonlinear equation of Rytov's *smooth perturbation method* (*SPM*)

$$\frac{\partial}{\partial x}\phi(x,\boldsymbol{R}) = \frac{i}{2k}\Delta_{\boldsymbol{R}}\phi(x,\boldsymbol{R}) + \frac{i}{2k}\left[\nabla_{\boldsymbol{R}}\phi(x,\boldsymbol{R})\right]^2 + i\frac{k}{2}\varepsilon(x,\boldsymbol{R}). \tag{13.23}$$

For the plane incident wave (in what follows, we will deal just with this case), we can set $u_0(\boldsymbol{R}) = 1$ and $\phi(0,\boldsymbol{R}) = 0$ without loss of generality.

Separating the real and imaginary parts in Eq. (13.23), we obtain

$$\frac{\partial}{\partial x}\chi(x,\boldsymbol{R}) + \frac{1}{2k}\Delta_{\boldsymbol{R}}S(x,\boldsymbol{R}) + \frac{1}{k}\left[\nabla_{\boldsymbol{R}}\chi(x,\boldsymbol{R})\right]\left[\nabla_{\boldsymbol{R}}S(x,\boldsymbol{R})\right] = 0, \tag{13.24}$$

$$\frac{\partial}{\partial x}S(x,\boldsymbol{R}) - \frac{1}{2k}\Delta_{\boldsymbol{R}}\chi(x,\boldsymbol{R}) - \frac{1}{2k}\left[\nabla_{\boldsymbol{R}}\chi(x,\boldsymbol{R})\right]^2 + \frac{1}{2k}\left[\nabla_{\boldsymbol{R}}S(x,\boldsymbol{R})\right]^2 = \frac{k}{2}\varepsilon(x,\boldsymbol{R}). \tag{13.25}$$

Using Eq. (13.24), we can derive the equation for wave intensity

$$I(x,\boldsymbol{R}) = e^{2\chi(x,\boldsymbol{R})}$$

in the form

$$\frac{\partial}{\partial x}I(x,\boldsymbol{R}) + \frac{1}{k}\nabla_{\boldsymbol{R}}\left[I(x,\boldsymbol{R})\nabla_{\boldsymbol{R}}S(x,\boldsymbol{R})\right] = 0. \tag{13.26}$$

If function $\varepsilon(x,\boldsymbol{R})$ is sufficiently small, then we can solve Eqs. (13.24) and (13.25) by constructing iterative series in field $\varepsilon(x,\boldsymbol{R})$. The first approximation of Rytov's SPM deals with the Gaussian fields $\chi(x,\boldsymbol{R})$ and $S(x,\boldsymbol{R})$, whose statistical characteristics are determined by statistical averaging of the corresponding iterative series. For example, the second moments (including variances) of these fields are determined from the linearized system of equations (13.24) and (13.25), i.e., from the system

$$\begin{aligned}\frac{\partial}{\partial x}\chi_0(x,\boldsymbol{R}) &= -\frac{1}{2k}\Delta_{\boldsymbol{R}}S_0(x,\boldsymbol{R}),\\ \frac{\partial}{\partial x}S_0(x,\boldsymbol{R}) &= \frac{1}{2k}\Delta_{\boldsymbol{R}}\chi_0(x,\boldsymbol{R}) + \frac{k}{2}\varepsilon(x,\boldsymbol{R}),\end{aligned} \tag{13.27}$$

while average values are determined immediately from Eqs. (13.24) and (13.25). Such amplitude–phase description of the wave filed in random medium was first used by A. M. Obukhov more than 50 years ago in paper [70] (see also [71]) where he pioneered considering diffraction phenomena accompanying wave propagation in random media using the perturbation theory. Before this work, similar investigations were based on the geometrical optics (acoustics) approximation. The technique suggested by Obukhov has been topical up untill now. Basically, it forms the mathematical apparatus of different engineering applications. However, as it was experimentally shown later in papers [72, 73], wavefield fluctuations rapidly grow with distance due to the effect of multiple forward scattering, and perturbation theory fails from a certain distance (region of strong fluctuations).

The liner system of equations (13.27) can be solved using the Fourier transform with respect to the transverse coordinate. Introducing the Fourier transforms of level, phase, and random field $\varepsilon(x, \boldsymbol{R})$

$$\chi_0(x, \boldsymbol{R}) = \int d\boldsymbol{q}\chi_{\boldsymbol{q}}^0(x)e^{i\boldsymbol{q}\boldsymbol{R}}, \quad \chi_{\boldsymbol{q}}^0(x) = \frac{1}{(2\pi)^2}\int d\boldsymbol{R}\chi_0(x, \boldsymbol{R})e^{-i\boldsymbol{q}\boldsymbol{R}};$$
$$S_0(x, \boldsymbol{R}) = \int d\boldsymbol{q}S_{\boldsymbol{q}}^0(x)e^{i\boldsymbol{q}\boldsymbol{R}}, \quad S_{\boldsymbol{q}}^0(x) = \frac{1}{(2\pi)^2}\int d\boldsymbol{R}S_0(x, \boldsymbol{R})e^{-i\boldsymbol{q}\boldsymbol{R}}; \tag{13.28}$$
$$\varepsilon(x, \boldsymbol{R}) = \int d\boldsymbol{q}\varepsilon_{\boldsymbol{q}}(x)e^{i\boldsymbol{q}\boldsymbol{R}}, \quad \varepsilon_{\boldsymbol{q}}(x) = \frac{1}{(2\pi)^2}\int d\boldsymbol{R}\varepsilon(x, \boldsymbol{R})e^{-i\boldsymbol{q}\boldsymbol{R}},$$

we obtain the solution to system (13.27) in the form

$$\chi_{\boldsymbol{q}}^0(x) = \frac{k}{2}\int_0^x d\xi\varepsilon_{\boldsymbol{q}}(\xi)\sin\frac{q^2}{2k}(x-\xi),$$
$$S_{\boldsymbol{q}}^0(x) = \frac{k}{2}\int_0^x d\xi\varepsilon_{\boldsymbol{q}}(\xi)\cos\frac{q^2}{2k}(x-\xi). \tag{13.29}$$

For random field $\varepsilon(x, \boldsymbol{R})$ described by the correlation and spectral functions (13.6) and (13.5), the correlation function of random Gaussian field $\varepsilon_{\boldsymbol{q}}(x)$ can be easily obtained by calculating the corresponding integrals.

Indeed, in the case of the delta-correlated approximation of random field $\varepsilon(x, \boldsymbol{R})$, the relationship between the correlation and spectral functions has the form

$$B_\varepsilon(x_1 - x_2, \boldsymbol{R}_1 - \boldsymbol{R}_2) = 2\pi\delta(x_1 - x_2)\int d\boldsymbol{q}\Phi_\varepsilon(0, \boldsymbol{q})e^{i\boldsymbol{q}(\boldsymbol{R}_1 - \boldsymbol{R}_2)}. \tag{13.30}$$

Multiplying Eq. (13.30) by $e^{-i(\boldsymbol{q}_1\boldsymbol{R}_1 + \boldsymbol{q}_2\boldsymbol{R}_2)}$, integrating the result over all $\boldsymbol{R}_1$, $\boldsymbol{R}_2$ and taking into account (13.28), we obtain the desired equality

$$\left\langle\varepsilon_{\boldsymbol{q}_1}(x_1)\varepsilon_{\boldsymbol{q}_2}(x_2)\right\rangle = 2\pi\delta(x_1 - x_2)\delta(\boldsymbol{q}_1 + \boldsymbol{q}_2)\Phi_\varepsilon(0, \boldsymbol{q}_1). \tag{13.31}$$

If field $\varepsilon(x, \boldsymbol{R})$ is different from zero only in layer $(0, \Delta x)$ and $\varepsilon(x, \boldsymbol{R}) = 0$ for $x > \Delta x$, then Eq. (13.31) is replaced with the expression

$$\left\langle\varepsilon_{\boldsymbol{q}_1}(x_1)\varepsilon_{\boldsymbol{q}_2}(x_2)\right\rangle = 2\pi\delta(x_1 - x_2)\theta(\Delta x - x)\delta(\boldsymbol{q}_1 + \boldsymbol{q}_2)\Phi_\varepsilon(0, \boldsymbol{q}_1). \tag{13.32}$$

If we deal with field $\varepsilon(x, \boldsymbol{R})$ whose fluctuations are caused by temperature turbulent pulsations, then the three-dimensional spectral density can be represented for a wide range of wave numbers in the form

$$\Phi_\varepsilon(\boldsymbol{q}) = AC_\varepsilon^2 q^{-11/3} \quad (q_{\min} \ll q \ll q_{\max}), \tag{13.33}$$

where $A = 0.033$ is a constant, C_ε^2 is the structure characteristic of dielectric permittivity fluctuations that depends on medium parameters. The use of spectral density (13.33) sometimes gives rise to the divergence of the integrals describing statistical characteristics of amplitude–phase fluctuations of the wavefield. In these cases, we can use the phenomenological spectral function

$$\Phi_\varepsilon(\boldsymbol{q}) = \Phi_\varepsilon(q) = AC_\varepsilon^2 q^{-11/3} e^{-q^2/\kappa_m^2}, \tag{13.34}$$

where κ_m is the wave number corresponding to the *turbulence microscale*.

Within the framework of the first approximation of Rytov's SPM, statistical characteristics of amplitude fluctuations are described by the variance of amplitude level, i.e., by the parameter

$$\sigma_0^2(x) = \left\langle \chi_0^2(x, \boldsymbol{R}) \right\rangle.$$

In the case of the medium occupying a layer of finite thickness Δx, this parameter can be represented by virtue of Eqs. (13.29) and (13.31) as

$$\sigma_0^2(x) = \int\int d\boldsymbol{q}_1 d\boldsymbol{q}_2 \langle \chi^0_{\boldsymbol{q}_1}(x) \chi^0_{\boldsymbol{q}_2}(x) \rangle e^{i(\boldsymbol{q}_1+\boldsymbol{q}_2)\boldsymbol{R}}$$
$$= \frac{\pi^2 k^2 \Delta x}{2} \int\limits_0^\infty dq q \Phi_\varepsilon(q) \left\{ 1 - \frac{k}{q^2 \Delta x} \left[\sin\frac{q^2 x}{k} - \sin\frac{q^2(x - \Delta x)}{k} \right] \right\}. \tag{13.35}$$

As for the average amplitude level, we determine it from Eq. (13.26). Assuming that incident wave is the plane wave and averaging this equation over an ensemble of realizations of field $\varepsilon(x, \boldsymbol{R})$, we obtain the equality

$$\langle I(x, \boldsymbol{R}) \rangle = 1.$$

Rewriting this equality in the form

$$\langle I(x, \boldsymbol{R}) \rangle = \left\langle e^{2\chi_0(x,\boldsymbol{R})} \right\rangle = e^{2\langle \chi_0(x,\boldsymbol{R})\rangle + 2\sigma_0^2(x)} = 1,$$

we see that

$$\langle \chi_0(x, \boldsymbol{R}) \rangle = -\sigma_0^2(x)$$

in the first approximation of Rytov's SPM.

The range of applicability of the first approximation of Rytov's SPM is restricted by the obvious condition

$$\sigma_0^2(x) \ll 1.$$

As for the wave intensity variance called also *flicker rate*, the first approximation of Rytov's SPM yields the following expression

$$\beta_0(x) = \left\langle I^2(x, \boldsymbol{R})\right\rangle - 1 = \left\langle e^{4\chi_0(x,\boldsymbol{R})}\right\rangle - 1 \approx 4\sigma_0^2(x). \tag{13.36}$$

Therefore, the one-point probability density of field $\chi(x, \boldsymbol{R})$ has in this approximation the form

$$P(x; \chi) = \sqrt{\frac{2}{\pi\beta_0(x)}} \exp\left\{-\frac{2}{\beta_0(x)}\left(\chi + \frac{1}{4}\beta_0(x)\right)^2\right\}.$$

Thus, the wavefield intensity is the logarithmic-normal random field, and its one-point probability density is given by the expression

$$P(x; I) = \frac{1}{I\sqrt{2\pi\beta_0(x)}} \exp\left\{-\frac{1}{2\beta_0(x)} \ln^2\left(Ie^{\frac{1}{2}\beta_0(x)}\right)\right\}. \tag{13.37}$$

The statistical analysis considers commonly two limiting asymptotic cases.

The first case corresponds to the assumption $\Delta x \ll x$ and is called the *random phase screen*. In this case, the wave first traverses a thin layer of fluctuating medium and then propagates in free space. The thin medium layer adds to the wavefield only phase fluctuations. In view of nonlinearity of Eqs. (13.24) and (13.25), the further propagation in free space transforms these phase fluctuations into amplitude fluctuations.

The second case corresponds to the *continuous medium*, i.e., to the condition $\Delta x = x$.

Consider these limiting cases in more detail assuming that wavefield fluctuations are weak.

13.2.1 Random Phase Screen ($\Delta x \ll x$)

In this case, the variance of amplitude level is given by the expression following from Eq. (13.35)

$$\sigma_0^2(x) = \frac{\pi^2 k^2 \Delta x}{2} \int_0^\infty dq\, q \Phi_\varepsilon(q) \left\{1 - \cos\frac{q^2 x}{k}\right\}. \tag{13.38}$$

If fluctuations of field $\varepsilon(x, \boldsymbol{R})$ are caused by turbulent pulsations of medium, then spectrum $\Phi_\varepsilon(q)$ is described by Eq. (13.33), and integral (13.35) can be easily calculated. The resulting expression is

$$\sigma_0^2(x) = 0.144 C_\varepsilon^2 k^{7/6} x^{5/6} \Delta x, \tag{13.39}$$

and, consequently, the flicker rate is given by the expression

$$\beta_0(x) = 0.563 C_\varepsilon^2 k^{7/6} x^{5/6} \Delta x. \tag{13.40}$$

As regards phase fluctuations, the quantity of immediate physical interest is the angle of wave arrival at point $(x, \boldsymbol{R})$,

$$\alpha(x, \boldsymbol{R}) = \frac{1}{k} |\nabla_{\boldsymbol{R}} S(x, \boldsymbol{R})|.$$

The derivation of the formula for its variance is similar to the derivation of Eq. (13.38); the result is as follows

$$\left\langle \alpha^2(x, \boldsymbol{R}) \right\rangle = \frac{\pi^2 \Delta x}{2} \int\limits_0^\infty dq\, q \Phi_\varepsilon(q) \left\{ 1 + \cos \frac{q^2 x}{k} \right\}.$$

13.2.2 Continuous Medium ($\Delta x = x$)

In this case, the variance of amplitude level is given by the formula

$$\sigma_0^2(x) = \frac{\pi^2 k^2 x}{2} \int\limits_0^\infty dq\, q \Phi_\varepsilon(q) \left\{ 1 - \frac{k}{q^2 x} \sin \frac{q^2 x}{k} \right\}, \tag{13.41}$$

so that parameters $\sigma_0^2(x)$ and $\beta_0(x)$ for turbulent medium pulsations assume the forms

$$\sigma_0^2(x) = 0.077 C_\varepsilon^2 k^{7/6} x^{11/6}, \quad \beta_0(x) = 0.307 C_\varepsilon^2 k^{7/6} x^{11/6}. \tag{13.42}$$

The variance of the angle of wave arrival at point $(x, \boldsymbol{R})$ is given by the formula

$$\left\langle \alpha^2(x, \boldsymbol{R}) \right\rangle = \frac{\pi^2 x}{2} \int\limits_0^\infty dq\, q \Phi_\varepsilon(q) \left\{ 1 + \frac{k}{q^2 x} \sin \frac{q^2 x}{k} \right\}. \tag{13.43}$$

We can similarly investigate the variance of the gradient of amplitude level. In this case, we are forced to use the spectral function $\Phi_\varepsilon(q)$ in form (13.33). Assuming that turbulent medium occupies the whole of the space and the so-called *wave parameter* $D(x) = \kappa_m^2 x / k$ (see, e.g., [33, 34]) is large, $D(x) \gg 1$, we can obtain for parameter

$$\sigma_q^2(x) = \left\langle [\nabla_{\boldsymbol{R}} \chi(x, \boldsymbol{R})]^2 \right\rangle$$

the expression

$$\sigma_q^2(x) = \frac{k^2 \pi^2 x}{2} \int\limits_0^\infty dq\, q^3 \Phi_\varepsilon(q) \left\{ 1 - \frac{k}{q^2 x} \sin \frac{q^2 x}{k} \right\} = \frac{1.476}{L_f^2(x)} D^{1/6}(x) \beta_0(x), \tag{13.44}$$

where we introduced the natural scale of length $L_f(x) = \sqrt{x/k}$ in plane $x = \text{const}$; this scale is independent of medium parameters and is equal in size to the first Fresnel zone that determines the size of the light–shade region in the problem on wave diffraction on the edge of an opaque screen (see, e.g., [33, 34]).

The first approximation of Rytov's SPM for amplitude fluctuations is valid in the general case under the condition

$$\sigma_0^2(x) \ll 1.$$

The region, where this inequality is satisfied, is called the *weak fluctuations region*. In the region, where $\sigma_0^2(x) \geq 1$ (this region is called the *strong fluctuations region*), the linearization fails, and we must study the nonlinear system of equations (13.24), (13.25).

As concerns the fluctuations of angle of wave arrival at the observation point $\alpha(x, \boldsymbol{R}) = \dfrac{1}{k}|\nabla_{\boldsymbol{R}} S(x, \boldsymbol{R})|$, they are adequately described by the first approximation of Rytov's SPM even for large values of parameter $\sigma_0(x)$.

Note that the approximation of the delta-correlated random field $\varepsilon(x, \boldsymbol{R})$ in the context of Eq. (13.1) only slightly restricts amplitude fluctuations; as a consequence, the above equations for moments of field $u(x, \boldsymbol{R})$ appear to be valid even in the region of strong amplitude fluctuations. The analysis of statistical characteristics in this case will be given later.

13.3 Method of Path Integral

Here, we consider statistical description of characteristics of the wavefield in random medium on the basis of problem solution in the functional form (i.e., in the form of the path integral).

As earlier, we will describe wave propagation in inhomogeneous medium starting from parabolic equation (13.1), page 355, whose solution can be represented in the operator form or in the form of path integral by using the method suggested by E. Fradkin in the quantum field theory [74–76].

To obtain such a representation, we replace Eq. (13.1) with a more complicated one with an arbitrary deterministic vector function $\boldsymbol{v}(x)$; namely, we consider the equation

$$\frac{\partial}{\partial x}\Phi(x, \boldsymbol{R}) = \frac{i}{2k}\Delta_{\boldsymbol{R}}\Phi(x, \boldsymbol{R}) + i\frac{k}{2}\varepsilon(x, \boldsymbol{R})\Phi(x, \boldsymbol{R}) + \boldsymbol{v}(x)\nabla_{\boldsymbol{R}}\Phi(x, \boldsymbol{R}),$$
$$\Phi(0, \boldsymbol{R}) = u_0(\boldsymbol{R}). \tag{13.45}$$

The solution to the original parabolic equation (13.1) is then obtained by the formula

$$u(x, \boldsymbol{R}) = \Phi(x, \boldsymbol{R})|_{\boldsymbol{v}(x)=0}. \tag{13.46}$$

In the standard way, we obtain the expression for the variational derivative $\dfrac{\delta\Phi(x,\boldsymbol{R})}{\delta\boldsymbol{v}(x-0)}$

$$\frac{\delta\Phi(x,\boldsymbol{R})}{\delta\boldsymbol{v}(x-0)} = \nabla_{\boldsymbol{R}}\Phi(x,\boldsymbol{R}), \tag{13.47}$$

and rewrite Eq. (13.45) in the form

$$\frac{\partial}{\partial x}\Phi(x,\boldsymbol{R}) = \frac{i}{2k}\frac{\delta^2\Phi(x,\boldsymbol{R})}{\delta\boldsymbol{v}^2(x-0)} + i\frac{k}{2}\varepsilon(x,\boldsymbol{R})\Phi(x,\boldsymbol{R}) + \boldsymbol{v}(x)\nabla_{\boldsymbol{R}}\Phi(x,\boldsymbol{R}). \tag{13.48}$$

We will seek the solution to Eq. (13.48) in the form

$$\Phi(x,\boldsymbol{R}) = e^{\frac{i}{2k}\int_0^{x-0} d\xi \frac{\delta^2}{\delta\boldsymbol{v}^2(\xi)}}\varphi(x,\boldsymbol{R}). \tag{13.49}$$

Because the operator in the exponent of Eq. (13.49) commutes with function $\boldsymbol{v}(x)$, we obtain that function $\varphi(x,\boldsymbol{R})$ satisfies the first-order equation

$$\frac{\partial}{\partial x}\varphi(x,\boldsymbol{R}) = i\frac{k}{2}\varepsilon(x,\boldsymbol{R})\varphi(x,\boldsymbol{R}) + \boldsymbol{v}(x)\nabla_{\boldsymbol{R}}\varphi(x,\boldsymbol{R}), \quad \varphi(0,\boldsymbol{R}) = u_0(\boldsymbol{R}), \tag{13.50}$$

whose solution as a functional of $\boldsymbol{v}(\xi)$ has the following form

$$\varphi(x,\boldsymbol{R}) = \varphi\left[x,\boldsymbol{R};\boldsymbol{v}(\xi)\right]$$
$$= u_0\left(\boldsymbol{R} + \int_0^x d\xi\boldsymbol{v}(\xi)\right)\exp\left\{i\frac{k}{2}\int_0^x d\xi\varepsilon\left(\xi,\boldsymbol{R} + \int_\xi^x d\eta\boldsymbol{v}(\eta)\right)\right\}. \tag{13.51}$$

As a consequence, taking into account Eqs. (13.49) and (13.46), we obtain the solution to the parabolic equation (13.1) in the operator form

$$u(x,\boldsymbol{R}) = \exp\left\{\frac{i}{2k}\int_0^x d\xi\frac{\delta^2}{\delta\boldsymbol{v}^2(\xi)}\right\}$$
$$\times u_0\left(\boldsymbol{R} + \int_0^x d\xi\boldsymbol{v}(\xi)\right)\exp\left\{i\frac{k}{2}\int_0^x d\xi\varepsilon\left(\xi,\boldsymbol{R} + \int_\xi^x d\eta\boldsymbol{v}(\eta)\right)\right\}\Bigg|_{\boldsymbol{v}(x)=0}. \tag{13.52}$$

In the case of the plane incident wave, we have $u_0(\boldsymbol{R}) = u_0$, and Eq. (13.52) is simplified

$$u(x, \boldsymbol{R}) = u_0 e^{\frac{i}{2k}\int\limits_0^x d\xi \frac{\delta^2}{\delta \boldsymbol{v}^2(\xi)}} \exp\left\{ i\frac{k}{2}\int\limits_0^x d\xi\, \varepsilon\left(\xi, \boldsymbol{R} + \int\limits_\xi^x d\eta \boldsymbol{v}(\eta)\right)\right\}\Bigg|_{\boldsymbol{v}(x)=0}. \tag{13.53}$$

Now, we formally consider Eq. (13.50) as the stochastic equation in which function $\boldsymbol{v}(x)$ is assumed the 'Gaussian' random vector function with the zero-valued mean and the imaginary 'correlation' function

$$\left\langle v_i(x)v_j(x')\right\rangle = \frac{i}{k}\delta_{ij}\delta(x - x'). \tag{13.54}$$

One can easily check that all formulas valid for the Gaussian random processes hold in this case, too.

Averaging Eq. (13.50) over an ensemble of realizations of 'random' process $\boldsymbol{v}(x)$, we obtain that average function $\langle \varphi(x, \boldsymbol{R})\rangle_{\boldsymbol{v}}$ satisfies the equation that coincides with Eq. (13.1). Thus the solution to parabolic equation (13.1) can be treated in the probabilistic sense; namely, we can formally represent this solution as the following average

$$u(x, \boldsymbol{R}) = \langle \varphi\left[x, \boldsymbol{R}; \boldsymbol{v}(\xi)\right]\rangle_{\boldsymbol{v}}. \tag{13.55}$$

This expression can be represented in the form of the *Feynman path integral*

$$u(x, \boldsymbol{R}) = \int D\boldsymbol{v}(x) u_0\left(\boldsymbol{R} + \int\limits_0^x d\xi \boldsymbol{v}(\xi)\right)$$
$$\times \exp\left\{ i\frac{k}{2}\int\limits_0^x d\xi \left[\boldsymbol{v}^2(\xi) + \varepsilon\left(\xi, \boldsymbol{R} + \int\limits_\xi^x d\eta \boldsymbol{v}(\eta)\right)\right]\right\}, \tag{13.56}$$

where the integral measure $D\boldsymbol{v}(x)$ is defined as follows

$$D\boldsymbol{v}(x) = \frac{\prod\limits_{\xi=0}^{x} d\boldsymbol{v}(\xi)}{\int \cdots \int \prod\limits_{\xi=0}^{x} d\boldsymbol{v}(\xi) \exp\left\{ i\frac{k}{2}\int\limits_0^x d\xi \boldsymbol{v}^2(\xi)\right\}}.$$

Representations (13.52) and (13.55) are equivalent. Indeed, considering the solution to Eq. (13.50) as a functional of random process $\boldsymbol{v}(\xi)$, we can reduce Eq. (13.55) to the following chain of equalities

$$u(x,\boldsymbol{R}) = \langle \varphi\left[x,\boldsymbol{R};\boldsymbol{v}(\xi)+\boldsymbol{y}(\xi)\right]\rangle_{\boldsymbol{v}}\Big|_{\boldsymbol{y}=0}$$

$$= \left\langle \exp\left\{\int\limits_0^x d\xi \boldsymbol{v}(\xi)\frac{\delta}{\delta\boldsymbol{y}(\xi)}\right\}\right\rangle_{\boldsymbol{v}} \varphi\left[x,\boldsymbol{R};\boldsymbol{y}(\xi)\right]\Bigg|_{\boldsymbol{y}=0}$$

$$= \exp\left\{\frac{i}{2k}\int\limits_0^x d\xi\frac{\delta^2}{\delta\boldsymbol{y}^2(\xi)}\right\}\varphi\left[x,\boldsymbol{R};\boldsymbol{y}(\xi)\right]\Bigg|_{\boldsymbol{y}=0},$$

and, consequently, to the operator form (13.52).

In terms of average field and second-order coherence function, the method of path integral (or the operator method) is equivalent to direct averaging of stochastic equations. However, the operator method (or the method of path integral) offers a possibility of obtaining expressions for quantities that cannot be described in terms of closed equations (among which are, for example, the expressions related to wave intensity fluctuations), and this is a very important point. Indeed, we can derive the closed equation for the fourth-order coherence function

$$\Gamma_4(x;\boldsymbol{R}_1,\boldsymbol{R}_2,\boldsymbol{R}_3,\boldsymbol{R}_4) = \left\langle u(x,\boldsymbol{R}_1)u(x,\boldsymbol{R}_2)u^*(x,\boldsymbol{R}_3)u^*(x,\boldsymbol{R}_4)\right\rangle$$

and then determine quantity $\left\langle I^2(x,\boldsymbol{R})\right\rangle$ by setting

$$\boldsymbol{R}_1 = \boldsymbol{R}_2 = \boldsymbol{R}_3 = \boldsymbol{R}_4 = \boldsymbol{R}$$

in the solution. However, this equation cannot be solved in analytic form; moreover, it includes many parameters unnecessary for determining $\left\langle I^2(x,\boldsymbol{R})\right\rangle$, whereas the path integral representation of quantity $\left\langle I^2(x,\boldsymbol{R})\right\rangle$ includes no such parameters. Therefore, the path integral representation of problem solution can be useful for studying asymptotic characteristics of arbitrary moments and–as a consequence–probability distribution of wavefield intensity. In addition, the operator representation of the field sometimes simplifies determination of the desired average characteristics as compared with the analysis of the corresponding equations. For example, if we would desire to calculate the quantity

$$\langle \varepsilon(y,\boldsymbol{R}_1)I(x,\boldsymbol{R})\rangle \quad (y < x),$$

then, starting from Eq. (13.1), we should first derive the differential equation for quantity $\varepsilon(y,\boldsymbol{R}_1)u(x,\boldsymbol{R}_2)u^*(x,\boldsymbol{R}_3)$ for $y < x$, average it over an ensemble of realizations of field $\varepsilon(x,\boldsymbol{R})$, specify boundary condition for quantity $\langle\varepsilon(y,\boldsymbol{R}_1)u(x,\boldsymbol{R}_2)u^*(x,\boldsymbol{R}_3)\rangle$ at $x = y$, solve the obtained equation with this boundary condition, and only then set $\boldsymbol{R}_2 = \boldsymbol{R}_3 = \boldsymbol{R}$. At the same time, the calculation of this quantity in terms of the operator representation only slightly differs from the above calculation of quantity $\langle\psi\psi^*\rangle$.

Now, we turn to the analysis of asymptotic behavior of plane wave intensity fluctuations in random medium in the region of strong fluctuations. In this analysis, we will adhere to works [77] (see also [2]).

13.3.1 Asymptotic Analysis of Plane Wave Intensity Fluctuations

Consider statistical moment of field $u(x,\boldsymbol{R})$

$$M_{nn}(x,\boldsymbol{R}_1,\ldots,\boldsymbol{R}_{2n}) = \left\langle \prod_{k=1}^{n} u(x,\boldsymbol{R}_{2k-1})u^*(x,\boldsymbol{R}_{2k}) \right\rangle. \tag{13.57}$$

In the approximation of delta-correlated field $\varepsilon(x,\boldsymbol{R})$, function $M_{nn}(x,\boldsymbol{R}_1,\ldots,\boldsymbol{R}_{2n})$ satisfies Eq. (13.10), page 357, for $n = m$. In the case of the plane incident wave, this is the equation with the initial condition, which assumes the form (in variables $\boldsymbol{R}_k$)

$$\begin{aligned}\left(\frac{\partial}{\partial x} - \frac{i}{2k}\sum_{l=1}^{2n}(-1)^{l+1}\Delta_{\boldsymbol{R}_l}\right) & M_{nn}(x,\boldsymbol{R}_1,\ldots,\boldsymbol{R}_{2n}) \\ &= \frac{k^2}{8}\sum_{l,j=1}^{2n}(-1)^{l+j}D(\boldsymbol{R}_l-\boldsymbol{R}_j)M_{nn}(x,\boldsymbol{R}_1,\ldots,\boldsymbol{R}_{2n}),\end{aligned} \tag{13.58}$$

where

$$D(\boldsymbol{R}) = A(0) - A(\boldsymbol{R}) = 2\pi\int d\boldsymbol{q}\Phi_\varepsilon(0,\boldsymbol{q})\left[1-\cos(\boldsymbol{q}\boldsymbol{R})\right] \tag{13.59}$$

and $\Phi_\varepsilon(0,\boldsymbol{q})$ is the three-dimensional spectrum of field $\varepsilon(x,\boldsymbol{R})$ whose argument is the two-dimensional vector $\boldsymbol{q}$.

Using the path integral representation of field $u(x,\boldsymbol{R})$ (13.56), and averaging it over field $\varepsilon(x,\boldsymbol{R})$, we obtain the expression for functions $M_{nn}(x,\boldsymbol{R}_1,\ldots,\boldsymbol{R}_{2n})$ in the form

$$\begin{aligned}M_{nn}(x,\boldsymbol{R}_1,\ldots,\boldsymbol{R}_{2n}) &= \int\ldots\int D\boldsymbol{v}_1(\xi)\ldots D\boldsymbol{v}_{2n}(\xi) \\ &\times\exp\left\{\frac{ik}{2x}\sum_{j=1}^{2n}(-1)^{j+1}\int_0^x d\xi\,\boldsymbol{v}_j^2(\xi)\right. \\ &\left.-\frac{k^2}{8}\sum_{j,l=1}^{2n}(-1)^{j+l+1}\int_0^x d\xi\,D\left(\boldsymbol{R}_j-\boldsymbol{R}_l+\int_\xi^x dx'\left[\boldsymbol{v}_j(x')-\boldsymbol{v}_l(x')\right]\right)\right\}.\end{aligned} \tag{13.60}$$

Another way of obtaining Eq. (13.60) consists in solving Eq. (13.58) immediately, by the method described earlier. We can rewrite Eq. (13.60) in the operator form

$$M_{nn}(x, \boldsymbol{R}_1, \ldots, \boldsymbol{R}_{2n}) = \prod_{l=1}^{2n} \exp\left\{\frac{i}{2k}(-1)^{l+1}\int\limits_0^x d\xi \frac{\delta^2}{\delta \boldsymbol{v}^2(\xi)}\right\}$$
$$\times \exp\left\{-\frac{k^2}{8}\sum_{j,l=1}^{2n}(-1)^{j+l+1}\right.$$
$$\left.\times \int\limits_0^x dx' D\left(\boldsymbol{R}_j - \boldsymbol{R}_l + \int\limits_{x'}^x d\xi \left[\boldsymbol{v}_j(\xi) - \boldsymbol{v}_l(\xi)\right]\right)\right\}_{\boldsymbol{v}=0}. \tag{13.61}$$

If we now superpose points $\boldsymbol{R}_{2k-1}$ and $\boldsymbol{R}_{2k}$ (i.e., if we set $\boldsymbol{R}_{2k-1} = \boldsymbol{R}_{2k}$), then function $M_{nn}(x, \boldsymbol{R}_1, \ldots, \boldsymbol{R}_{2n})$ will grade into the function $\left\langle\prod_{k=1}^{n} I(x, \boldsymbol{R}_{2k-1})\right\rangle$ that describes correlation characteristics of wave intensity. Further, if we set all $\boldsymbol{R}_l$ identical ($\boldsymbol{R}_l = \boldsymbol{R}$), then function

$$M_{nn}(x, \boldsymbol{R}, \ldots, \boldsymbol{R}) = \Gamma_{2n}(x, \boldsymbol{R}) = \langle I^n(x, \boldsymbol{R})\rangle$$

will describe the n-th moment of wavefield intensity.

Prior to discuss the asymptotics of functions $\Gamma_{2n}(x, \boldsymbol{R})$ for the continuous random medium, we consider a simpler problem on wavefield fluctuations behind the random phase screen.

Random Phase Screen

Suppose that we deal with the inhomogeneous medium layer whose thickness is so small that the wave traversed this layer acquires only random phase incursion

$$S(\boldsymbol{R}) = \frac{k}{2}\int\limits_0^{\Delta x} d\xi \varepsilon(\xi, \boldsymbol{R}), \tag{13.62}$$

the amplitude remaining intact. As earlier, we will assume that random field $\varepsilon(x, \boldsymbol{R})$ is the Gaussian field delta-correlated in the x-direction. After traversing the inhomogeneous layer, the wave is propagating in homogeneous medium where its propagation is governed by the equation (13.1) with $\varepsilon(x, \boldsymbol{R}) = 0$. The solution to this problem is given by the formula

$$u(x, \boldsymbol{R}) = e^{i\frac{x}{2k}\Delta_{\boldsymbol{R}}} e^{iS(\boldsymbol{R})} = \frac{k}{2\pi i x}\int d\boldsymbol{v} \exp\left\{\frac{ik}{2x}\boldsymbol{v}^2 + iS(\boldsymbol{R} + \boldsymbol{v})\right\}, \tag{13.63}$$

which is the finite-dimensional analog of Eqs. (13.53) and (13.56).

Consider function $M_{nn}(x, \boldsymbol{R}_1, \ldots, \boldsymbol{R}_{2n})$. Substituting Eq. (13.63) in Eq. (13.57) and averaging the result, we easily obtain the formula

$$M_{nn}(x, \boldsymbol{R}_1, \ldots, \boldsymbol{R}_{2n})$$
$$= \left(\frac{k}{2\pi x}\right)^{2n} \int \ldots \int d\boldsymbol{v}_1 \ldots d\boldsymbol{v}_{2n} \exp\left\{\frac{ik}{2x}\sum_{j=1}^{2n}(-1)^{j+1}\boldsymbol{v}_j^2\right.$$
$$\left. - \frac{k^2\Delta x}{8}\sum_{j,l=1}^{2n}(-1)^{j+l+1}D\left(\boldsymbol{R}_j - \boldsymbol{R}_l + \boldsymbol{v}_j - \boldsymbol{v}_l\right)\right\} \tag{13.64}$$

which is analogous to Eq. (13.60).

First of all, we consider in greater detail the case with $n = 2$ for superimposed pairs of observation points

$$\boldsymbol{R}_1 = \boldsymbol{R}_2 = \boldsymbol{R}', \quad \boldsymbol{R}_3 = \boldsymbol{R}_4 = \boldsymbol{R}'', \quad \boldsymbol{R}' - \boldsymbol{R}'' = \boldsymbol{\rho}.$$

In this case, function

$$\Gamma_4(x; \boldsymbol{R}', \boldsymbol{R}', \boldsymbol{R}'', \boldsymbol{R}'') = \left\langle I(x, \boldsymbol{R}')I(x, \boldsymbol{R}'')\right\rangle$$

is the covariation of intensities $I(x, \boldsymbol{R}) = |u(x, \boldsymbol{R})|^2$. In the case of $n = 2$, we can introduce new integration variables in Eq. (13.64),

$$\boldsymbol{v}_1 - \boldsymbol{v}_2 = \boldsymbol{R}_1, \quad \boldsymbol{v}_1 - \boldsymbol{v}_4 = \boldsymbol{R}_2, \quad \boldsymbol{v}_1 - \boldsymbol{v}_3 = \boldsymbol{R}_3, \quad \frac{1}{2}(\boldsymbol{v}_1 + \boldsymbol{v}_2) = \boldsymbol{R},$$

and perform integrations over $\boldsymbol{R}$ and $\boldsymbol{R}_3$ to obtain the simpler formula

$$\left\langle I(x, \boldsymbol{R}')I(x, \boldsymbol{R}''\right\rangle$$
$$= \left(\frac{k}{2\pi x}\right)^2 \int\int d\boldsymbol{R}_1 d\boldsymbol{R}_2 \exp\left\{\frac{ik}{x}\boldsymbol{R}_1(\boldsymbol{R}_2 - \boldsymbol{\rho}) - \frac{k^2\Delta x}{4}F(\boldsymbol{R}_1, \boldsymbol{R}_2)\right\}, \tag{13.65}$$

where $\boldsymbol{\rho} = \boldsymbol{R}' - \boldsymbol{R}''$and function $F(\boldsymbol{R}_1, \boldsymbol{R}_2)$ is determined from Eq. (13.22),

$$F(\boldsymbol{R}_1, \boldsymbol{R}_2) = 2D(\boldsymbol{R}_1) + 2D(\boldsymbol{R}_2) - D(\boldsymbol{R}_1 + \boldsymbol{R}_2) - D(\boldsymbol{R}_1 - \boldsymbol{R}_2),$$
$$D(\boldsymbol{R}) = A(0) - A(\boldsymbol{R}).$$

The integral in Eq. (13.65) was studied in detail (including numerical methods) in many works. Its asymptotics for $x \to \infty$ has the form

$$
\begin{aligned}
\langle I(x,\boldsymbol{R}')I(x,\boldsymbol{R}'')\rangle &= 1 + \exp\left\{-\frac{k^2\Delta x}{2}D(\boldsymbol{\rho})\right\} \\
&+ \pi k^2 \Delta x \int d\boldsymbol{q}\Phi_\varepsilon(\boldsymbol{q})\left[1-\cos\frac{q^2 x}{k}\right]\exp\left\{i\boldsymbol{q}\boldsymbol{\rho} - \frac{k^2\Delta x}{2}D\left(\frac{\boldsymbol{q}x}{k}\right)\right\} \\
&+ \pi k^2 \Delta x \int d\boldsymbol{q}\Phi_\varepsilon(\boldsymbol{q})\left[1-\cos\left(\boldsymbol{q}\boldsymbol{\rho} - \frac{q^2 x}{k}\right)\right] \\
&\times \exp\left\{-\frac{k^2\Delta x}{2}D\left(\boldsymbol{\rho} - \frac{\boldsymbol{q}x}{k}\right)\right\} + \cdots.
\end{aligned}
\tag{13.66}
$$

Note that, in addition to spatial scale ρ_{coh}, the second characteristic spatial scale

$$
r_0 = \frac{x}{k\rho_{\text{coh}}} \tag{13.67}
$$

appears in the problem.

Setting $\boldsymbol{\rho} = 0$ in Eq. (13.66), we can obtain the expression for the intensity variance

$$
\begin{aligned}
\beta^2(x) &= \left\langle I^2(x,\boldsymbol{R})\right\rangle - 1 \\
&= 1 + \pi\Delta x \int d\boldsymbol{q}\, q^4 \Phi_\varepsilon(\boldsymbol{q}) \exp\left\{-\frac{k^2\Delta x}{2}D\left(\frac{\boldsymbol{q}x}{k}\right)\right\} + \cdots.
\end{aligned}
\tag{13.68}
$$

If the turbulence is responsible for fluctuations of field $\varepsilon(x,\boldsymbol{R})$ in the inhomogeneous layer, so that spectrum $\Phi_\varepsilon(\boldsymbol{q})$ is given by Eq. (13.33), page 364, then Eq. (13.68) yields

$$
\beta^2(x) = 1 + 0.429\beta_0^{-4/5}(x), \tag{13.69}
$$

where $\beta_0(x)$ is the intensity variance calculated in the first approximation of Rytov's smooth perturbation method applied to the phase screen (13.40).

The above considerations can be easily extended to higher moment functions of fields $u(x,\boldsymbol{R})$, $u^*(x,\boldsymbol{R})$ and, in particular, to functions $\Gamma_{2n}(x,\boldsymbol{R}) = \langle I^n(x,\boldsymbol{R})\rangle$. In this case, Eq. (13.64) assumes the form

$$
\begin{aligned}
\langle I^n(x,\boldsymbol{R})\rangle &= \left(\frac{k}{2\pi x}\right)^{2n} \int \ldots \int d\boldsymbol{v}_1 \ldots d\boldsymbol{v}_{2n} \\
&\times \exp\left\{\frac{ik}{2x}\sum_{j=1}^{2n}(-1)^{j+1}\boldsymbol{v}_j^2 - F(\boldsymbol{v}_1,\ldots,\boldsymbol{v}_{2n})\right\},
\end{aligned}
\tag{13.70}
$$

where

$$F(\boldsymbol{v}_1,\ldots,\boldsymbol{v}_{2n}) = \frac{k^2\Delta x}{8}\sum_{j,l=1}^{2n}(-1)^{j+l+1}D\left(\boldsymbol{v}_j-\boldsymbol{v}_l\right). \tag{13.71}$$

Function $F(\boldsymbol{v}_1,\ldots,\boldsymbol{v}_{2n})$ can be expressed in terms of random phase incursions $S(\boldsymbol{v}_i)$ (13.62) by the formula

$$F(\boldsymbol{v}_1,\ldots,\boldsymbol{v}_{2n}) = \frac{1}{2}\left\langle\left[\sum_{j=1}^{2n}(-1)^{j+1}S(\boldsymbol{v}_j)\right]^2\right\rangle \geq 0.$$

This formula clearly shows that function $F(\boldsymbol{v}_1,\ldots,\boldsymbol{v}_{2n})$ vanishes if each odd point $\boldsymbol{v}_{2l+1}$ coincides with certain pair point among even points, because the positive and negative phase incursions cancel in this case. It becomes clear that namely the regions in which this cancellation occur will mainly contribute to moments $\langle I^n(x,\boldsymbol{R})\rangle$ for $\sqrt{x/k} \gg \rho_{\text{coh}}$. It is not difficult to calculate that the number of these regions is equal to $n!$. Then, replacing the integral in (13.70) with $n!$ multiplied by the integral over one of these regions A_1 in which

$$|\boldsymbol{v}_1-\boldsymbol{v}_2| \sim |\boldsymbol{v}_3-\boldsymbol{v}_4| \sim \cdots \sim |\boldsymbol{v}_{2n-1}-\boldsymbol{v}_{2n}| < \rho_{\text{coh}},$$

we obtain

$$\begin{aligned}\left\langle I^n(x,\boldsymbol{R})\right\rangle \approx n!\left(\frac{k}{2\pi x}\right)^{2n}\int_{A_1}\cdots\int d\boldsymbol{v}_1\ldots d\boldsymbol{v}_{2n}\\ \times\exp\left\{\frac{ik}{2x}\sum_{j=1}^{2n}(-1)^{j+1}\boldsymbol{v}_j^2 - F(\boldsymbol{v}_1,\ldots,\boldsymbol{v}_{2n})\right\}.\end{aligned} \tag{13.72}$$

Terms of sum (13.71)

$$\frac{k^2\Delta x}{8}D(\boldsymbol{v}_1-\boldsymbol{v}_2),\quad \frac{k^2\Delta x}{8}D(\boldsymbol{v}_3-\boldsymbol{v}_4),\quad \text{and so on}$$

ensure the decreasing behavior of the integrand with respect to every of variables $\boldsymbol{v}_1-\boldsymbol{v}_2$, $\boldsymbol{v}_3-\boldsymbol{v}_4$, and so on. We keep these terms in the exponent and expand the exponential function of other terms in the series to obtain the following approximate expression

$$\langle I^n(x,\boldsymbol{R})\rangle \approx n!\left(\frac{k}{2\pi x}\right)^{2n}\int\limits_{A_1}\cdots\int d\boldsymbol{v}_1\ldots d\boldsymbol{v}_{2n}$$

$$\times\exp\left\{\frac{ik}{2x}\sum_{j=1}^{2n}(-1)^{j+1}\boldsymbol{v}_j^2-\frac{k^2\Delta x}{4}\sum_{l=1}^{n}D\left(\boldsymbol{v}_{2l-1}-\boldsymbol{v}_{2l}\right)\right\}$$

$$\times\left\{1+\frac{k^2\Delta x}{8}\sum_{j,l=1}^{2n}{}'(-1)^{j+l+1}D\left(\boldsymbol{v}_j-\boldsymbol{v}_l\right)+\cdots\right\}. \tag{13.73}$$

Here, the prime of the sum sign means that this sum excludes the terms kept in the exponent. Because the integrand is negligible outside region A_1, we can extend the region of integration in Eq. (13.73) to the whole space. Then the multiple integral in Eq. (13.73) can be calculated in analytic form, and we obtain for $\langle I^n(x,\boldsymbol{R})\rangle$ the formula

$$\langle I^n(x,\boldsymbol{R})\rangle = n!\left[1+n(n-1)\frac{\beta^2(x)-1}{4}+\cdots\right], \tag{13.74}$$

where quantity $\beta^2(x)$ is given by Eq. (13.69). We discuss this formula a little later, after considering wave propagation in continuous random medium, which yields a very similar result.

Continuous Medium

Consider now the asymptotic behavior of higher moment functions of the wavefield $M_{nn}(x,\boldsymbol{R}_1,\ldots,\boldsymbol{R}_{2n})$ propagating in random medium. The formal solution to this problem is given by Eqs. (13.60) and (13.61). They differ from the phase screen formulas only by the fact that ordinary integrals are replaced with path integrals.

We consider first quantity

$$\langle I(x,\boldsymbol{R}')I(x,\boldsymbol{R}'')\rangle = M_{22}(x,\boldsymbol{R}_1,\ldots,\boldsymbol{R}_4)|_{\boldsymbol{R}_1=\boldsymbol{R}_2=\boldsymbol{R}',\ \boldsymbol{R}_3=\boldsymbol{R}_4=\boldsymbol{R}''}\cdot$$

In the case of the plane wave ($u_0(\boldsymbol{R})=1$), we can use (13.61) and introduce new variables similar to those for the phase screen to obtain

$$\langle I(x,\boldsymbol{R}')I(x,\boldsymbol{R}'')\rangle = \exp\left\{\frac{i}{k}\int\limits_0^x d\xi\frac{\delta^2}{\delta\boldsymbol{v}_1(\xi)\delta\boldsymbol{v}_2(\xi)}\right\}$$

$$\times\exp\left\{-\frac{k^2}{4}\int\limits_0^x dx'\left[2D\left(\boldsymbol{\rho}+\int\limits_{x'}^x d\xi\boldsymbol{v}_1(\xi)\right)+2D\left(\int\limits_{x'}^x d\xi\boldsymbol{v}_2(\xi)\right)\right.\right.$$

$$-D\left(\rho+\int\limits_{x'}^{x}d\xi\,[\boldsymbol{v}_1(\xi)+\boldsymbol{v}_2(\xi)]\right)-D\left(\rho+\int\limits_{x'}^{x}d\xi\,[\boldsymbol{v}_1(\xi)-\boldsymbol{v}_2(\xi)]\right)\Bigg]\Bigg\}_{\boldsymbol{v}_i=0},\tag{13.75}$$

where $\boldsymbol{\rho}=\boldsymbol{R}'-\boldsymbol{R}''$.

Using Eq. (13.60), formula (13.75) can be represented in the form of the path integral; however, we will use here the operator representation. As in the case of the phase screen, we can represent $\langle I(x,\boldsymbol{R}')I(x,\boldsymbol{R}'')\rangle$ for $x\to\infty$ in the form

$$B_I(x,\boldsymbol{\rho})=\langle I(x,\boldsymbol{R}')I(x,\boldsymbol{R}'')\rangle-1=B_I^{(1)}(x,\boldsymbol{\rho})+B_I^{(2)}(x,\boldsymbol{\rho})+B_I^{(3)}(x,\boldsymbol{\rho}),\tag{13.76}$$

where

$$B_I^{(1)}(x,\boldsymbol{\rho})=\exp\left\{-\frac{k^2x}{2}D(\boldsymbol{\rho})\right\},$$

$$B_I^{(2)}(x,\boldsymbol{\rho})=\pi k^2\int\limits_0^x dx'\int d\boldsymbol{q}\Phi_\varepsilon(\boldsymbol{q})\left[1-\cos\frac{\boldsymbol{q}^2}{k}(x-x')\right]$$

$$\times\exp\left\{i\boldsymbol{q}\boldsymbol{\rho}-\frac{k^2x'}{2}D\left(\frac{\boldsymbol{q}}{k}(x-x')\right)-\frac{k^2}{2}\int\limits_{x'}^{x}dx'D\left(\frac{\boldsymbol{q}}{k}(x-x'')\right)\right\},$$

$$B_I^{(3)}(x,\boldsymbol{\rho})=\pi k^2\int\limits_0^x dx'\int d\boldsymbol{q}\Phi_\varepsilon(\boldsymbol{q})\left[1-\cos\left(\boldsymbol{q}\boldsymbol{\rho}-\frac{\boldsymbol{q}^2}{k}(x-x')\right)\right]$$

$$\times\exp\left\{-\frac{k^2x'}{2}D\left(\boldsymbol{\rho}-\frac{\boldsymbol{q}}{k}(x-x')\right)-\frac{k^2}{2}\int\limits_{x'}^{x}dx'D\left(\boldsymbol{\rho}-\frac{\boldsymbol{q}}{k}(x-x'')\right)\right\}.$$

Setting $\boldsymbol{\rho}=0$ and taking into account only the first term of the expansion of function

$$1-\cos\frac{\boldsymbol{q}^2}{k}(x-x'),$$

we obtain that intensity variance

$$\beta^2(x)=\left\langle I^2(x,\boldsymbol{R})\right\rangle-1=B_I(x,0)-1$$

is given by the formula similar to Eq. (13.68)

$$\beta^2(x) = 1 + \pi \int_0^x dx'(x - x') \int d\mathbf{q}\, q^4 \Phi_\varepsilon(\mathbf{q})$$
$$\times \exp\left\{-\frac{k^2 x'}{2} D\left(\frac{\mathbf{q}}{k}(x - x')\right) - \frac{k^2}{2} \int_{x'}^{x} dx' D\left(\frac{\mathbf{q}}{k}(x - x'')\right)\right\} + \cdots . \tag{13.77}$$

If we deal with the turbulent medium, then Eq. (13.77) yields

$$\beta^2(x) = 1 + 0.861\,(\beta_0(x))^{-2/5}, \tag{13.78}$$

where $\beta_0(x)$ is the wavefield intensity variance calculated in the first approximation of Rytov's smooth perturbation method (13.42).

Expression (13.77) remains valid also in the case when functions $\Phi_\varepsilon(\mathbf{q})$ and $D(\boldsymbol{\rho})$ slowly vary with x. In this case, we can easily reduce Eq. (13.77) to Eq. (13.68) by setting $\Phi_\varepsilon(\mathbf{q}) = 0$ outside layer $0 \le x' \le \Delta x \ll x$.

Concerning correlation function $B_I(x, \boldsymbol{\rho})$, we note that the main term $B_I^{(1)}(x, \boldsymbol{\rho})$ in Eq. (13.76) is the squared modulus of the second-order coherence function (see, e.g., [77]).

Now, we turn to the higher moment functions $\langle I^n(x, \mathbf{R})\rangle = \Gamma_{2n}(x, 0)$. Similarly to the case of the phase screen, one can easily obtain that this moment of the wavefield in continuous medium is represented in the form of the expansion

$$\left\langle I^n(x, \mathbf{R})\right\rangle = n!\left[1 + n(n-1)\frac{\beta^2(x) - 1}{4} + \cdots\right], \tag{13.79}$$

which coincides with expansion (13.74) for the phase screen, excluding the fact that parameter $\beta^2(x)$ is given by different formulas in these cases.

Formula (13.79) specifies two first terms of the asymptotic expansion of function $\langle I^n(x, \mathbf{R})\rangle$ for $\beta_0^2(x) \to \infty$. Because $\beta^2(x) \to 1$ for $\beta_0(x) \to \infty$, the second term in Eq. (13.79) is small in comparison with the first one for sufficiently great $\beta_0(x)$. Expression (13.79) makes sense only if

$$n(n-1)\frac{\beta^2(x) - 1}{4} \ll 1. \tag{13.80}$$

However, we can always select numbers n for which condition (13.80) will be violated for a fixed $\beta_0(x)$. This means that Eq. (13.79) holds only for moderate n. It should be noted additionally that the moment can approach the asymptotic behavior (13.79) for $\beta_0(x) \to \infty$ fairly slowly.

Formula (13.79) yields the singular probability density of intensity. To avoid the singularities, we can approximate this formula by the expression (see, e.g., [78])

$$\left\langle I^n(x,\boldsymbol{R})\right\rangle = n!\exp\left\{n(n-1)\frac{\beta^2(x)-1}{4}\right\}, \tag{13.81}$$

which yields the probability density (see, e.g., [78, 79])

$$P(x,I) = \frac{1}{\sqrt{\pi}\,(\beta(x)-1)}\int\limits_0^\infty dz\exp\left\{-zI - \frac{\left[\ln z - \dfrac{\beta(x)-1}{4}\right]^2}{\beta(x)-1}\right\}. \tag{13.82}$$

Note that probability distribution (13.82) is not applicable in a narrow region $I \sim 0$ (the width of this region the narrower the greater parameter $\beta_0(x)$). This follows from the fact that Eq. (13.82) yields infinite values for the moments of inverse intensity $1/I(x,\boldsymbol{R})$. Nevertheless, moments $\langle 1/I^n(x,\boldsymbol{R})\rangle$ are finite for any finite-valued parameter $\beta_0^2(x)$ (arbitrarily great), and the equality $P(x,0) = 0$ must hold. It is clear that the existence of this narrow region around the point $I \sim 0$ does not affect the behavior of moments (13.81) for larger $\beta_0(x)$.

Asymptotic formulas (13.81) and (13.82) describe the transition to the region of *saturated* intensity fluctuations, where $\beta(x) \to 1$ for $\beta_0(x) \to \infty$. In this region, we have correspondingly

$$\left\langle I^n(x,\boldsymbol{R})\right\rangle = n!, \quad P(x,I) = e^{-I}. \tag{13.83}$$

The exponential probability distribution (13.83) means that complex field $u(x,\boldsymbol{R})$ is the Gaussian random field. Recall that

$$u(x,\boldsymbol{R}) = A(x,\boldsymbol{R})e^{iS(x,\boldsymbol{R})} = u_1(x,\boldsymbol{R}) + iu_2(x,\boldsymbol{R}), \tag{13.84}$$

where $u_1(x,\boldsymbol{R})$ and $u_2(x,\boldsymbol{R})$ are the real and imaginary parts, respectively. As a result, the wavefield intensity is

$$I(x,\boldsymbol{R}) = A^2(x,\boldsymbol{R}) = u_1^2(x,\boldsymbol{R}) + u_2^2(x,\boldsymbol{R}).$$

From the Gaussian property of complex field $u(x,\boldsymbol{R})$ follows that random fields $u_1(x,\boldsymbol{R})$ and $u_2(x,\boldsymbol{R})$ are also the Gaussian statistically independent fields with variances

$$\left\langle u_1^2(x,\boldsymbol{R})\right\rangle = \left\langle u_2^2(x,\boldsymbol{R})\right\rangle = \frac{1}{2}. \tag{13.85}$$

It is quite natural to assume that their gradients

$$\boldsymbol{p}_1(x,\boldsymbol{R}) = \nabla_{\boldsymbol{R}} u_1(x,\boldsymbol{R}) \text{ and } \boldsymbol{p}_2(x,\boldsymbol{R}) = \nabla_{\boldsymbol{R}} u_2(x,\boldsymbol{R})$$

are also statistically independent of fields $u_1(x, \boldsymbol{R})$ and $u_2(x, \boldsymbol{R})$ and are the Gaussian homogeneous and isotropic (in plane $\boldsymbol{R}$) fields with variances

$$\sigma_p^2(x) = \left\langle p_1^2(x, \boldsymbol{R}) \right\rangle = \left\langle p_2^2(x, \boldsymbol{R}) \right\rangle. \tag{13.86}$$

With these assumptions, the joint probability density of fields $u_1(x, \boldsymbol{R})$, $u_2(x, \boldsymbol{R})$ and gradients $\boldsymbol{p}_1(x, \boldsymbol{R})$ and $\boldsymbol{p}_2(x, \boldsymbol{R})$ has the form

$$P(x; u_1, u_2, \boldsymbol{p}_1, \boldsymbol{p}_2) = \frac{1}{\pi^3 \sigma_p^4(x)} \exp\left\{ -u_1^2 - u_2^2 - \frac{\boldsymbol{p}_1^2 + \boldsymbol{p}_2^2}{\sigma_p^2(x)} \right\}. \tag{13.87}$$

Consider now the joint probability density of wavefield intensity $I(x, \boldsymbol{R})$ and amplitude gradient

$$\boldsymbol{\kappa}(x, \boldsymbol{R}) = \nabla_{\boldsymbol{R}} A(x, \boldsymbol{R}) = \frac{u_1(x, \boldsymbol{R})\boldsymbol{p}_1(x, \boldsymbol{R}) + u_2(x, \boldsymbol{R})\boldsymbol{p}_2(x, \boldsymbol{R})}{\sqrt{u_1^2(x, \boldsymbol{R}) + u_2^2(x, \boldsymbol{R})}}.$$

We have for this probability density the expression

$$\begin{aligned} P(x; I, \boldsymbol{\kappa}) &= \langle \delta\left(I(x, \boldsymbol{R}) - I\right) \delta\left(\boldsymbol{\kappa}(x, \boldsymbol{R}) - \boldsymbol{\kappa}\right) \rangle_{u_i, \boldsymbol{p}_i} \\ &= \frac{1}{\pi^3 \sigma_p^4(x)} \int\limits_{-\infty}^{\infty} du_1 \int\limits_{-\infty}^{\infty} du_2 \int d\boldsymbol{p}_1 \int d\boldsymbol{p}_2 \exp\left\{ -u_1^2 - u_2^2 - \frac{\boldsymbol{p}_1^2 + \boldsymbol{p}_2^2}{\sigma_p^2(x)} \right\} \\ &\quad \times \delta\left(u_1^2 + u_2^2 - I\right) \delta\left(\frac{u_1 \boldsymbol{p}_1 + u_2 \boldsymbol{p}_2}{\sqrt{u_1^2 + u_2^2}} - \boldsymbol{\kappa} \right) \\ &= \frac{1}{2\pi \sigma_p^2(x)} \exp\left\{ -I - \frac{\boldsymbol{\kappa}^2}{2\sigma_p^2(x)} \right\}. \end{aligned} \tag{13.88}$$

Consequently, the transverse gradient of amplitude is statistically independent of wavefield intensity and is the Gaussian random field with the variance

$$\left\langle \boldsymbol{\kappa}^2(x, \boldsymbol{R}) \right\rangle = 2\sigma_p^2(x). \tag{13.89}$$

We note that the transverse gradient of amplitude is also independent of the second derivatives of wavefield intensity with respect to transverse coordinates.

At the end of this section, we note that the path integral representation of field $u(x, \boldsymbol{R})$ makes it possible to investigate the applicability range of the approximation of delta-correlated random field $\varepsilon(x, \boldsymbol{R})$ in the context of wave intensity fluctuations. It turns out that all conditions restricting applicability of the approximation of delta-correlated random field $\varepsilon(x, \boldsymbol{R})$ for calculating quantity $\langle I^n(x, \boldsymbol{R}) \rangle$ coincide with those obtained for quantity $\langle I^2(x, \boldsymbol{R})\rangle$. In other words, the approximation of delta-correlated random field $\varepsilon(x, \boldsymbol{R})$ does not affect the shape of the probability distribution of wave-field intensity.

In the case of turbulent temperature pulsations, the approximation of the delta-correlated random field $\varepsilon(x, \boldsymbol{R})$ holds in the region of weak fluctuations under the conditions

$$\lambda \ll \sqrt{\lambda x} \ll x,$$

where $\lambda = 2\pi/k$ is the wavelength.

As to the region of strong fluctuations, the applicability range of the approximation of delta-correlated random field $\varepsilon(x, \boldsymbol{R})$ is restricted by the conditions

$$\lambda \ll \rho_{\text{coh}} \ll r_0 \ll x,$$

where ρ_{coh} and r_0 are given by Eqs. (13.21), page 360, and (13.67), page 374. The physical meaning of all these inequalities is simple. The delta-correlated approximation remains valid as long as the correlation radius of field $\varepsilon(x, \boldsymbol{R})$ (its value is given by the size of the first Fresnel zone in the case of turbulent temperature pulsations) is the smallest longitudinal scale of the problem. As the wave approaches at the region of strong intensity fluctuations, a new longitudinal scale appears; its value $\sim \rho_{\text{coh}}\sqrt{kx}$ gradually decreases and can become smaller than the correlation radius of field $\varepsilon(x, \boldsymbol{R})$ for sufficiently large values of parameter $\beta_0(x)$. In this situation, the delta-correlated approximation fails.

We can consider the above inequalities as restrictions from above and from below on the scale of the intensity correlation function. In these terms, the delta-correlated approximation holds only if all scales appearing in the problem are small in comparison with the length of the wave path.

13.4 Elements of Statistical Topography of Random Intensity Field

The above statistical characteristics of wavefield $u(x, \boldsymbol{R})$, for example, the intensity correlation function in the region of strong fluctuations, have nothing in common with the actual behavior of the wavefield propagating in particular realizations of the medium (see Figs. 13.1 and 13.2). In the general case, wave field intensity satisfies Eq. (13.26), page 362,

$$\begin{aligned} &\frac{\partial}{\partial x} I(x, \boldsymbol{R}) + \frac{1}{k}\nabla_{\boldsymbol{R}}\left[I(x, \boldsymbol{R})\nabla_{\boldsymbol{R}} S(x, \boldsymbol{R})\right] = 0, \\ &I(0, \boldsymbol{R}) = I_0(\boldsymbol{R}), \end{aligned} \qquad (13.90)$$

from which follows that the power of wave field in plane $x = \text{const}$ remains intact for arbitrary incident wave beam:

$$E_0 = \int I(x, \boldsymbol{R}) d\boldsymbol{R} = \int I_0(\boldsymbol{R}) d\boldsymbol{R}.$$

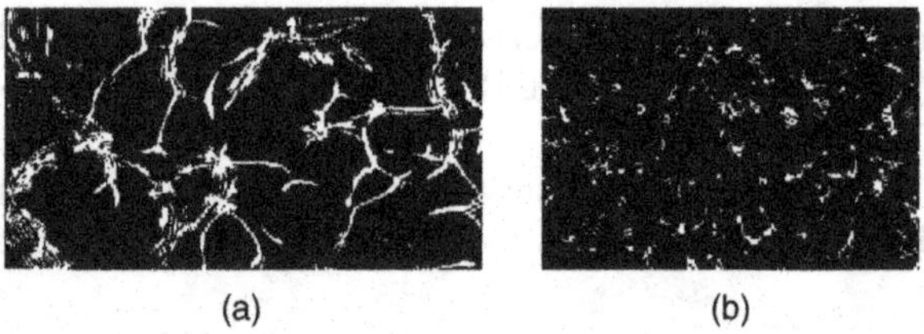

Figure 13.1 Transverse section of laser beam propagating in turbulent medium in the regions of (a) strong focusing and (b) strong (saturated) fluctuations. Experiment in laboratory conditions.

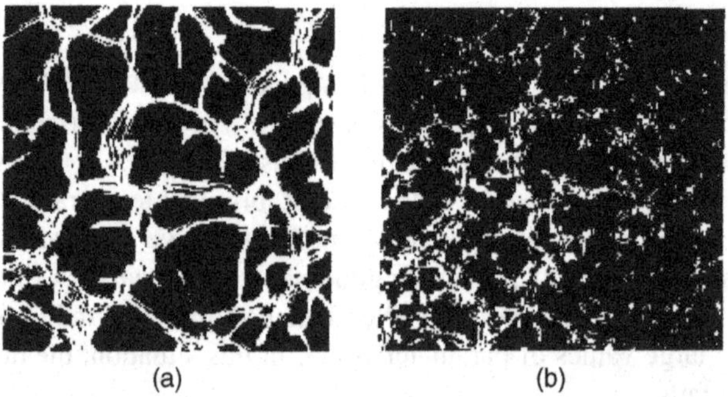

Figure 13.2 Transverse section of laser beam propagating in turbulent medium in the regions of (a) strong focusing and (b) strong (saturated) fluctuations. Numerical simulations.

Eqs. (13.90) can be treated as the equation of conservative tracer transfer in the potential velocity field in the conditions of tracer clustering. As a consequence, realizations of the wave intensity field also must show cluster behavior. In the case under consideration, this phenomenon appears in the form of caustic structures produced due to random processes of wave field focusing and defocusing in random medium. For example, Fig. 13.1 shows photos of the transverse sections of laser beam propagating in turbulent medium [80] for different magnitudes of dielectric permittivity fluctuations (laboratory investigations). Figure 13.2 shows similar photos borrowed from [81]. These photos were simulated numerically [82, 83]. Both figures clearly show the appearance of caustic structures in the wave field.

In order to analyze the detailed structure of wavefield, one can use methods of statistical topography [69]; they provide an insight into the formation of the wavefield caustic structure and make it possible to ascertain the statistical parameters that describe this structure.

If we deal with the plane incident wave, all one-point statistical characteristics, including probability densities, are independent of variable $\boldsymbol{R}$ in view of spatial homogeneity. In this case, a number of physical quantities that characterize cluster structure of wavefield intensity can be adequately described in terms of specific (per unit area) values.

Among these quantities are:

- specific average total area of regions in plane $\{\boldsymbol{R}\}$, which are bounded by level lines of intensity $I(x, \boldsymbol{R})$ inside which $I(x, \boldsymbol{R}) > I$,

$$\langle s(x, I) \rangle = \int_I^\infty dI' P(x; I'),$$

where $P(x; I)$ is the probability density of wavefield intensity;
- specific average field power within these regions

$$\langle e(x, I) \rangle = \int_I^\infty I' dI' P(x; I');$$

- specific average length of these contours per *first Fresnel zone* $L_f(x) = \sqrt{x/k}$

$$\langle l(x, I) \rangle = L_f(x) \left\langle |\boldsymbol{p}(x, \boldsymbol{R})| \delta \left(I(x, \boldsymbol{R}) - I \right) \right\rangle,$$

where $\boldsymbol{p}(x, \boldsymbol{R}) = \nabla_{\boldsymbol{R}} I(x, \boldsymbol{R})$ is the transverse gradient of wavefield intensity;
- estimate of average difference between the numbers of contours with opposite normal orientation per first Fresnel zone

$$\langle n(x, I) \rangle = \frac{1}{2\pi} L_f^2(x) \left\langle \kappa(x, \boldsymbol{R}; I) |\boldsymbol{p}(x, \boldsymbol{R})| \delta \left(I(x, \boldsymbol{R}) - I \right) \right\rangle,$$

where $\kappa(x, \boldsymbol{R}; I)$ is the curvature of the level line,

$$\kappa(x, \boldsymbol{R}; I) = \frac{-p_y^2(x, \boldsymbol{R}) \dfrac{\partial^2 I(x, \boldsymbol{R})}{\partial z^2} - p_z^2(x, \boldsymbol{R}) \dfrac{\partial^2 I(x, \boldsymbol{R})}{\partial y^2}}{p^3(x, \boldsymbol{R})} + 2 \frac{p_y(x, \boldsymbol{R}) p_z(x, \boldsymbol{R}) \frac{\partial^2 I(x, \boldsymbol{R})}{\partial y \partial z}}{p^3(x, \boldsymbol{R})}.$$

The first Fresnel zone plays the role of the natural medium-independent length scale in plane $x =$ const. It determines the size of the transient light–shadow zone appeared in the problem on diffraction by the edge of an opaque screen.

Consider now the behavior of these quantities with distance x (parameter $\beta_0(x)$).

13.4.1 Weak Intensity Fluctuations

The region of weak intensity fluctuations is limited by inequality $\beta_0(x) \leq 1$. In this region, wavefield intensity is the lognormal process described by probability distribution (13.37).

The typical realization curve of this logarithmic-normal process is the exponentially decaying curve

$$I^*(x) = e^{-\frac{1}{2}\beta_0(x)},$$

and statistical characteristics (moment functions $\langle I^n(x,\boldsymbol{R})\rangle$, for example) are formed by large spikes of process $I(x,\boldsymbol{R})$ relative this curve.

In addition, various majorant estimates are available for lognormal process realizations. For example, separate realizations of wavefield intensity satisfy the inequality

$$I(x) < 4e^{-\frac{1}{4}\beta_0(x)}$$

for all distances $x \in (0,\infty)$ with probability $p = 1/2$. All these facts are indicative of the onset of cluster structure formations in wavefield intensity.

As we have seen earlier, the knowledge of probability density (13.37) is sufficient for obtaining certain quantitative characteristics of these cluster formations. For example, the average area of regions within which $I(x,\boldsymbol{R}) > I$ is

$$\langle s(x,I)\rangle = \Pr\left(\frac{1}{\sqrt{\beta_0(x)}}\ln\left(\frac{1}{I}e^{-\beta_0(x)/2}\right)\right), \tag{13.91}$$

and specific average power confined in these regions is given by the expression

$$\langle e(x,I)\rangle = \Pr\left(\frac{1}{\sqrt{\beta_0(x)}}\ln\left(\frac{1}{I}e^{\beta_0(x)/2}\right)\right), \tag{13.92}$$

where probability integral function $Pr(z)$ is defined as (4.20), page 94.

The character of cluster structure spatial evolution versus parameter $\beta_0(x)$ essentially depends on the desired level I. In the most interesting case of $I > 1$, the values of these functions in the initial plane are $\langle s(0,I)\rangle = 0$ and $\langle e(0,I)\rangle = 0$. As $\beta_0(x)$ increases, small cluster regions appear in which $I(x,\boldsymbol{R}) > I$; for certain distances, these regions remain almost intact and actively absorb a considerable portion of total energy. With further increasing $\beta_0(x)$, the areas of these regions decrease and the power within them increases, which corresponds to an increase of regions's average brightness. The cause of these processes lies in radiation focusing by separate portions of medium.

Figure 13.3 shows functions $\langle s(x,I)\rangle$ and $\langle e(x,I)\rangle$ for different parameters $\beta_0(x)$ from a given range. The specific average area is maximum at $\beta_0(x) = 2\ln(I)$, and

$$\langle s(x,I\rangle_{\max} = \Pr\left(-\sqrt{2\ln I}\right).$$

The average power at this value of $\beta_0(x)$ is $\langle e(x,I)\rangle = 1/2$.

In the region of weak intensity fluctuations, the spatial gradient of amplitude level $\nabla_{\boldsymbol{R}}\chi(x,\boldsymbol{R})$ is statistically independent of $\chi(x,\boldsymbol{R})$. This fact makes it possible both to calculate specific average length of contours $I(x,\boldsymbol{R}) = I$ and to estimate specific average number of these contours. Indeed, in the region of weak fluctuations, probability density of amplitude level gradient $\boldsymbol{q}(x,\boldsymbol{R}) = \nabla_{\boldsymbol{R}}\chi(x,\boldsymbol{R})$ is the Gaussian density

$$P(x;\boldsymbol{q}) = \langle\delta\left(\nabla_{\boldsymbol{R}}\chi(x,\boldsymbol{R}) - \boldsymbol{q}\right)\rangle = \frac{1}{\sigma_q^2(x)}\exp\left\{-\frac{\boldsymbol{q}^2}{\sigma_q^2(x)}\right\}, \tag{13.93}$$

where $\sigma_q^2(x) = \langle\boldsymbol{q}^2(x,\boldsymbol{R})\rangle$ is the variance of amplitude level gradient given by Eq. (13.44), page 366.

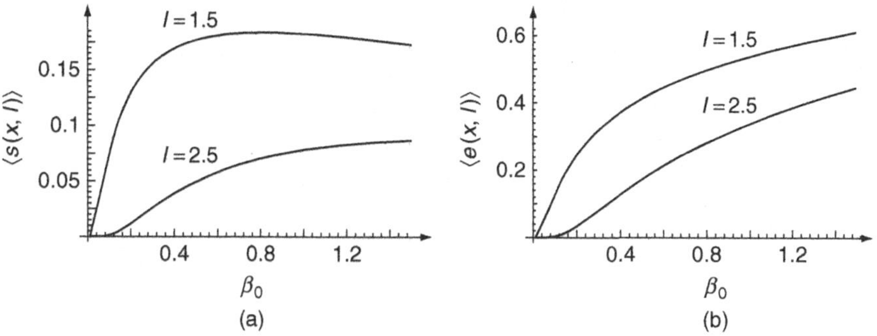

Figure 13.3 (a) Average specific area and (b) average power versus parameter $\beta_0(x)$.

As a consequence, we obtain that the specific average length of contours is described by the expression

$$\langle l(x, I)\rangle = 2L_f(x)\langle |q(x, \boldsymbol{R})|\rangle IP(x; I) = L_f(x)\sqrt{\pi\sigma_q^2(x)}IP(x; I). \tag{13.94}$$

For the specific average number of contours, we have similarly

$$\begin{aligned}\langle n(x, I)\rangle &= \frac{1}{2\pi}L_f^2(x)\langle \kappa(x, \boldsymbol{R}, I)|\boldsymbol{p}(x, \boldsymbol{R})\delta(I(x, \boldsymbol{R}) - I)\rangle \\ &= -\frac{1}{2\pi}L_f^2(x)I\langle \Delta_{\boldsymbol{R}}\chi(x, \boldsymbol{R})\delta(I(x, \boldsymbol{R}) - I)\rangle \\ &= -\frac{1}{\pi}L_f^2(x)\left\langle \boldsymbol{q}^2(x, \boldsymbol{R})\right\rangle I\frac{\partial}{\partial I}IP(x; I) \\ &= \frac{L_f^2(x)\sigma_{\boldsymbol{q}}^2(x)}{\pi\beta_0(x)}\ln\left(Ie^{\frac{1}{2}\beta_0(x)}\right)IP(x; I).\end{aligned} \tag{13.95}$$

We notice that Eq. (13.95) vanishes at $I = I_0(x) = \exp\{-\frac{1}{2}\beta_0(x)\}$. This means that this intensity level corresponds to the situation in which the specific average number of contours bounding region $I(x, \boldsymbol{R}) > I_0$ coincides with the specific average number of contours bounding region $I(x, \boldsymbol{R}) < I_0$. Figures 13.4 show functions $\langle l(x, I)\rangle$ and $\langle n(x, I)\rangle$ versus parameter $\beta_0(x)$.

Dependence of specific average length of level lines and specific average number of contours on turbulence microscale reveals the existence of small-scale ripples imposed upon large-scale random relief. These ripples do not affect the redistribution of areas and power, but increases the irregularity of level lines and causes the appearance of small contours.

As we mentioned earlier, this description holds for $\beta_0(x) \leq 1$. With increasing parameter $\beta_0(x)$ to this point, Rytov's smooth perturbation method fails, and we must consider the nonlinear equation in wavefield complex phase. This region of fluctuations called the *strong focusing region* is very difficult for analytical treatment. With

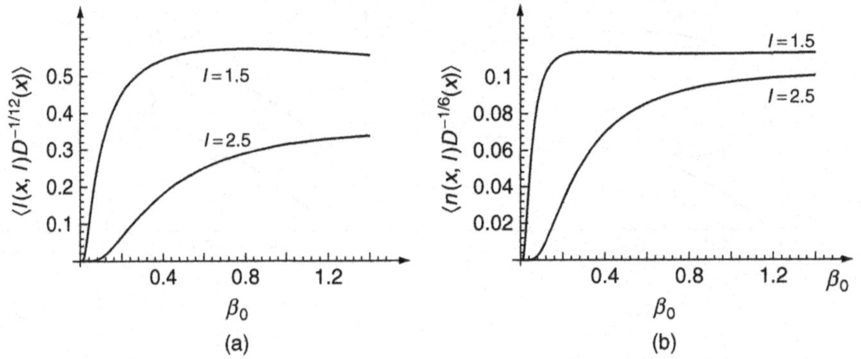

Figure 13.4 (a) Average specific contour length and (b) average contour number versus parameter $\beta_0(x)$.

further increasing parameter $\beta_0(x)$ ($\beta_0(x) \geq 10$) statistical characteristics of intensity approach the saturation regime, and this region of parameter $\beta_0(x)$ is called the *region of strong intensity fluctuations*.

13.4.2 Strong Intensity Fluctuations

From Eq. (13.82) for probability density follows that specific average area of regions within which $I(x, \boldsymbol{R}) > I$ is

$$\langle s(x, I)\rangle = \frac{1}{\sqrt{\pi\,(\beta(x)-1)}} \int\limits_0^\infty \frac{dz}{z} \exp\left\{-zI - \frac{\left[\ln z - \dfrac{\beta(x)-1}{4}\right]^2}{\beta(x)-1}\right\}, \tag{13.96}$$

and specific average power concentrated in these regions is given by the expression

$$\langle e(x, I)\rangle = \frac{1}{\sqrt{\pi\,(\beta(x)-1)}} \int\limits_0^\infty \frac{dz}{z}\left(I+\frac{1}{z}\right) \exp\left\{-zI - \frac{\left[\ln z - \dfrac{\beta(x)-1}{4}\right]^2}{\beta(x)-1}\right\}. \tag{13.97}$$

Figure 13.5 shows functions (13.96) and (13.97) versus parameter $\beta(x)$. We note that parameter $\beta(x)$ is a very slow function of $\beta_0(x)$. Indeed, limit process $\beta_0(x) \to \infty$ corresponds to $\beta(x) = 1$ and value $\beta_0(x) = 1$ corresponds to $\beta(x) = 1.861$.

Asymptotic formulas (13.96) and (13.97) adequately describe the transition to the region of saturated intensity fluctuations ($\beta(x) \to 1$). In this region, we have

$$P(I) = e^{-I}, \quad \langle s(I)\rangle = e^{-I}, \quad \langle e(I)\rangle = (I+1)\,e^{-I}. \tag{13.98}$$

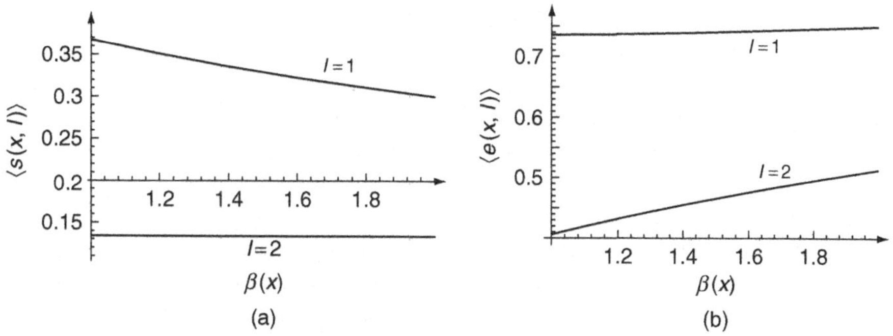

Figure 13.5 (a) Average specific area and (b) average contour number in the region of strong intensity fluctuations versus parameter $\beta(x)$.

Moreover, we obtain the expression for specific average contour length

$$\begin{aligned}\langle l(x, I)\rangle &= L_f(x)\,\langle|\boldsymbol{p}(x, \boldsymbol{R})|\delta\,(I(x, \boldsymbol{R}) - I)\rangle \\ &= 2L_f(x)\sqrt{I}\,\langle|\boldsymbol{\kappa}(x, \boldsymbol{R})|\delta\,(I(x, \boldsymbol{R}) - I)\rangle \\ &= 2L_f(x)\sqrt{I}\,\langle|\boldsymbol{q}(x, \boldsymbol{R})|\rangle\, P(x; I) = L_f(x)\sqrt{2\pi\sigma_q^2(x)I}\,P(x; I),\end{aligned} \tag{13.99}$$

where the variance of amplitude level gradient in the region of saturated fluctuations coincides with the variance calculated in the first approximation of Rytov's smooth perturbation method. Specific average contour length (13.99) is maximum at $I = 1/\sqrt{2}$.

In the region of saturated intensity fluctuations, the estimator of specific average number of contours is given by the following chain of equalities

$$\begin{aligned}\langle n(x, I)\rangle &= \frac{L_f^2(x)}{2\pi}\,\langle\kappa(x, \boldsymbol{R}, I)|\boldsymbol{p}(x, \boldsymbol{R})|\delta\,(I(x, \boldsymbol{R}) - I)\rangle \\ &= -\frac{L_f^2(x)}{2\pi}\sqrt{I}\,\langle\Delta_{\boldsymbol{R}}A(x, \boldsymbol{R})\delta\,(I(x, \boldsymbol{R}) - I)\rangle \\ &= -\frac{L_f^2(x)}{\pi}\left\langle\boldsymbol{\kappa}^2(x, \boldsymbol{R})\right\rangle\sqrt{I}\frac{\partial}{\partial I}\sqrt{I}P(x; I) \\ &= -\frac{2L_f^2(x)\sigma_q^2(x)}{\pi}\sqrt{I}\frac{\partial}{\partial I}\sqrt{I}e^{-I} = \frac{2L_f^2(x)\sigma_q^2(x)}{\pi}\left(I - \frac{1}{2}\right)e^{-I}.\end{aligned} \tag{13.100}$$

Expression (13.100) is maximum at $I = 3/2$, and the level at which specific average number of contours bounding region $I(x, \boldsymbol{R}) > I_0$ is equal to specific average number of contours corresponding to $I(x, \boldsymbol{R}) < I_0$ is $I_0 = 1/2$.

We note that Eq. (13.100) fails in the narrow region around $I \sim 0$. The correct formula must vanish for $I = 0$ ($\langle n(x, 0)\rangle = 0$).

As may be seen from Eqs. (13.99) and (13.100), average length of level lines and average number of contours keep increasing behaviors with parameter $\beta_0(x)$ in the region of saturated fluctuations, although the corresponding average areas and powers remain fixed. The reason of this behavior consists in the fact that the leading (and defining) role in this regime is played by interference of partial waves arriving from different directions.

Behavioral pattern of level lines depends on the relationship between processes of focusing and defocusing by separate portions of turbulent medium. Focusing by large-scale inhomogeneities becomes apparent in random intensity relief as high peaks. In the regime of maximal focusing ($\beta_0(x) \sim 1$), the narrow high peaks concentrate about a half of total wave power. With increasing parameter $\beta_0(x)$, radiation defocusing begins to prevail,which results in spreading the high peaks and forming highly idented (interference) relief characterized by multiple vertices at level $I \sim 1$.

In addition to parameter $\beta_0(x)$, average length of level lines and average number of contours depend on wave parameter $D(x)$; namely, they increase with decreasing microscale of inhomogeneities. This follows from the fact that the large-scale relief is distorted by fine ripples appeared due to scattering by small-scale inhomogeneities.

Thus, in this section, we attempted to qualitatively explain the cluster (caustic) structure of the wavefield generated in turbulent medium by the plane light wave and to quantitatively estimate the parameters of such a structure in the plane transverse to the propagation direction. In the general case, this problem is multiparametric. However, if we limit ourselves to the problem on the plane incident wave and consider it for a fixed wave parameter in a fixed plane, then the solution is expressed in terms of the sole parameter, namely, the variance of intensity in the region of weak fluctuations $\beta_0(x)$. We have analyzed two asymptotic cases corresponding to weak and saturated intensity fluctuations. It should be noted that applicability range of these asymptotic formulas most likely depends on intensity level I. It is expected naturally that this applicability range will grow with decreasing the level.

As regards the analysis of the intermediate case corresponding to the developed caustic structure (this case is the most interesting from the standpoint of applications), it requires the knowledge of the probability density of intensity and its transverse gradient for arbitrary distances in the medium. Such an analysis can be carried out either by using probability density approximations for all parameters [79], or on the basis of numerical simulations (see, e.g., [82–84]).

Problems

Problem 13.1

Find the second-order coherence function and average intensity at the axis of parabolic waveguide in the problem formulated in terms of the dynamic equation

$$\frac{\partial}{\partial x}u(x,\boldsymbol{R}) = \frac{i}{2k}\Delta_{\boldsymbol{R}}u(x,\boldsymbol{R}) + \frac{ik}{2}\left[-\alpha^2\boldsymbol{R}^2 + \varepsilon(x,\boldsymbol{R})\right]u(x,\boldsymbol{R})$$

assuming that $\varepsilon(x, \boldsymbol{R})$ is the homogeneous isotropic Gaussian field delta-correlated in x and characterized by the parameters

$$\langle \varepsilon(x, \boldsymbol{R}) \rangle = 0, \quad B_\varepsilon(x, \boldsymbol{R}) = B_\varepsilon^{eff}(x, \boldsymbol{R}) = A(\boldsymbol{R})\delta(x).$$

Solution

$$\Gamma_2(x, \boldsymbol{R}, \boldsymbol{\rho}) = \frac{e^{-i\alpha \boldsymbol{R}\boldsymbol{\rho} \tan(\alpha x)}}{\cos^2(\alpha x)} \int d\boldsymbol{q} \gamma_0 \left(\boldsymbol{q}, \frac{1}{\cos(\alpha x)} \boldsymbol{\rho} - \frac{\boldsymbol{q}}{\alpha k} \tan(\alpha x) \right)$$
$$\times \exp \left\{ i \frac{1}{\cos(\alpha x)} \boldsymbol{q}\boldsymbol{R} - \frac{k^2}{4} \int_0^x d\xi D \left(\frac{\cos(\alpha \xi)}{\cos(\alpha x)} \boldsymbol{\rho} - \frac{1}{\alpha k} \frac{\sin\left[\alpha (x - \xi)\right]}{\cos(\alpha x)} \boldsymbol{q} \right) \right\},$$

where

$$\gamma_0 (\boldsymbol{q}, \boldsymbol{\rho}) = \frac{1}{(2\pi)^2} \int d\boldsymbol{R} \Gamma_2(0, \boldsymbol{R}, \boldsymbol{\rho}) e^{-i\boldsymbol{q}\boldsymbol{R}}$$

and $D(\boldsymbol{\rho}) = A(0) - A(\boldsymbol{\rho})$.

Setting $\boldsymbol{R} = 0$ and $\boldsymbol{\rho} = 0$, we obtain the average intensity at waveguide axis as a function of longitudinal coordinate x

$$\langle I(x, \boldsymbol{0}) \rangle = \frac{1}{\cos^2(\alpha x)} \int d\boldsymbol{q} \gamma_0 \left(\boldsymbol{q}, -\frac{\boldsymbol{q}}{\alpha k} \tan(\alpha x) \right)$$
$$\times \exp \left\{ -\frac{k^2}{4} \int_0^x d\xi D \left(\frac{1}{\alpha k} \frac{\sin(\alpha \xi)}{\cos(\alpha x)} \boldsymbol{q} \right) \right\}.$$

Problem 13.2

Within the framework of the diffusion approximation, derive equations for average field $\langle u(x, \boldsymbol{R}) \rangle$ and second-order coherence function

$$\Gamma_2(x; \boldsymbol{R}, \boldsymbol{\rho}) = \left\langle u \left(x, \boldsymbol{R} + \frac{1}{2}\boldsymbol{\rho} \right) u^* \left(x, \boldsymbol{R} - \frac{1}{2}\boldsymbol{\rho} \right) \right\rangle$$

starting from parabolic equation (13.1) [55, 56].

Solution

$$\left(\frac{\partial}{\partial x} - \frac{i}{2k} \Delta_{\boldsymbol{R}} \right) \langle u(x, \boldsymbol{R}) \rangle$$
$$= -\frac{k^2}{4} \int_0^x dx' \int d\boldsymbol{R}' B_\varepsilon(x', \boldsymbol{R} - \boldsymbol{R}') e^{\frac{ix'}{2k} \Delta_{\boldsymbol{R}}} \left[\delta(\boldsymbol{R} - \boldsymbol{R}') e^{-\frac{ix'}{2k} \Delta_{\boldsymbol{R}}} \langle u(x, \boldsymbol{R}) \rangle \right],$$

$$\left(\frac{\partial}{\partial x}-\frac{i}{k}\nabla_{\boldsymbol{R}}\nabla_{\boldsymbol{\rho}}\right)\Gamma_2(x,\boldsymbol{R},\boldsymbol{\rho})$$
$$=-\frac{k^2}{4}\int\limits_0^x dx_1\int d\boldsymbol{R}_1\left[B_\varepsilon\left(x_1,\boldsymbol{R}-\boldsymbol{R}_1+\frac{\boldsymbol{\rho}}{2}\right)-B_\varepsilon\left(x_1,\boldsymbol{R}-\boldsymbol{R}_1-\frac{\boldsymbol{\rho}}{2}\right)\right]$$
$$\times e^{\frac{i}{k}x_1\nabla_{\boldsymbol{R}}\nabla_{\boldsymbol{\rho}}}\left[\delta\left(\boldsymbol{R}-\boldsymbol{R}_1+\frac{1}{2}\boldsymbol{\rho}\right)-\delta\left(\boldsymbol{R}-\boldsymbol{R}_1-\frac{1}{2}\boldsymbol{\rho}\right)\right]$$
$$\times e^{-\frac{i}{k}x_1\nabla_{\boldsymbol{R}}\nabla_{\boldsymbol{\rho}}}\Gamma_2(x-x_1,\boldsymbol{R},\boldsymbol{\rho}).$$

If we introduce now the two-dimensional spectral density of inhomogeneities

$$B_\varepsilon(x,\boldsymbol{R})=\int d\boldsymbol{q}\Phi_\varepsilon^{(2)}(x,\boldsymbol{q})e^{i\boldsymbol{q}\boldsymbol{R}},$$
$$\Phi_\varepsilon^{(2)}(x,\boldsymbol{q})=\frac{1}{(2\pi)^2}\int d\boldsymbol{R}B_\varepsilon(x,\boldsymbol{R})e^{-i\boldsymbol{q}\boldsymbol{R}}$$

and Fourier transforms of wave field $u(x,\boldsymbol{R})$ and coherence function $\Gamma_2(x,\boldsymbol{R},\boldsymbol{\rho})$ with respect to transverse coordinates

$$u(x,\boldsymbol{R})=\int d\boldsymbol{q}\widetilde{u}(x,\boldsymbol{q})e^{i\boldsymbol{q}\boldsymbol{R}},\quad \widetilde{u}(x,\boldsymbol{q})=\frac{1}{(2\pi)^2}\int d\boldsymbol{R}u(x,\boldsymbol{R})e^{-i\boldsymbol{q}\boldsymbol{R}},$$
$$\Gamma_2(x,\boldsymbol{R},\boldsymbol{\rho})=\int d\boldsymbol{q}\widetilde{\Gamma}_2(x,\boldsymbol{q},\boldsymbol{\rho})e^{i\boldsymbol{q}\boldsymbol{R}},$$
$$\widetilde{\Gamma}_2(x,\boldsymbol{q},\boldsymbol{\rho})=\frac{1}{(2\pi)^2}\int d\boldsymbol{R}\Gamma_2(x,\boldsymbol{R},\boldsymbol{\rho})e^{-i\boldsymbol{q}\boldsymbol{R}},$$

then we obtain that average field is given by the expression

$$\langle u(x,\boldsymbol{R})\rangle=\frac{1}{(2\pi)^2}\int d\boldsymbol{q}\int d\boldsymbol{R}'u_0(\boldsymbol{R}')$$
$$\times\exp\left\{i\boldsymbol{q}(\boldsymbol{R}-\boldsymbol{R}')-i\frac{\boldsymbol{q}^2x}{2}-\frac{k^2}{2}\int\limits_0^x dx'D(x',\boldsymbol{q})\right\},$$

where

$$D(x,\boldsymbol{q})=\int\limits_0^x d\xi\int d\boldsymbol{q}'\Phi_\varepsilon^{(2)}(\xi,\boldsymbol{q}')\exp\left\{-\frac{i\xi}{2k}\left(\boldsymbol{q}'^2-2\boldsymbol{q}'\boldsymbol{q}\right)\right\},$$

and function $\widetilde{\Gamma}_2(x, \boldsymbol{q}, \boldsymbol{\rho})$ satisfies the integro-differential equation

$$\left(\frac{\partial}{\partial x} + \frac{1}{k}\boldsymbol{q}\nabla_{\boldsymbol{\rho}}\right)\widetilde{\Gamma}_2(x, \boldsymbol{q}, \boldsymbol{\rho}) = -\frac{k^2}{4}\int\limits_0^x dx_1 \int d\boldsymbol{q}_1 \Phi_\varepsilon^{(2)}(x_1, \boldsymbol{q}_1)$$

$$\times \left\{\cos\left[\frac{x_1}{2k}\boldsymbol{q}_1(\boldsymbol{q}_1 - \boldsymbol{q})\right] - \cos\left[\boldsymbol{q}_1\boldsymbol{\rho} - \frac{x_1}{2k}\boldsymbol{q}_1(\boldsymbol{q}_1 - \boldsymbol{q})\right]\right\}\widetilde{\Gamma}_2(x, \boldsymbol{q}_1, \boldsymbol{\rho}).$$

If distances a wave passes in medium satisfy the condition $x \gg l_\parallel$, where $l_\parallel$ is the longitudinal correlation radius of field $\varepsilon(x, \boldsymbol{R})$, then we obtain

$$\langle u(x, \boldsymbol{R})\rangle = \frac{1}{(2\pi)^2}\int d\boldsymbol{q}\int d\boldsymbol{R}' u_0(\boldsymbol{R}')\exp\left\{i\boldsymbol{q}(\boldsymbol{R} - \boldsymbol{R}') - i\frac{\boldsymbol{q}^2 x}{2} - \frac{k^2}{2}xD(\boldsymbol{q})\right\},$$

where

$$D(\boldsymbol{q}) = \int\limits_0^\infty d\xi \int d\boldsymbol{q}' \Phi_\varepsilon^{(2)}(\xi, \boldsymbol{q}')\exp\left\{-\frac{i\xi}{2k}\left(\boldsymbol{q}'^2 - 2\boldsymbol{q}'\boldsymbol{q}\right)\right\}.$$

For the plane incident wave, we have $u_0(\boldsymbol{R}) = 1$, and, consequently, average intensity is independent of $\boldsymbol{R}$,

$$\langle u(x, \boldsymbol{R})\rangle = e^{-\frac{1}{2}k^2 xD(0)}.$$

The applicability range of this expression is described by the condition

$$\frac{k^2}{2}D(0)l_\parallel \ll 1.$$

Problem 13.3

Derive operator expression for average field of plane incident wave in the general case of the Gaussian field $\varepsilon(x, \boldsymbol{R})$ with correlation function $B_\varepsilon(x, \boldsymbol{R})$ and calculate average field in the delta-correlated approximation.

Solution Averaging Eq. (13.53), page 369, over an ensemble of realizations of field $\varepsilon(x, \boldsymbol{R})$, we obtain in the general case the expressionv

$$\langle u(x, \boldsymbol{R})\rangle = u_0 \exp\left\{\frac{i}{2k}\int\limits_0^x d\xi \frac{\delta^2}{\delta \boldsymbol{v}^2(\xi)}\right\}$$

$$\times \exp\left\{-\frac{k^2}{8}\int\limits_0^x d\xi_1 \int\limits_0^x d\xi_2 B_\varepsilon\left(\xi_1 - \xi_2, \int\limits_{\xi_1}^{\xi_2} d\eta\, \boldsymbol{v}(\eta)\right)\right\}\Bigg|_{\boldsymbol{v}(x)=0}. \quad (13.101)$$

According to Eq. (13.6), page 356, the correlation function of the Gaussian delta-correlated field is $B_\varepsilon(x, \boldsymbol{R}) = A(\boldsymbol{R})\delta(x)$ and, consequently, we obtain the expression

$$\langle u(x, \boldsymbol{R})\rangle = u_0 \exp\left\{-\frac{k^2}{8}A(0)x\right\}.$$

Problem 13.4

Derive operator expression for the second-order coherence function of plane incident wave in the general case of the Gaussian field $\varepsilon(x, \boldsymbol{R})$ with correlation function $B_\varepsilon(x, \boldsymbol{R})$ and calculate the second-order coherence function in the delta-correlated approximation.

Solution On the case of plane incident wave, the second-order coherence function is given by the operator expression

$$\Gamma_2(x, \boldsymbol{R}_1 - \boldsymbol{R}_2)$$
$$= \left\langle u(x, \boldsymbol{R}_1)u^*(x, \boldsymbol{R}_2)\right\rangle = |u_0|^2 \exp\left\{\frac{i}{2k}\int\limits_0^x d\xi \left[\frac{\delta^2}{\delta \boldsymbol{v}_1^2(\xi)} - \frac{\delta^2}{\delta \boldsymbol{v}_2^2(\xi)}\right]\right\}$$
$$\times \exp\left\{-\frac{k^2}{4}\int\limits_0^x d\xi\, D\left(\boldsymbol{R}_1 - \boldsymbol{R}_2 + \int\limits_\xi^x d\eta\,[\boldsymbol{v}_1(\eta) - \boldsymbol{v}_2(\eta)]\right)\right\}_{\boldsymbol{v}_i=0},$$

where $D(\boldsymbol{R}) = A(0) - A(\boldsymbol{R})$.
Changing functional variables

$$\boldsymbol{v}_1(x) - \boldsymbol{v}_2(x) = \boldsymbol{v}(x), \quad \boldsymbol{v}_1(x) + \boldsymbol{v}_2(x) = 2\boldsymbol{V}(x)$$

and introducing new space variables

$$\boldsymbol{R}_1 - \boldsymbol{R}_2 = \boldsymbol{\rho}, \quad \boldsymbol{R}_1 + \boldsymbol{R}_2 = 2\boldsymbol{R},$$

we can rewrite last expression in the form

$$\Gamma_2(x, \boldsymbol{R}_1 - \boldsymbol{R}_2) = |u_0|^2 \exp\left\{\frac{i}{k}\int\limits_0^x d\xi\, \frac{\delta^2}{\delta \boldsymbol{v}(\xi)\delta \boldsymbol{V}(\xi)}\right\}$$
$$\times \exp\left\{-\frac{k^2}{4}\int\limits_0^x d\xi\, D\left(\boldsymbol{\rho} + \int\limits_\xi^x d\eta\, \boldsymbol{v}(\eta)\right)\right\}_{\boldsymbol{v}_i=0} = |u_0|^2 \exp\left\{-\frac{k^2 x D(\boldsymbol{\rho})}{4}\right\}.$$

References

[1] Novikov, E.A., 1965, *Functionals and the random force method in turbulence theory*, Sov. Phys. JETP **20**(5), 1290–1294.

[2] Klyatskin, V.I., 2005, *Stochastic Equations Through the Eye of the Physicist*, Elsevier, Amsterdam.

[3] Stratonovich, R.L., 1963, 1967, *Topics in the Theory of Random Noise*, Vol. 1, 2, Gordon and Breach, New York; London.

[4] Klyatskin, V.I. and A.I. Saichev, 1997, *Statistical theory of the diffusion of a passive tracer in a random velocity field*, JETP **84**(4), 716–724.

[5] Klyatskin, V.I. and A.I. Saichev, 1992, *Statistical and dynamical localization of plane waves in randomly layered media*, Soviet Physics Usp. **35**(3), 231–247.

[6] Klyatskin, V.I. and D. Gurarie, 1999, *Coherent phenomena in stochastic dynamical systems*, Physics-Uspekhi **42**(2), 165–198.

[7] Nicolis, G. and I. Prigogin, 1989, *Exploring Complexity, an Introduction*, W.H. Freeman and Company, New York.

[8] Isichenko, M.B., 1992, *Percolation, statistical topography, and transport in random media*, Rev. Modern Phys. **64**(4), 961–1043.

[9] Ziman, J.M., 1979, *Models of Disorder, The Theoretical Physics of Homogeneously Disordered Systems*, Cambridge University Press, Cambridge.

[10] Klyatskin, V.I., 2000, *Stochastic transport of passive tracer by random flows*, Atmos. Oceanic Phys. **36**(2), 757–765.

[11] Klyatskin, V.I. and K.V. Koshel', 2000, *The simplest example of the development of a cluster-structured passive tracer field in random flows*, Physics-Uspekhi **43**(7), 717–723.

[12] Klyatskin, V.I., 2003, *Clustering and diffusion of particles and tracer density in random hydrodynamic flows*, Physics-Uspekhi, **46**(7), 667–688.

[13] Klyatskin, V.I., 2008, *Dynamic stochastic systems, typical realization curve, and Lyapunov's exponents*, Atmos. Oceanic Phys. **44**(1), 18–32.

[14] Klyatskin, V.I., 2008, *Statistical topography and Lyapunov's exponents in dynamic stochastic systems*, Physics-Uspekhi, **53**(4), 395–407.

[15] Klyatskin, V.I., 2009, *Modern methods for the statistical description of dynamical stochastic systems*, Physics-Uspekhi, **52**(5), 514–519.

[16] Klyatskin, V.I. and O.G. Chkhetiani, 2009, *On the Diffusion and Clustering of a magnetic Field in Random Velocity Fields*, JETP **109**(2), 345–356.

[17] Koshel', K.V. and O.V. Aleksandrova, 1999, *Some resullts of numerical modeling of diffusion of passive scalar in randon velocity field*, Izvestiya, Atmos. Oceanic Phys. **35**(5), 578–588.

[18] Zirbel, C.L. and E. Çinlar, 1996, *Mass transport by Brownian motion*, In: *Stochastic Models in Geosystems*, eds. S.A. Molchanov and W.A. Woyczynski, IMA Volumes in Math. and its Appl., **85**, 459–492, Springer-Verlag, New York.

[19] Aref, H., 2002, *The development of chaotic advection*, Physics of Fluids, **14**, 1315–1325.

Lectures on Dynamics of Stochastic Systems. DOI: 10.1016/B978-0-12-384966-3.00014-3

[20] Mesinger, F. and Y. Mintz, 1970, *Numerical simulation of the 1970–1971 Eole experiment, Numerical simulation of weather and climate*, Technical Report No. 4, Department of Meteorology. University of California, Los Angeles.

[21] Mesinger, F. and Y. Mintz, 1970, *Numerical simulation of the clustering of the constant-volume balloons in the global domain, Numerical simulation of weather and climate*, Technical Report No. 5, Department Meteorology, University of California, Los Angeles.

[22] Mehlig, B. and M. Wilkinson, 2004, *Coagulation by random velocity fields as a Kramers Problem*, Phys. Rev. Letters, **92**(25), 250602-1–250602-4.

[23] Wilkinson, M. and B. Mehlig, 2005, *Caustics in turbulent aerosols*, Europhys. Letters, **71**(2), 186–192.

[24] Dolzhanskii, F.V., V.I. Klyatskin, A.M. Obukhov and M.A. Chusov, 1974, *Nonlinear Hydrodynamic Type Systems*, Nauka, Moscow [in Russian].

[25] Obukhov, A.M., 1973, *On the problem of nonlinear interaction in fluid dynamics*, Gerlands Beitr. Geophys. **82**(4), 282–290.

[26] Lorenz, E.N., 1963, *Deterministic nonperiodic flow*, J. Atmos. Sci. **20** (3), 130–141.

[27] Landau, L.D., and E.M. Lifshitz, 1984, *Course of Theoretical Physics*, Vol. 8: *Electrodynamics of Continuous Media*, Butterworth-Heinemann, Oxford.

[28] Batchelor, G.K., 1953, *Theory of Homogeneous Turbulence*, Cambridge University Press, London.

[29] Maxey, M.R., 1987, *The Gravitational Settling of Aerosol Particles in Homogeneous turbulence and random flow field*, J. Fluid Mech. **174**, 441–465.

[30] Lamb, H., 1932, *Hydrodynamics*, Sixth Edition, Dover Publications, New York.

[31] Klyatskin, V.I. and T. Elperin, 2002, *Clustering of the Low-Inertial Particle Number Density Field in Random Divergence-Free Hydrodynamic Flows*, JETP **95**(2), 328–340.

[32] Pelinovskii, E.N., 1976, *Spectral analysis of simple waves*, Radiophys. Quantum Electron. **19**(3), 262–270.

[33] Rytov, S.M., Yu.A. Kravtsov and V.I. Tatarskii, 1987–1989, *Principles of Statistical Radiophysics*, **1–4**, Springer-Verlag, Berlin.

[34] Tatarskii, V.I., 1977, *The Effects of the Turbulent Atmosphere on Wave Propagation*, National Technical Information Service, Springfield, VA.

[35] Monin, A.S. and A.M. Yaglom, 1971, 1975, *Statistical Fluid Mechanics*, **1**, **2**, MIT Press, Cambridge, MA.

[36] Ambartsumyan, V.A., 1943, *Diffuse reflection of light by a foggy medium*, Comptes Rendus (Doklady) de l'USSR **38**(8), 229–232.

[37] Ambartsumyan, V.A., 1944, *On the problem of diffuse reflection of light*, J. Phys. USSR **8**(1), 65.

[38] Ambartsumyan, V.A., 1989, *On the principle of invariance and its some applications*, in: *Principle of Invariance and its Applications*, 9–18, eds. M.A. Mnatsakanyan, H.V. Pickichyan, Armenian SSR Academy of Sciences, Yerevan.

[39] Casti, J. and R. Kalaba, 1973, *Imbedding Methods in Applied Mathematics*, Addison-Wesley, Reading, MA.

[40] Kagiwada, H.H. and R. Kalaba, 1974, *Integral Equations Via Imbedding Methods*, Addison-Wesley, Reading, MA.

[41] Bellman, R. and G.M. Wing, 1992, *An Introduction to Invariant Imbedding*, Classics in Applied Mathematics **8**, SIAM, Philadelphia.

[42] Golberg, M.A., 1975, *Invariant imbedding and Riccati transformations*, Appl. Math. and Com. **1**(1), 1–24.

[43] Meshalkin, L.D. and Ya.G. Sinai, 1961, *Investigation of the stability of a stationary solution of a system of equations for the plane movement of an incompressible viscous flow*, J. Appl. Math. and Mech. **25**(6), 1700–1705.

[44] Yudovich, V.I., 1965, *Example of the birth of secondary steady-state or periodic flow at stability failure of the laminar flow of viscous noncompressible liquid*, J. Appl. Math. and Mech. **29**(3), 527–544.

[45] Klyatskin, V.I., 1972, *To the nonlinear theory of stability of periodic flow*, J. Appl. Math. and Mech. **36**(2), 263–271.

[46] Furutsu, K., 1963, *On the statistical theory of electromagnetic waves in a fluctuating medium*, J. Res. NBS. **D-67**, 303.

[47] Shapiro, V.E. and V.M. Loginov, 1978, *'Formulae of differentiation' and their use for solving stochastic equations*, Physica, **91A**, 563–574.

[48] Bourret, R.C., U. Frish and A. Pouquet, 1973, *Brownian motion of harmonical oscillator with stochastic frequency*, Physica **A 65**(2), 303–320.

[49] Klyatskin, V.I. and T. Elperin, 2002, *Diffusion of Low-Inertia Particles in a Field of Random Forces and the Kramers Problem*, Atmos. Oceanic Phys. **38**(6), 725–731.

[50] Tatarskii, V.I., 1969, *Light propagation in a medium with random index refraction inhomogeneities in the Markov process approximation*, Sov. Phys. JETP **29**(6), 1133–1138.

[51] Klyatskin, V.I., 1971, *Space-time description of stationary and homogeneous turbulence*, Fluid Dynamics **6**(4), 655–661.

[52] Kuzovlev, Yu.E. and G.N. Bochkov, 1977, *Operator methods of analysing stochastic non-gaussian processes systems*, Radiophys. Quantum Electron. **20**(10), 1036–1044.

[53] Landau, L.D. and E.M. Lifshitz, 1957, *On hydrodynamic fluctuations*, JETP, **32**(3), 210.

[54] Lifshits, E.M. and L.P. Pitaevskii, 1980, *Statistical Physics*, Part 2, Elsevier Science & Technology, UK.

[55] Saichev, A.I. and M.M. Slavinskii, 1985, *Equations for moment functions of waves propagating in random inhomogeneous media with elongated inhomogeneities*, Radiophys. Quantum Electron. **28**(1), 55–61.

[56] Virovlyanskii, A.L., A.I. Saichev and M.M. Slavinskii, 1985, *Moment functions of waves propagation in waveguides with elongated random inhomogeneities of the refractive index*, Radiophys. Quantum Electron. **28**(9), 794–803.

[57] Moffatt, H.K., 1967, *The interaction of turbulence with strong wind shear*, in: *Atmospheric Turbulence and Radio Wave Propagation*, eds. A.M. Yaglom and V.I. Tatarskii, 139–154, Nauka, Moscow.

[58] Lighthill, J., 1952, *On sound generated aerodynamically. 1. General theory*, Proc. Roy. Soc. **A.211**(1107), 564–587.

[59] Lighthill, J., 1954, *On sound generated aerodynamically. II. Turbulence as a source of sound*, Proc. Roy. Soc. **A.222**(1148), 1–32.

[60] Lighthill, J., 1962, *Sound generated aerodynamically*, Proc. Roy. Soc. **A.267**(1329), 147–182.

[61] Klyatskin, V.I., 1966, *Sound radiation from vortex system*, Izvestiya AN SSSR, Mekhanika Zhidkosti i Gaza No. 6, 87–92.

[62] Milne-Thomson, L.M., 1996, *Theoretical Hydrodynamics* 5th edition, Dover Publications, Inc., New York.

[63] Landau, L.D. and E.M. Lifshitz, 1987, *Fluid Mechanics*, Second edition, Pergamon Press, London.

[64] Klyatskin, V.I., 2009, *Integral one-point statistical characteristics of vector fields in stochastic magnetohydrodynamic flows*, JETP **109**(6), 1032–1034.

[65] Balkovsky, E., G. Falkovich and A. Fouxon, 2000, *Clustering of inertial particles in turbulent flows*, http://arxiv.org/abs/chao-dyn/9912027.

[66] Balkovsky, E., G. Falkovich and A. Fouxon, 2001, *Intermittent Distribution of Inertial Particles in Turbulent Flows*, Phys. Rev. Letters **86**(13), 2790–2793.

[67] Guzev, M.A., V.I. Klyatskin and G.V. Popov, 1992, *Phase fluctuations and localization length in layered randomly inhomogeneous media*, Waves in Random Media **2**(2), 117–123.

[68] Klyatskin, V.I., 2004, *Propagation of electromagnetic waves in randomly inhomogeneous media as a problem of statistical mathematical physics*, Physics–Uspekhi **47**(2), 169–186.

[69] Klyatskin, V.I. and I.G. Yakushkin, 1997, *Statistical theory of the propagation of optical radiation in turbulent media*, JETP **84**(6), 1114–1121.

[70] Obukhov A.M., 1953, *On the effect of weak atmospheric inhomogeneities on propagation of sound and light*, Izv. AN SSSR, Geofizika, No. 2, P. 155–165.

[71] Obukhov A.M., 1988, *Turbulens and Atmospheric Dynamics*, Gidrometeoizdat, Leningrad [in Russian].

[72] Gracheva, M.E. and A.S. Gurvich, 1965, *On strong intensity fluctuation of light propagating in atmospheric layer near the Earth,* Radiophys. Quantum Electron. **8**(4), 511–515.

[73] Gracheva, M.E., A.S. Gurvich, S.O. Lomadze, V.V. Pokasov and A.S. Khrupin, 1974, *Probability distribution of 'strong' fluctuations of light intensity in the atmosphere*, Radiophys. Quantum Electron. **17**(1), 83–87.

[74] Fradkin, E.S., 1965, *Method of Green's Functions in the Theory of Quantized Fields and Quantum Statistics*, Trudy FIAN, **29**(7), 65.

[75] Fradkin, E.S., 1966, *Applications of functional methods in a quantum field theory and quantum statistics, II*, Nucl. Phys. **11**, 588.

[76] Klyatskin, V.I. and V.I. Tatarskii, 1970, *The parabolic equation approximation for propagation of waves in a random medium with random inhomogeneities*, Sov. Phys. JETP **31**(2), 335–339.

[77] Zavorotnyi, V.U., V.I. Klyatskin and V.I. Tatarskii, 1977, *Strong fluctuations of the intensity of electromagnetic waves in randomly inhomogeneous media*, Sov. Phys. JETP **46**(2), 252–260.

[78] Dashen R., 1984, *Distribution of intensity in a multiply scattering medium*, Opt. Lett. **9**, 110–112.

[79] Churnside, J.H. and S.F. Clifford, 1987, *Log-normal Rician probability–density function of optical scintillations in the turbulent atmosphere*, JOSA **A4**(10), 1923–1930.

[80] Gurvich, A.S., M.A. Kallistratova and F.E. Martvel, 1977, *An investigation of strong fluctuations of light intensity in a turbulent medium at a small wave parameter*, Radiophys. Quantum Electron. **20**(7), 705–714.

[81] Kravtsov, Yu.A., 1992, *Propagation of electromagnetic waves through a turbulent atmosphere*, Rep. Prog. Phys. **55**, 39–112.

[82] Flatté, S.M., G.-Y. Wang and J. Martin, 1993, *Irradiance variance of optical waves through atmospheric turbulence by numerical simulation and comparison with experiment*, JOSA **A10**, 2363–2370.

[83] Flatté, S.M., C. Bracher and G.Y. Wang, 1994, *Probability–density functions of irradiance for waves in atmospheric turbulence calculated by numerical simulation*, JOSA **A11**, 2080–2092.

[84] Martin, J.M. and S.M. Flatté, 1988, *Intensity images and statistics from numerical simulation of wave propagation in 3-D random media*, Appl. Opt. **27**(11), 2111–2126.

Printed in the United States
By Bookmasters